PLANT NEMATODE CONTROL

Plant Nematode Control

A.G. WHITEHEAD
PhD, ARCS

formerly Plant Nematologist,
East Africa Agriculture and Forestry Research Organization and Rothamsted Experimental Station, Harpenden, UK

CAB INTERNATIONAL

CAB INTERNATIONAL
Wallingford
Oxon OX10 8DE
UK

Tel: +44 (0)1491 832111
Fax: +44 (0)1491 833508
E-mail: cabi@cabi.org

CAB INTERNATIONAL
198 Madison Avenue
New York, NY 10016–4314
USA

Tel: +1 212 726 6490
Fax: +1 212 686 7993
E-mail: cabi-nao@cabi.org

A catalogue record for this book is available from the British Library, London, UK

Library of Congress Cataloging-in-Publication Data
Whitehead, A.G. (Alan G.)
Plant nematode control / A.G. Whitehead.
p. cm.
Includes bibliographical references and index.
ISBN 0-85199-188-2 (alk. paper)
1. Plant nematodes. 2. Plant nematodes—Control. I. Title.
SB998.N4W47 1997 97-22222
632′.6257—dc21 CIP

ISBN 0 85199 188 2

Typeset in Photina 10/12pt by Columns Design Ltd, Reading
Printed and bound in the UK at the University Press, Cambridge

Contents

Preface

The important role of plant-parasitic nematodes as pests of world crops has been clearly established over the last fifty years. Research on the taxonomy, biology and control of these minute animals increased greatly in this period and the literature generated by it is extensive. The object of this book is to bring together useful information from many sources on successful control measures for the principal pest species. To control a pest, the first essential is to recognize it and understand its way of life. To provide this essential background for the non-specialist and to put the control measures detailed in context, important nematode species and their effects on crop plants are illustrated and the main features of their lifestyles are described. The nematodes are grouped according to their position in host plants (Chapters 2–8). The final chapter gives an overview of successful control of the nematodes referred to and considers potential advances in nematode control. The contents of this book are indexed by nematode species and by crop plant to make them readily accessible to all concerned with the control of nematode–crop plant interactions.

Alan G. Whitehead
Harpenden, UK
January 1997

Acknowledgements

The author wishes to acknowledge the help he received from colleagues in the preparation of this book. Special thanks are due to Professor Brian Kerry for facilities made available in the Entomology and Nematology Department, Rothamsted Experimental Station, and to Mr Gordon Higgins and his staff for photographic assistance. Many colleagues in the UK, USA and Japan kindly loaned slides for reproduction in this book and their contributions are indicated in the figures and plates. Mr Michael Whitehead is thanked for guidance in computerization of the text. The assistance and contribution of Rhône-Poulenc Agriculture to the production of this publication is gratefully acknowledged.

Plate 1.
Pine wilt disease due to *Bursaphelenchus xylophilus*.
A, dieback of pine foliage.
B, hillside with some healthy pines among many dead and dying pines.
C, dead pines in front of healthy pines.
D, *B. xylophilus* extracted from affected pine tissue.
(Photos: Y. Mamiya.)

RHÔNE-POULENC

The assistance and contribution of Rhône-Poulenc Agriculture to the production of the colour plates section is gratefully acknowledged.

A

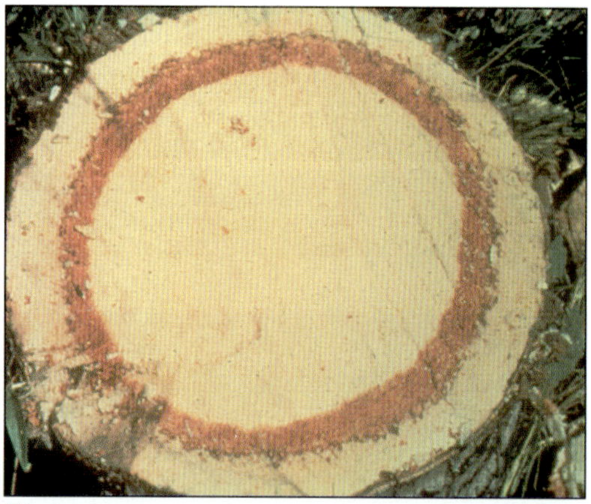
B

C

D

Plate 2.

A, coconut palms affected by the red ring nematode, *Rhadinaphelenchus cocophilus*. (Photo: W.F. Mai.)

B, coconut palm trunk in transverse section showing the 'red ring'. (Photo: W.F. Mai.)

C, patch of rice plants stunted by rice stem nematode, *Ditylenchus angustus*. (Photo: R. Plowright.)

D, rice shoots with symptoms of 'ufra' disease due to *D. angustus*. (Photo: J. Bridge.)

A

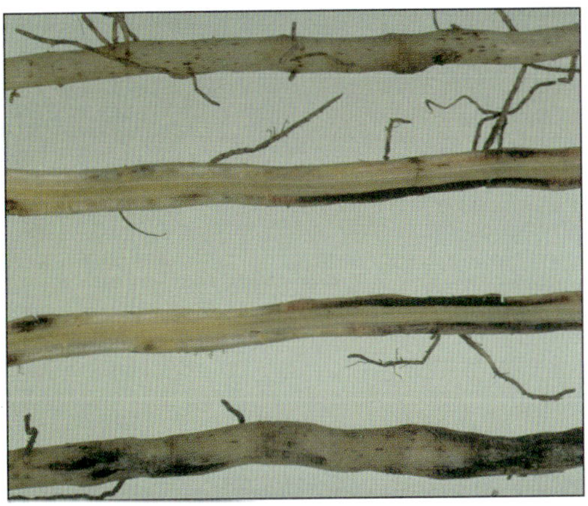

B

C

Plate 3.
A, uprooting of bananas following burrowing nematode root-rot.
B, cord roots of banana split longitudinally to show rotting of the cortex due to *Radopholus similis*. (Photo: D.J. Hooper.)
C, paring *R. similis*-infected corm tissue from suckers prior to replanting.

Plate 4.

A, patchy growth of potatoes due to infestation with potato cyst-nematodes, *Globodera rostochiensis* and *G. pallida*.

B, symptoms of potato cyst-nematode injury.

C, *G. pallida*, pale females and brown cysts.

D, *G. rostochiensis*, yellow females and brown cysts.

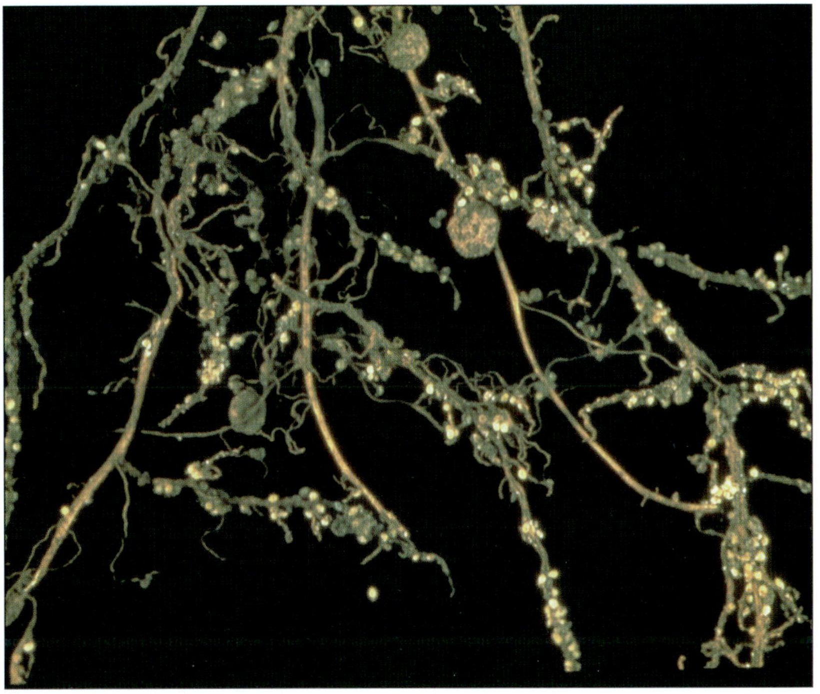

A

B

C

Plate 5.
A, soybean roots infested with soybean cyst-nematode, *Heterodera glycines*. (Photo: R.A. Motsinger.)
B, better performance of soybeans grown in rotation with non-host crops (left) than grown continuously in soil infested with *H. glycines*. (Photo: D. Schmitt.)
C, improved growth of resistant (tolerant) cultivars of soybeans (upper left) compared with susceptible (intolerant) cultivars (foreground) in soil infested with *H. glycines*. (Photo: J.P. Ross.)

A B C D

Plate 6.

A, 'beet sick' patch in sugarbeet crop, showing stunted, wilting plants in soil heavily infested with beet cyst-nematode, *Heterodera schachtii*.

B and C, affected sugarbeet with numerous white females on lateral roots.

D, white females and brown cysts of *H. schachtii* on lateral roots.

(Photos: D.A. Cooke.)

A

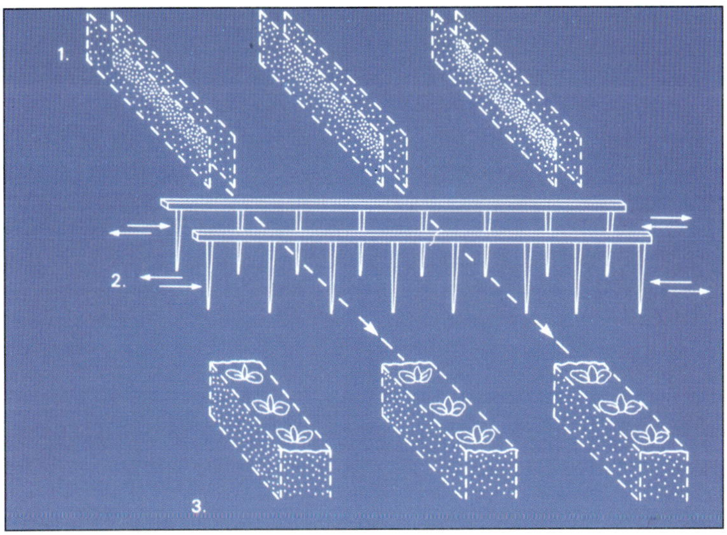

B

Plate 7.

A, response of sugarbeet to control of *Heterodera schachtii* by soil fumigation with 1,3-dichloropropene and application of aldicarb to the seedbed (left-hand rows) compared with no control of the nematode (right-hand rows), California. (Photo: F.G.W. Jones.)

B, vertical band-reciprocating harrow technique for incorporating granular nematicide in bands 15 cm wide and 15 cm deep into which sugarbeet or another row crop is sown.

Plate 8.

A, tobacco crop destroyed by *Meloidogyne javanica* and *M. incognita,* Tanzania.

B, tomato crop similarly attacked, Kenya.

C, pyrethrum affected by *M. hapla,* Kenya.

D, chlorotic tea seedlings infested with *M. javanica*, Tanzania.

1 Plant-parasitic Nematodes, their Importance and Control

Nematodes (roundworms) are found in virtually all 'watery' environments worldwide, in as diverse situations as cold oceans, hot springs, vinegar, soil, as well as in animals and plants. Cultivated soils usually contain a wide range of nematodes, differing in their food requirements. Some feed on different components of the microflora, some feed on soil animals, including other nematodes, some are parasites of invertebrates or vertebrates and some parasitize plants. This book is concerned with the last mentioned group and their control.

Worldwide Pests

At present, 24 genera of plant-parasitic nematodes include species that are economically important pests of crop plants. Other species of these and other genera of plant-parasitic nematode may assume importance in the future as their host–parasite relationships become better understood. It has been estimated that some 10% of world crop production is lost as a result of plant nematode damage, which is one third of the losses attributed to pests and diseases generally. Obviously, losses vary greatly from negligible to total depending on the circumstances. What is clear, however, is that, left unchecked, nematode pests can cause severe financial losses to the grower.

Nematode Form and Size

All plant-parasitic nematodes pass at least part of the life cycle in the soil, as eggs, juveniles, adults or any combination of these stages. Most plant-parasitic nematodes are vermiform in shape and vary in length from about 0.2 to 5 mm or more, depending on the species and life stage (Fig. 1.1). The females of some sedentary root parasites and some of the large root ectoparasites, when they aggregate on roots, are just visible to the naked eye but most species are not seen in the soil or in plant tissues because they are translucent as well as small.

(a)

Fig. 1.1. (*and opposite*) (a) Different genera of vermiform plant-parasitic nematode (Photo: D.J. Hooper). (b) Adults of a vermiform nematode (*Ditylenchus dipsaci*) showing major body features (Photo: D.J. Hooper after J.B. Goodey).

Habitats and Life Cycle

Most plant-parasitic nematodes attack underground plant tissues, especially roots. Other species are predominantly shoot parasites, attacking stems, leaves, flowers, seeds or any combination thereof. The nematodes may feed ecto-, semi-endo- or endo-parasitically on host tissues and the important species considered in this book are grouped accordingly. The nematode cuticle, though flexible, has limited elasticity, so the juveniles can only grow to adult dimensions by a series of four moults, the first of which often occurs within the egg, so that the hatching juvenile is already the second of the four juvenile stages. Males are active and vermiform. Reproduction may be asexual (parthenogenetic) or sexual. In some species the females are really self-fertilizing hermaphrodites and in others reproduction may be first parthenogenetic and then sexual, as in aphids. The number of new generations per crop or per annum can vary from one to many depending on climate, nematode and host plant. As multiplication rates per crop or cropping season can vary from as little as five fold to 1000-fold or more for some nematodes, it is not surprising that numbers of some plant-parasitic nematodes may exceed 10^3 g^{-1} of soil and 10^4 g^{-1} of plant tissue. Although nematode damage increases with nematode numbers, very small numbers of some nematodes, especially those that are vectors of pathogenic viruses, may cause serious yield losses.

(b)

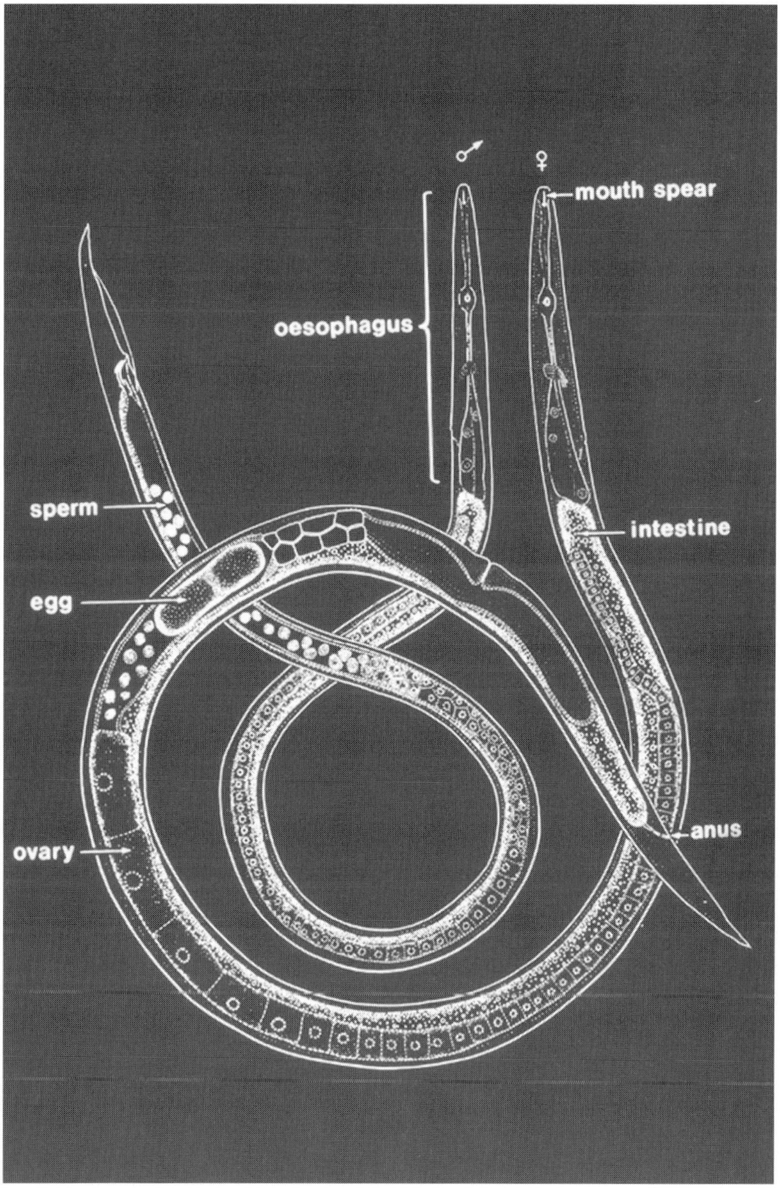

Fig. 1.1. *Continued*

Feeding

All nematode pests of plants feed on plant tissues with a narrow mouth spear or stylet. Most species belong to the order Tylenchida, characterized by the possession of a hollow, feeding stylet with basal knobs. Muscle fibres attached to these knobs and the basal plate of the head, when they contract, push the stylet tip through the stoma and into a plant cell. Digestive enzymes produced in the base of the oesophagus are pumped forward by the muscular median bulb of the

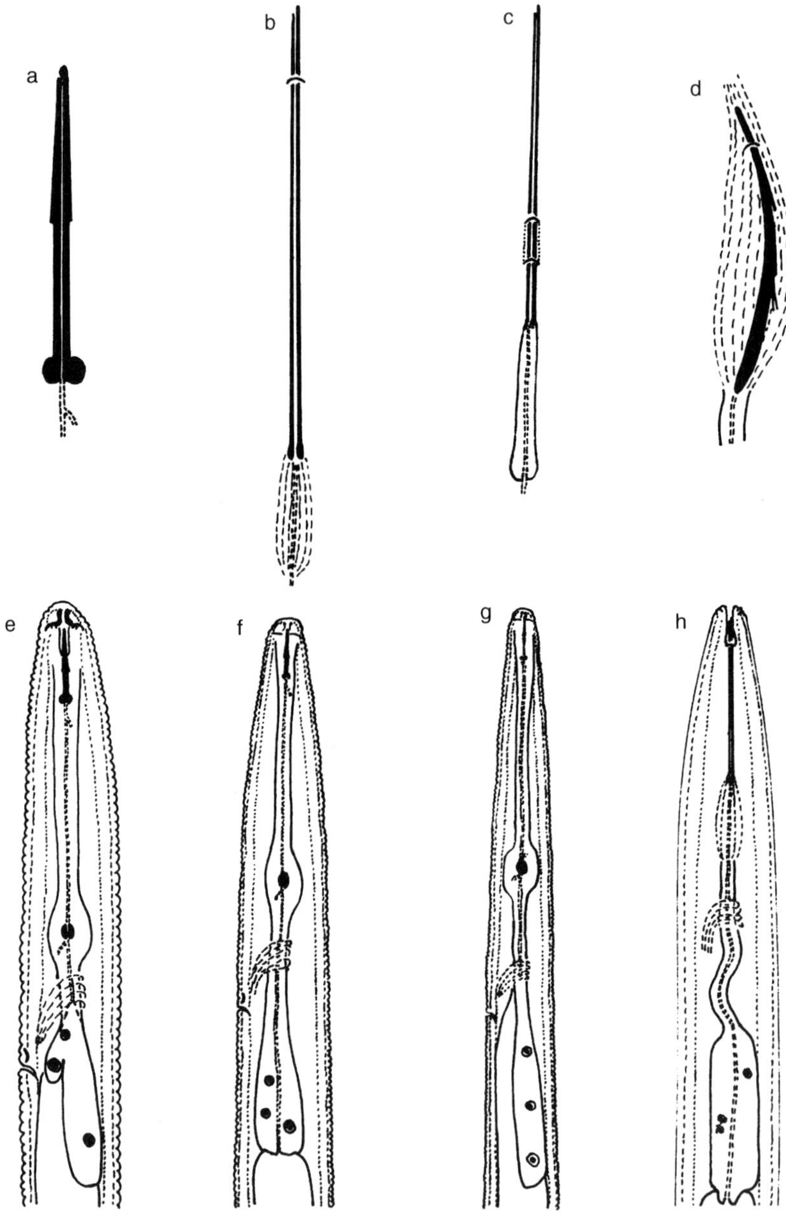

Fig. 1.2. Main types of plant nematode feeding stylet: a, tylenchid; b, *Longidorus*; c, *Xiphinema*; and d, trichodorid. Main types of plant nematode oesophagus: e, tylenchid (overlapping intestine); f, tylenchid (abutting intestine); g, aphelenchid; and h, dorylaimid.

oesophagus, through the stylet and into the plant cell. Predigested cell contents are withdrawn through the oesophagus into the intestine, which acts as a storage organ. Other species, belonging to the order Dorylaimida, have a different type of stylet, which in trichodorids (species of *Trichodorus* and *Paratrichodorus*) is curved. Trichodorid nematodes may feed on predigested cell contents, as do other plant-parasitic nematodes, but may also graze roots, using the stylet like a rasp. These three types of stylet are shown in Fig. 1.2. Most species of tylenchid nematode pest belong to the Tylenchina, as distinct from a small number of species that belong to the Aphelenchina. These two groups are clearly differentiated by the form of the oesophagus and the disposition of the oesophageal gland duct openings into the oesophageal lumen (Fig. 1.2). The drawings of the different nematode species covered in this book, although made from photographs of actual specimens, are intended purely for illustration. Reliable identification of the species requires full descriptions and measurements and the eyes of an experienced taxonomist and is outside the scope of this account.

Plant Damage

The damage done to a plant increases with the number of nematodes feeding on it. Determination of actual loss of plant yield or crop quality in the field is often difficult and time consuming. Although loss of yield and quality of produce may increase with the number of nematodes attacking or feeding on the plant, in the field there may be additional organisms involved. When specific symptoms, caused by the nematode alone, such as lesions or galls, increase as crop yield decreases, one can be fairly confident that the loss is due, at least in part, to the nematode. The symptoms of plant nematode attack are varied and often of a non-specific nature, so they may be confused with nutrient deficiencies or injury due to other pests or diseases. Not surprisingly, a reduced root system may result in the shoot displaying symptoms of drought stress and mineral deficiences, even when both requirements are in adequate supply for a healthy plant. Most plant-parasitic nematodes are obligate parasites and at least some of them form disease complexes with pathogenic bacteria and fungi. Some plant-parasitic dorylaimids are also vectors of important plant viruses.

Nematode Survival

Plant-parasitic nematodes may survive unfavourable environments in several ways. The eggs may resist low temperatures or drought that juveniles or adults cannot. Conversely, some juvenile stages may be very resistant to drought. Eggs may be retained in the dead body of the female, which becomes a tanned, protective envelope or 'cyst' around them. Nematodes without very resistant stages may have very wide host ranges and some of the hosts may offer better protection from a hostile environment than others. Migration into the subsoil may protect some vulnerable species from desiccation in the topsoil. The production of large numbers of eggs also increases the chances of the population surviving. Nematodes

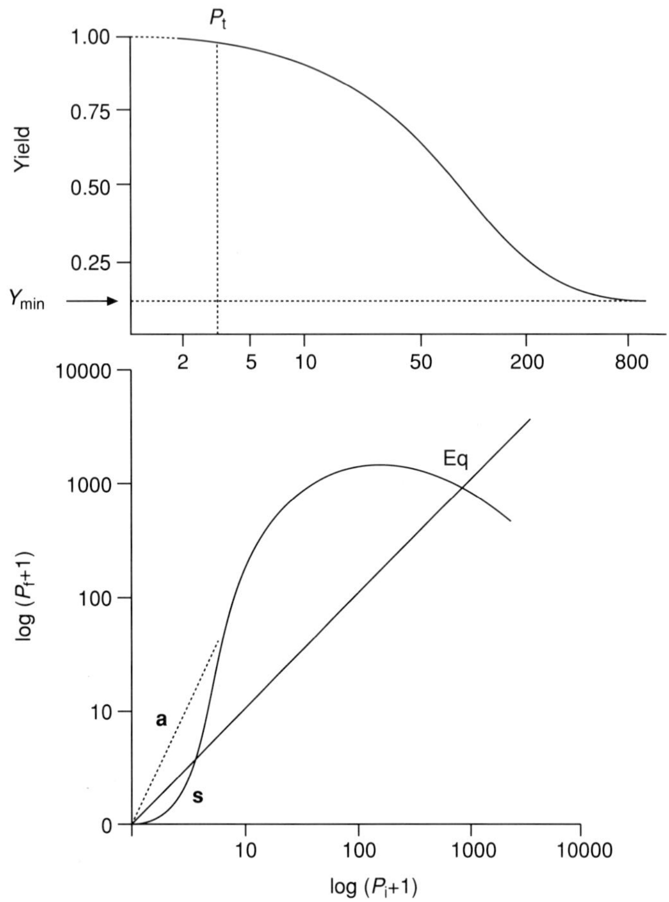

Fig. 1.3. Relationship between plant yield and number of viable nematodes at planting (upper graph). Relationship between initial and final infestation of soil (lower graph). P_t = nematode population density above which yield is less than the maximum (1.00); Y_{min} = minimum yield; P_f = viable nematodes g^{-1} soil after harvest; P_i = viable nematodes g^{-1} soil at planting; Eq = equilibrium population, at which $P_f/P_i = 1$; **a** = asexual nematodes; **s** = sexually reproducing nematodes.

with very drought resistant stages may be transported by wind to a more favourable environment. Self-dispersal of nematodes is slow, rarely more than a metre a year, but they may be carried over considerable distances by water, on farm implements, on vehicles and by animals.

Aims of Control

The object of nematode control is to prevent significant loss of yield and quality in vulnerable crop plants and, in the longer term, to keep nematode populations small.

To meet the first objective it is necessary to decrease the actual population density (P_i) to the damage threshold density (P_t) relevant to the tolerance of the specific crop plant and the environment in which it is grown (Fig. 1.3, upper graph). For a virus-vector nematode, P_t may be 0 or close to 0. Obviously the larger the value of P_i the greater the potential damage to the plant, the greater the yield loss and the greater the percentage of the nematodes that must be inactivated or killed to prevent yield loss. Whereas yield loss due to nematode attack is *proportional* to nematode population density, nematode multiplication is *inversely proportional* to population density (Fig. 1.3, lower graph). When P_i is small, multiplication is great, because the nematodes have a non-limiting source of food. As P_i increases, multiplication decreases as the available food is limited. When P_i is sufficiently large (the equilibrium density, P_e), there is no population increase, so population density at harvest, $P_f = P_i$. When P_i exceeds P_e , P_f is less than P_i . When P_i is very small, a proportion of the females may remain unfertilized, so multiplication is less for a sexually reproducing nematode than for a parthenogenetic nematode.

Except to prevent virus transmission, when the nematode vector must be virtually eradicated from the soil, prevention of crop loss due to nematode attack is usually easier to achieve than control of population increase on susceptible crop plants. To prevent damage, the percentage of the nematodes (K) that must be killed or immobilized is given by the equation

$$K = 100\,(P_i - P_t)\,/\,P_i \qquad (1.1)$$

To prevent population increase, i.e. to achieve full control, K is given by the empirical equation

$$K = 100\,(1 - 1\,/\,a) \qquad (1.2)$$

where a is the increase of those nematodes that hatch. The actual multiplication that occurs is

$$(P_f - P_u)\,/\,P_x \qquad (1.3)$$

P_u is the number of nematodes that did not hatch when the host plant was grown and P_x is the number of nematodes remaining to attack the plant after the use of a control measure. The difficulty of obtaining full population control is proportional to a. Values of a can vary from as little as 10 or less, for some root-ectoparasitic nematodes in cool soils, to 50, for some cyst nematodes and to 1000 or more for some stem nematodes and root-knot nematodes. The corresponding values of K needed for full population control of these nematodes would therefore be 90%, 98% and 99.9% or more, respectively. Because the equation from which these percentages have been derived is empirical, ignoring the fact that a decreases as initial population density approaches the equilibrium density, they overestimate somewhat the values of K required. More explanation is given elsewhere (Whitehead and Bridge, 1985; McSorley and Philipps, 1993). It is clear, however, that population control is often difficult to achieve for susceptible crop plants. When a suitable resistant cultivar is available, i.e. one which prevents population increase (fully resistant) or lessens it (partially resistant), a level of population control is achieved.

Methods of Control

The old adage 'prevention is better than cure' certainly applies to plant-parasitic nematodes. Every effort should be made to exclude damaging exotic nematodes from uninfested areas. Although such controls are expensive to implement, they may cost far less than the control measures that have to be followed if the pest becomes established. The 'golden nematode program', for example, though very expensive, has been very successful in preventing the pest becoming established in mainland potato regions of the USA. The potato race of the false root-knot nematode, endemic in parts of South America, is at present unknown in Europe and is recognized as a potentially serious pest here, which must be kept out. Quarantine measures may succeed in delaying establishment of exotic nematodes in new areas but they are unlikely to remain 100% effective indefinitely.

Comparatively few measures are effective in the control of nematode infestations. These may be classified as physical, biological or chemical. Any combination of control measures is called integrated control.

Physical Control Measures

Nematodes in soil may be killed by heating the soil water to a lethal temperature. This may be done for small areas by electrically heating the soil or by passing steam through it, though both measures are very expensive in energy. An alternative procedure is to cover the soil with one or two layers of polyethylene film and allow the sun to raise the soil temperature. This is known as soil solarization and it is effective in hot climates in the topsoil but not usually to depths of more than about 20 cm. Infested plant tissue – roots, seeds, bulbs, tubers and plantlets – may be freed of nematodes by dipping it in water at a carefully chosen and controlled temperature, long enough to kill viable nematodes. There is often a risk of plant damage from hot water treatment, especially of non-dormant tissues, and for these, dipping in cold solutions or suspensions of nematicides may be preferable. There is some evidence that some nematodes are harmed mechanically (e.g. some trichodorids) by soil cultivations, but passing an electric current through the soil, once claimed effective in the control of root-knot nematodes, was later shown to be ineffective. Some nematodes are controlled by flooding the land for several weeks or months but others seem to be unaffected by flooding.

Biological Control Measures

Nematodes with narrow host ranges may be controlled by growing host crops infrequently in rotation with non-host crops. Such crop rotation is more difficult for nematodes with wide host ranges, where the choice of non-host crops may be very limited and uneconomic. The success of control by crop rotation may be reduced if weed hosts are not adequately controlled. Crop rotation is not an option for long-term perennial crops. Cultivars or rootstocks which resist nematode invasion and/or multiplication can give long-term population control. In time, however, they may select virulent biotypes from field populations, losing their resistance. It is important that resistant plants are not intolerant of attack. If they are, they must be grown in lightly infested soils or be used in conjunction with an effective nematicide. Soil populations may sometimes be decreased by growing 'trap crops', which are invaded by the nematodes but which are destroyed before the nematodes have multiplied. Natural biological control of some nematodes is well documented but the commercial use of nematode predators and pathogens is in its infancy.

Chemical Control Measures

Chemicals which paralyse or kill nematodes are referred to collectively as nematicides. Fumigant nematicides have large vapour pressures and diffuse rapidly through the network of soil pores in the vapour phase. They kill nematodes by diffusing slowly through the soil water films around the worms. All currently available fumigant nematicides are phytotoxic and, to prevent plant damage, must be applied to soil days, weeks or months before planting, according to compound and soil temperatures during fumigation. Most currently used compounds are either halogenated aliphatic hydrocarbons or methyl isothiocyanate precursor compounds or mixtures. Limited use is now made of chloropicrin and dibromochloropropane has been withdrawn from use in most countries. The structural formulae, water solubilities and vapour pressures of these fumigants are in Table 1.1. Seeds may sometimes be disinfested successfully by carefully controlled fumigation with methyl bromide but seed moisture content is critical both to successful eradication and to impairment of germination and seedling vigour.

A range of non-fumigant, non-phytotoxic nematicides has been used to control plant-parasitic nematodes. The most widely effective of these are organophosphates and carbamates, especially oximecarbamates. The structural formulae of non-fumigant nematicides frequently mentioned in this book along with their major features are in Table 1.1.

Treating all the topsoil with a nematicide is expensive and economic only for high value crops. In annual crop plants, yield loss is usually greater from nematode injury to young than to older, established plants. Yield losses may therefore be adequately and economically controlled by treating the rows in which the seeds are sown or the soil around the holes in which transplants are set. Damage may also be limited by applying nematicides to the seeds or to the bare roots of nursery stock before transplanting. Such treatments may have little or no effect on nematode multiplication. In perennial crops, non-fumigant nematicides must usually be applied several times per annum to prevent nematode injury and limit population increase. Details of successful application techniques are given in the succeeding chapters.

Non-hosts and resistant plants decrease nematode populations deep in soil to the depth of their root systems. They do this much more cheaply and without risk of contaminating ground water compared with the large amounts of fumigant nematicides needed to achieve the same result.

Table 1.1. Widely used nematicides.

Common name (trade names)	Structural formula	Vapour pressure mmHg (°C)	Solubility in water % (°C)	Acute oral LD_{50} rats ($mg\ kg^{-1}$)
SOIL FUMIGANTS				
Methyl bromide (Embafume C, Brozone, Dowfume MC 2 and MC 33)	$CH_3\ Br$	1420 (20)	1.8 (20)	514
Ethylene dibromide (Dowfume W 85, Soilbrom, etc.)	$Br\ CH_2\ CH_2\ Br$	8 (20)	0.3 (20)	146
Chloropicrin	$Cl_3\ C\ NO_2$	24 (25)	0.2 (0)	toxic gas
1,3-Dichloropropene (Telone, Telone II)	$Cl\ CH_2CH.CHCl$ (*cis* (*Z*) and *trans* (*E*))	55 (20)	0.1 (20)	250–500
D-D	1,3-dichloropropene plus 1,2-dichloropropane		0.1 (20)	140
Dazomet (Basamid, Mylone)	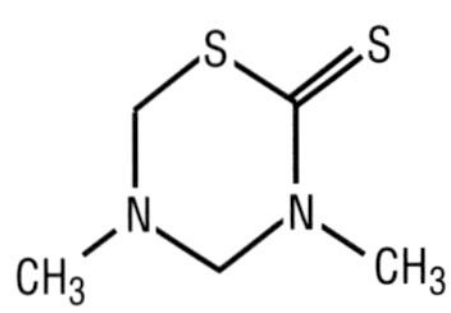	21 (20)	0.8 (20)	500
Metham-sodium (Vapam, Sistan)	$(C_2H_4NNaS_2)$		72 (20)	1750
CARBAMATES				
Aldicarb (Temik)	$CH_3S.C(CH_3)_2.CH=NO.CONHCH_3$	1×10^{-4} (25)	0.6 (25)	0.9
Oxamyl (Vydate)	$(CH_3)_2NCO.C(SCH_3)=NO.CONHCH_3$	2.3×10^{-4} (25)	28 (25)	5.4
Carbofuran (Furadan, Yaltox)	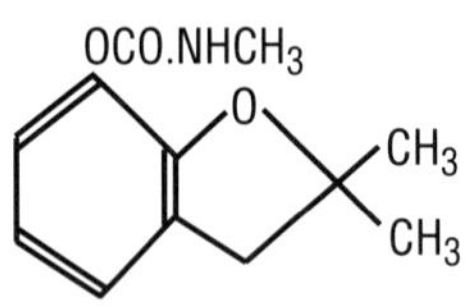	2×10^{-5} (33)	0.07 (25)	8–14
ORGANOPHOSPHATES				
***O,O*-diethyl *O*-phosphorothioates**				
Fensulphothion (Dasanit, Terracur P)	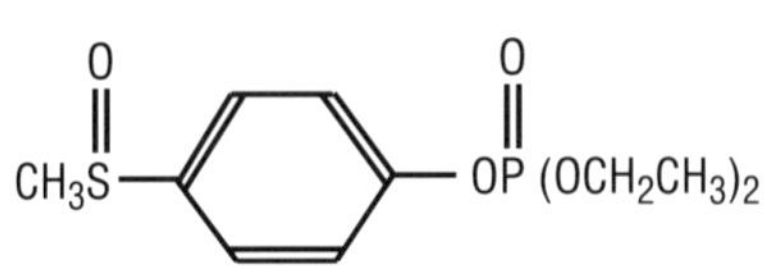		0.15 (25)	1.9–10.5

Isazophos (Miral)	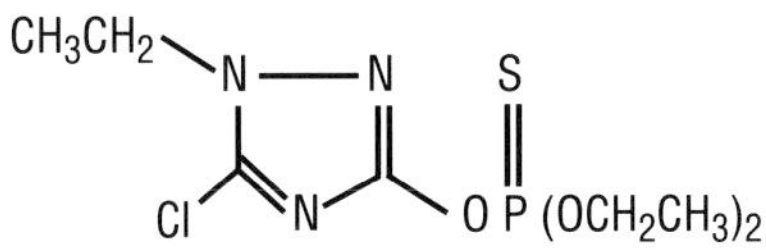	3.2×10^{-4} (20)	0.00025 (20)	40–60
Thionazin (Nemaphos, Zinophos, Cynem)	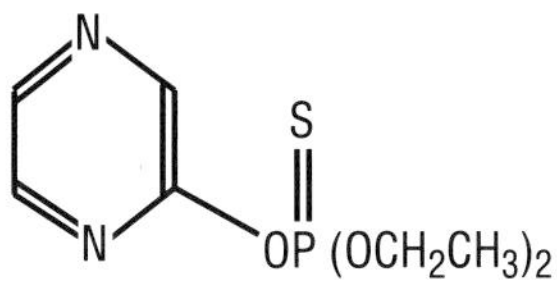	3×10^{-3} (30)	0.11 (27)	12
***O,O*-diethyl phosphorodithioates**				
Disulphoton (Di-Syston, etc.)	$CH_3CH_2SCH_2CH_2SP(=S)(OCH_2CH_3)_2$	1.8×10^{-4}(20)	0.000025 (22)	2.6–12
Phorate (Thimet)	$CH_3CH_2SCH_2SP(=S)(OCH_2CH_3)_2$	8.4×10^{-4}(20)	0.00005	1.6–3.7
Terbufos (Counter)	$(CH_3)_3CSCH_2SP(=S)(OCH_2CH_3)_2$	2.6×10^{-4} (25)	0.0000125	1.6
Other organophosphates				
Ethoprophos (Mocap, Prophos)	$CH_3CH_2OP(=O)(SCH_2CH_2CH_3)_2$	3.5×10^{-4} (26)	0.00075	62
Fenamiphos (Nemacur)	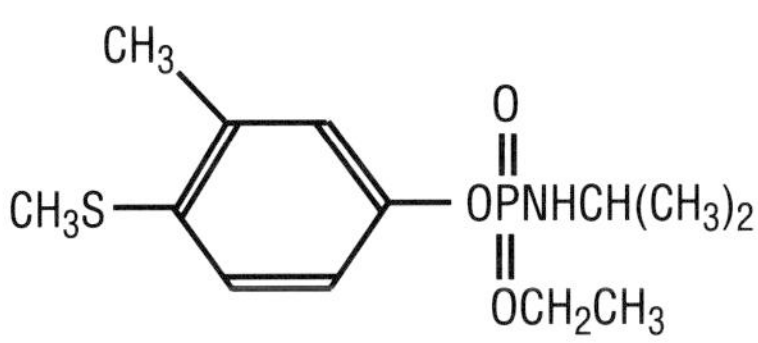	1×10^{-6} (30)	0.0007	15–19

Integrated Control

Plant-parasitic nematodes are usually controlled better by a combination of effective measures than by relying on a single measure. Frequent use of a resistant cultivar may select virulent biotypes from field populations. Frequent use of a nematicide may accelerate its microbial degradation or select nematodes tolerant or resistant to it. There are, of course, fewer integrated control options for perennial than for annual crop plants.

Problems in Implementing Control Measures

Each control measure may fail if it is not applied appropriately or if it is not executed correctly. Crop rotations may not succeed if the chosen non-host crops are exposed to populations to which they are susceptible. Similarly, repeated use of resistant cultivars or rootstocks may select virulent biotypes that break resistance. Trap cropping is ineffective if the invading nematodes are allowed to complete their life cycle in the roots. Hot water treatment of planting stock may be phytotoxic. Inadequate sealing of the soil surface or the presence of large organic matter fractions in the soil may render soil fumigation ineffective, as will very large or very small soil moisture contents. Insufficient aeration of the soil after fumigation may leave phytotoxic residues harmful to the new crop. There are pitfalls too with non-fumigant nematicides. Failure can often be traced to inappropriate application techniques, excessive leaching in sandy soils, high soil pH reducing the half-life of carbamates and organophosphates or low soil moisture content, which inactivates the compound. Despite these pitfalls, there are many reports of successful control, worldwide, and these are reviewed in the following chapters.

Seed-gall and Leaf Nematodes (*Anguina* and *Aphelenchoides*) 2

Anguina tritici (Steinbuch, 1799) Chitwood, 1935 (Fig. 2.1)

The wheat seed-gall nematode, *A. tritici*, was the first plant-parasitic nematode to be recognized. It was reported by Needham (1744), who found large numbers of the worms inside galled wheat seed. Roffredi (1775, 1776) showed that these worms were responsible for the formation of the seed galls or 'cockles' in wheat ears (Fig. 2.2). Their life cycle was elucidated by Davaine (1857). Although the nematode has been reported in many countries, where its principal hosts – wheat, emmer (*Triticum dicoccum*), spelt (*T. spelta*) and rye – have been grown, it is now limited to those areas where infested seed stocks are still used.

Control

Although juvenile nematodes can survive for many years inside desiccated seed galls, in the field the nematodes do not survive more than a year in the absence of a host plant. Oats and barley are virtually immune to the nematode, so growing them or other non-host crops for a year can rid land of the pest. Only uninfested seed stocks of host crops should be sown. Infested stocks can be freed of the nematode by mechanical seed-cleaning techniques. Most of the seed galls pass through a 3 mm mesh sieve (Paruthi and Bhatti, 1992). As the galls are less dense than healthy grain, they can be blown away in a winnowing machine (Kort, 1972). Where such machinery is not available, the galls may be separated from the denser, healthy grain in small amounts of seed by stirring it thoroughly into 10, 15 or 20% salt solutions and allowing the galls to float to the surface. The supernatant is discarded and the healthy grain rinsed in water and dried (Goodey, 1933; Paruthi and Bhatti, 1992).

Good control of *A. tritici* has been obtained by treating contaminated seed with 600 or 800 mg neem kernel powder (*Azadirachta indica*) 100 g^{-1} (Gotke and Swarup, 1989), or by treating infested soil with leaf extracts of neem or *Catharanthus roseus* (Qamar *et al.*, 1989). Ethyl parathion spray has eliminated the galls from wheat (Singh and Prasad, 1973), but this is a dangerous proce-

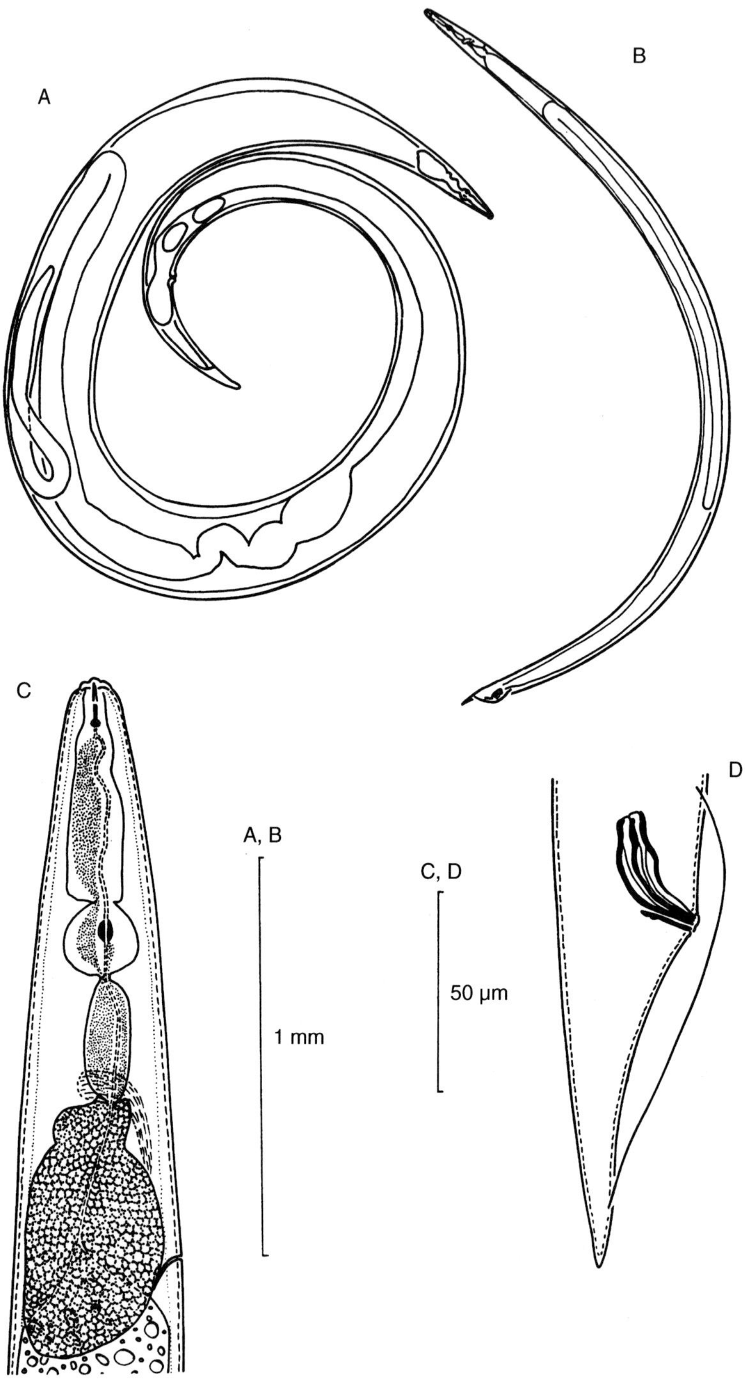

Fig. 2.1. *Anguina tritici*: A, female; B, male; C, anterior end of female; D, male tail.

Fig. 2.2. Wheat ears and grain – healthy (left), affected by wheat seed-gall nematode, *Anguina tritici* (right) (Photo: D.J. Hooper).

dure. The most practical of these methods of control are growing non-host crops for a year and the use of uncontaminated seed stocks of host crops. Provided they are efficient, dry decontamination techniques are preferable to flotation in water or salt solutions.

Aphelenchoides besseyi Christie, 1942 (Fig. 2.3)

This is a worldwide pest causing 'white tip' disease of rice (Fig. 2.4) and 'summer crimp' disease of strawberry. It does not, however, occur in cool, temperate soils. The nematode is ectoparasitic on plant shoots, moving about on plant surfaces in moisture films. It infests bud tissues and leaf surfaces of strawberry. In rice it infests the first leaf sheaths of seedlings, then in larger numbers the developing panicles, where it occurs first beneath the glumes and later beneath the paleae. The nematode is active at high air temperatures, the optimum for egg-laying and hatching being 30°C, at which temperature the life cycle takes only 8–12 days. Tikhonova (1966) reported ten generations per annum on rice in Tadzhik SSR, 13 in Uzbek SSR and eight in Kazakh SSR. Yield losses of up to 60% have been reported in rice and of up to 75% in strawberry. Currently such losses are probably rare, as the diseases provoked are now easily controlled. As rice grain ripens, the nematodes, mostly females, aggregate and coil up in the glume axis, passing into an anabiotic state, in which they remain viable in dry-stored seed for up to three years. The nematode usually dies within a few months in moist soil, so it is spread primarily in infested seed or plants.

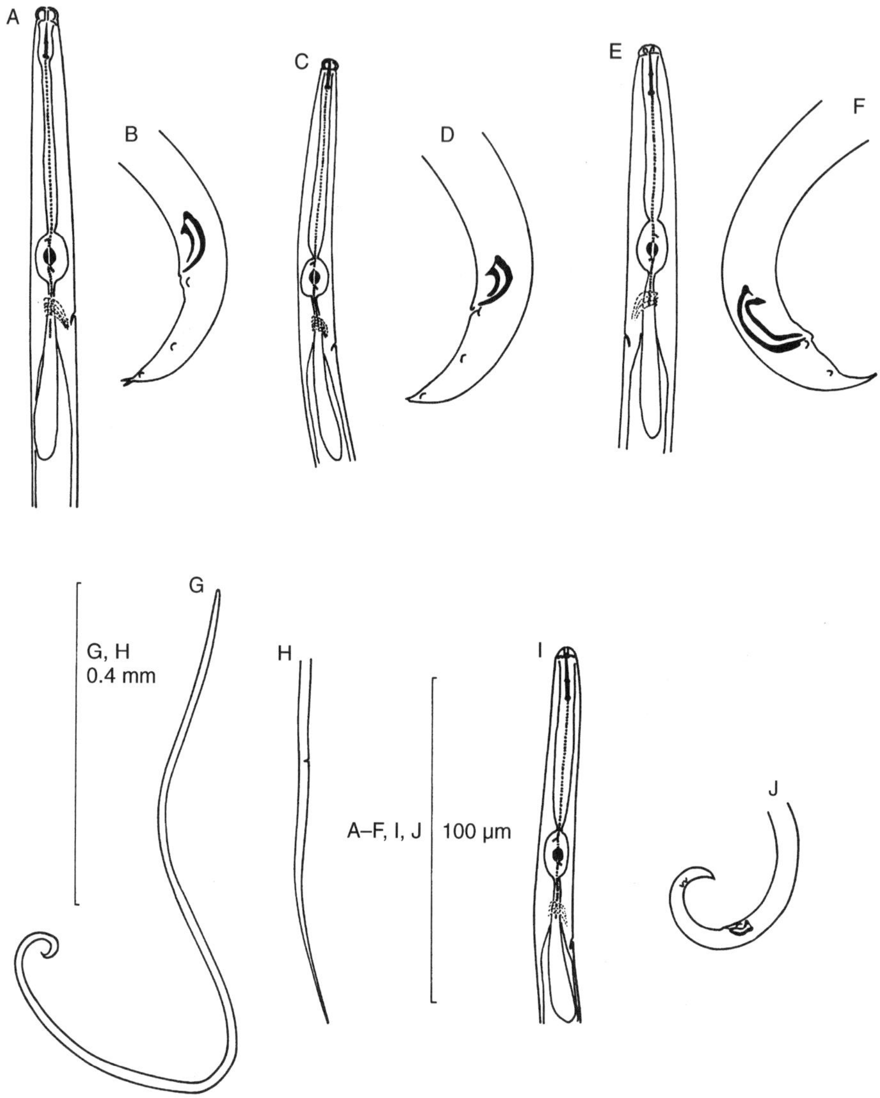

Fig. 2.3. Aphelenchid nematodes. *Aphelenchoides besseyi*: A, female, anterior end; B, male tail. *A. fragariae*: C, female anterior end; D, male tail. *Bursaphelenchus xylophilus*: E, female anterior end; F, male tail. *Rhadinaphelenchus cocophilus*: G, male; H, female posterior end; I, female anterior end; J, male tail.

Rice

Seed treatments

SEED FUMIGATION

Most of the nematodes in rice seed can be killed by fumigating it with methyl bromide. Germination and seedling vigour may be harmed if the seed is too moist or

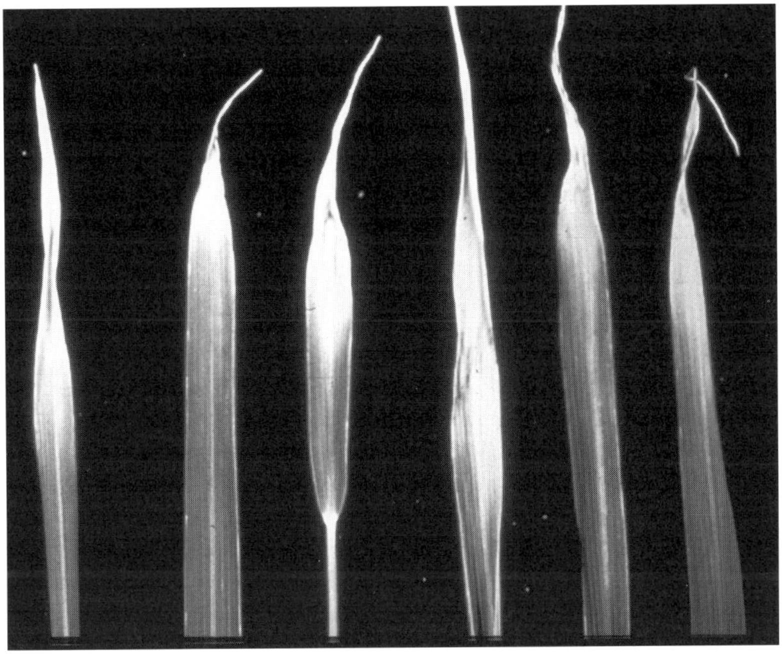

Fig. 2.4. Rice 'white tip' leaf symptom of *Aphelenchoides besseyi* (Photo: R.D. Riggs).

if the concentration × time product is too large. Konova and Vinnichuck (1959) found that in rice seed at 14% moisture, 98.8% of the nematodes were killed by exposure to 20.2 g methyl bromide m^{-3} for 12 h. Although 100% were killed by 100 g m^{-3} at 18–20°C for 72 h, seed germination was decreased by 12% and, if this treatment was made at 27–28°C, all the seeds were killed.

GAMMA RADIATION

Exposing rice seeds at 14% moisture to gamma radiation (at 30 Grey) reduced infestation from 29% to 2–6% (Aleksandrova, 1985).

SEED DRESSINGS

Dry coating infested seeds with effective nematicides can protect rice from 'white tip'. At 80 g (50 l)$^{-1}$ of seed, 10% 3-*p*-chlorophenyl-5-methyl rhodanine controlled *A. besseyi* satisfactorily on an experimental scale (Todd and Atkins, 1959). Coating rice seeds with a 10% dust of DSP (*O*, *O*-dimethyl-*O*-(4-dimethyl sulphamoyl-phenyl) phosphorothioate) at 10% (w/w) was very effective in controlling the nematode (Uebayashi *et al.*, 1971). The nematode was eliminated from rice seed by applying 2 kg etaphos a.i. t^{-1} seed (Chumakova *et al.*, 1985). Similarly effective was carbofuran applied at 667 g Furadan 75 PM (100 kg)$^{-1}$ seed (Martins *et al.*, 1976). Benomyl at 80 or 160 g (50 l)$^{-1}$ seed controlled the disease and increased rice yields (Templeton *et al.*, 1971). Benomyl applied at 0.3% (w/w) to rice seed plus spraying the rice plants 1–15 days after transplanting (at 2.5 g dm^{-2}) can prevent infestation with *A. besseyi* in the field and ensure new seed is free of infestation (Gergon and Prot, 1993).

NEMATICIDAL SEED DIPS

Dipping infested rice seed in nematicidal liquids can also be very effective against 'white tip' disease. Effective treatments include isofenfos and carbosulphone dips (Prasad *et al.*, 1986a), soaking for 24 h in REE-200 (an ester of rhodaneacetic acid) at a dilution of 1:600 in water, which eliminated the disease (Tanaka, 1959), and soaking seed for 12–24 h in aqueous suspensions (1:500) of malathion (80% e.c.), Sumithion (50% e.c.) or Lebaycid (50% e.c.) (Lee *et al.*, 1972). Penetration of the seed by the nematicide is increased by the use of ultrasound or vacuum infiltration. In this way, the efficacy of dipping in oxamyl solution was improved. Ultrasound, used in addition to oxamyl applied to the water in which the seeds were placed, killed 94% of the nematodes. Vacuum infiltration doubled the efficacy of oxamyl solutions (Nagy, 1987). Heavily infested rice seed was disinfested by soaking it first in 0.2% mancozeb plus monocrotophos, followed by either vacuum fumigation for 24 h at 30°C with about 32 g methyl bromide m^{-3}, or atmospheric fumigation with about 9.3 g aluminium phosphide m^{-3} (Prasad and Varaprasad, 1992).

HEAT THERAPY

Drying infested rice seed in the sun for 6 h at 40–41°C killed 87% of the nematodes. This was increased to 95–97.6% when the seed was first soaked in water or 1% sodium or potassium chloride and then sun dried to its original moisture content (Sivakumar, 1987). Soaking infested seeds in hot water also kills the nematodes, but temperature and immersion time must be controlled carefully to achieve the greatest kill without harming the seed. Although submersion of infested seed for 15 min in water at 54–57°C has been effective, the germination of some cultivars was decreased (Rahim, 1988). Presoaking millet seeds in hydrogen peroxide for 3 h , then immersing them in water at 50°C for 15 min killed all *A. besseyi* infesting them (Gotke and Mathur, 1993).

SEED TREATMENT EFFICACY

Whichever seed treatment is used, a high percentage kill is needed to eliminate 'white tip' disease, because the nematode can complete many generations on the rice plant. Small initial numbers can increase greatly, especially in the hotter rice-growing areas.

Seedbed treatments

'White tip' was prevented by treating rice seedbeds with Ekatin TD at 15 kg $(1000\ m)^{-2}$ at sowing plus irrigation, or diazinon at 8 or 12 kg $(1000\ m)^{-2}$, or 'Sumithion plus S 12927A' at 12 kg $(1000m)^{-2}$ (Yoshida *et al.*, 1967). A 0.1% solution of Cartap (1,3-bis(carbamoylthio)-2-(*N, N*-dimethylamino) propane hydrochloride) applied to the seedbed was equally effective (Uebayashi *et al.*, 1971).

Seedling dips

Infested rice seedlings can be freed of *A. besseyi* by dipping them in Folidol for 24 h (Nishizawa, 1953a) or in DBCP, Sumithion or Lebaycid (Lee *et al.*, 1972).

Field application of non-fumigant nematicides

Although effective, the field application of fensulphothion, ethoprophos or carbofuran granules (Jung and Iwaya, 1971; Lee *et al.*, 1976) is expensive. Choi *et al.* (1989) found disinfestation of seed with non-fumigant nematicides was more effective than applying the same compounds to paddy fields. However, transplanting rice seedlings raised in seedbeds treated with MEP, fenitrothion or Dasuzin, to paddy fields treated with 3% carbofuran granules the day before, was successful (Cho *et al.*, 1987).

Resistant cultivars

Cralley (1952) listed seven rice cultivars resistant to *A. besseyi* – Arkansas, Fortuna, Nira 43, Bluebonnet, Improved Bluebonnet, Century 23 and Century 52. Among 48 cultivars assessed in Japan, 33 were susceptible, nine were symptomless (tolerant) and six were resistant. Cralley (1954) reported six cultivars resistant and nine susceptible to the nematode (Nishizawa, 1953b). Other cultivars that have been found resistant include Bella Patha, Bluebonnet 50, Bluebelle, Century Patha, Vegold, Bonnet 73, Nira, Rexoro and Kuban. In assessing the susceptibility of over 1000 rice cultivars to the nematode, Popova *et al.* (1994) found Bluebonnet, Bluebonnet 50 and Starbonnet immune and ten other cultivars highly resistant to the nematode. Cultivars that are uninfested in infested soil may be considered immune, those very lightly infested resistant but those that are symptomless may or may not be infested. Symptomless cultivars which allow the nematode to multiply to some extent are strictly tolerant susceptible cultivars, not resistant cultivars.

Strawberry

'Summer crimp' disease of strawberry is best prevented by planting uninfested stock in uninfested soil. Three or four months of bare fallow in winter should minimize soil infestation, though *A. besseyi* can feed on certain fungi and may survive in small numbers in warm soil. Dipping infested transplants in thionazin (300 ppm) for 5 min has controlled the pest (Locascio *et al.*, 1967). Alternatively, aldicarb applied to the soil at or soon after planting is translocated to the young shoots and stolons and has given good control of *Ditylenchus dipsaci* in Britain (Lewis, 1979) and could protect first year strawberries from 'summer crimp'. EPPO recommends dipping actively growing strawberry runners, from which fully formed leaves have been removed, in water at 30°C for 30 min or at 40°C for 10 min and then transferring them to water at room temperature for 10 min (Anon., 1994).

Aphelenchoides fragariae (Ritzema-Bos, 1891) Christie, 1932 (Fig. 2.3)

Like *A. besseyi*, *A. fragariae* is primarily ectoparasitic on the leaves and buds of strawberry on which it causes 'spring crimp' or 'dwarf crimp' disease (Fig. 2.5).

(a)

(b)

(c)

Fig. 2.5. Effects of *Aphelenchoides fragariae* on strawberry: (a) severely injured strawberry beds; (b) crown dieback and leaf puckering due to nematode infestation; (c) infested leaflet (left) and uninfested leaflet (right) (Photos: D. McNamara).

Unlike *A. besseyi*, *A. fragariae* requires cool, moist conditions to enable it to move on plant surfaces, where it feeds and reproduces. The symptoms are most obvious in spring, when the young leaves are puckered or 'crimped'. In severe cases the central crown may be aborted, resulting in the formation of secondary crowns and delayed flowering. As flowering and fruiting are often affected, yields can be decreased. In Ireland, where strawberries were infested with both *A. fra-*

gariae and *A. ritzemabosi* (Schwartz, 1911) Steiner & Buhrer, 1932, yields were reduced by up to 60% (Duggan, 1969) and in Poland yield losses in seven cultivars due to these nematodes were 32–61% (Szczygiel, 1967).

Control

Although *A. fragariae* is a pest of mint, ferns and various flowers, e.g. violets, lilies and begonias (Anon., 1984), in soil it survives only for a few months in the absence of a host plant, so 'spring crimp' of strawberry is not a problem if uninfested stock is planted after a non-host crop. Any affected plants that appear in spring should be rogued and burnt to stop the spread of the disease to adjacent, healthy plants. Runners from infested mother plants should not be replanted.

Infestation of plants can be minimized, even eradicated, by hot water treatment of planting stock, dipping plants in nematicidal liquids, applying non-fumigant nematicides to the soil at planting or by spraying them several times with non-phytotoxic nematicides.

Hot water therapy

Healthy, vigorous strawberry plants were obtained by Staniland (*in litt.* to T. Goodey, 1933) by immersing them for 20 min in water at 43.3°C and then plunging them in cold water. Szczygiel (1969a) found August or early September the best time for hot water treatment in Poland and 10 min at 46–46.5°C was satisfactory. Damage to the plants was lessened if they were without roots and were first prewarmed to 25°C before immersing them for 10 min in water at 46.1°C. Not all the nematodes were killed but their numbers continued to decline when the treated plants were set out in the field (Hirling, 1974). In Britain, the recommended treatment is to immerse the runners in water at 46°C for 10 min and then to plunge them into cold water immediately afterwards. Good planting conditions and proper firming in are essential for success (Anon., 1984). Qiu, *et al.* (1994a) found that five cultivars of strawberry were unharmed when they were immersed for 8–10 min in water at 47.7°C, 10–15 min at 46.1°C or 20–30 min at 44.4°C.

Nematicidal dips

Although good control of the nematode in strawberry plants was obtained by dipping them in 0.07% ethyl parathion or in 0.1 or 0.18% methyl parathion (Szczygiel, 1969b), a 5 min dip in 0.14% thionazin was preferred, because it was less toxic to operators (Szczygiel, 1969c). Plants 99.5% free of the nematode were obtained by dipping them in 0.2% thionazin 2E (Tacconi, 1979). Szczygiel (1980a) found all *A. fragariae* were killed when plants were dipped in 0.1% thionazin or 0.2% fenamiphos, though they suffered some damage. In contrast, dipping in 0.2% oxamyl was less effective but did not harm the plants.

Field applied nematicides

Strawberry plants were freed of the nematode by applying fenamiphos or thionazin granules at 0.2 g per plant on three occasions (Szczygiel, 1980b). Nematode control was increased when strawberry plants, which had been dipped for 10 min in water at 46°C, were planted out in soil treated with heterophos, etaphos, diazinon or oxamyl (Metlitski and Pokhodenko, 1986). Good results have also been obtained from applying aldicarb to infested plants. Very effective control resulted from the application of 0.15 g aldicarb per plant or 0.2 g per plant over one year old (Szczygiel, 1970). At 0.2 g aldicarb m^{-1} row, both *A. fragariae* and *A. ritzemabosi* were so well controlled in runner beds that nematode-free plants were available for transplanting ten months later (Heide, 1976). One year old plants were undamaged and *A. fragariae* was controlled by 0.2 g aldicarb per plant (Szczygiel, 1980b) and the same dose applied in autumn and again the following spring, before regrowth, freed autumn-planted strawberries of the nematode (Szczygiel, 1980c). Spring application of granular nematicides is more effective than autumn application and 0.05 g aldicarb per plant in the spring has given good nematode control (Szczygiel, 1981).

In infested land, 80–90% of strawberry plants were freed of infestation by spraying the foliage with 5 l thionazin 20E or 3 l isazophos 50 ha^{-1} in September, October, November and February (Tacconi, 1979) or in September, October and once or twice in February or March (Tacconi, 1980).

3 Stem and Trunk Nematodes (*Bursaphelenchus*, *Rhadinaphelenchus* and *Ditylenchus*)

Bursaphelenchus xylophilus (Steiner & Buhrer, 1934) Nickle, 1970

In Japan, a lethal wilt disease of pine trees has been known since 1905. It is closely associated with infestation by *B. xylophilus*, the pine wilt nematode (Plate 1; Fig. 2.3). Inoculated with 30,000 nematodes per tree, *Pinus densiflora* and *P. thunbergii* were killed, whereas trees inoculated with sterilized water were unaffected (Kiyohara and Tokushige, 1971). Similar results were obtained by Mamiya (1972) on *P. densiflora*. There seems little doubt that *B. xylophilus*, or some other organism carried by it, is responsible for this disease. Virulent and non-virulent forms of the nematode exist. The virulent form causes widespread cavitation in tracheids, denaturation of cytoplasm in xylem parenchyma, leading on to water deficiency and death of seedlings (*P. thunbergii*). The non-virulent form causes a limited amount of cavitation but the cambium and xylem parenchyma are unaffected, so the seedlings are not subject to moisture deficit (Fukuda *et al.*, 1992). Within two weeks of inoculation with virulent *B. xylophilus*, oleoresin ceases to exude from artificial wounds to the trunk of pine trees, transpiration is reduced and the sapwood dries out. The pine needles turn yellow and die and the trees die 30–40 days after the first visible symptoms of the disease. Very large numbers (tens of millions) of *B. xylophilus* are found throughout the tree. The disease is much more severe in hot, dry summers, when the trees are suffering moisture stress. There is also circumstantial evidence that atmospheric pollution may weaken *P. densiflora* and make it more prone to disease (Nakane and Kimura, 1992). Kuroda and Ito (1992) found that the nematode migrated from inoculated branches of *P. thunbergii* into the main stem at about 1.5 m day^{-1} and that the fungal flora was the same in inoculated as in uninoculated trees. Five weeks after inoculation, when there was already widespread cavitation and necrosis of the cambium of the stem, *Ceratocystis* sp., a blue stain fungus, and bacteria were also found in the stem. All this strongly indicates that *B. xylophilus* is the cause of pine wilt disease. One female nematode lays about 79 eggs in a 28-day period and, as the life cycle (observed on *Botrytis cinerea*) took only four to five days at 25°C, it could give rise to 263,000 nematodes in 15 days. The nematode enters the tree via wounds made by its cerambycid vectors (species of *Monochamus*) during maturation feeding on pine twigs. The dauer juveniles moult to the adult stage within 24 h inside the tree and rapidly disperse throughout the trunk, branches and roots. Nematode reproduction is sexual. Third (dispersal) stage juveniles migrate to the outside of the pupal chamber of the beetle. There they become dauer juveniles attached to blue stain fungi, *Ceratocystis* spp. The adult beetles become contaminated with the nematodes on emergence from the pupal chambers.

The pine wilt nematode is widespread in Japan, where some 17% of the 2.6 million ha of pine forests is affected by the disease. Forty five of the 47 prefectures of Japan are affected by the disease and losses reached a peak of 2.4 million m^3 of timber in 1979 (Mamiya, 1984). The disease is also very damaging on the island of Okinawa. In the USA, *B. xylophilus* affects several species of commercial and ornamental pine trees. It is commoner in the eastern than in the western half of the USA, where it occurs in California, northern Texas and southern Arizona. Although *B. xylophilus* is unknown in Europe and the former USSR and is a quarantine pest in the EU, it appears to be conspecific with a German isolate of *B. mucronatus* with which it successfully hybridized and the resulting F_1 generation was able to multiply (Braasch, 1994).

In Japan, the principal vector of the nematode is *Monochamus alternatus*, the pine sawyer, though seven other species of cerambycid beetle can also transmit the dauer juveniles. In the USA, *M. alternatus* is absent and the known vectors are *M. carolinensis*, *M. scutellaris* and *M. titallator*.

Control of the Beetle Vectors

Destruction of the beetle vectors, before they can emerge from dead wood, can stop the spread of the disease to healthy trees. This can be done by felling dead trees late in summer or in autumn, when the beetle larvae are still beneath the bark, then peeling the bark, burning infested wood, or spraying unpeeled timber with an effective nematicide in emulsion form or in fuel oil. All the bark must be thoroughly wetted, so this is an expensive operation. Alternatively, aerial spraying of the foliage with a 3% fenitrothion emulsion at 60 l ha^{-1}, in late May to mid-June in Japan, prevents maturation feeding of *M. alternatus* on healthy twigs. Spraying the crowns from the ground with 2–3 l per tree of an 0.5% emulsion of fenitrothion, fenthion or carbaryl has also been effective, though only suitable on a small scale and not to tall trees (Mamiya, 1984). *Beauveria bassiana*, extracted from dead *M. alternatus*, is a highly effective biological control agent against the beetle. *M. alternatus* larvae were decreased by 70–80% by spraying dead stems with the nematode *Neoaplectana carpocapsae*. The emergence of *M. alternatus* from dead stems was prevented by covering them with vinyl sheet not less than 0.2 mm. thick (Ogawa, 1988).

Control of the Nematode

The nematode can be controlled by heat treatment of wood for export, by applying nematicides to living trees and by planting pines resistant to the nematode.

Wood chips for export can be freed of *B. xylophilus* by steam heating them to 85°C for 2 min, or 60°C for 8 min, or dipping them in hot water (55°C for 2 min or 50°C for 6 min) (Kim, 1986). In Georgia, USA, pinewood nematodes were killed in a 71°C kiln, when the greenwood temperature was 60°C. This point was reached in 1 h (5 cm × 10 cm timber) or 4 h (15 cm × 15 cm timber). *Monochamus* spp. were also killed at 60°C (Dwinell, 1990). The nematode was similarly killed in *Pinus* green sawnwood imported into Finland from Canada, by placing the timber in a sawmill kiln at 70°C for a period dependent on the dimensions of the wood (Tomminen and Nuorteva, 1992).

A single application of 50 g fensulphothion injected into the trunk of a tree 25 cm in diameter about 1.5 m above ground level protected it from infestation by *B. xylophilus* for at least three years (Matsuura *et al.*, 1975, 1976). The nematode has also been excluded by trunk injection of mesulfenfos, by incorporating disulphoton in the soil (Takase *et al.*, 1982) or by injecting the trunk, trunk base or roots with fensulphothion three weeks before inoculation with *B. xylophilus* (Matsuura, 1984). Mesulfenfos (50%) injected into stands of *Pinus* prevented pine wilt for up to two years. Injection with 4% levermyzol [levamisole?] or 12.5% morantel (anthelmintics) protected them for up to three years (Ogawa, 1988). Sprayed on the crowns of pine trees, fenitrothion, deltamethrin or cypermethrin reduced nematode damage and killed adult beetles, but three sprays a year were needed to ensure all beetles were killed because they emerge over a long period. The beetle larvae were killed and the nematodes were controlled when omethoate, monocrotophos, acephate or carbofuran was injected into the trunks of diseased pines. Sprayed on diseased logs, the first three of these compounds retarded emergence (Li *et al.*, 1994). The antibiotic Jietacin A (Giencke *et al.*, 1989) and *N*-octadecylamine (Nagase *et al.*, 1982) were very toxic to the nematode. Injecting pine trunks, two weeks after inoculating them with *B. xylophilus*, with mesulfenfos, fensulphothion or thionazin promoted good recovery of the trees (Matsuura, 1984), so treating trees, when the first symptoms of wilt disease are detected, is worthwhile. Pine tree mortality due to *B. xylophilus* was decreased in China when 200 g 5% fenamiphos granules was applied to the soil around the roots (Zhu *et al.*, 1994).

Pinus spp. found resistant to *B. xylophilus* include *P. taeda*, *P. rigida*, *P. elliottii*, *P. taiwanensis*, *P. excelsa*, *P. strobus*, *P. massoniana*, *P. thunbergii* × *P. massoniana*, *P. resinosa*, *P. tabulaeformis*, *P. banksiana*, *P. contorta*, *P. echinata*, *P. palustris*, *P. pungens*, *P. bungeana*, *P. caribaea*, *P. armandii*, *P. jeffreyi* (intolerant) and pine cv. Virginia. *Pinus densiflora* (Japanese red pine) and *P. thunbergii* (Japanese black pine) are highly susceptible to the disease and very important in Japan. However, resistant trees of both species have been identified by inoculating 40–80 year old trees with *c.* 2000 *B. xylophilus* annually for three years. Seedlings two to five years old from the resistant trees were also highly resistant (Kasuya *et al.*, 1990). Open-pollinated *P. thunbergii* seedlings which survived inoculation in July with 10,000 *B. xylophilus* mostly survived a retest with 10,000 nematodes per plant in the following year (Nagahama, 1987). Inoculated with 10,000 nematodes per tree, there was no extensive cavitation of tracheae or xylem and no significant blockage of sap ascent in resistant *P. taeda*. Similarly, inoculated resistant *P. strobus* showed some transient wilting in some branches (Kuroda *et al.*, 1991). Resistance to *B. xylophilus* was induced in *P. thunbergii* by prior inoculation with non-pathogenic isolates of the nematode. A 10–30 day interval between inducer and challenger inoculation resulted in a high level of resistance, which lasted at least 120 days. The induced resistance was species specific, not isolate specific (Kiyohara, 1984).

The resistance of *P. massoniana* is thought due to the presence of the nematode repellent α-humulene and the nematicides pinosylvin monomethylether and (−)-nortrachelogenin in the heartwood and to the nematicidal compounds methyl ferulate and (+)-pinoresinol in the bark. The resistance of *P. strobus* and *P. palustris* may also be linked to the presence of pinosylvin monomethyl ether in the heartwood and bark of *P. strobus* and in the bark of *P. palustris*. Pinosylvin monomethyl ether had little anticholinesterase activity but was the most nematotoxic of these compounds with an LD_{50} of 4 ppm (Suga *et al.*, 1993).

Rhadinaphelenchus cocophilus (Cobb, 1919) J.B. Goodey, 1960

'Red ring' disease of coconut and other palms is caused by infestation of roots, trunk and leaves by *R. cocophilus*, the red ring nematode (Plate 2; Fig. 2.3). It occurs in the West Indies, except for northern Caribbean islands, and in Central and South America. It is unknown in Cuba or Florida but has been reported in the Dominican Republic. Christie (1959) mentions a report of Schuurmans Stekhoven that this nematode was found in oil palm in Zaire and Malaguti (1953) found the nematode in oil palms imported into Venezuela from Africa. The nematode is otherwise unknown in the Old World at present. In the New World, *R. cocophilus* is of major importance as a pest of coconut and oil palm and it is known to multiply in 15 other species of palm.

In coconut, red ring disease affects plants 2.5–10 years old, especially those 4–7 years old. It is rarely found in plants less than 2 or more than 20 years old. Blair (1969) pointed out that coconut palms are most susceptible to the disease when the internodes are longest, i.e. at 5–7 years old crowns are 3.6–7.2 m long, whereas at 1–2 years old and at 20 years old crowns are only 0.45–0.6 m long. Long internodes are thought favourable to the transmission of the nematode by its principal insect vector, the palm weevil, *Rhynchophorus palmarum*. The nematode can also be damaging to oil palms, though the disease takes longer to develop and the tissues are less heavily infested than those of coconut. The nematode can cause a form of 'little leaf' disease of oil palms in Surinam (Van Hoof and Seinhorst, 1962). Symptoms of red ring disease in coconuts are first the premature shedding of nuts, then the older leaves turn yellow, then brown and hang downwards. The crown may be destroyed later by palm weevils feeding on the diseased tissues. The cortex of infested roots turns orange to faint red and the core of infested petioles turns dull red then brown up to 75 cm from the leaf axil. The most characteristic symptom, however, is the 'red ring' of diseased tissue, full of *R. cocophilus*, 2–4 cm wide, located 2.5–5 cm beneath the trunk surface. This red ring extends from the roots to the crown of the tree. Adult nematodes are usually located outside the red ring tissue. Males are common and sexual reproduction seems likely. The life cycle is short and multiplication is great. One gram of infested tissue may harbour 5000–10,000 nematodes. The tree dies 3–4 months after the first symptoms have appeared. Infested oil palm tissues harbour less than one tenth of the nematodes found in the corresponding coconut tissues and the disease takes ten times as long to kill the oil palm. Diseased coconut palms take up less water than healthy palms, probably from the poisoning effect on the leaves of a toxin produced in the red ring tissue and/or occlusion of the xylem vessels by tyloses.

There has been much discussion about the transmission of *R. cocophilus* to coconut. Fenwick (1962) showed that the nematode could penetrate undamaged roots or any part of the crown of the coconut palm. Infested palm tissue dropped on moist soil resulted in infestation of the roots and red ring disease (Fenwick, 1969). Similarly, the nematode can be transferred to crown tissues on the bodies of the palm weevil. Third stage (infective) juveniles are also found in the bodies of weevil larvae and adults and some are deposited with the eggs of the weevil, which are usually injected into a leaf axil. Weevil transmission of the nematode is of greater concern than root transmission, because it is much faster. However, there is more chance of controlling weevil transmission than root transmission of the disease.

Coconut

Attempts to cure diseased coconut palms of 'red ring' disease with nematicides have been largely unsuccessful, or are impractical, and attention has been concentrated on preventing rapid spread of the disease by control of its principal vector, the palm weevil, *R. palmarum*. It is most important to recognize the early symptoms of the disease, quickly confirm that the nematode is infesting the trunk by analysing bore samples and then rapidly killing the tree with an arboricide, e.g. cacodylic acid, having first sprayed the crown with an insecticide to kill the weevils. Only when the tree is totally dead should it be cut down, because chips of infested tissue on the soil surface can infest the roots of neighbouring, healthy palms, especially when the soil is moist (Fenwick, 1969). All tissues should be burnt *in situ*. If the palm is in too advanced a stage of the disease, it cannot be killed with herbicides but must be cut down, sprayed with insecticide (usually methomyl) and burnt two weeks later. The weevils are attracted to fermenting, damaged, coconut tissue 'bait' soaked in methomyl, contained in wire mesh baskets (two ha^{-1}) left on the ground (Griffith and Koshy, 1990).

Oil Palm

A local wild palm, *Oenocarpus distichus*, was found to play host to both *R. cocophilus* and *R. palmarum* on an oil palm estate in Pará, Brazil and the nematode was carried on the weevil to infest the oil palms. As for coconut, early removal of infested oil palms, prevention of wounding and killing all the diseased tissues limited the spread of infestation, without use of chemical treatments

(Schuiling and van Dinther, 1981). Similar procedures have worked well in Pará and Bahia provinces of Brazil since 1975, and trapping the weevils in baits of oil palm cubes limited the spread of the disease (Morin *et al.*, 1986).

Ditylenchus angustus (Butler, 1913) Filipjev, 1936

(Fig. 3.1)

This is an ectoparasite of the shoots of rice, other *Oryza* spp. and the weeds *Leersia hexandra, Echinochloa colona* and *Sacciolepis interrupta* (Plate 2). Being wholly ectoparasitic on shoot tissues, it needs a relative humidity of 75% or more for it to move over plant surfaces. It is, therefore, not surprising that, as a rice parasite, it is found mainly in deep water rice areas and in lowland rice in river deltas – the Ganges Delta in Bangladesh and India, Uttar Pradesh in India, the Mekong Delta in Vietnam, the Irrawaddy Delta in Burma, the Nile Delta of Egypt, Malaysia, south Thailand and the Philippines. In Madagascar, *D. angustus* is widespread, except on the west coast, and is particularly damaging on the central plateaux around Tananarive.

The nematode feeds on young shoot tissues but is restricted to the growing point of deep water rice. As a result, the tissues become twisted and distorted, with white patches, later brown lesions, on the leaf bases. The panicle may not emerge, or only emerge partially, bearing few, shrivelled grains. The disease is called 'Ufra' or 'Dak Pora' in Bangladesh, 'Akhet-pet' in Burma, 'Yad-ngo' in south Thailand and 'Tiem Dot San' in Vietnam. The disease occurs erratically, even in the same field, as the nematode is sensitive to environmental conditions. It is greater in wet years and in areas with high water tables. It does not, however, survive submersion in water for long periods. Conversely, in a coiled, anabiotic state on rice grains it is very resistant to desiccation. In severe infestations, yield loss can approach 100%. The nematode infests some 200,000 ha in Bangladesh and about 100,000 ha in the Mekong Delta of Vietnam. As *D. angustus* can pass three generations on the rice plant, serious yield losses can result from transplanting seedlings of which only 4–10 % are infested. As the rice plants mature, 'nematode wool' (masses of drought resistant, coiled nematodes) forms on the shoots and inside the glumes. In this state, the nematodes can survive many months of drought, reviving and infesting the next crop, when moisture returns. Infestations do not seem to result from sowing infested seed but from transfer from infested stubbles or volunteers of the previous rice crop or from wild rice plants. Ibrahim and Perry (1994) suggest that the fourth stage juveniles are the main survivors between rice crops.

Control

Seed treatment

Sun drying rice seeds to less than 12% moisture killed *D. angustus* on them (Cuc and Giang, 1982).

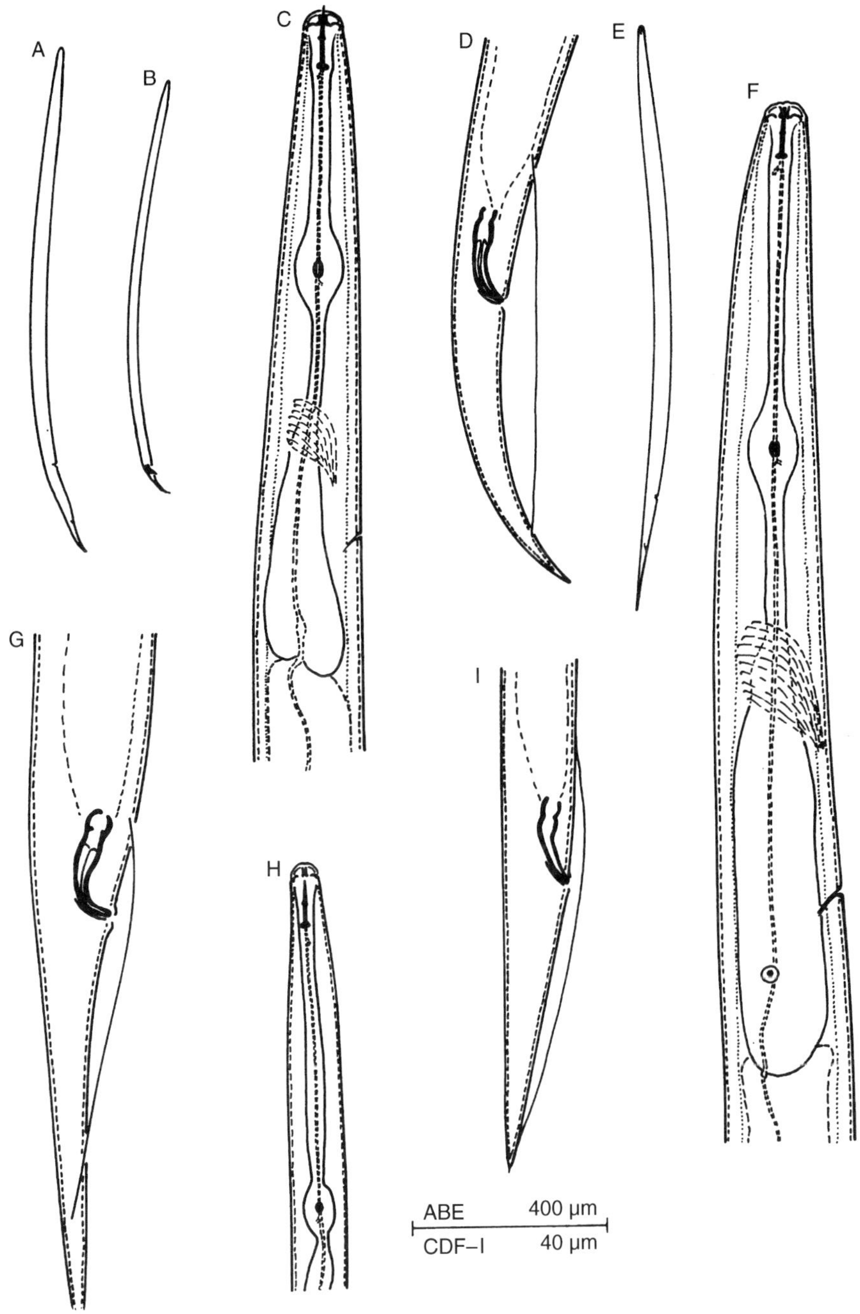

Fig. 3.1. Species of *Ditylenchus*. *D. destructor*: A, female; B, male; C, female anterior end; D, male tail. *D. dipsaci*: E, female; F, female anterior end; G, male tail. *D. angustus*: H, female anterior end; I, male tail.

Cultural control

As infestations seem to originate from infested stubbles, volunteer or wild rice and perhaps the three weeds referred to above, their elimination can prevent disease in newly sown or transplanted healthy rice. Stubble burning or ploughing under the stubbles can greatly diminish disease incidence, where practicable. Delaying planting until after wet periods or using early maturing cultivars can greatly lessen disease. Rotating rice with other (non-host) crops can also lessen the risk of yield loss. Flood control measures can slow the spread of the pest (Butler, 1919; Cox, 1980; Rahman and Taylor, 1983). In Assam, Rayada 16–06 and late sown susceptible cv. Rangabao suffered little or no infestation, while the early cv. Padmapani escaped infestation (Bhagwati and Bora, 1993).

Chemical control

Although hexadrin at one part in 700 in water sprayed twice at 2 week intervals was said to be 100% effective in eliminating the nematode from rice plants (Pal, 1970), as a persistent aromatic chlorinated hydrocarbon its use cannot be recommended. Applying 1 kg carbofuran ha^{-1} (as 3% granules) to the soil before transplanting lightly infested rice (cv. IR 8) into it, decreased the disease and the spread of the nematode and increased grain yields (Rahman *et al.*, 1981). Treating ufra-affected soil in Bangladesh with 2.5 kg ethoprophos or 1 kg carbofuran ha^{-1} at transplanting lessened infestation of the panicles and increased rice yields by 0.9 and 0.8 t ha^{-1}, respectively (Mondal *et al.*, 1990). Widespread chemical control of *D. angustus* in rice seems unlikely in view of the costs of the nematicides and the risks of exposure to them in unprotected workers.

Resistant cultivars

In Assam, India, Rayada 16-06, a deep water rice line, is very resistant to the nematode (Bora and Medhi, 1992) and in Bangladesh nine Rayada 16 series deep water rice lines were highly resistant (Rahman, 1987). From these and other lines, new resistant cultivars may become available.

Ditylenchus destructor Thorne, 1945 (Fig. 3.1)

Although it attacks a wide range of crops and weeds and can feed on many soil fungi, *D. destructor* is most important as a pest of potatoes (Fig. 3.2). It can also be troublesome in iris, garlic and groundnut. On potato it is currently of importance in western and central–southern areas of the former USSR and of some importance in Germany and the Netherlands. On potato it causes a dry rot of the tubers, which are invaded from infested soil via lenticels, eyes or stolons (Fig. 3.2). In contrast to *D. dipsaci*, *D. destructor* has no drought-resistant stage, so it cannot survive in very dry soils. It is, however, able to withstand a wide range of soil temperatures (−28°C to + 34°C). In potato tubers, it feeds and multiplies just below the skin, which is not attacked, producing lesions, which are at first whitish, then yellowish and finally black, as secondary invading bacteria and fungi lead to a dry rot. In contrast to *D. dipsaci*, this rot is located just below the skin and does not extend throughout the tuber, which can occur with *D. dipsaci* infestation.

Fig. 3.2. Potato dry rot symptoms due to *D. destructor* (Photo: D.J. Hooper).

Potato

Crop rotation

Kiryanova and Krall (1971) controlled *D. destructor* satisfactorily by growing potatoes in three-, or four-course rotations with small grains, vetch and lupins. Three years monoculture of non-hosts, such as buckwheat, carrot or lupins cleared soil of infestation, so that a crop of uninfested potatoes was obtained in the fourth year (Efremenko and Burshtein, 1975). *Ditylenchus destructor* was greatly decreased when healthy seed potatoes were planted late in spring as the fourth crop after rice, lucerne and winter rye and the daughter tubers were harvested early (Abylova and Vasilevskii, 1983). Control by crop rotation may not succeed if the rotation includes host crops, if many (infested) volunteer potatoes are present, or if weed control is inadequate. Success may also be limited if the nematode survives on soil fungi.

Seed tubers

Use of uninfested seed tubers or seed tubers selected for low levels of infestation can minimize or lessen tuber dry rot (Ivanova, 1971). As it does not attack the stems, nematode-free tubers can be obtained from infested tubers by raising plants from stem cuttings (Bashkardin, 1971). Tubers were freed of infestation, without phytotoxicity, by dipping them in autumn for 1 h in a 1% suspension of Nemafos 46 EC (Wilski, 1972). Solutions of 0.025–0.05% heterophos applied to seed tubers lessened infestation of the daughter tubers and increased their yield (Chukantseva, 1976, 1983). Infestation of seed tubers was reduced from 25% in untreated soil to 2.7% in soil treated with 50 g heterophos m^{-2} before planting and the yield of seed tubers was increased from 400 to 700 g per plant.

Application of 100 kg heterophos ha^{-1} to soil infested with the nematode reduced tuber infestation from 36% to 0% (Vorona, 1984).

Soil treatment

Since its discovery in Wisconsin, USA in 1953, the nematode has been eliminated there by soil fumigation, strict quarantine of infested tubers and supervised disposal of potatoes from infested fields. Autumn plough-sole application of 37 l ethylene dibromide ha^{-1}, followed by reploughing to bury the top layer of the soil plus a further 18.5 l ha^{-1}, gave such good control that not one infested tuber could be found after ten subsequent potato crops (Darling, 1959; Darling *et al.*, 1983). At 750 kg ha^{-1}, metham-sodium applied to the soil before planting decreased infestation of harvested tubers from 27.5% to 4% (Abylova and Vasilevskii, 1983).

Resistance

Resistance to *D. destructor* is the exception among potato cultivars, but partial resistance has been found in some cultivars. In Poland, Belg, Grom, Pimpernel, Robijn and Rode Star had only 5% of tubers infested with up to 10% of surface damage. In Belorussia, Cardi, Franzi, Scutella, Kardia and Patroness have been reported resistant. Okopnyĭ *et al.* (1983a) found that potatoes resistant to *D. destructor* had a greater total alkaloid content than susceptible potatoes. Phenolic compounds, notably chlorogenic acid, occurred in larger amounts in resistant than in susceptible potatoes. Chlorogenic acid and the alkaloid solanine fully inhibited nematode pectinase and cellulase, which may explain the resistance of such cultivars to *D. destructor*. Two phytoalexins, rishitin and lubimin, are produced in potato tissue infested with *D. destructor* and *D. dipsaci*. The amount of rishitin produced increases with the number of nematodes in the tissues and the extent of necrosis and is also greater in resistant than in susceptible cultivars. Rishitin may therefore play a part in cultivar resistance to these nematodes (Zinovieva and Chalova, 1987).

Iris

Ditylenchus destructor was well controlled when iris bulbs were immersed in water at 53°C for 3 h, or in 0.5% formalin solution (Nakanishi, 1979). Infestation at harvest was minimized by applying fenamiphos in a 30 cm wide band directly over iris bulbs during planting (Haglund, 1983). On Wedgewood iris, subsoil bands of 16.8 kg of 10% phorate granules ha^{-1} controlled the nematode, without phytotoxicity. Applied at planting, oxamyl drench (900 ppm) or granules of oxamyl (9 kg 10% granules ha^{-1}), fenamiphos (8.4 kg 15% granules ha^{-1}) or carbofuran (2.8 kg 10% granules ha^{-1}) were also effective, despite the comparatively small amounts of active ingredient applied by the granules. Dipping the bulbs in fenamiphos (250 ppm) for 20 min before planting them was also effective (Maggenti and Hart, 1975).

Garlic

Applied to garlic seed at 1% (w/w), a wettable powder thiram-benomyl mixture prevented rotting of garlic in the growing season. Dry storage of harvested garlic at 34–36°C for 12–17 days greatly decreased the number of *D. destructor* in the tissues (Fujimura *et al.*, 1989).

Groundnut

Advancing groundnut harvest by 15 days lessened loss of seed quality due to *D. destructor* in S. Africa (Venter *et al.*, 1992). The cultivars BP 1, Buffelspoort, Kimberley Choice, Late Harvest, Sackfiller, Up-to-date and Vanderplant were poor hosts of this race.

Ditylenchus dipsaci (Kuhn, 1857) Filipjev, 1936 (Fig. 3.1)

Stem nematode, *D. dipsaci*, is a damaging pest of many crop plants (see Figs 3.3, 3.4), which can also infest many species of weeds and other uncultivated plants (Goodey *et al.*, 1965). The nematode occurs worldwide in temperate soils, where cool, moist conditions favour its invasion of plants (e.g. Griffin, 1987a). Serious yield losses may result from attacks in cereals (oats, rye, maize), legumes (field beans, peas, lucerne, red clover, white clover), root crops (sugarbeet, fodder beet, mangels, turnip, carrot and potato), strawberry and flower bulbs (narcissus, tulip and hyacinth). All shoot tissues and sometimes roots may be infested, i.e. stems, leaves, flowers, seeds, bulbs and the crowns of storage roots. Although the nematode may invade roots, it is usually in much smaller numbers than in shoot tissues.

In susceptible plants, stem nematodes secrete pectolytic enzymes, which cause dissolution of the middle lamellae between parenchyma cells (Chitwood and Krusberg, 1977), so the cells separate and hypertrophy or collapse as a result of the nematodes feeding on them. The tissues often become swollen or 'bloated' and the nematodes are then able to move freely through the intercellular spaces created by separation and death of cells. However, Hesling (1972) found that although it multiplied greatly in a range of crop plants, the tulip race of the nematode often did so without inducing obvious symptoms. The life cycle is quite short, e.g. 19–23 days in onion plants at 15°C and a female can lay up to 500 eggs (laid singly) and persist for two or three generations (Yuksel, 1960), so the nematode can easily multiply 1000-fold in a season on a suitable host plant. Indeed, it may greatly exceed this. Hesling (1970) reported narcissus stem nematode increasing 18,000-fold in narcissus bulbs in the growing season. The nematodes reproduce sexually and the hatching juveniles are second stage, as for many other plant-parasitic nematodes. The fourth stage juveniles are very resistant to desiccation in plant tissues that dry out gradually. In such tissue, they aggregate and coil up like watch springs. As this 'eelworm wool', they can survive in an anabiotic state for many years in a dry atmosphere (Fielding, 1951).

(a)

(b)

(c)

Fig. 3.3. (*and opposite*) (a) Crop injury by *Ditylenchus dipsaci*. Patch of onions killed by *D. dipsaci*. (b) Healthy onion (right) and infested (bloated) onions (left) (Photo: D.J. Hooper). (c) Crown rot of sugarbeet (Photo: R.A. Dunning). (d) Healthy oat seedling (right) and infested (bloated) oat seedling (left) (Photo: D.J. Hooper).

(d)

Fig. 3.3. *Continued*

Heavily infested plant tissues may be destroyed, partly as a result of nematode feeding and partly or largely by secondary invasions of bacteria and fungi, which cause a generalized rotting of affected tissues, as, for example, in 'crown rot' of sugarbeet. *Dityenchus dipsaci* has also been shown to exacerbate *Erwinia* soft rot of the crown buds of rhubarb (Metcalfe, 1940) and *Clavibacter michiganense* ssp. *insidiosum* wilt of lucerne (Hawn, 1963). In the latter disease, the nematodes were found to carry the bacteria on their bodies into the crown buds (Hawn, 1971), as well as lowering plant resistance to the bacterium (Hawn and Hanna, 1967). Hijinck (1963) found an almost perfect correlation between the numbers of the nematode in the soil and the percentage of potato plants infected by *Phoma solanicola*, yields of tubers being negatively correlated with increase in incidence of *D. dipsaci* and *Phoma*. The nematode has also increased the severity of wilt in lucerne due to *Verticillium albo-atrum* in susceptible cv. Vernal but not in resistant cv. Maris Kabul (Vrain, 1987). Griffin (1990) found a synergistic relationship between the nematode and *Fusarium oxysporum* f.sp. *medicaginis* in lucerne susceptible to both organisms.

More than 20 'preferred-host races' of *D. dipsaci* have been recognized (Sturhan, 1971), which often differ in their host ranges e.g. oat–onion, narcissus, tulip, giant, red clover, white clover, lucerne, rye and hyacinth races. However, there are many reports that different isolates of a race can also differ in

(a)

(b)

Fig 3.4.(*and opposite*) (a) Crop injury by *Ditylenchus dipsaci.* Patch of lucerne stunted by *D. dipsaci.* (b) Healthy lucerne plant (right) and infested (bloated) plants middle and left. (c) *D. dipsaci* controlled in lucerne by carbofuran (left) compared with untreated rows (right). (d) *D. dipsaci* controlled by aldicarb in the onion rows (left) compared with untreated rows (right).

host range (e.g. Sturhan, 1965), a race may multiply by different amounts on different cultivars of the preferred host (Hesling, 1966a,1972), a cultivar resistant to one or more populations of a race may be susceptible to other populations of that race (Whitehead, 1984) and field or wild populations of the nematode may be mixtures of races. Despite all this variation, the races can interbreed (Sturhan,1964a; Webster, 1964; Eriksson, 1965) and so belong to the same nematode species, but it is doubtful if the races can ever be properly defined (Sturhan, 1969). Although lucerne only allowed lucerne race nematodes to multiply in its tissues, other 'preferred hosts' allowed multiplication of two or more races in their tissues. A reliable differential host test for identification of the races therefore seems unattainable (Whitehead *et al.*, 1987a). Viglierchio (1971) reported an isolate of *D. dipsaci* from garlic able to multiply greatly in lucerne seedlings and moderately in red clover and white clover seedlings but perhaps this was a mixture of races.

Ditylenchus dipsaci lives in the upper layers of the soil, migrating to the soil surface after rain (Wallace, 1962) or to narcissus bulbs (Hesling, 1967).

(c)

(d)

Fig. 3.4. *Continued*

Numbers of the nematode decrease dramatically after a susceptible crop has been harvested. This is especially so in sandy soils, whereas in clay soils the nematode can persist, even in small numbers, for many years (Seinhorst, 1956). As pointed out by Hesling (1966b), the nematodes may survive between host crops by feeding and multiplying on cotyledonary leaves and petioles (Courtney, 1962), because these are susceptible even when true leaves and stems are resistant (Grundbacher, 1962). They may even survive by feeding on seedlings which germinate but fail to emerge from the soil. Viglierchio (1971) found that an isolate from garlic was able to feed and reproduce slightly on two species of fungus. Very small numbers of the nematode, which may not be detected by normal sampling and extraction procedures, may cause serious damage to sensitive crops like onion and sugarbeet, given cool, moist conditions in spring. This makes risk assessment by soil sampling impractical for such crops.

Control

Control of *D. dipsaci* may be obtained to a greater or lesser extent by management techniques, crop rotations, use of resistant crop cultivars, hot water treatment of seeds and bulbs, by the use of nematicides and by integrated control programmes.

Cultural techniques

Damage to very sensitive crops can sometimes be avoided by planting seeds or sets when soil conditions are less favourable to the nematodes. In north-west France, for example, maize drilled in May is little affected, whereas maize drilled in April may be badly affected by 'maize toppling' disease, caused by as few as 50 nematodes per plant (Caubel, 1973a,b). Soil solarization has given excellent control of *D. dipsaci* and increased yield of garlic in Israel (Siti *et al.*, 1982) and in Italy, soil solarization for 2–8 weeks controlled the nematode equally as well as 300 l D-D ha^{-1}. After eight weeks solarization, only 1.6% of the viable nematodes remained in the soil (Greco *et al.*, 1985). Flooding was found to decrease numbers of *D. dipsaci* in sandy soils in Dutch bulb fields in proportion to its duration. After nine weeks all were killed (Muller and van Aartrijk, 1989).

Crop rotation

As a general rule, losses due to stem nematode are less likely when susceptible and intolerant crop plants are grown infrequently than when they are grown in short rotations, or as monocultures. However, as the nematodes may persist for many years in the soil in the absence of host crops and may multiply rapidly when one is grown, crop rotations cannot be relied on to prevent losses due to *D. dipsaci*. Although their complete host ranges are unknown, some races have very wide host ranges – oat–onion race (Hooper, 1972), narcissus race (Hesling, 1966a), tulip race (Hesling, 1972) and lucerne race (Whitehead *et al.*, 1987a). Other races seem to have narrower host ranges – red clover race (Courtney, 1962), giant race (Hooper, 1984) and white clover race (Whitehead *et al.*, 1987a) and some are intermediate, such as race 'X' of carrot (Goodey and Brown, 1956) and the south-west Ontario race (Sayre and Mountain, 1962), both of which seem to be oat–onion race type but with more restricted host ranges. As a field may be infested with a mixture of races, effective rotations may be difficult or impossible to designate. Nevertheless, some preferred hosts are non-hosts of other races. Lucerne was susceptible to 13 English isolates of lucerne race but resistant to 54 isolates of *D. dipsaci* from other crops (Whitehead *et al.*, 1987a). Similarly, red clover, white clover, carrot, tulip, hyacinth and narcissus are selectively susceptible to stem nematodes, so they could also be useful where their preferred races are absent. For example, narcissus race does not multiply in tulip and hyacinth race does not multiply in narcissus or tulip. Tulip race, however, multiplies in narcissus and hyacinth (Southey, 1957). The efficacy of an apparently non-host crop plant needs to be determined before it is widely used. Sainfoin (*Onobrychis viciaefolia*) is an alternative to lucerne in soil infested with lucerne race but in the USA a cultivar of sainfoin was nearly as good a host as susceptible lucerne (Griffin *et al.*, 1975). Control of stem nematode through crop rotation is often not an option because non-host crops may not be marketable.

Some crop plants are susceptible to several or many races – broad bean (*Vicia faba*), sugarbeet and strawberry, for example, so their inclusion in a rotation could be counter-productive. Crop rotations alone are, therefore, unlikely to control stem nematodes, though they may be of some benefit when linked to the use of resistant cultivars and/or nematicides.

Resistance

Among the principal crop plants susceptible to *D. dipsaci*, resistance to it has been bred into oat (*Avena sativa* and *A. byzantina*), lucerne and red clover. Some cultivars of white clover also have a degree of resistance. There are no cultivars of onion, sugarbeet, peas or field beans resistant to the nematode but some cultivars of garlic, strawberry and sweet potato are resistant (Table 3.1). Some 'resistant' cultivars are really tolerant–susceptible because although they grow well, despite infestation, they allow the nematodes to multiply in their tissues. As breeding for resistance is usually lengthy and hence costly, it cannot be justified for crops in which attacks are infrequent or of minor importance. Control of such problems must be left to other *ad hoc* measures.

OATS

Resistance to *D. dipsaci*, which induces 'tulip root' disease in oats was first studied by T. Goodey (1937), who found resistance to the nematode in Grey Winter and Aberystwyth S 81 oats. Subsequently, resistance was demonstrated in the commercial cultivars Milford (S 225) (McLeod *et al.*, 1954), and in S 172, Early Grey Winter and Unique (Jones *et al.*, 1955), all of which are winter oats. The spring oat Manod (S 235) was reported resistant to *D. dipsaci* by Davies (1960). The source of its resistance is in doubt. Goodey and Hooper (1962) found resistance to oat stem nematode in the spring oat Milford and in the winter oats S 172, Grey Winter ON 3, Picton ON 425 and S 231. Of the six isolates used, one induced mild tulip root in Grey Winter at panicle emergence, another similarly in S 172 and a third isolate induced tulip root in Grey Winter and Picton oats. These results were clearly not due to differences in aggressiveness between the isolates, but to cultivar × isolate interactions. In Britain, at that time, five winter oats

Table 3.1. Some cultivars resistant to *Ditylenchus dipsaci.*

Crop plant	Cultivar
Garlic	Grodvensk, G 109 (Parus), Voronezhskii 240, Grodnenskii K 109, Tambovskii 22, Tambovskii 13, Belgorodskii 10.
Lucerne	Prime, WL 457, WL 317, 5364, Rayen INIA, Palihue INIA, Huinca INIA, Nematol, IV 532, IV 631, IV 684, Sechine Hamedan, Mohajeran, Zuzana, Resistar, Nemastan, Hamedan Huinca INIA, Alfa II, Fortier Pergamino M.A.G., Deseret (= Kayseri), Sitel, Kara, Vertus, Resis, San Martin, Nemasint, Sverre, Euver, Lahontan, Trek, M 5, Talent, Vernal, Mv Nematol, Lesina, Washoe, Resistador II, Pioneer 581, Falkiner, Cordura 73, Nova, WL 310, Nevada Synthetic XX, Aquarius, '5246'.
Oat	Resistant to 'oat race' ... Elan, Curt, Moregrain, Algerian, Avon, Cassis, Solva, Craig, Image, Gerald, Maris Tabard, Milford, Pennant, Peniarth, S 172, Barnwell, Milns Prophet, Panema.
Rye	Heetvelder.
Red clover	Resistant ... Bjorn, Bambi, Britta, Disa, Hedda, Hermes II, Kora, Molly, Resistenta (improved), Norseman, Sabtoron, Pearce's Broad Red. Moderately resistant ... Essi II, Fanny, Perenta, Reko, Sally, Triton, Lucrum, Temara, Renova, Rittinova, Quin.
Strawberry	Voskhod (Ascent), Rannyaya Plotnaya, Pamyatnaya.
Sweet potato	Yu Shu 3, Kogane Sungau (Norin 31), Mianfen No. 1, Lanshu 88.
White clover	Pitau, English Wild White, Milka, Lodi, Otofte, Zero, Regal Ladino.

(Barnwell, Milns Prophet, Pennant (S 227), Peniarth (S238) and S172) were resistant to *D. dipsaci* and two spring oats (Manod (S235) and Milford (S225)) were similarly resistant, while the spring oat Early Miller was tolerant of attack (Anon.,1963). Blake (1962a) showed that the nematode multiplied less in Manod and Pennant than in susceptible Sun II and Powys and the life cycle took twice as long in the former as in the latter cultivars, so that there were only two or three generations in Manod but twice as many in Sun II. Although the nematodes caused little separation and hypertrophy of parenchyma cells, they did withdraw cytoplasm from them and caused cavitation of the tissue, as in Sun II, though without provoking hyperplasia or necrosis (Blake, 1962b). Currently, four winter cultivars Gerald, Solva, Image and the naked oat Harpoon are resistant to oat stem nematode but no spring oat currently grown in Britain is resistant. Rivoal *et al.* (1978) have pointed out that the tulip root symptom in the susceptible cv. Naine de Moyencourt was not an infallible guide to infestation, as plants not showing tulip root could contain up to 4000 nematodes. Furthermore, of plants with tulip root, 16% had up to 2000, 28% up to 4000 and 56% up to 10,000 per plant.

Against an oat race population at Rothamsted, England, Manod proved fully susceptible and Maris Tabard, a putatively susceptible oat, was as resistant as were Peniarth, Panema, Pennant, Maris Quest and Milford, all of which derived their resistance from Grey Winter oats (Whitehead *et al.*, 1983). The source of Maris Tabard's resistance is in doubt, but it may have come from one of its parents – cv. Tama, or from one of its great grandparents – cv. Victoria (R. Cook, personal communication).

WHEAT AND BARLEY

Stem nematode has been known to attack wheat and barley, as well as rye and oats (e.g. Savary, 1952; Belloni, 1954). Their status as true hosts, i.e. susceptibility to *D. dipsaci*, is unclear, however, despite the characteristic symptoms in young plants. In England, Goodey and Hooper (1958) could not reproduce the symptoms reported in the two cultivars of wheat in Italy by Belloni, even though they applied up to 100,000 *D. dipsaci* from oats per 18 cm pot. Further, wheat or barley can be grown in infested fields in England without damage by the nematode. It is possible that the English oat–onion race differs from that that attacks wheat and barley, occasionally, elsewhere. The susceptibility/resistance of modern wheat and barley cultivars to *D. dipsaci* has not been studied, no doubt because stem nematode attacks on them are uncommon.

LUCERNE

Stem nematode is a major pest of lucerne in cool, moist temperate areas and in hotter regions that are irrigated.

In a plot heavily infested with lucerne stem nematode in Utah, USA, Thorne (1938) reported Turkestan 19304 lucerne had 'almost perfect' immunity to stem nematode. Turkestan 19304 was subsequently increased and named Nemastan. By polycross breeding from Nemastan, progenies were obtained which had resistance to the nematodes and to bacterial wilt, as well as good seed and forage yields. A new synthetic variety was produced by growing together in isolation mother plants of the five best of these progenies. It is called Lahontan and is almost immune to the nematode and yields as well in uninfested soil as cv. Ranger. It is adapted to western lucerne areas of the USA. It yielded poorly compared with cv. Grimm in south Sweden (60°N), not being sufficiently winter hardy. Though less resistant than Lahontan, cv. Kayseri from Turkey yielded as well as Grimm (Bingefors, 1961). In terrines of seedlings, percentages of plants not showing characteristic swelling at the hypocotyl–cotyledon junction, indicative of susceptibility, were 83 (cv. Vertus), 75 (cv. Alfa II), 72 (cv. Kayseri) and 68 (cv. Netun) but only 8 for the susceptible cv. Maris Kabul (Caubel *et al.*, 1977). Bingefors (1985) stated that all lucerne seed sold in Sweden was resistant to stem nematode. He classified cultivars as a) susceptible (less than 15% of plants stunted in seedling tests with *D. dipsaci*), b) moderately susceptible / resistant (c. 30% stunted – cv. Lesina (1976) and c) resistant (40% or more stunted) to very resistant (more than 60% stunted) – Alfa II (1967), Sverre (1974) and Vertus (1969).

Elgin *et al.* (1977) found that for two susceptible cultivars (Ranger and Moapa) and two resistant cultivars (Lahontan and Caliverde 65) severity of symptoms, plant stand at 12 wk and number of plants with slight or no symptoms at 12 wk were significantly affected by which USA or Canadian nematode isolate was used. However, they considered the results sufficiently similar to justify the use of a single isolate in breeding for resistance to lucerne stem nematode and thought that the resistant plants would be resistant in other lucerne growing regions. Whitehead (1984) reported differences in the virulence of 11 English populations of lucerne race to cv. Vertus, because although it was generally more resistant than cv. Sverre or cv. Europe, some populations of the nematode multiplied equally as well in Vertus as in Sverre or Europe. The results suggested at least two pathotypes among the 11 isolates used. In selection for resistance it was pointed out that early hypertrophy at the seedling apex is a reaction to invasion of the tissues and not necessarily indicative of susceptibility, because some early hypertrophy in Vertus seedlings decreased or disappeared as the seedlings aged. It was also suggested that a mixture of isolates should be used in selection for resistance, rather than a single isolate, which can only include a tiny fraction of the nematode genotypes to which selected plants will be subjected in the field. Furthermore, assessments should continue to plant maturity before the resistance/susceptibility status can be reliably determined, as age-related resistance is not uncommon in plants. Resistance may break down at high temperatures, so a cultivar bred for use in a cool climate may not be resistant in a hotter climate. In further studies it was shown that cv. Euver was as susceptible to English and French isolates of lucerne race as was cv. Europe and that, despite differences between isolates in their fecundity on lucerne, some isolates multiplied on Vertus as much or more than on Europe or Euver (Whitehead *et al.*, 1987a; Whitehead, 1992a).

Because a cultivar of lucerne is a mixture of resistant and susceptible genotypes, the proportion of resistant types may decrease when the cultivar is grown repeatedly for seed, so it needs to be reselected repeatedly for resistance (Lundin, 1969). If this is not done, then new seed stocks may be less resistant than the originally selected ones. This may explain why some cultivars, putatively resistant, prove to be susceptible in later testing. Leclercq and Caubel (1991) found that a single selection from a susceptible cultivar resulted in resistance equivalent to that in Vertus. The gain in resistance percentage (Y) was according to the equation

$$Y = -0.64\,(X) + 54.5 \qquad (3.1)$$

where X was the initial percentage resistance. So, when the percentage resistance before selection (X) is > 1 but < 85 the equation holds. In 72 trials, which included cultivars Europe, Vertus and Cinna, as well as many lucerne families, no nematode pathotypes able to increase in resistant genotypes were observed.

Other species of *Medicago* have shown good resistance to lucerne stem nematode (Bingefors,1967; Whitehead, 1992a). Unfortunately these are annual species which makes this resistance very difficult to transfer to perennial lucerne by traditional breeding techniques. Perennial *Medicago* spp. are generally susceptible to the nematode, though resistant plants were found in four species (Whitehead, 1992a). Some of the annual *Medicago* spp. resistant to lucerne stem nematode are grown commercially in Australia, but they are not frost hardy (Gintzberger and Prosperi, 1987).

RED CLOVER

The breeding of red clover for resistance to stem nematode started in Sweden, where the pest was widespread and damaging. The early resistant cultivars Merkur and Resistenta, adapted to south Sweden, were followed by the winter-hardy resistant cultivars Ulva, Disa and Reko. Generally, resistance in red clover has been very reliable in Sweden and Denmark (Bingefors, 1969). However, in an earlier publication (Bingefors, 1967), the percentage of Merkur plants with eggs of *D. dipsaci* varied from 0 to 17.5, according to which of 14 Swedish isolates they were exposed. None of the isolates attacked Merkur 'to such an extent that [it] could be characterised as susceptible' and there were greater differences in resistance between seed samples of Merkur and Resistenta than between one stock of seed and different isolates of the nematode. With less resistant red clover cultivars, he points out, differences in virulence between different isolates of the nematode are of greater practical significance. In Denmark, Frandsen (1966) found differences between populations of the nematode in their ability to attack red clover cultivars, although significant interaction between nematode populations and clover cultivars was very infrequent. Currently in Sweden, only one of 17 approved red clovers is susceptible to the red clover race of stem nematode – cv. Bjursele, introduced in 1962 and now grown only in north Sweden (Bingefors, 1985). The 16 resistant cultivars are either moderately resistant (Essi II, Fanny, Perenta, Reko, Sally and Triton) or resistant to very resistant (Bjorn, Bombi, Britta, Disa, Hedda, Hermes II, Kora, Molly, Resistenta (improved) and Sara).

The multiplication of eight populations of *D. dipsaci* from red clover on resistant and susceptible red clover cultivars varied, both with cultivar and with nematode isolate (Whitehead *et al.*, 1987a). There was little variation in the levels of multiplication on the most resistant cultivars (Norseman and Sabtoron) but on other cultivars (Quin, Britta, Rittinova, Lucrum, Temara, Leisi) multiplication

Table 3.2. Multiplication of six isolates of red clover race on five cultivars of red clover in two pot experiments.

Cultivar	Multiplication (P_f/P_i)	(range)	Average % of Redhead
Redhead	364	110–761	100
Britta	77	13–133	21
Quin	28	1–180	8
Norseman	12	0.9–48	3
Sabtoron	7.5	0.7–16	2

varied greatly between isolates, which all multiplied well on susceptible cv. Redhead (Table 3.2). Such interactions suggest that, as Bingefors (1967) remarked, cultivars only moderately resistant would not be reliably resistant in the field.

Bingefors (1967) listed 15 species of *Trifolium* resistant or possibly resistant to the red clover race stem nematode and Whitehead (1992a) listed 26 species resistant or very resistant to a mixture of four isolates of the nematode. Bingefors found *T. alpestre* and *T. spumosum* resistant and *T. echinatum* susceptible, whereas Whitehead found *T. echinatum* very resistant and the other two species susceptible. The differences in results may be due to differences between the nematodes used as inoculum or to genetic differences between the seed batches used. The results show, nevertheless, that there is plenty of resistance awaiting exploitation among wild species of *Trifolium*.

WHITE CLOVER

White clover (*Trifolium repens*) is of great value in temperate pastures, fixing atmospheric nitrogen, to the benefit of companion grasses, so reducing dependence on and cost of applying larger amounts of nitrogen fertilizer to the soil. The management of white clover in mixed swards, however, is not easy. Careful choice of cultivar appropriate to the intensity of grazing or cutting practised and strictly controlled application of small amounts of fertilizer N to avoid the grasses out-competing the clover are required. A number of pests and diseases can limit white clover growth. Of these, stem nematode is often most important. In England and Wales, 40% of white clover in mixed swards was found to be infested (Cook *et al.*, 1992b) and the pest is one of several, which affect the predictability and persistence of white clover in the UK (Cook and York, 1985). The nematode is also important in New Zealand and probably elsewhere in cool, temperate pastures. The symptoms of infestation, though characteristic, are difficult to see in mixed swards in the field, as the damage is often at or just below soil surface and often hidden by the grasses.

Establishment and early yields of S 184 and four partially resistant white clover cultivars (each with about 50% resistant plants) in mixed swards with perennial ryegrass (*Lolium perenne*) were inversely related to the numbers of *D. dipsaci* in the soil. Although the resistant cultivars (Katrina, Donna, Alice and Grasslands Huia) were initially less infested than S 184, infestations of the nematode developed on all of them after the establishment phase. Treating the seedbed with aldicarb or growing the white clovers after a sequence of non-host crops increased yields nine months after sowing 3.5- and 8.9-fold, respectively (Cook *et al.*, 1992a). In an earlier study, the cultivar Aran, although the most resistant to white clover stem nematode of 53 *T. repens* accessions inoculated, eventually became infested and allowed the nematodes to multiply, either through latent infestations or by sequential inoculations (Cook and Evans, 1988). This is probably less serious than in lucerne or red clover, because after about 15 months it may be offset by rapid compensatory growth of daughter plants from stolons, though the balance between pest and host is probably affected by subsequent environmental conditions.

Nematicides and hot water therapy

As stem nematodes can multiply so rapidly on host plants, thereby injuring sensitive plants, much harm may be avoided by ensuring that all planting material is free of infestation. Widespread dissemination of the nematodes to land, which may be uninfested by some or all races of the nematode, is also avoided. Applying fumigant or non-fumigant nematicides to the soil or plants in infested land may lessen or prevent yield losses but nematode multiplication is rarely, if ever, prevented. Infestation of planting material (seeds, bulbs or transplants) can be minimized, even eliminated, by applying nematicides to them or immersing them in hot water before planting.

TREATMENT OF SEEDS, SEEDLINGS AND BULBS

Methyl bromide fumigation. Stem nematodes can be eliminated from seeds by exposing them to an appropriate dose of methyl bromide for an appropriate length of time, i.e. a certain concentration × time product (CTP) (Goodey, 1945). As the gas is very phytotoxic, the seed must be dormant. Fumigation of seedlings or bulbs causes too much plant injury to be practical. The nematodes occur beneath, as well as upon, the seed coat (e.g. Southey, 1965) and not all of them will be killed if the seed is too dry. Conversely, if the seed is too moist, germination and seedling vigour may suffer (e.g. Powell, 1975).

At a CTP not less than 600 mg h l^{-1} and with seed at no more than 12% moisture, T. Goodey (1945) eliminated *D. dipsaci* from onion seed and J.B. Goodey (1949) did likewise with red clover and teazle seed, without significantly affecting germination. Fumigation lasted 18 or 20 h. At a CTP of 800 mg h l^{-1}, methyl bromide was similarly effective for infested shallot seed (Hague, 1968). Page *et al.* (1959) stated that a CTP of 1000 should suffice to control the nematode in lucerne seed and that some variation around the recommended 12% moisture level did not significantly harm the seed. Gostick (1963) eliminated the nematode from lucerne seed and the accompanying plant debris at a fixed CTP for 15–30 h at 10–30°C, without affecting germination. He suggested CTPs of 1200 (20°C), 1000 (25°C) and 750 (30°C). The nematode was eliminated from lucerne seed by CTPs of 1280 (16 h at atmospheric pressure) and 1000 (20 h in partial vacuum) (Caubel *et al.*, 1985). Currently in the EU, lucerne seed fumigation approved for seed imports to member states is not less than 1200 g h m^{-3} at 10–20°C for 20–30 h. Methyl bromide fumigation is less successful for field bean seed. In heavily infested seed, 3000 mg h l^{-1} killed all the nematodes but seriously reduced germination. In lightly infested seed, 800–2000 mg h l^{-1} killed all the nematodes with little reduction in germination at a seed moisture of not more than 12% (Powell, 1974). Other workers have found efficacy, without phytotoxicity, difficult to achieve. The nematode was eliminated from whole garlic bulbs at CTPs < 800 (Caubel *et al.*, 1985).

Non-fumigant nematicides. Several non-phytotoxic nematicides have been applied to seeds or sets or used as bulb dips to protect young plants from injury by stem nematodes. Such treatments have been used most on strawberry, garlic and tulip.

Vik (1968) disinfested onion seed by dipping it for 24 or 48 h in 4% 'Polybarit' (20% barium polysulphides). Yakimenko and Efremenko (1973) found 0.5% formaldehyde solution at 44°C for 2 h eliminated *D. dipsaci* from onion seed, whereas a 1% solution of ammonium polysulphide for 5 min reduced infestation by 67%. Organophosphate nematicides were less effective and phytotoxic.

Many experiments on the efficacy of dipping garlic and flower bulbs in nematicidal solutions have been reported. Infestation of garlic cloves (bulbils) in the field was greatly decreased by dipping them or the bulbs in 0.1% thionazin for 15–30 min (Brzeski, 1973a), 0.4% heterophos for 30 min (Mel'nik and Knaub, 1977), 1% phoxim for 2 h (Caubel, 1983), 0.05% triazophos or 0.15% thionazin for 15 min (Cíndea, 1982) or 0.35% of 70% ethoprophos for 1 h (del Toro *et al.*, 1988). Cíndea (1979, 1981) found that in contrast to hot water treatment of garlic, which adversely affected subsequent growth, excellent control of *D. dipsaci* (over 98%) was obtained by vacuum infiltration of infested garlic with 0.15% thionazin or 0.2% oxamyl for 15 min at 25 torr, which did not affect subsequent sprouting. Good results were also obtained by soaking garlic cloves in 0.2% thionazin for 4 h at 40°C.

Stem nematode in tulips has been controlled by dipping the bulbs in suspensions of thionazin or phorate before planting (Oliff *et al.*, 1969). Windrich (1970) found that the leaves of tulip bulbs dipped in 0.5% aqueous thionazin (5000 ppm) were only protected from attack for 4 weeks. Winfield (1972) found numbers of *D. dipsaci* in tulip bulbs reduced, without phytotoxicity, when they had been dipped in thionazin suspensions of 2300 or 4600 ppm in July or September. However, when thionazin concentration has decreased sufficiently and soil temperatures have increased in late spring or early summer, surviving nematodes multiply rapidly, so when the bulbs are lifted at the end of the season they need to be retreated to prevent serious losses during storage. The withdrawal of thionazin from commercial availability stimulated the search for effective alternative cold dips for tulip bulbs, as hot water treatment usually harms tulip bulbs. Phoxim, carbofuran and thionazin were shown to be equally effective in controlling the nematode and in preventing them laying eggs in the bulbs. Eight months after treatment with 2500 ppm dips of thionazin, carbofuran or phoxim there were no eggs in the bulbs, compared with 210 per untreated bulb. After carbofuran or phoxim there were no nematodes in the bulbs and only 20 per bulb after thionazin, compared with 5800 per untreated bulb. Phoxim had the advantage of lower vertebrate toxicity than thionazin or carbofuran (Damadzadeh and Hague, 1979). The treatments did not prevent the nematodes leaving the bulbs (Hague *et al.*, 1983). Other possible alternatives to thionazin for tulips are 0.2 or 0.4% oxamyl dips (30 min), which gave good control (Rasmussen and Lindhardt, 1980) and, for light infestations, 0.25% triazophos (Lane and Saynor, 1983).

Whereas tulip bulbs are left in the ground for only one season, narcissus bulbs remain in the ground for two or three seasons. Chemical treatment of narcissus bulbs may keep stem nematode under control for one season or a part of a season but not longer. Purnell and Hague (1965) found that thionazin or diethyl chlorovinyl phosphate was as effective as hot water treatment in controlling the nematode in the bulbs. Similarly, dipping narcissus bulbs in thionazin was found as effective as hot water treatment in controlling the nematode and in promoting vigorous plant growth (Oliff, 1966). Winfield (1971) controlled the nematode by dipping narcissus bulbs in thionazin suspension (2300 ppm) for 1 or 2 h. Hesling (1971) pointed out that stem nematodes that survived in bulbs treated with thionazin could not resume their activities until the concentration of the chemical had decreased greatly, i.e. late in the growing season, whereas the nematodes that survived hot water treatment could pursue their lives immediately after treatment or planting. As with tulips, carbofuran, phoxim or phorate can be equally as effective as thionazin as bulb dips against stem nematode (Damadzadeh and Hague, 1979; Hague *et al.*, 1983; Oliff *et al.*, 1969).

Early control of stem nematode on lucerne has been obtained in outdoor plots by applying phorate dust at 1% (w/w) to the seed (Gostick, 1963), soaking seed for 10 h in 0.1% carbofuran, fenamiphos or oxamyl in acetone (Gray and Soh, 1989), or applying furathiocarb to the seed (Adamová and Rotnekl, 1991), the latter proving more effective than carbofuran. These treatments improved early growth of lucerne. All of them have high vertebrate toxicity, so it is doubtful if they will be approved for use on lucerne seed. Field bean seeds, each bearing a dressing of about 0.5 mg aldicarb or carbofuran, sown at 0.5–1.0 million seeds ha^{-1} (0.25–0.5 kg a.i. ha^{-1}), significantly reduced but did not eliminate *D. dipsaci* in the harvested seed. A number of organophosphate seed dressings, similarly applied were ineffective (Whitehead and Tite, 1987). Incorporated in pelleted field bean seed, carbofuran at 3 mg per seed or carbosulphan at 8 mg per seed was effective against stem nematode for three months and increased seed yields by 6–12%.. The pelleting technique was said to ensure slower release than from granules applied to the soil near the seeds (Schiffers *et al.*, 1984). The dosages were equivalent to 1.2 kg carbofuran ha^{-1} or 3.2 kg carbosulphan ha^{-1}, assuming a seed rate of 400,000 seeds ha^{-1} (50 cm rows, seeds 5 cm apart).

Hot water treatments. Garlic cloves, whose protective, outer leaf and abscission zone were infested with *D. dipsaci*, were freed of infestation by soaking them in a solution of 1% formalin plus 0.1% detergent at 24°C and then transferring them for 20 min to 1% formalin plus 0.1% detergent solution at 49°C (Lear and Johnson, 1962). Complete control was also achieved by immersing garlic in water at 45°C for 20 min (del Toro and Mavrich, 1977). Immersion of shallot sets in water at 44.5°C for 1–2 h eliminated the nematode and increased yield (Caubel, 1968).

Tulips differ in sensitivity to hot water treatment but warm storage at 34°C for seven days beforehand lessens bulb sensitivity. Warm storage, however, acclimatizes *D. dipsaci* to higher temperatures, necessitating higher tank temperatures. Few nematodes survived 2 h at 45.6 or 46.7°C and none survived 3 h at these temperatures, though the price of eradication was some plant damage (Millar, 1981). Hot water treatment is standard practice for narcissus bulbs, which are less sensitive to it than tulip bulbs.

To minimize damage, fully dormant narcissus bulbs are warm stored (e.g. 7 days at 30°C), then presoaked in water for 3 h, then put in a bath of 0.5% formalin plus wetter at up to 46.7°C for 3 h (Tomsett, 1975; Lane, 1984). In Lisse, the Netherlands, narcissus bulbs were stored for 1 or 2 weeks at 25–30°C, soaked in cold water for 24 h, then placed in water at 45°C for 4 h or 47°C for 4 h without soaking (Windrich, 1973). Two weeks storage of narcissus bulbs at 18°C, prior to hot water treatment, especially if the bulbs had been lifted in summer, largely prevented damage to several narcissus cultivars (Hanks, 1995). Qiu *et al.* (1994b) controlled the nematode in daffodil bulbs by dipping them for 150 min in 0.37% formaldehyde in water at 44°C, or for 240 min in water alone at 44°C. Hot water treatment is ineffective for heavily infested bulbs, because control is rarely 100% effective and the survivors would give rise to populations large enough to ruin the next crop (Winfield, 1970). Annual treatment of two cultivars of narcissus (2 h at 43.5°C) controlled *D. dipsaci* in the Netherlands (Vreeburg, 1981). The increased costs of annual treatment, except for very valuable bulb stocks, would be hard to justify.

Strawberry plants have been successfully freed of the nematode by hot water treatments. In Britain, strawberry runners are immersed for 10 min in water at 46°C, then plunged into cold water to minimize or prevent plant injury (Anon., 1984). In the former USSR, the nematode has been controlled by dipping strawberry plants in water at 48°C (Bobrysheva, 1972). In the Crimea, plants were selected from the field in March and dipped for 15 min at 48°C (Eremin and Bobrysheva, 1972) and in Kazakhstan, 17 min at 48°C gave 100% control with little loss of plants (Vladimirova, 1975). There is, however, variation in varietal sensitivity to hot water treatment.

TREATMENT OF SOILS AND PLANTS

Fumigant nematicides. At rates up to 1025 l ha^{-1}, D-D prevented damage to onions by *D. dipsaci* in organic soils in southern New York and increased onion bulb yields (Lewis and Mai, 1958). D-D and its active ingredient, 1,3-dichloropropene, were more effective in organic soil kept hot and dry than in the same soil kept wetter and cooler. In test bags of soil buried 12.5 cm deep, maximum kill with *c.* 500 l D-D ha^{-1} was 98.9%. Kills were greatest at 26.7°C and at soil moisture close to the wilting point. These conditions favour rapid diffusion of the gases and low adsorption on soil organic matter (Lewis and Mai, 1963). However, unless the fumigant is applied in two doses, one before and one after ploughing, few nematodes are likely to be killed near the soil surface and in organic soils adsorption of the gases by organic matter necessitates very large doses. Soil fumigation with D-D was greatly improved when the soil was covered with plastic sheeting (Schenk, 1972). Dazomet worked into the soil at 500 kg ha^{-1} killed most onion stem nematodes (Tikhornik, 1966). In sandy loam inoculated with *D. dipsaci* from field beans infested with oat–onion race of the nematode, dazomet at 220–880 kg ha^{-1} increased onion yields 3- to 4-fold and decreased numbers of the nematodes left in the soil after harvest from 1200 l^{-1} soil in untreated plots to 100 l^{-1} soil (220 kg $ha)^{-1}$ down to 0 l^{-1} soil (880 kg $ha)^{-1}$ and greatly reduced losses of stored onions due to rotting (Whitehead and Tite, 1972). In Ireland, however, in soil in which earlier onion crops had failed due to stem nematode attack, neither D-D at 448 or 673 kg ha^{-1} nor dazomet at 610 or 915 kg ha^{-1} prevented eventual infestation and rotting of the bulbs (Duggan, 1971). In Israel, very good control of the nematode and increased yields of garlic were obtained by fumigating the soil with 500 kg methyl bromide ha^{-1} (Siti *et al.*, 1982). Treating infested soil in Italy, 30 days before planting out strawberries in August, with 750 kg methyl bromide, 500 kg dazomet or 500 l D-D or Di-Trapex ha^{-1}, decreased numbers of *D. dipsaci* found in the leaves in the following year and increased fruit yields (Vovlas and Inserra, 1980).

Non-fumigant nematicides. A number of organophosphate and oximecarbamate nematicides have been more or less effective in controlling *D. dipsaci* in bulb crops, legumes and strawberry.

Applied to sandy loam infested with 1–800 stem nematodes (500 $g)^{-1}$ soil three weeks after sowing onions, 10 kg thionazin ha^{-1} inactivated 98% (at low densities) to > 99% of the nematodes (at high densities), thereby preventing significant loss of onion sets harvested in July and minimizing rotting in cool storage over eight months. The treatment was effective when applied in 60,000 l water ha^{-1} or when applied in 1000 l ha^{-1} followed over the next three days by 17.7 mm rain, but it was ineffective when applied in 1000 l water ha^{-1} and followed by only 3.6 mm rain (Kaai and Pronk, 1965; Kaai, 1968). Brzeski (1973b) found 3–4 kg thionazin ha^{-1}, applied to onion rows in 10% granules at sowing controlled stem nematode well, especially at the 4 kg rate, which decreased storage losses even further than the 3 kg rate. Applied in a 5 cm wide band to the soil in which onion seeds were sown, with or without a subsequent top dressing, aldicarb greatly increased yields of onions (Winfield *et al.*, 1971). Applied over the seed furrows at sowing, 1.4–5.6 kg aldicarb or 1.3–5.2 kg oxamyl ha^{-1} prevented much of the damage due to stem nematode attack in spring-sown onions and sometimes reduced rotting of bulbs in store (Table 3.3). Applied at 2.2 kg ha^{-1} to the seedbed in two consecutive years, aldicarb permitted good crops of onions to grow in the second year, when untreated plots yielded no harvestable

Table 3.3. Yield response in spring-sown onions (cv. Robusta) to control of stem nematode in three soils by aldicarb or oxamyl applied to the rows at sowing.

	Onions (t ha^{-1})		
Treatment	Peaty loam	Sandy loam	Clay loam
Untreated	3.7	11.5	31.8
Aldicarb (2.8 kg ha^{-1})	31.1	27.7	39.2
Oxamyl (2.6 kg ha^{-1})	19.0	30.1	41.6
L.S.D. (5%)	15.5	10.7	5.9

onions (Fig. 3.4). In summer-sown onions, applying half the nematicide at sowing and half as a top dressing in the following spring was sometimes more effective than applying the whole dose at sowing (Whitehead *et al.*, 1979d). Applying aldicarb to the plant rows at sowing, in oat–onion race infested soil, lessened attacks and greatly increased the yields of a number of crop plants, some of which were susceptible to the nematodes and others not and lessened multiplication of the nematodes in host plants (Whitehead *et al.*, 1983). In heavily infested soil, oxamyl at 50, 100 or 150 mg m^{-1} row or triazophos at 200 or 300 mg m^{-1} row prevented significant yield loss in onions but did not eliminate losses in stored onions (Baksik, 1989). Oxamyl or aldicarb applied to onion seed crops lessened the number of flower shoots infested by *D. dipsaci* and increased 1000 seed weights (Brzeski, 1986). Only 2.5 kg aldicarb or less ha^{-1} can be applied to onions without leaving unacceptable residues in harvested bulbs (i.e. > 0.15 μg g^{-1}) (Whitehead *et al.*, 1979d).

Stem nematode in garlic cannot be controlled by methyl bromide fumigation or hot water treatment of the cloves without plant damage. In former Czechoslovakia, infestation of garlic seedlings and cloves was decreased by up to 95% by soaking them in solutions of barium polysulphides or lime sulphur (Janýška, 1958). In Roumania, 3 kg aldicarb or thionazin ha^{-1} applied to the soil before or at planting controlled *D. dipsaci* in garlic (Romaşcu, 1977). Calcium polysulphide (8%) or thionazin (0.1 or 0.2%), applied as preplanting treatments to garlic, were very effective (Romaşcu *et al.*, 1977). Lime sulphur sprays have also controlled the nematode in garlic (Mel'nik and Knaub, 1976). Applied around garlic cloves at planting, 2.5 or 5.0 kg aldicarb or fenamiphos ha^{-1} suppressed nematode infestation, but, as with hot water treatment, did not prevent infestation, which 5 kg carbofuran ha^{-1} did, though not without phytotoxicity (Roberts and Greathead, 1986). Aldicarb residues were found in garlic in soil given 8 kg but not 4 or 2 kg ha^{-1} . No residues of carbofuran were found in garlic following treatment of the soil with up to 6 kg ha^{-1} (Becker, 1988).

Costs of treatment and nematicide residue limits are less restricting for valuable ornamental plants like narcissus and tulip. In a sandy loam in England, thionazin drenched onto plots of narcissus at dosages varying from about 8 to about 98 kg ha^{-1} gave up to 99% control of the nematode and prevented significant damage for three years after treatment (Oliff, 1966). Applied to infested peaty loam in much smaller amounts of water than Oliff had used, thionazin at up to 44 kg ha^{-1} had little effect on the nematode in narcissus (Winfield, unpublished report 1966). Thionazin is known to work poorly when not leached into the soil. Its efficacy is also reduced by adsorption on soil organic matter. Of a number of treatments for *D. dipsaci*, Hart and Maggenti (1974) found 5.6 kg aldicarb ha^{-1} distributed over narcissus bulbs in the planting furrows gave the best control. Infestation of tulip bulbs by the nematode in plots with one infested bulb planted in the centre was greatly reduced by 6 kg aldicarb ha^{-1} applied mid-November or in late December and again in early May the next year. Oxamyl had a similar but lesser effect, when applied at 2.8 kg ha^{-1} in late April, mid-May and in early June (8.4 kg ha^{-1} in all) (Windrich, 1985).

In strawberries, *D. dipsaci* has been controlled by non-fumigant nematicides applied to the soil and /or to the foliage. In Italy, isazophos, oxamyl or thionazin sprayed on the plants in spring and autumn controlled the nematode and increased yields (Tacconi, 1976). In other experiments, nematode infestations were decreased in three-year-old strawberries by foliar sprays of thionazin (55 l ha^{-1}), oxamyl (5 l ha^{-1}) or carbofuran (24 kg ha^{-1}) in November and again in February–March and strawberry yields were increased 60–90% (Vovlas *et al.*, 1978). The nematode was controlled more effectively in the field by aldicarb granules than by D-D, oxamyl or thionazin applied to the soil. Light attacks were controlled by 26 mg aldicarb per plant, applied at the base of the plant, but whereas applications within 21 days of planting suppressed attacks, applications 15 or 30 weeks after planting were ineffective (Lewis, 1979). Dipping strawberry plants in water at 48°C for 15 min before planting them improved control achieved by aldicarb, oxamyl, heterophos, etaphos, fenamiphos or isazophos applied to the soil (Metlitskii and Pokhodenko, 1986), presumably by lessening infestation in the transplants.

Crown rot of sugarbeet and related crops, initiated by *D. dipsaci*, has been a serious problem in cool, moist soils. It is currently controlled by non-fumigant nematicides. As very small numbers of the nematode may cause great damage and their detection is too unreliable for predictive soil sampling, routine applications of nematicides are necessary in areas prone to the disease, e.g the northern plateau of Switzerland. Thionazin or parathion applied on two or three occasions in the growing season along the sugarbeet rows protected the plants from crown rot (Graf *et al.*, 1964). Parathion was more effective and cheaper than thionazin in other trials and, if used at not more than 6 kg ha^{-1}, did not leave unacceptably large residues in the harvested sugarbeet (Clausen *et al.*, 1964). Two row treatments of 2 g of 5% granules of parathion m^{-1} row gave good economic control, but when *D. dipsaci* was abundant, a third treatment was necessary (Graf and Meyer, 1973). Applications were made at seedling emergence and at about three week intervals thereafter to compensate for the relatively short persistence of parathion in the soil. Aldicarb and carbofuran were also effective as row treatments against *D. dipsaci* of sugarbeet (Küthe, 1978) and in Switzerland can be used instead of parathion only at seedling emergence, subsequent treatments being with parathion. Terbufos as a 2% granule applied at 0.5 g m^{-1} row controlled the nematode up to 60 days after sowing as effectively as aldicarb in 5% granules. Although beet yield was increased more by aldicarb, terbufos was cheaper and it controlled aphids and leaf miner beetle (*Atomaria linearis*) as effectively as aldicarb (Pfister and Mittnacht, 1992). Row application of granular nematicides at sowing, emergence and later in the growing season is easily done but the cost can be 10% of the value of the harvested crop.

Field beans are subject to attack by 'oat–onion', 'giant' and other races of *D. dipsaci*, of which the giant race is the most damaging. In spring-sown beans, 4–5 kg aldicarb ha^{-1} applied around the seeds at sowing , controlled both seed-borne and soil-borne infestations of stem nematode (Hooper, 1984). In soil heavily infested with oat race, 1.5 kg aldicarb ha^{-1} applied at sowing to spring bean rows 38 cm apart greatly decreased invasion of the stems, greatly increased seed yields and decreased numbers of the nematode left in the soil after harvest (Whitehead *et al.*, 1983). In further experiments (Whitehead and Tite, 1987), giant race was well controlled by up to 5 kg aldicarb or 2.5 kg carbofuran ha^{-1}. In two of three

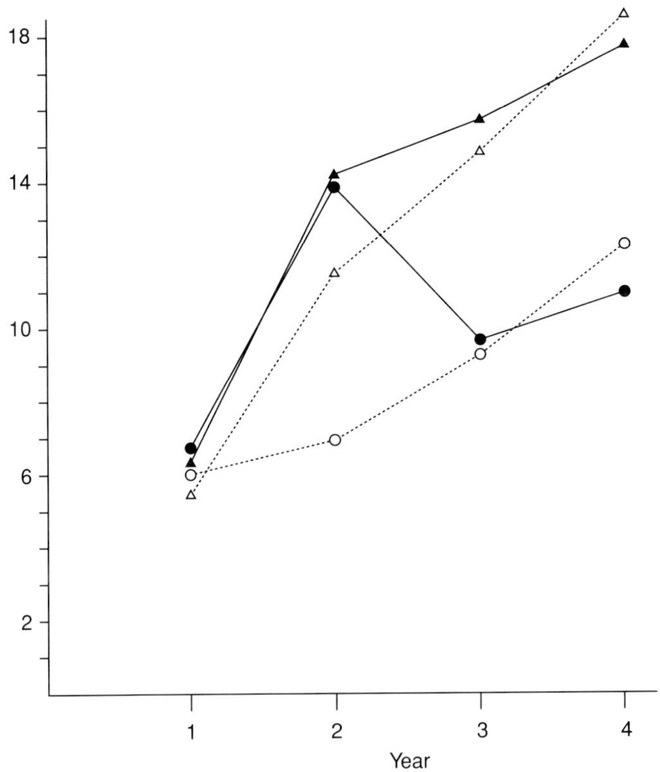

Fig. 3.5. Dry matter yields (t ha^{-1}) of two lucerne cultivars with and without carbofuran applied at 1.65 kg ha^{-1} to the rows at sowing in year 1 in soil infested with lucerne race of *D. dipsaci*. Broken lines, untreated; unbroken lines, treated; circles, cv. Europe (susceptible); triangles, cv. Vertus (resistant).

experiments, aldicarb at 4.4 or 5 kg ha^{-1} reduced percentage of infested stems by about 95 and 97%, respectively, and decreased seed infestation at harvest by 99.7%. Control was not improved by applying half the dose at sowing and half to the rows of young plants but 2.5 kg phoxim or aldicarb ha^{-1} over the rows of plants as a supplementary treatment improved control in one experiment. The nematode was less controlled by aldicarb or carbofuran in autumn-sown beans, though split applications were rather more effective than applications made in autumn alone, no doubt as a result of poor protection in spring following winter leaching of the nematicides. In one experiment, in which aldicarb or oxamyl had been applied to the seed furrows, phoxim or aldicarb over the plant rows mid-season enhanced nematode control.

Applied around the seeds at sowing, 1.5 kg carbofuran ha^{-1} can improve the establishment of lucerne, control *Sitona* weevil damage and increase herbage yields into the second year, both in soil infested with lucerne race stem nematode (which was controlled) and in uninfested soil. In infested soil, yields were further increased by spraying established plants with thiodicarb hypophosphite or carbendazim (Whitehead and Tite, 1988). In soil inoculated with both *D. dipsaci* and

Aphelenchoides ritzemabosi, carbofuran granules around lucerne seeds at sowing in the spring greatly decreased stunting associated with *D. dipsaci* and increased herbage yields of two resistant and two susceptible cultivars, on average, by 17% in the first year and by 59% in the second year of a four year experiment, in which the stem nematode was controlled best and herbage yields were largest where carbofuran granules at sowing had been combined with the use of the resistant cultivar Vertus (Fig. 3.5). Small amounts of carbofuran applied to the rows at sowing also increased yields of summer-sown lucerne in infested soil (Whitehead, 1994).

Carbofuran at 0.5, 1, 2 or 5 kg ha^{-1} or aldicarb at 2 kg ha^{-1} applied to the seed furrows at sowing in the spring, in soil infested with 'white clover race' stem nematode, dramatically increased the percentage ground cover and herbage yields in the first two harvests of four white clover cultivars. In untreated plots there was less than 1% ground cover compared with 96% in plots given 2 kg and 99% in plots given 5 kg carbofuran ha^{-1} .The improvement is thought to be due in part to control of the stem nematode (Lewis *et al.*, 1985).

Integrated control

With its great capacity for multiplication on susceptible plants and ability to infest many crops and weeds, its ability to survive long periods in a dried state or in moist soil without host crops, *D. dipsaci* remains a difficult pest to control, especially for those crops without resistant cultivars and for those that remain for several years in the ground. The limitations of each control measure make it imperative to use a combination of them to limit crop losses. Crop rotations and efficient weed control must be supplemented by the use of uninfested seeds, bulbs or transplants, resistant cultivars, if available and marketable, and, when appropriate, chemical control measures to limit injury to crops. When feasible, planting in cold moist soils should be avoided, as such conditions favour *D. dipsaci*. Early maturing crops may also lessen the impact of *D. dipsaci*, allowing a shorter period in which the nematode can multiply.

4 Ectoparasitic Nematodes of Roots (*Belonolaimus*, *Criconemella*, *Hoplolaimus*, *Longidorus*, *Trichodorus*, *Paratrichodorus*, *Tylenchorhynchus* and *Xiphinema*)

Belonolaimus longicaudatus Rau, 1958 (Fig. 4.1)

This 'sting nematode' is a major pest of many crops and ornamental plants on the Atlantic coastal plains of the USA, from Virginia to Florida. It has also been reported in Alabama, Kansas and Oklahoma and was probably included in earlier records (as *B. gracilis* Steiner, 1949) in Louisiana, Texas, Connecticut and New Jersey. *Belonolaimus longicaudatus* occurs mainly in the top 30 cm of light, sandy soils, where it feeds ectoparasitically on the roots of many plant species. It is very damaging to cabbage, carrot, cereals, *Citrus*, *Chrysanthemum*, clovers, cotton, grapevine, grasses, groundnut, onion, potato, soybean, sunflower, strawberry, tomato, turnip and a range of trees and ornamentals (Fig. 4.2). Notable poor or non-hosts are American jointvetch (*Aeschynomene americana*), asparagus, sunn hemp (*Crotalaria spectabilis*), hairy indigo (*Indigofera hirsuta*), okra (*Hibiscus esculentus*) and marigold (*Tagetes minuta*). Three races of the nematode with differing host ranges have been recognized in Florida. Differences in host range have also been found between North Carolina and Georgia populations (Robbins and Barker, 1973). The nematode can break resistance in cotton to wilt disease (*Fusarium oxysporum* f.sp. *vasinfectum*) and it interacts with *Pythium aphanidermatum* to the detriment of chrysanthemums (Johnson and Littrell, 1970).

This nematode can be controlled by fumigant or non-fumigant nematicides, because it occurs in light, sandy soils, containing relatively little organic matter, in which fumigant adsorption by organic matter is not a serious problem and in which lipophilic as well as hydrophilic nematicides are effective. As neither DBCP nor EDB may now be used in the USA, control of *B. longicaudatus* depends on dichloropropene-dichloropropane mixtures (D-D, Vidden D), 1,3-dichloropropene mixture (Telone II) or non-fumigant nematicides. Dichloropropene mixtures kill most nematodes in sandy soils to 30 or more cm deep, except for the top 5 cm. Non-fumigant nematicides can paralyse most nematodes in the top 15 cm of the soil if well mixed with this layer but, as it is usually impractical to incorporate them deeper than 15 cm, they can only affect sting nematodes 15–30 cm deep if leached there by heavy rainfall or irrigation.

The nematode is most active in soil at 20–34°C (Boyd *et al.*, 1972) but it can

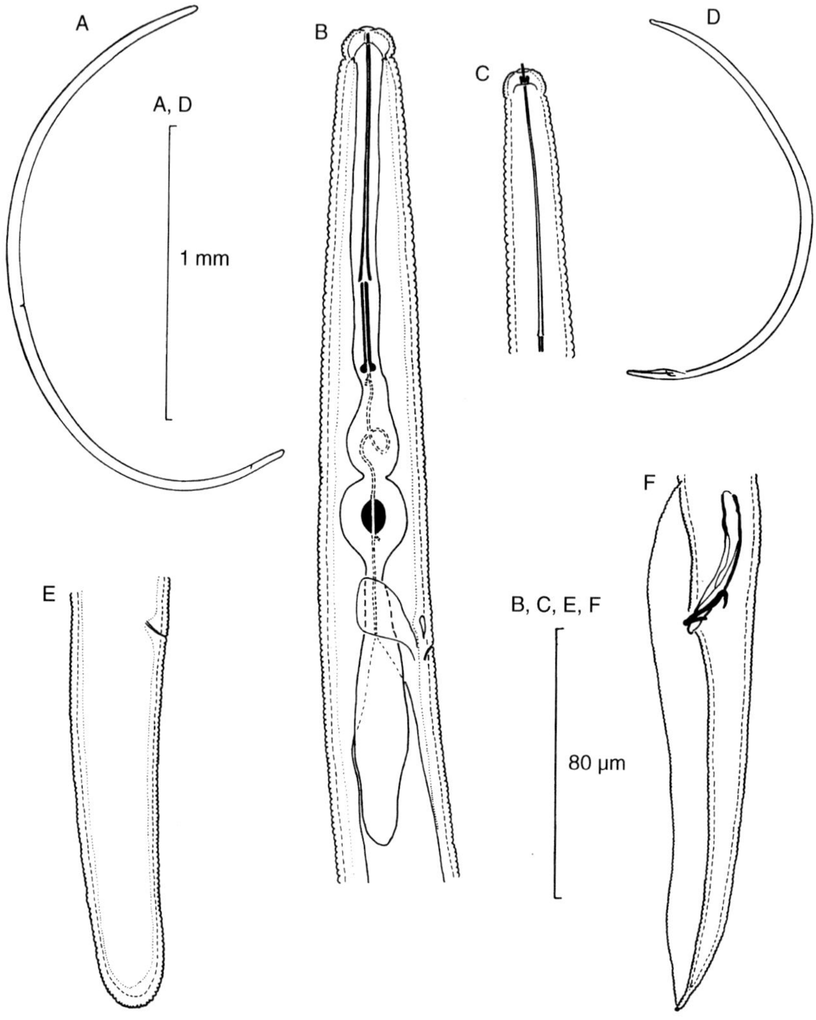

Fig. 4.1. Sting nematode, *Belonolaimus longicaudatus.* A, female; B, female anterior end; E, female tail; D, male; C, male anterior end; F, male tail.

still reproduce at 35°C and damage roots at 39.4°C (Boyd and Perry, 1970). It is most abundant 15–30 cm deep in July in Florida but it increases greatly in the top 15 cm in September and declines dramatically in the top 30 cm in winter.

Effective control measures for selected plants are described below.

Grasses

Belonolaimus longicaudatus causes stunting and yellowing of the leaves of bermuda grass (*Cynodon dactylon* cv. Ormond) but 18 or 36 kg fensulphothion ha^{-1} in liquid or granular formulation applied to the land controlled the nema-

(a)

(b)

Fig. 4.2. (a) Sting nematode injury to cotton roots (Photo: C.M. Heald). (b) Response of cabbage to chemical control of *Belonolaimus longicaudatus*; two untreated rows on the right (Photo: H.L. Rhoades).

tode and prevented the disease (Winchester and Burt, 1964). Fensulphothion and diazinon (Smart, 1971) and fenamiphos and oxamyl (Smart and Perry, 1971) were also effective against *B. longicaudatus* on bermuda grass. In Florida, bermuda, centipede and *Zoysia* grasses were seriously injured by *B. longicaudatus* (and by *Hoplolaimus galeatus* and *Dolichodorus heterocephalus*) but in commercial turf, such as on golf courses, the nematodes were controlled by applying 17–22 kg ethoprophos, fenamiphos or fensulphothion ha^{-1}, 6.5–11 kg aldicarb or 11–17 kg carbofuran ha^{-1} to the land. Such treatments were best made between June and September or in the rainy season, as control was not achieved unless the soil was moist (Perry, 1974). 'Tifton 328' bermuda grass grew badly in soil containing 500 *B. longicaudatus* $(500\ ml)^{-1}$ soil but 21.5 kg fenamiphos ha^{-1} greatly improved turf quality and even 40 weeks later the number of nematodes was still depressed (Lucas, 1979). In a subsequent report (Lucas, 1982), turf quality was shown to be negatively correlated with the numbers of the nematode in the soil and was best after 21.5 kg fenamiphos ha^{-1}, the residual benefits of which were apparent up to two years after treatment. The same treatment was equally effective on St. Augustine grass (*Stenotaphrum secundatum* FX-313) (Busey *et al.*, 1991).

Maize and Sweetcorn

The yields of maize cobs and grain were negatively correlated with the numbers of *B. longicaudatus* in the soil seven weeks after planting (Johnson and Dickson, 1973). The nematode was well controlled and maize yields were increased when 2.24 kg ethoprophos, fensulphothion, fenamiphos or carbofuran ha^{-1} was applied to the soil in bands 38 cm wide, 5–8 cm deep before maize seeds were sown in the middle of them. Rotary cultivating the granules into the 38 cm wide bands ahead of the planter controlled the nematode better than seed furrow application of nematicide (Rhoades, 1974). Treatments at planting were also more effective than treatments after planting (Rhoades, 1979). Yields of sweetcorn were significantly increased by treating infested soil with 1,3-dichloropropene, metham-sodium, aldoxycarb, fenamiphos or terbufos, all of which significantly decreased numbers of *B. longicaudatus* in the soil (Rhoades, 1985). In Georgia, nematode numbers were decreased and yields were increased by 2.24 kg aldicarb or 1.12 kg terbufos ha^{-1} (Johnson and Wright, 1988).

Cabbage

Applying 2.24 kg granular aldicarb, carbofuran, fenamiphos, ethoprophos or fensulphothion ha^{-1} in the rows, or in-row D-D soil fumigation reduced numbers of active *B. longicaudatus* and greatly increased the yield of direct-seeded cabbage (Rhoades, 1971) (Fig. 4.2). In transplanted cabbage, 2.24 kg ha^{-1} carbofuran, ethoprophos, fensulphothion, oxamyl or fenamiphos in liquid or granular form applied in the rows, just before transplanting cabbage into them, greatly decreased numbers of the nematode and significantly increased yields in years when the nematode was abundant (Rhoades, 1977). Aldicarb, carbofuran, ethoprophos, oxamyl, fenamiphos or terbufos, all at 2.24 kg ha^{-1} in-row, controlled the nematode and increased cabbage yields (Rhoades, 1981a).

Hairy indigo, sown as a summer cover crop, greatly decreased numbers of *B. longicaudatus* and increased yields of cabbage, compared with plots sown to summer cover crops of *Sorghum bicolor* or *Sesbania macrocarpa*, on both of which the nematode fed and multiplied (Rhoades, 1976).

Sweet Pepper (*Capsicum annuum*)

As a row treatment two weeks before planting, 75 l D-D ha^{-1} gave excellent control of the nematode and significantly increased sweet pepper yields. Similarly, at 2.24 or 3.36 kg ha^{-1}, aldicarb, carbofuran, ethoprophos, oxamyl, fenamiphos or terbufos decreased the numbers of active nematodes and usually increased yields. Applying the nematicides as granules or in transplanting water was usually equally effective (Rhoades, 1981b).

Snap or Dwarf Beans (*Phaseolus vulgaris*)

The 38 cm band technique (see above) at planting with 2.24 kg ha^{-1} of fenamiphos, fensulphothion, ethoprophos or carbofuran controlled the nematode and greatly increased snap bean yields (Rhoades, 1974). The nematode was controlled equally well and bean yields were increased equally as much by a preceding cover crop of hairy indigo as by fumigating the soil with 260 l D-D ha^{-1} (Rhoades, 1976). Numbers of *B. longicaudatus* were kept small by a three month cover crop of hairy indigo or American jointvetch, by bare fallow or by mulching the soil with clippings of cowpea or *Paspalum notatum* (Rhoades and Forbes, 1986).

Potato

In Florida, 3.36 kg aldicarb ha^{-1} decreased *B. longicaudatus* on potatoes and significantly increased tuber yields (Weingartner *et al.*, 1978). Combining soil fumigation with 56 l 1,3-dichloropropene ha^{-1} with 3.36 kg aldicarb ha^{-1} applied at planting gave best control of the nematode and increased tuber yield from 21.8 t ha^{-1} in untreated plots to 30.6 t ha^{-1} (Weingartner and Shumaker, 1981).

Soybean

Soybean yield was also negatively correlated with numbers of *B. longicaudatus* in the soil (Schmitt, 1989). At 2.24 kg ha^{-1} aldicarb, oxamyl, fenamiphos, carbofuran or ethoprophos controlled the nematode equally well when applied to the seed furrows or in 38 or 25 cm bands over the rows. Yields were thereby increased, except for ethoprophos, which interacted phytotoxically with the metribuzin herbicide (Rhoades, 1980). The efficacy of seed-furrow treatment contrasts with its poorer performance on maize. This may be due to the differing rooting habits of the two crops, soybean having a tap-root system, whereas maize has a fibrous root system.

Strawberry

Numbers of *B. longicaudatus* and strawberry yields were inversely correlated and nematode numbers were suppressed by treating the soil with phorate or fensulphothion (Overman, 1964). Season-long control of the nematode was given by 6.7 or 13.4 kg fenamiphos ha^{-1} or four, weekly sprays of oxamyl (Overman and Howard, 1984). Solar soil heating decreased the nematode for the first half of the season (Overman *et al.*, 1987).

Groundnut

The nematode was controlled and groundnut yields were increased by 1.1–6.7 kg ha^{-1} of Cynem (diethyl pyrazinyl phosphorothioate) (Cooper and Sasser, 1962).

Criconemella xenoplax (Raski, 1962) Luc and Raski, 1981 (Fig. 4.3)

This is a root ectoparasite of temperate and subtropical soils. Although it prefers a sandy soil, it is vulnerable to desiccation. It parasitizes almond, apple, apricot, cherry, grapefruit, guava, peach, plum, carnation, lettuce and species of *Acer*, *Juniperus* and *Pinus*. The association of this nematode with peach has been closely studied in the USA. Under glass, *C. xenoplax* caused stunting, chlorosis and premature leaf loss in peach (Lownsbery *et al.*, 1973; Wehunt, 1984). In the field, loss of feeder roots was associated with *C. xenoplax* infestation of the soil. The nematode feeds at root tips and along the sides of roots. It feeds mostly on the outer layers of peach root cells and in the USA has been found in all orchards examined in which 'peach tree short life' (PTSL) syndrome occurs. In affected orchards, trees die young in late winter or early in spring, following cold injury and/or bacterial canker (*Pseudomonas syringae*) (Nyczepir, 1989), to which *C. xenoplax* predisposes trees (Lownsbery *et al.*, 1973) (Fig. 4.4). The yield and mortality of peach trees were, respectively, negatively and positively correlated with the number of the nematodes in the soil (Wehunt *et al.*, 1980; Wehunt and Weaver, 1982). PTSL may be linked to the withdrawal of non-structural carbohydrate from the shoots into the roots of an intolerant rootstock (such as Nemaguard), which did not occur with a tolerant rootstock (BY 520–9) (Olien *et al.*, 1995). A decline of peach trees in New Jersey was associated with soil infestation by *C. curvata* (Hung and Jenkins, 1969), the nematode causing extensive pits and lesions in peach roots. This species also injures carnation (Streu *et al*, 1961; Varo Alcala *et al.*, 1970). The pathogenicity of *C. xenoplax* to carnation was demonstrated by Sher (1959).

Control

Criconemella xenoplax affecting peach is controlled by management practices, choice of rootstock and the use of fumigant and non-fumigant nematicides.

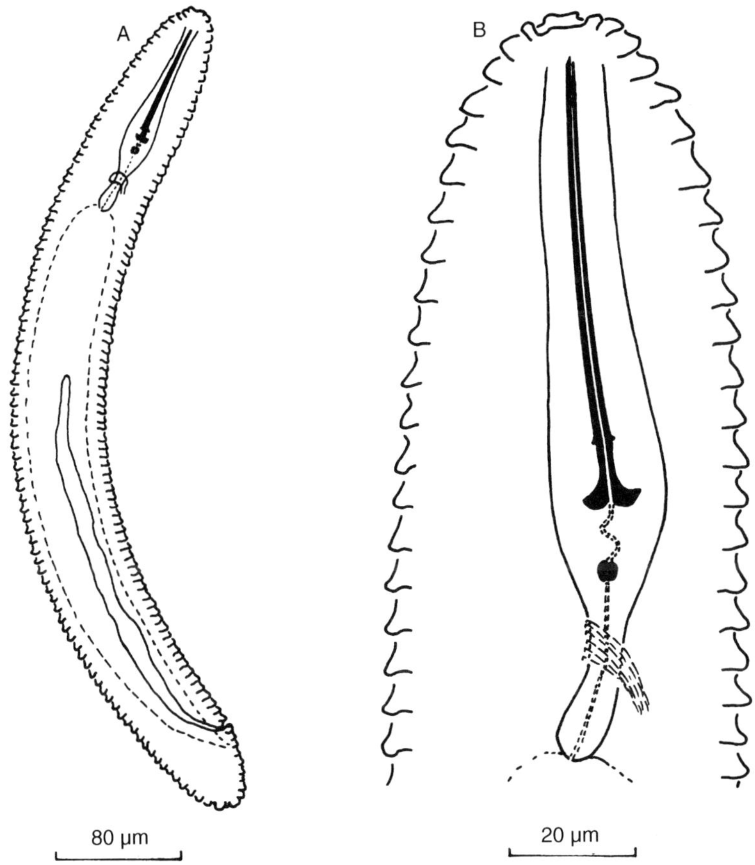

Fig. 4.3. *Criconemella xenoplax.* A, female; B, female anterior end.

Management

On sites affected by PTSL, Reighard *et al.*(1990) recommend that own-rooted peaches should be screened for susceptibility to *C. xenoplax*, bacterial canker and other relevant factors, before they are planted. Meyer *et al.* (1992) found that a short perennial grass, *Muhlenbergia schreberi* was a suitable ground cover plant for peach orchards in North and South Carolina, where it suppressed *C. xenoplax*, smothered out weeds without out-competing the peach trees and did not harbour several other peach pests.

Rootstocks

Susceptibility to PTSL was lessened when the trees were grafted on Lovell, rather than on Elberta or Nemaguard rootstocks (Zehr *et al.*, 1976). Although the nematode reproduced less on Lovell than on Elberta or Nemaguard rootstocks, resistance to *C. xenoplax* was not found in more than 160 seedling genotypes (Wehunt, 1984).

Fig. 4.4. (a) Root damage to Nemaguard peach rootstock by *Criconemella xenoplax* – top left to bottom right: 0, 112, 448, 1792, 7168, 14000 per pot of 1500 ml. (b) Peach tree dieback ('short life') initiated by *C. xenoplax* injury to the roots (Photos: A.P. Nyczepir).

Fumigant nematicides

In soil infested primarily with *C. xenoplax*, preplant fumigation with 374 l of D-D ha^{-1} improved the growth of peaches (Lownsbery, 1959), but fumigating the topsoil with 1,3-dichloropropene did not prevent severe loss of peach trees six years later (Zehr and Golden, 1986). The incidence of PTSL in south-east USA was lessened by applying DBCP to the soil (Nesmith and Dowler, 1974) and large amounts of lime applied to the soil suppressed *C. xenoplax* (Weaver and Wehunt, 1976). In South Carolina, preplant soil fumigation with 1,3-D or DBCP coupled with the use of Lovell rootstocks prevented tree losses over a three year period. However, if Elberta or Nemaguard rootstocks were used, some trees died of cold injury or bacterial canker, while in unfumigated soil tree losses were sometimes severe when Nemaguard rootstocks were used (Zehr *et al.*, 1976). Fumigating a PTSL site with methyl bromide before replanting it decreased numbers of *C. xenoplax* and, over a three year period, greatly increased growth and vigour of the trees, which were much less susceptible to cold injury. Cold injury was also reduced by pruning the trees in March, instead of in December and by using Lovell rootstocks (Sharpe *et al.*, 1989). Application of DBCP before and after planting controlled *C. xenoplax* and prevented tree loss but, although post-plant applications at reduced dosages of EDB or 1,3-D lessened numbers of the nematode, they did not lessen peach tree losses (Zehr *et al.*, 1982).

Non-fumigant nematicides

As the only non-phytotoxic soil fumigant (DBCP) is no longer available for control of *C. xenoplax* in established peaches or grapevines, attention has switched to the efficacy of non-phytotoxic, non-fumigant nematicides, especially fenamiphos.

Numbers of *C. xenoplax* and death of peach trees were decreased most by fenamiphos, when it was applied twice in autumn and once in spring (Ritchie and Bennett, 1986). In Louisiana, 62 l fenamiphos ha^{-1} per annum decreased numbers of the nematode in all five test years (Johnson *et al.*, 1987). At 6.7 kg fenamiphos ha^{-1} per annum, applied in two autumn and one spring doses, *C. xenoplax* was suppressed and 88–100% of the trees were still productive after eight years, the best results being obtained when post-plant closely followed preplant treatment (Ritchie, 1989). In contrast, Johnson *et al.* (1988) found 62 l Nemacur 3 (fenamiphos) ha^{-1} to peach orchards in the USA over the period 1982–1987 did not always decrease the nematode significantly and did not benefit the trees. Presumably, presence of *C. xenoplax* in the soil is no guarantee that trees will succumb to PTSL, though the nematode is invariably present at PTSL sites. As already mentioned, in soils infested with *C. xenoplax* the risk of tree death is greatly decreased if the land is fumigated before planting and if Lovell rootstocks are used. In grapevine, 62 l. fenamiphos ha^{-1} decreased numbers of the nematode up to harvest time (Santo *et al.*, 1985a). Control of *C. xenoplax* with fenamiphos, which is lipophilic, is inversely related to the adsorptive capacity of the soil.

Hoplolaimus von Daday, 1905 (Fig. 4.5)

'Lance' nematodes (*Hoplolaimus* spp.) are migratory parasites of plant roots, feeding ecto-, endo-, or semi-endoparasitically, according to plant species and/or location in the root system. Economically, the most important species are *H. columbus* Sher, 1963, *H. galeatus* (Cobb, 1913) Thorne, 1935, *H. indicus* Sher, 1963 and, to a lesser extent, *H. pararobustus* (Schuurmans Stekhoven and Teunissen, 1938) Sher 1963, *H. seinhorsti* Luc, 1958 and *H. aegypti* Shafiee and Koura, 1970. Lance nematodes have wide host ranges, including annual and perennial crop plants, grasses and woody perennials. *Hoplolaimus columbus* is important in the Coastal Plain region of the south-eastern USA, mainly as a pest

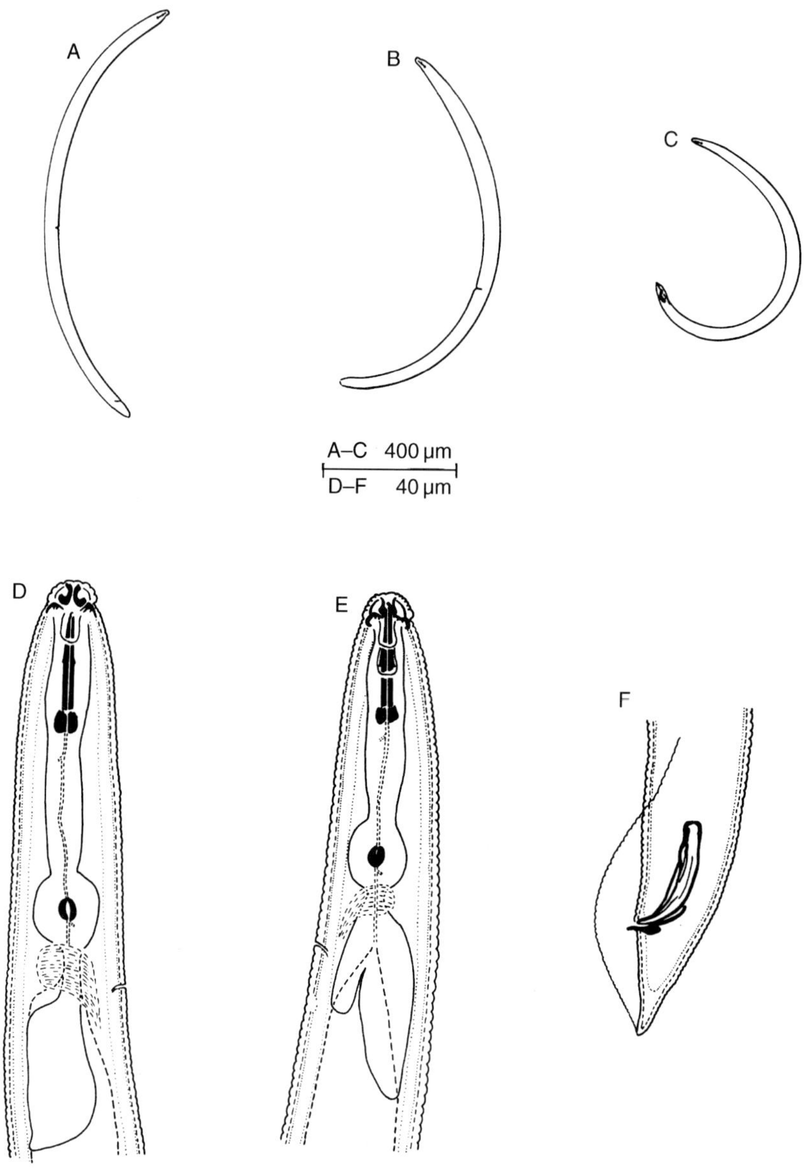

Fig. 4.5. Species of *Hoplolaimus. H. columbus*: A, female; D, female anterior end. *H. indicus*: B, female; C, male; E, female anterior end; F, male tail.

of soybean and cotton (Fig. 4.6). *Hoplolaimus galeatus* is widespread in the USA and is also known in Canada, Central America and India. *Hoplolaimus indicus* is widespread in India and has a very wide host range, including many important crop plants. *Hoplolaimus pararobustus* is a pest of banana in Africa but is also associated with the roots of many other crop plants. *Hoplolaimus seinhorsti* attacks cowpea, cotton and other crop plants and is known from widely separated countries in the tropics. *Hoplolaimus aegypti* is a pest of cotton in Egypt.

Fig. 4.6. Stunting of cotton by *Hoplolaimus columbus* (Photo: C.M. Heald).

In *H. seinhorsti*, reproduction is parthenogenetic, males being unknown. Parthenogenesis is also the norm in *H. columbus*, though males have been found in some populations. Males are more frequent or common in *H. galeatus*, *H. indicus* and *H. pararobustus*, so sexual reproduction may also occur in these species.

The pathogenicity of *H. columbus* to soybean and cotton was demonstrated by Lewis *et al.* (1976). The nematode is endoparasitic in soybean roots, penetrating the endodermis, pericycle and phloem of Hampton 266 soybeans but feeding mainly in the cortex of Bragg soybeans. In Deltapine 16 cotton, the nematode is endo- and semi-endoparasitic but does not penetrate to the root endodermis. Stunting, yellowing and partial defoliation of cotton in North Carolina is associated with *H. galeatus* in the roots, feeding on the cortex and especially on the phloem. This species is capable of destroying roots of *Pinus* seedlings (Ruehle, 1972) and reducing the weight of bermuda grass (*Cynodon dactylon*) by 50% over a period of six months (Di Edwardo, 1963). *Hoplolaimus indicus* causes stunting and chlorosis in rice, millet (*Eleusine coracana*), sugarcane, *Citrus* spp., maize and pearl millet (*Pennisetum typhoides*). *Hoplolaimus pararobustus* can cause severe damage to banana roots, as a result of its endo- and semi- endoparasitic feeding. The feeding of lance nematodes results in necrosis of cells around the heads of the nematodes and through the tissues which they invade. The damage may remain within the cortex or the stele but lesions may appear on the surface of roots and these may rupture or the root cortex may die and disintegrate.

Fassuliotis (1971, 1976) found that *H. columbus* can survive for several years in very dry soil (0.2–0.3% moisture) and subjection in very dry soil to 70°C for one hour. The wide host ranges of lance nematodes and their ability to injure plants at low population densities also makes them difficult to control. Resistance to *Hoplolaimus* spp. has not yet been detected in plants. Although lance nematodes can be controlled to a greater or lesser extent by fumigant or non-fumigant nematicides, only *H. columbus* has been studied sufficiently to warrant review here.

Soybean

As many as 42 *H. columbus* ml^{-1} soil were found around the roots of soybean plants that were severely stunted, chlorotic and with few pods (Fassuliotis *et al.*, 1968). Such an infestation may be the result both of aggregation of nematodes on the roots and nematode multiplication, because the damage threshold for the nematode on soybeans varied from as few as 4 up to 100 $(100\ ml)^{-1}$ soil (Rickard and Barker, 1982). Although the nematode has a wide host range, sweet potato, tomato, pepper, tall fescue and bahia grass were non-hosts and groundnut was a poor host (Lewis and Fassuliotis, 1982). Some control may therefore be possible, given good weed control, by adequately researched crop rotations. At present, however, control relies heavily on the use of nematicides.

In South Carolina, *H. columbus* was controlled better in four years out of five by D-D or DBCP than by carbofuran or fenamiphos and in severely infested fields the use of nematicides increased soybean yields (Blackmon and Musen, 1974). Soybean yields were increased more, where the soil had a compacted ('hard pan') layer, by combining the application of a nematicide with subsoiling 36 cm deep than by the application of nematicide alone. Similarly, in Georgia, Parker *et al.* (1975) found that in an infested sandy loam, subsoiling to 46 cm deep in addition to application of DBCP increased yields more than application of DBCP alone and resulted in fairly uniform distribution of *H. columbus* to 46 cm deep, whereas in soil not subsoiled most nematodes were found to 20 cm deep. In the second year of the experiment, a second subsoiling did not increase yields but DBCP did, presumably because nematode numbers were decreased by DBCP, not by subsoiling alone. In North Carolina, cv. Coker 317 was significantly more tolerant of attack by *H. columbus* than cvs Coker 156, Centennial, Deltapine 105 or Gordon. In infested soil, 1,3-dichloropropene, aldicarb, carbofuran and fenamiphos significantly increased soybean yields (Schmitt and Imbriani, 1987). Combined use of a tolerant cultivar and an effective nematicide could, therefore, prevent or minimize yield losses. However, Noe (1993) has shown that although a yield loss of nearly 10% at 60 nematodes $(100\ ml)^{-1}$ soil could be corrected by a relatively low cost nematicide treatment, the low price of soybeans (4 cts kg^{-1} seed) made the treatment uneconomic, whatever the soil population density and despite a potential maximum yield loss of 48%.

Cotton

Annual subsoiling to 35 cm deep under the planting rows maximized cotton seed yields and decreased *H. columbus* in the top 20 cm of the soil by encouraging deeper rooting. Although aldicarb or DBCP treatment of the seedbed decreased numbers of the nematode, it did not increase yields more than subsoiling alone (Hussey, 1977). In Egypt, aldicarb at 0.3 g per plant at sowing decreased numbers of *H. columbus* and increased yield of Egyptian cotton (*Gossypium barbadense*) by 88% (Koura and El-Sharkawi, 1978). In South Carolina, numbers of *H. columbus* in the soil were significantly decreased by 28 l Telone II ha^{-1}, with or without 3.9 kg aldicarb ha^{-1} (Mueller *et al.*,

1992). Mueller and Sullivan (1988) detected neither resistance, nor tolerance to *H. columbus* in a range of cotton cultivars.

Longidorus (Micoletzky, 1922) Thorne and Swanger, 1936 (Fig. 4.7)

Species of *Longidorus* ('needle' nematodes) are long, vermiform ectoparasites of subterranean plant tissues that feed principally on root tips, which are usually galled as a result. Young tap and lateral roots or fibrous roots may be very stunted if there are numerous nematodes feeding on them. Needle nematodes can feed on a wide range of woody and herbaceous plants. Most of these

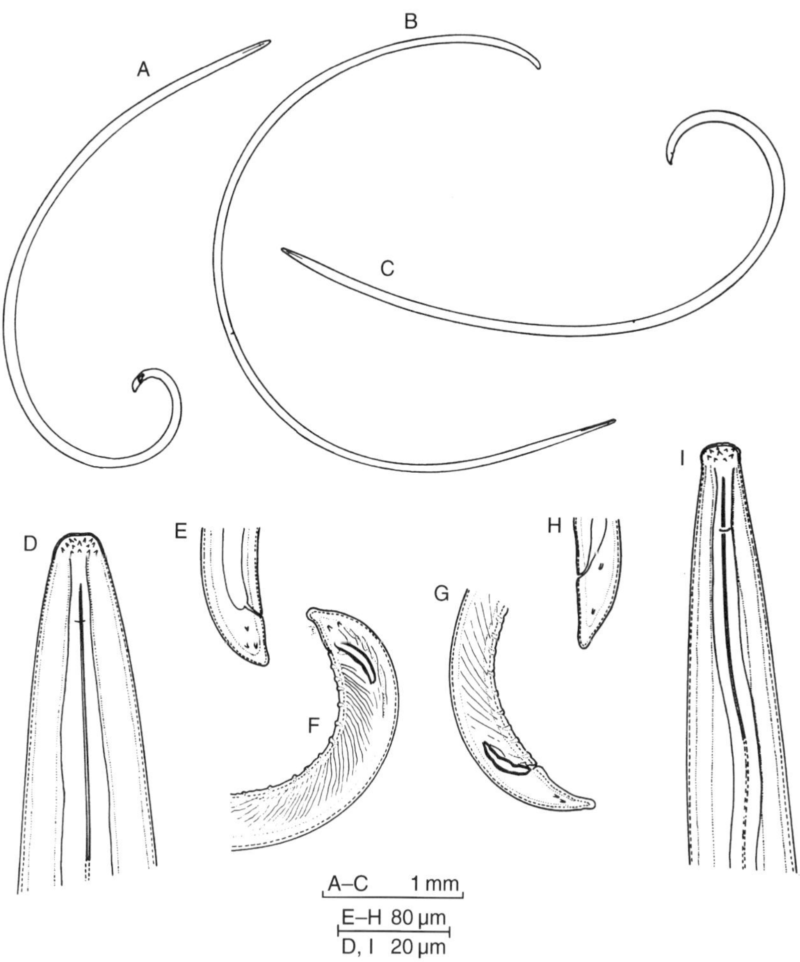

Fig. 4.7. Species of *Longidorus*. *L. elongatus*: A, male; C, female; D, female anterior end; E, female tail; F, male tail. *L. attenuatus*: B, female; G, male tail; H, female tail; I, female anterior end.

(a)

(b)

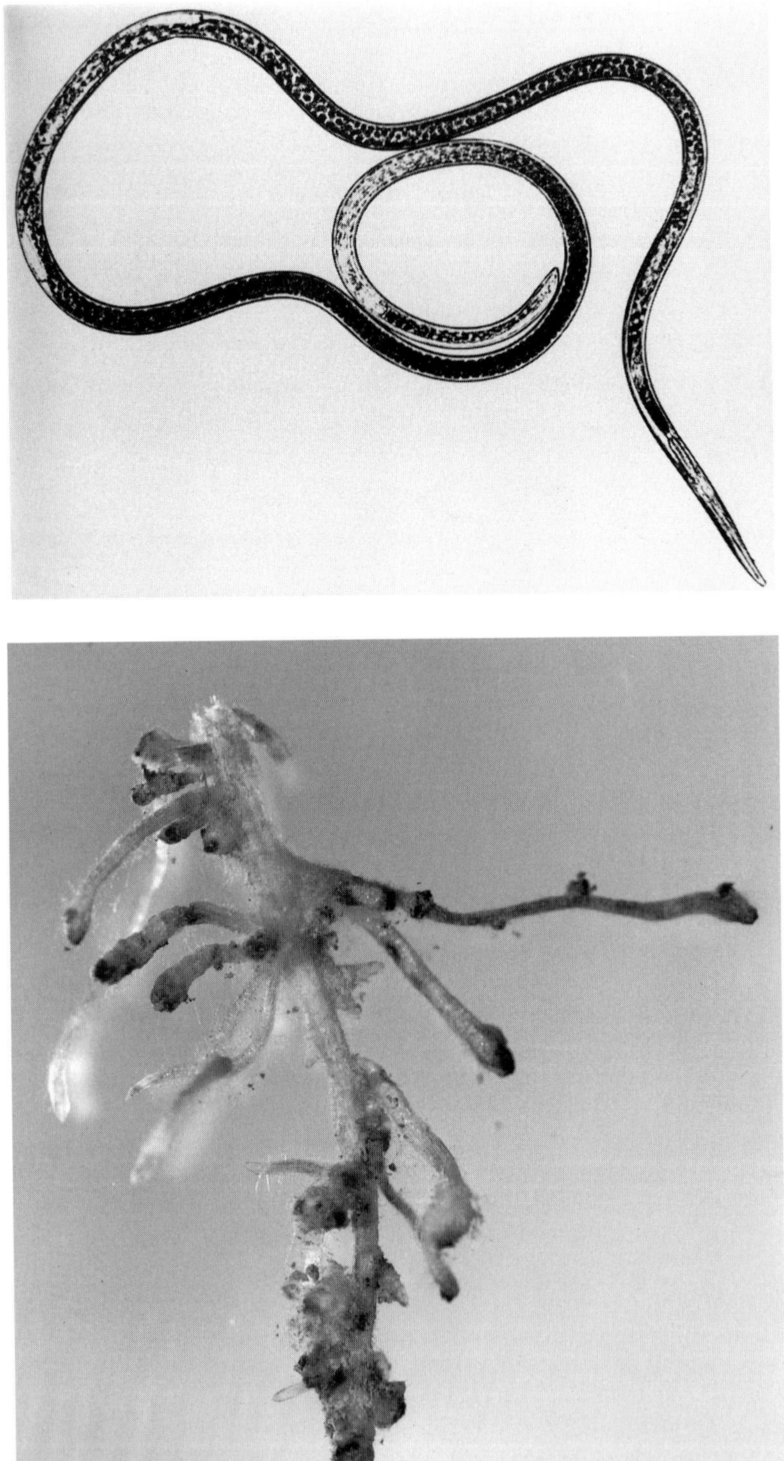

Fig. 4.8. (*and opposite*) (a) *Longidorus elongatus* female (Photo: D.J. Hooper). (b) Root tip galling of sugarbeet seedling due to feeding of *L. elongatus*. (c) Stunting of three sugarbeet seedlings in heavily infested soil (left) compared with two seedlings from lightly infested soil (Photos: C.C. Doncaster).

nematodes can acquire and transmit specific plant viruses, which may be of greater importance than the direct damage resulting from nematode feeding. When the nematodes are numerous, however, feeding damage may be marked (Figs 4.8, 4.9).

Several species of *Longidorus* have life cycles that last more than one year and the adults may live for more than one year. The adults are several mm in length. The best known species is *L. elongatus* (de Man, 1876) Thorne and Swanger, 1936, which is widespread in temperate areas of the world. It is usually parthenogenetic, males being very rare or absent in most populations, though common in others. Multiplication is slow. There is probably only one generation per annum and although the nematode may multiply 2- to 4-fold on some crops, it may increase as much as 20-fold on strawberry (Yassin, 1969). The depth distribution of *Longidorus* spp. can vary considerably – it may be linked to the depth distribution of the plant roots on which the nematodes feed, though ploughing and cultivations may affect its abundance in the topsoil. Apart from *L. elongatus*, four other species described by Hooper (1961) occur in Europe. *L. attenuatus*,

(a)

(b)

Fig. 4.9. (*and opposite*) Crop injury by *Longidorus elongatus*. (a) Uneven growth of carrots due to attack by *L. elongatus*; (b) Ryegrass ley badly injured by *L. elongatus*; (c) Roots of carrot seedling from heavily infested soil (left) and lightly infested soil (right) (Photo: C.C. Doncaster). (d) Ryegrass plants from lightly infested soil (left) and heavily infested soil (right) (Photo: C.C. Doncaster).

which occurs locally in Europe and the former USSR and *L. leptocephalus*, which occurs in north-west Europe, are parthenogenetic, whereas *L. macrosoma*, occurring in temperate areas of Europe and *L. caespiticola*, occurring in western Europe, would seem to multiply sexually, as males are commonly found in field populations. Most control studies have been on *L. elongatus*. The control of other species has received less attention.

(c)

(d)

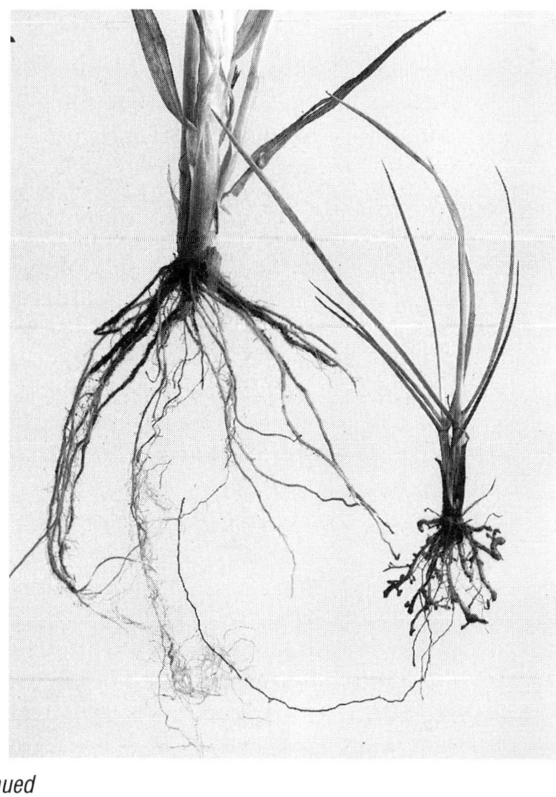

Fig. 4.9. *Continued*

Control of *L. elongatus*

The decline of peppermint due to *L. sylphus* (= *L. elongatus*) was controlled by fumigating the soil with D-D (189 l ha^{-1}), EDB (26.5 l ha^{-1}) or DBCP (11 l ha^{-1}) (Jensen and Horner, 1957). D-D at 151 l ha^{-1} or quintozene at 67 kg ha^{-1} killed in excess of 95% of *L. elongatus*, thereby minimizing transmission of raspberry ring spot and tomato black ring viruses to strawberry, and greatly increased fruit yields (Taylor and Murant, 1968). Dazomet at 336 kg ha^{-1} or D-D at 280 l ha^{-1} controlled the nematode and transmission of raspberry ringspot virus in raspberry plantations in Scotland. Quintozene at 90 kg ha^{-1} decreased nematode numbers slowly and controlled virus transmission (Trudgill and Alphey, 1976).

Applying 207 l 1,3-dichloropropene ha^{-1} to soil killed 90% of *L. elongatus* and, under a grass–clover sward, numbers were still only 70% of those in untreated soil five years later (Boag, 1984). Nematostatic control afforded by organophosphate or carbamate nematicides, although adequate for control of the nematode on some annual crop plants, is too brief for perennials, unless they can be treated repeatedly without leaving unacceptable residues in the harvested crop. Even if this is theoretically possible, foliar applications of these compounds are hazardous and only a tiny fraction of what is applied to the foliage is secreted from the roots to paralyse the nematodes in the adjacent soil (Alphey *et al.*, 1985).

Control of Other Species of *Longidorus*

Longidorus vineacola Sturhan and Weischer, 1964 has been controlled in onions and celery in Israel by preplant fumigation of the soil with EDB and EDB or methyl bromide, respectively. Celery planted in late summer, when nematode numbers were small, rather than in late autumn, when nematode numbers were larger, avoided injury and obviated soil fumigation (Cohn *et al.*, 1968b; Cohn and Auscher, 1971). *Longidorus africanus*, which injured lettuce in California at temperatures above 24°C, was greatly decreased by injecting 1,3-dichloropropene into the soil 20–25 cm deep at 53 l ha^{-1} in the centre of the prepared seedbed, four days before lettuce was sown (Radewald *et al.*, 1969). In southern Germany, where *L. maximus* attacks most crops, other than cereals and some *Cruciferae*, D-D, metham-sodium or dazomet cannot fully control the nematode, because many live deep in the soil and are not reached by fumigant gases (Sturhan, 1964b). Over 99% of *L. attenuatus*, a pest of sugarbeet in England, were killed by 371 l D-D or chloropicrin injected 15 cm deep into sandy soils in winter (Whitehead *et al.*, 1970a). Most of the nematodes in the top 20 cm of the soil were killed by 134 or 268 l D-D ha^{-1}, similarly injected in autumn or winter. Injected beneath sugarbeet rows before sowing the seeds in them, 50 l D-D or 75% 1,3-dichloropropene (Telone) killed half the *L. attenuatus* in the rows and 100 l ha^{-1} of either fumigant, similarly applied, killed three-quarters of the nematodes in the top 20 cm of the soil. The treatments also increased sugar yields significantly, especially when applied 15 or 22.5 cm deep (Whitehead *et al.*, 1970b). As little as 67 l D-D or 50 l 75% 1,3-dichloropropene under the rows at or before sowing killed over

90% of the nematodes in the rows but only 25% of them 12.5 cm from the rows and none 25 cm from the rows (Cooke *et al.*, 1970). D-D at 66 l ha^{-1} in the rows killed 74–92% of *Longidorus* 0–23 cm deep in the treated row, 32–48% 13 cm from the row and none 25 cm from the row (Cooke *et al.*, 1973). Despite this localization of nematode control, sugar beet grew and yielded better in treated than in untreated rows (Table 4.1). In sandy soils, 1,3-dichloropropene injected 15 cm deep, 15 and 30 cm deep or 15 and 45 cm deep controlled *L. attenuatus*, *L. leptocephalus* and *L. macrosoma* to 60 cm deep in the soil (Chapman, 1983).

Small amounts of non-fumigant nematicides, notably aldicarb, applied to the seed furrows or around the seeds at sowing also protect sugarbeet seedlings from injury by ectoparasitic nematodes, including *L. attenuatus*, in the critical early weeks of seedling establishment. Dosages of 0.6–1.1 kg aldicarb ha^{-1} in the sugarbeet rows have largely prevented root ectoparasitic nematode damage and greatly improved beet yields, particularly in years with cold, wet springs favourable to the movement and feeding of these nematodes (Dunning and Winder, 1969).

Trichodorus Cobb, 1913 and *Paratrichodorus* Siddiqi, 1974
(Figs 4.10, 4.11)

Several species of these 'trichodorid' nematodes injure crop plants directly and may also transmit tobacco rattle virus (TRV) and pea early browning virus (PEBV) to roots during feeding. Christie and Perry (1951) were the first to recognize damage due to a trichodorid nematode, named *Trichodorus christiei* by Allen (1957), on several crop plants in the Sanford area of Florida. Trichodorids are migratory, root ectoparasites, which feed on root hairs, on epidermal cells in the root elongation zone and on root tips. They sometimes feed on cortical cells exposed by root cracks and may also feed on the apical meristems of roots, devitalizing the root tips. They may feed on epidermal cells by a rasping action of the onchiostyle (stylet) but also feed by piercing cell walls and withdrawing cytoplasm (Russell and Perry, 1966). Wyss (1971) showed that, in feeding on tobacco roots, *T. similis* Seinhorst, 1963 formed a feeding tube in the stoma which was attached to the hole in the epidermal cell. Cytoplasm was withdrawn through this tube, which remained in place after the nematode had moved away after feeding. As a result of nematode feeding, the root systems of young plants may be stunted, with sparse, lateral roots, or they may exhibit discoloured or necrotic areas, where the nematodes have been feeding and bear numerous 'stubby' roots – hence the common name 'stubby root' nematode. Trichodorids are polyphagous and have wide host ranges.

Paratrichodorus minor (Colbran, 1956) Siddiqi, 1974 (syn. *Trichodorus christiei* Allen, 1957) is widespread in the USA and is known in Puerto Rico, Brazil, South Africa, the Philippines, New Zealand and Afghanistan (Fig. 4.12). It is a pest of warm soils, its life cycle on tomato taking 21–22 days at 22°C but only 16–17 days at 30°C (Rhode and Jenkins, 1957a). The temperature optimum for this nematode is around 25°C and it does not complete its life cycle at 20 or 35°C (Malek and Jenkins, 1964). Reproduction is parthenogenetic, males being

Table 4.1. Effect of D-D or Telone applied in sugarbeet rows at or before sowing on yields of beet sugar (t ha^{-1}) in fields infested with *Longidorus attenuatus* and trichodorid nematodes in west Norfolk, UK.

		Site			
		Gayton A		Gayton B	
Longidorus l^{-1} soil :		70		130	
Trichodorids l^{-1} soil :		1200		600	
Treatment					
Untreated		6.9		7.1	
D-D*l ha^{-1}	48	7.9		8.1	
	65	7.8		8.2	
	96	8.0		8.2	
L.S.D. (5%)		0.52		0.33	
		Gayton C†		E. Winch‡	
Untreated			7.5		6.6
D-D, l ha^{-1} before sowing		56	8.4	65	7.4
		79	8.7	91	7.3
		101	8.2	112	7.3
D-D, l ha^{-1} at sowing		56	8.4	67	7.6
Telone, l ha^{-1} at sowing		56	8.5	–	–
L.S.D. (5%)			0.29		0.45

* Shallow or deep tines alone in the sugarbeet rows had no significant effect on sugar yield in either experiment, nor in the E. Winch experiment.
† Yields determined from two rows × 440 m.
‡ Yields determined from two rows × 343 m.

extremely rare. On tomato and other good hosts it may multiply 10-fold or more. Seven species of trichodorid are fairly common in northern Europe – *P. pachydermus* (Seinhorst, 1954) Siddiqi, 1974, *P. teres* (Hooper, 1962) Siddiqi, 1974, *P. anemones* (Loof, 1965) Siddiqi, 1974, *T. primitivus* (de Man, 1880) Micoletzky, 1922, *T. similis* Seinhorst, 1963, *T. viruliferus* Hooper, 1963 and *T. cylindricus* Hooper 1962. They too feed on root tips and along the sides of roots (Pitcher and Flegg, 1965; Whitehead and Hooper, 1970) (Fig. 4.13). Like *P. minor*, *P. teres* is parthenogenetic, males being absent, rare or non-functional. In contrast, *P. anemones* and *P. pachydermus* are obligately amphimictic and *T. primitivus*, *T. similis*, *T. cylindricus* and *T. viruliferus* are bisexual and probably reproduce amphimictically. *P. pachydermus*, *T. primitivus* and *T. viruliferus* can transmit TRV and PEBV and *T. similis* is known to transmit TRV.

Trichodorids live in the moisture films around soil particles. Their activity is dependent on adequate soil moisture to allow them to move on and between roots. Transmission of TRV, for example, depended on at least 15% soil moisture, with maximum transmission at 30% soil moisture (Cooper and Harrison, 1973). Stunting of sugarbeet by trichodorid and longidorid nematodes in sandy soils in

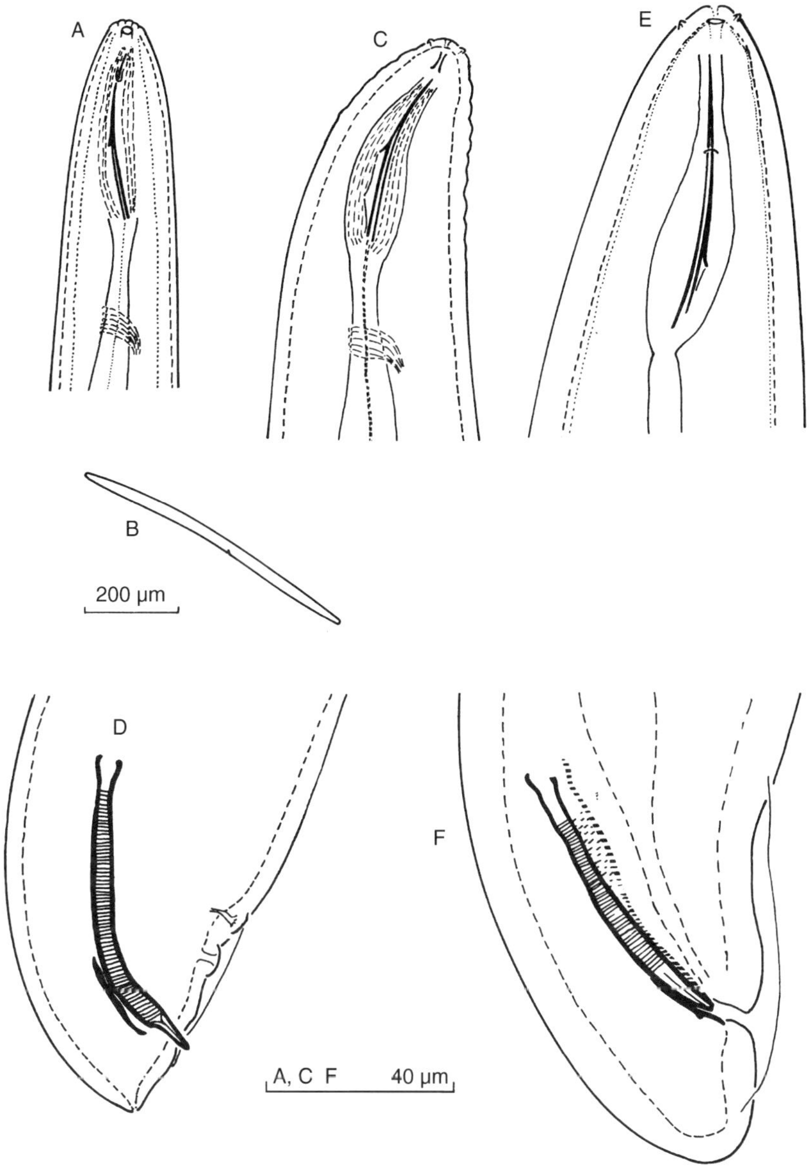

Fig. 4.10. Species of *Paratrichodorus*. *P. minor*: B, female; A, female anterior end. *P. teres*: C, female anterior end; D, male tail. *P. pachydermus*: E, female anterior end; F, male tail.

England is dependent on cool, wet spring weather after the seeds have been sown (e.g. Whitehead *et al.*, 1971).

Because they have wide host ranges and crops resistant to them have not been bred, control of trichodorids depends largely on the use of nematicides. *P. minor* differs from other trichodorids in several respects relevant to control and is therefore dealt with separately.

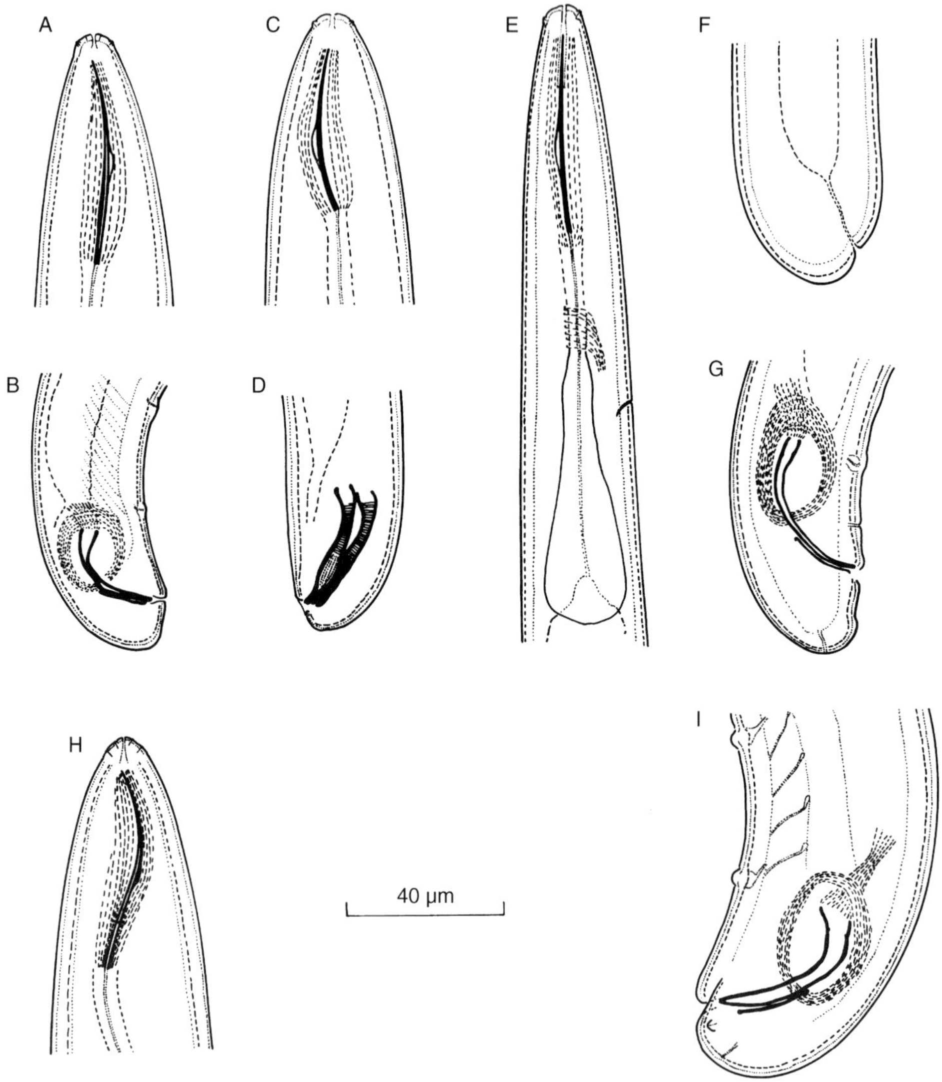

Fig. 4.11. Species of *Trichodorus. T. viruliferus*: A, female anterior end; B, male tail. *T. cylindricus*: C, female anterior end; D, male tail. *T. primitivus*: E, female anterior end; F, female tail; G, male tail. *T. similis*: H, female anterior end; I, male tail.

Control of *P. minor*

Although it has a wide host range, *P. minor* may multiply little on a number of crop plants, including cucumber, strawberry, tobacco, garden pea, radish, rye and spinach and not at all on asparagus and sunn hemp (*Crotalaria spectabilis*) (Rhode and Jenkins, 1957b). Crop rotations might therefore be devised to reduce numbers of the nematode in the soil. Such an approach is, however, complicated

Fig. 4.12. Stubby root injury to maize by *Paratrichodorus minor* (Photo: D.W. Dickson).

by the existence of physiological races of the nematode with differing host preferences, so the benefit of a rotation could be short-lived because the nematode may multiply rapidly on a susceptible crop plant. Unlike the other trichodorids mentioned below, *P. minor* declines rapidly in fallowed soil and none survived a four month fallow in sandy loam (Thomason, 1959). As other trichodorids, the nematode is little affected by flooding of the soil. Although methyl bromide gas under polyethylene soil cover, D-D or EDB kills nearly all nematodes in the soil, *P. minor* multiplies more rapidly than other plant-parasitic nematodes when host plants are sown (Perry, 1953). How this is achieved is unclear, perhaps the eggs are not killed by the fumigants but hatch more slowly than in untreated soil, because damage to onions was prevented and onion yields were increased when organic soil was fumigated with 374 l 1,3-dichloropropene ha^{-1} (Hoff and Mai, 1962). Rhoades (1969) found the nematode was controlled better on onions by several non-fumigant nematicides (aldicarb, carbofuran, fensulphothion, ethoprophos, thionazin or phorate) than by soil fumigants (D-D, EDB or DBCP). Similarly, aldicarb, carbofuran, fensulphothion, fenamiphos or ethoprophos controlled the nematode better on sweetcorn (*Zea mays* var. *saccharai*) than did D-D, EDB or DBCP (Johnson and Chalfont, 1972). Applied at planting in Florida, 10% granules of aldicarb or carbofuran improved potato tuber quality, increased tuber yields and lessened corky ring spot (TRV) transmitted by *P. minor* (Weingartner *et al.*, 1976).

Control of Other Trichodorid Species

Although two years of bare fallow decreased numbers of active *T. cylindricus* and *P. pachydermus* in sandy soils in England, it did not eliminate them from the soil (Whitehead *et al.*, 1970a). The numbers of common species of trichodorid in the

(a)

(b)

Fig. 4.13. *Caption opposite*

(c)

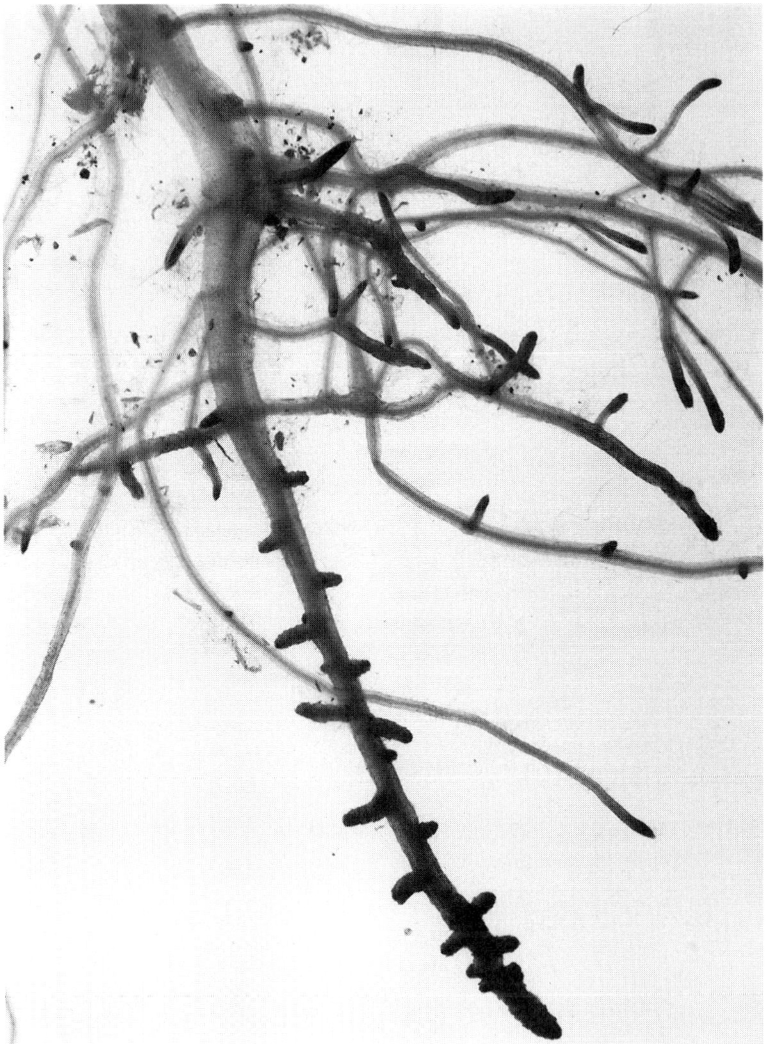

Fig. 4.13. (*and opposite*) (a) Uneven growth of sugarbeet due to trichodorid ('stubby root') nematodes. (b) Healthy sugarbeet seedling (left) and two seedlings stunted by *Paratrichodorus teres*. (c) Close-up of stubby root symptoms. (Photo a: Broom's Barn Experimental Station. Photos b and c: C.C. Doncaster.)

soil can be quickly and greatly decreased by soil fumigation. Taconis and Kuiper (1964) found 600 l D-D eliminated *P. pachydermus* and *T. primitivus* from soil 20–60 cm deep but did not kill all of them near the soil surface. At 771 l ha^{-1}, D-D or chloropicrin killed over 90% of trichodorids in the top 20 cm of four sandy soils in England. In two of these soils, there were only 5–20% of pre-fumigation numbers of the nematodes in the soil after two crops of sugarbeet, on which the nematodes fed. In a third soil, in which 99% of *T. cylindricus* had been killed, there was little nematode increase after two crops of sugarbeet, whereas in a fourth soil, in which 93% of *P. anemones* had been killed, there were nearly as

(a)

(b)

(c)

Fig. 4.14. (*and opposite*) Chemical control of stubby root and needle nematode injury to sugarbeet. (a) land-wheel controlled fumigation of sugarbeet rows with 1,3-dichloropropene just before sowing; (b) fumigation tine, slit fillers and press wheel; (c) fumigated rows (left and extreme right), unfumigated rows (right); (d) effect of granular aldicarb applied to a sugarbeet row (left) compared with two untreated rows (right) (Photo: R.A. Dunning).

(d)

Fig. 4.14. *Continued.*

many nematodes in fumigated as in unfumigated soil after the second sugarbeet crop (Whitehead *et al.*, 1970a). Cooke and Draycott (1969) found the same dosage of D-D killed 95% or more of trichodorids in sandy soils in England. In the Netherlands, a disease of sugarbeet and potatoes, called 'T' disease after the island of Texel where it was first encountered, was closely associated with the feeding of *Paratrichodorus teres* and/or 'sheath' nematodes (*Hemicycliophora* spp.). Soil fumigation greatly decreased numbers of the nematodes and increased crop yields (Bockel, 1977).

Injected beneath predetermined crop rows in spring, 72 l D-D or 45 l 1,3-dichloropropene killed over 80% of trichodorids harmful to sugarbeet in sandy soils in a wet spring and greatly increased sugar yields (Whitehead *et al.*, 1970b).

More conveniently, and without risk of phytotoxicity, granular non-fumigant nematicides can be applied to the rows of sugarbeet during sowing (Dunning and Winder, 1969) (Fig. 4.14). In sandy soils, however, leaching of nitrate N (and aldicarb) from the seedbed in a very wet spring may lessen yield response to aldicarb treatment, compared with D-D or 1,3-dichloropropene row treatment (Cooke *et al.*, 1970). In a soil infested with *T. viruliferus*, sugarbeet yields responded to aldicarb row treatment in a wet spring, where the nematodes were abundant, but not where they were less numerous, nor in a dry spring (Cooke, 1984a). Other non-fumigant nematicides are also effective against trichodorids as row treatments at sowing, i.e. carbofuran, carbosulfan, benfuracarb and oxamyl (Cooke, 1989).

Trichodorids are important as vectors of TRV to potatoes, especially in light, sandy soils. Although some cultivars of potato are not susceptible to some strains of TRV, experiments have shown that transmission of the virus to susceptible cultivars can be greatly decreased, if not always prevented, by applying fumigant or non-fumigant nematicides to the soil. Joubert *et al.* (1971) obtained excellent control of *P. pachydermus* and the TRV it transmitted to potatoes, by fumigating the soil with 400 l D-D ha^{-1} or 1000 l metham-sodium ha^{-1}. Soil fumigation with similarly large amounts of D-D or metham-sodium greatly decreased the incidence of tuber necrosis due to TRV in the three consecutive potato crops that followed it and almost eradicated the nematode from the soil (Joubert and Dalmasso, 1974). In Scotland, D-D at 186 l ha^{-1} applied in autumn controlled the spread of TRV for two years in treated soil planted with potatoes susceptible to the virus (Cooper and Thomas, 1971). In the Netherlands, where regionally differing strains of TRV made use of 'resistant' cultivars of potato impractical, the trichodorid vectors of the virus were eliminated from the soil by applying 500 l D-D ha^{-1} in two equal doses, the second after the soil had been ploughed three weeks after the first dose had been applied (Maas, 1975).

Sprinkled into open potato drills, 2.24 kg aldicarb or 5.6 kg oxamyl or ethoprophos ha^{-1} controlled 'spraing' (TRV) in Pentland Dell potatoes in England in soil lightly infested with *P. pachydermus* and *T. primitivus* (Brown and Sykes, 1973). In Scotland, in soil infested with both these species and *T. cylindricus*, spraing was controlled best on Pentland Dell potatoes by 3.4 kg aldicarb or 5.6 or 7.8 kg oxamyl ha^{-1} applied to the soil at planting (Alphey *et al.*, 1975). None of these treatments eliminated the disease, but aldicarb, oxamyl, fenamiphos or carbofuran controlled spraing better in northern England, when they were incorporated in the top 10–15 cm of the soil before planting, than when they were applied in the seed furrows at planting (French and Wilson, 1976). To be fully effective in preventing virus transmission, when applied as a row treatment, a non-fumigant nematicide would need to be fully systemic in the root system and fully inhibit nematode feeding on the roots. In-furrow application of aldicarb was more effective than side-band application 7–8 weeks after planting (Alphey *et al.*, 1986). In Belgium, oxamyl applied to experimental plots infested with trichodorids prevented or minimized transmission of TRV to cvs Resy and Bintje, both of which are susceptible to TRV (Pelsmaeker, 1987).

Tylenchorhynchus Cobb,1913 and *Merlinius* Siddiqi, 1970

Several species of *Tylenchorhynchus* and *Merlinius* ('stunt' nematodes) may injure plant roots. They are migratory root parasites that feed on root hairs, epidermal cells and sometimes on cortical cells. Compared with many other nematodes that feed on crop plants, they do not usually cause obvious above-ground symptoms of injury but they are widely distributed and may be responsible for unsuspected plant growth retardation. Of the many recognized species, only four warrant mention here.

Because of their wide host ranges, 'stunt' nematodes are not readily controlled by crop rotations, nor are resistant crop cultivars available. These nematodes are not, however, resistant to desiccation and their numbers may be greatly reduced by dry season bare fallowing, where this is practical. As they are mostly ectoparasitic on roots, they are easily killed by soil fumigation or inactivated by non-fumigant nematicides applied to the soil.

Tylenchorhynchus annulatus (Cassidy, 1930) Golden, 1971 (syn. *T. martini* Fielding, 1956)

Widespread in tropical and subtropical soils, this parthenogenetic species feeds on a wide range of crop plants, especially *Gramineae*. It has been reported to injure sugarcane, sorghum, rice and Bermuda grass (*Cynodon dactylon*) and may retard the growth of other plants on which it feeds by stunting lateral roots, thereby lessening uptake of water and soil nutrients.

Control

Tylenchorhynchus annulatus has been controlled by methyl bromide (Atkins and Fielding, 1956), hydrogen sulphide in flooded rice fields (Rodriguez-Kábana *et al.*, 1965) and by DBCP, D-D, EDB and fensulphothion (Winchester, 1968).

Tylenchorhynchus claytoni Steiner, 1937

This amphimictic species is widespread in eastern states of the USA. It is injurious to young maize, turf grasses, azaleas and pine seedlings. It is also parasitic on the roots of wheat, oats, tobacco, soybean and several trees.

Control

Tylenchorhynchus claytoni numbers were reduced by 80% and 40–60%, respectively, when seeds of wheat and rye were dipped for 5 min before sowing in 1.25% (w/v) oxamyl, carbofuran or fenamiphos (Rodriguez-Kábana *et al.*, 1977). In tobacco, the nematode was controlled by fumigating the soil with a mixture of 75% 1,3-dichloropropene and 19% EDB (Miller and Taylor, 1958). In potatoes, the nematode was controlled by applying aldicarb, disulphoton or fensulphothion to the soil (Miller and Kring, 1970).

Tylenchorhynchus brassicae Siddiqi, 1961

This nematode has been found in India, Oman and Egypt, where it is a pest of crucifers, especially cabbage and cauliflower, the growth of which is inversely related to nematode numbers. *Tylenchorhynchus brassicae* invades the root cortex and occasionally the stele. Cabbage and cauliflower may be severely stunted by root infestations of this nematode. The optimum temperature for its multiplication is 30°C and a soil moisture of 25–30% favours its activity in the soil. *Tylenchorhynchus brassicae* can survive about eight months in soil devoid of host plants.

Control

Tylenchorhynchus brassicae on cabbage and cauliflower was well controlled by D-D or metham-sodium (Siddiqui and Khan, 1975) and in rice nurseries by 1.5 kg carbofuran, 5 kg fensulphothion or 3 kg aldicarb ha^{-1} applied broadcast to the seedbeds one week after sowing (Chhabra *et al.*, 1974). More recent research in India and Pakistan has concentrated on the control of this pest with oilcakes, sawdusts and plant extracts. Treating seeds of cabbage or cauliflower before sowing with latexes of *Calotropis gigantea*, *C. procera*, *Euphorbia milii*, *E. neriifolia* or *E. tirucalli* lessened nematode multiplication and improved plant growth, but did not prevent all damage to roots (Siddiqui and Alam, 1988). Dipping the roots of cabbage or cauliflower seedlings in leaf extracts of *Azadirachta indica* (neem) or *Melia azadarach* (Persian lilac) before transplanting them prevented root damage and lessened nematode increase according to the concentration × dipping-time product (Siddiqui and Alam, 1989a). Similarly effective were bare root dips in leaf extracts of *Eichhornia crassipes* (water hyacinth) (Siddiqui and Alam, 1990a). Added to infested soil, especially with the addition of ammonium sulphate, neem sawdust decreased numbers of *T. brassicae* (Siddiqui and Alam, 1990b). Finally, oilcakes of neem, groundnut, mustard or castor applied to soil at dosages equivalent to 100 kg N ha^{-1} decreased numbers of *T. brassicae* in plots of cabbage or cauliflower by 60–70% at harvest (Siddiqui *et al.*, 1979).

Tylenchorhynchus dubius (Butschli, 1873) Filipjev, 1936 (Fig. 4.15)

This nematode is widespread in Europe and has been found in parts of the USA and India. It has a wide host range, including both temperate and tropical crops, though it has not been reported as a pest in the tropics. *Tylenchorhynchus dubius* is amphimictic and is capable of damaging field beans, sorghum, rice, wheat,

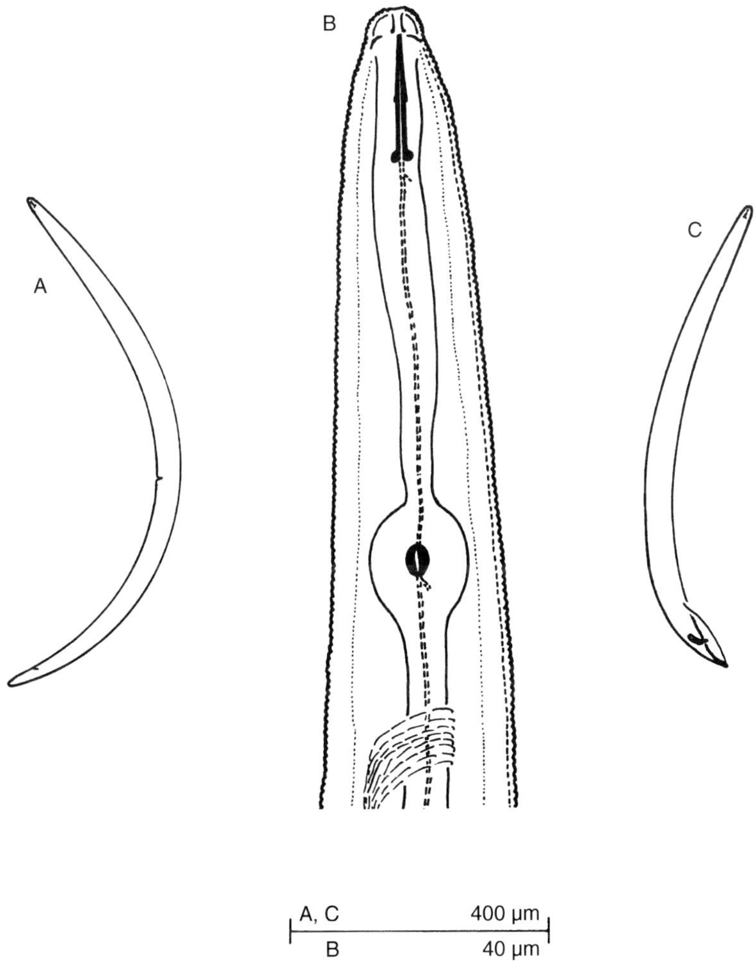

Fig. 4.15. *Tylenchorhynchus dubius*: A, female; B, female anterior end; C, male.

millet, cotton, pea, turnip, *Phaseolus radiatus* and seedlings of Sitka spruce. In the field, the nematode has been associated with damage to oats, *Poa pratensis*, *Agrostis palustris* and carnation. It interacts with *Phoma medicaginis* var. *pinodella* on peas, resulting in blackened stem bases and premature ripening, and with *Fusarium roseum* on *Poa pratensis*, severely stunting roots and shoots. Interactions between *T. dubius* (and other 'stunt' nematodes) and other soil microflora probably await discovery.

Control

Tylenchorhynchus dubius was greatly decreased, when a number of *Compositae* were grown in infested soil (Hijinck and Suatmadji, 1967). *Tagetes patula* grown as a green manure crop also greatly decreased *T. dubius*. The nematode was controlled by methyl bromide, D-D or metham-sodium (Adams and True, 1962) and by 1,3-dichloropropene injected in sugarbeet rows or 0.6–2.2 kg aldicarb ha^{-1} applied to a 5 cm band over sugarbeet rows (Whitehead *et al.*, 1970b). Thionazin lessened multiplication of *T. dubius* (den Ouden and Seinhorst, 1964) and many other non-fumigant nematicides, especially acetylcholinesterase inhibitors, are likely to inactivate the nematode, as they do other species of *Tylenchorhynchus*, preventing injury to young plants.

Xiphinema Cobb, 1913 (Fig. 4.16)

These 'dagger' nematodes are migratory, root ectoparasites, which feed at root tips and/or along the sides of roots. Like longidorids, they live in the large cracks and spaces in soil, adults varying from 1.5 to 5 mm in length, depending on species, and they also have long stylets, which may penetrate to the vascular elements of the thinner roots or to the apical meristems of root tips during feeding. *Xiphinema* spp. cause direct injury through their feeding, often resulting in sparser lateral feeder roots, which may be swollen, necrotic and with swollen, necrotic, root tips (Fig. 4.17). However, they have greater importance as vectors of the polyhedral, isodiametric viruses (nepoviruses), as distinct from the rod-like tobraviruses, which are transmitted by trichodorids (*Paratrichodorus* spp. and *Trichodorus* spp.). Although there are numerous species of *Xiphinema* that are vectors of nepoviruses, this account is restricted to three species known to be pathogenic to crop plants, as well as being nepovirus vectors. The first report of virus transmission by a plant-parasitic nematode was that of Hewitt *et al.* (1958), who showed that *X. index* Thorne and Allen, 1950 transmitted fanleaf virus to grapevines. Each of the three species discussed below may be made up of several physiological races.

Xipinema diversicaudatum (Micoletzky, 1927) Thorne, 1939

This nematode has been reported from Europe, USA, the former USSR, Canada, Australia, New Zealand and the island of Guam (Pacific). Its wide host range includes strawberry, rose, hops, grapevine, raspberry, apple, asparagus, cabbage, carrot, cherry, red clover, peach, tamarillo and *Pinus ponderosa*. It is a serious pest of strawberry and roses but is also a vector of arabis mosaic virus of strawberry and hops, strawberry latent ringspot virus, raspberry ringspot virus, carnation ringspot virus and (perhaps) cherry leaf-roll virus.

Flegg (1968) found the life cycle could take up to three years to complete, with individual females living for more than five years, mating several times. Males are as abundant as females. The longevity of the nematode compensates

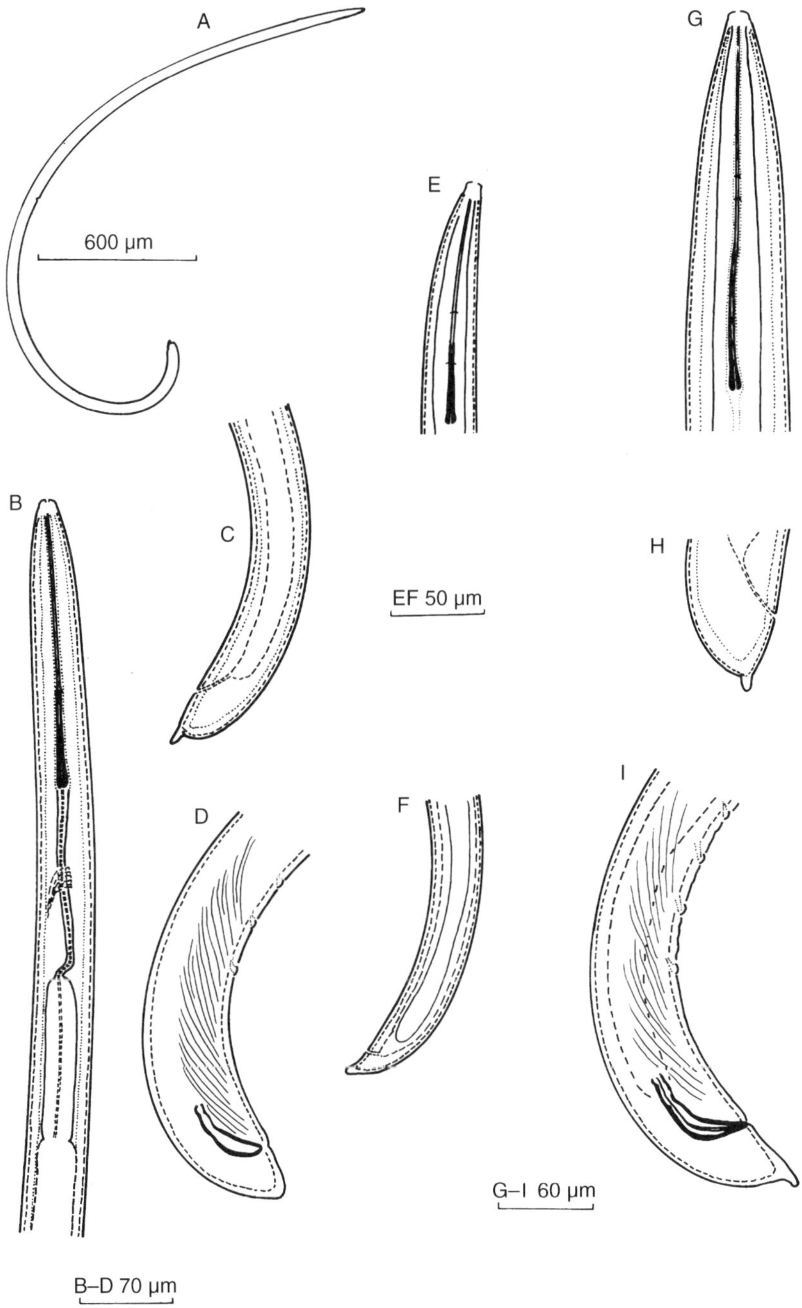

Fig. 4.16. Species of *Xiphinema*. *X. index*: A, female; B, female anterior end; C, female tail; D, male tail. *X. americanum*: E, female anterior end; F, female tail. *X. diversicaudatum*: G, female anterior end; H, female tail; I, male tail.

Fig. 4.17. Galling of rose roots by *Xiphinema diversicaudatum* (Photo: D.J. Hooper).

for its slow rate of multiplication. It can survive in soil devoid of higher plants for up to three years and is most numerous where roots are most abundant in the soil. Eggs are killed at 5°C, whereas juveniles and adults are not and usually survive the winter in Europe.

Control

For the reasons given above, neither bare fallowing, nor crop rotations are practical methods of controlling this nematode. Chemical control of the nematode can prevent direct crop injury. Soil fumigation can exert control for several years after its application, because the nematode population increases slowly on host plants. To prevent virus transmission, more than 99% of the nematodes must be killed or inactivated.

Harrison *et al.* (1963) found that in three different soil types 815 l D-D ha^{-1} killed *X. diversicaudatum* to 71 cm deep in soil. As the kill achieved exceeded 99%, arabis mosaic virus transmission to strawberry was almost eliminated for two years after planting and was much less in the third year than in untreated soil. They found that if there was less than one viable nematode $(2\ l)^{-1}$ soil after treatment, retreatment was unnecessary for several years. In south-west England, 388 l D-D ha^{-1} controlled the virus on strawberry (G.T. Morgan, 1964), whereas in Cheshire, England, neither this treatment, nor 318 l chloropicrin ha^{-1} killed sufficient nematodes to control virus transmission. Repeating these treatments 18 months later, however, greatly reduced virus spread in strawberries (Dale and Hendy, 1967). In a glasshouse soil, 749 l D-D ha^{-1} controlled the nematode on roses and increased plant growth and bloom production (Peachey and Brown, 1965). Although the nematode was well controlled by D-D at 560 or 840 l ha^{-1}, Di-Trapex at 560 l ha^{-1} or dazomet at 370 kg ha^{-1} in hop garden soil, such soil fumigation was less effective in controlling the hop strain of arabis mosaic virus to hops than a two year hop-free fallow. Although fallowing had little effect on nematode numbers, they were then free of virus (Pitcher and McNamara, 1973).

Xiphinema americanum Cobb, 1913

This nematode is widespread in the USA and has been reported from Hawaii, Poland, Australia, New Zealand, Chile, Mexico, Egypt, Canada, north India, Sri Lanka, Pakistan, Guatemala and Belize. Its wide host range includes strawberry, raspberry, maize, soybean, oats, wheat, sugarcane, flax, cherry, lucerne, red clover, grapevine, rose, citrus, mango, arabica coffee, walnut and *Pinus echinata*. The life cycle lasts at least one year and males are extremely rare. Unlike *X. diversicaudatum*, it overwinters mainly as eggs, but even these may not survive long periods in frozen soil. However, in South Dakota, USA, all stages overwintered and the eggs were more affected by low winter temperatures than active stages of the nematode (Malek, 1969). The nematode damages strawberry, apple, coffee, citrus, apricot, bird cherry (*Prunus avium* cv. Mazzard), lucerne, red clover, young spruce (*Pinus echinata*) and reduces the vigour and yield of peach trees. Grapevine root systems are reduced in size and have fewer feeder roots. *Xiphinema americanum* is a vector of tobacco ringspot, tomato ringspot, peach yellow bud mosaic, peach rosette mosaic and cherry rasp leaf viruses. *Xiphinema americanum* belongs to a complex of closely related species. Clear morphometric differences have been found between populations raised from ten single females of a field population of *X. americanum* (Halbrendt and Brown, 1994).

Control

Control of this nematode by crop rotation is difficult because of its wide host range. Brodie *et al.* (1970) found the nematode did not multiply on *Crotalaria* or Sudan grass and it decreased 94% under chicory (Boldyrev and Borzykh, 1979). In soil with 3500 *X. americanum* $(l)^{-1}$, the nematode decreased by 89–95% in plots of rye, barley, wheat and millet, compared with the numbers persisting in fallow plots (Boldyrev and Borzykh, 1983). Such population decreases would lessen damage to a subsequent crop plant but would not prevent virus transmission. Some brassica crops, notably oilseed rape, suppressed the nematode in orchard soil, probably because they were sensitive to the glucosinolate-thioglucosidase system of the plants (Halbrendt and Jing, 1994).

The nematode was controlled for at least two years on strawberry, when soil containing 6400 $(l)^{-1}$ was fumigated before planting with 327 l Vorlex (D-D plus methyl isothiocyanate) ha^{-1}. Strawberry yields were also increased (Adams and Hickman, 1970). In Michigan, USA, large amounts of Telone II, Vorlex, D-D–chloropicrin mixture (70% : 30%, respectively) or methyl bromide-chloropicrin mixture (67% : 32%, respectively) injected 20 cm and 91 cm deep into infested soil protected *Vitis labrusca* cv. Concord grapevines from transmission of peach rosette mosaic virus for at least six years after planting (Ramsdell *et al.*, 1983). Numbers of the nematode on apple were significantly decreased by 18.5 kg oxamyl ha^{-1}, or two applications of 6.7 kg carbofuran ha^{-1} to the infested soil (Thompson *et al.*, 1984). Numbers of the nematode on blueberry (*Vaccinium corymbosum*) were decreased by 70–80% by applying fenamiphos to soil annually, either at 37.4 l of Nemacur 3 ha^{-1} in spring or autumn or at 18.7 l ha^{-1} in spring and autumn (Olsen and Nelson, 1988).

Xiphinema index Thorne and Allen, 1950

This species is a pest of grapevines worldwide. It has been reported in the USA, Europe, South Africa, Victoria, Australia, North Africa, Iraq, Tadzhik SSR and South America. Its hosts include grapevine, fig, rose, mulberry, *Citrus aurantium*, strawberry, walnut, *Prunus* spp., pear, tomato and pistachio nut. It is very damaging to grapevine seedlings. Roots attacked by *X. index* are necrotic, lack laterals, show root tip swellings, cellular hypertrophy and multinucleate cells near nematode feeding sites. The result is an impoverished root system. Males are extremely rare and reproduction is by meiotic parthenogenesis. In California, the life cycle is completed in 22–27 days at 24°C, whereas in Israel it takes 7–9 months at 20–23 °C and 3–5 months at 28°C on a good host plant. In California, the nematode did not survive for more than 10 months without host tissue to feed on but it survived for over 4 years if grapevine roots were left in the soil, after grapevines had been uprooted. It can survive freezing in winter.

Control

Because of the persistence of grapevine roots on which the nematode can survive in soil, after grapevines have been grubbed, replanting is not recommended for five years in California and six or seven years in France. As wheat is not a host of *X. index* or grapevine fanleaf virus, growing it for several years can eliminate both nematode and virus, so the land can then be replanted with virus-free stock. This assumes, of course, that all grapevine roots have been removed or destroyed and weed seeds, which may be infected with the virus, are either destroyed by fumigating the soil with methyl bromide or dazomet or are killed with herbicides. Philis (1994) found that 40–52 months of fallow, or a fallow–barley rotation, after uprooting infested grapevines, decreased *X. index* to non-detected numbers in Cyprus.

Large amounts of soil fumigants can prevent direct injury to roots by killing most *X. index* in the soil, but long-term protection of grapevines from infection with fanleaf virus requires very thorough soil preparation before applying large amounts of methyl bromide, D-D or 1,3-dichloropropene.

Boubals and Dalmasso (1968) found that although large amounts of D-D injected into shallow vineyard soil in southern France appeared to eliminate the nematode and quadrupled grape yield, the virus reappeared after the first grape harvest from the replanted vines. In California, split depth application of 2339 l 1,3-dichloropropene ha^{-1} virtually eliminated the nematode to 2.4 m deep in the soil and no nematodes or fanleaf virus were detected in the first five growing seasons thereafter (Raski and Schmitt, 1972). In Germany, soil heavily infested with fanleaf-infected *X. index* was fumigated with 1000 l D-D ha^{-1} and replanted with Traminer/504 and Traminer/5C vines. After the fourth year, the costs of treatment were repaid by the increased grape yields and by the eleventh year, when nearly all the vines in untreated soil were infected with fanleaf virus, only 22.4% of vines in treated plots were infected with the virus (Rüdel, 1978). The efficacy of methyl bromide in killing *X. index* and *X. pachtaicum*, both vectors of fanleaf virus, was increased by applying it under gas-proof Saranex covers, instead of 45 μm thick polyethylene sheeting, and by applying it 'hot'. The nematodes were eradicated from soil to 90 cm deep by 200 kg hot methyl bromide ha^{-1} applied under Saranex soil covers (Basile *et al.*, 1986). In California, however, after only two grapevine growing seasons, following 448 kg methyl bromide ha^{-1} applied under soil covers or 1560 kg 1,3-dichloropropene ha^{-1}, the nematode was found in half the samples from methyl bromide-treated plots and in all the samples from 1,3-dichloropropene-treated plots. Six years after replanting, the methyl bromide plots had 5% of vines infected with fanleaf and the 1,3-dichloropropene plots had 36% of vines infected (Raski and Goheen, 1988). Soil fumigation, therefore, may only protect grapevines from fanleaf infection for a number of years, not permanently. Retreatment, after a 5–7 year non-host break, and replanting with virus-free stock, is then necessary.

Non-fumigant nematicides appear less effective in controlling virus transmission by *X. index* than soil fumigants, for they must be applied repeatedly and cannot readily be distributed deeply down the soil profile. Virus-resistant rootstocks, if they can be produced, would seem to offer a better long-term solution to nematode transmitted viruses of perennial crop plants than heavy doses of soil fumigants.

Semi-endoparasitic Nematodes of Roots (*Helicotylenchus*, *Rotylenchulus* and *Tylenchulus*)

5

This chapter deals with two species of 'spiral' nematode belonging to the genus *Helicotylenchus* Steiner 1945, the 'reniform' nematode *Rotylenchulus reniformis* Linford and Oliveira, 1940 and the 'citrus' nematode *Tylenchulus semipenetrans* Cobb, 1913.

Helicotylenchus dihystera (Cobb, 1893) Sher, 1961 (Fig. 5.1)

This sedentary, semi-endoparasite of roots is of worldwide distribution, multiplying on a wide range of crop plants. It may cause small superficial lesions on the roots and on some plants the nematodes may penetrate more deeply into the roots and may be completely embedded in the cortex. Cells may become brown and necrotic around the nematodes, as a result of their feeding. The nematode reproduces by mitotic parthenogenesis, males being very rare. *Helicotylenchus dihystera* has been associated with injury to grasses (Sumner, 1967; Wallace, 1971) and injures olive seedlings (Diab and El-Eraki, 1968). Wilting in tomatoes due to *Pseudomonas solanacearum* and in carnation due to *P. caryophylli* were greatly increased when the soil was also infested with *H. dihystera* (Stewart and Schindler, 1956; Libman *et al.*, 1964). In general, *H. dihystera* is considered to be a mild pathogen, though at 4 g^{-1} soil it caused chlorosis, sparse root development and stunted shoots in aubergine, tomato and wheat (Firoza and Maqbool, 1995).

Control

Nitrogen fertilizers differ in their toxicity to *Helicotylenchus* spp. At rates up to 240 kg N ha^{-1} in water, urea was more toxic than ammonium sulphate, which in turn was more toxic than calcium nitrate (Gupta, 1987). In soil also infested with *Meloidogyne incognita* and *Pratylenchus coffeae*, neem cake at 2 t ha^{-1} and sugarcane filter cake at 25 t ha^{-1} decreased nematode numbers under sugarcane by about 80% after 90 days and by about 60% after 180 days. Sugar yields were

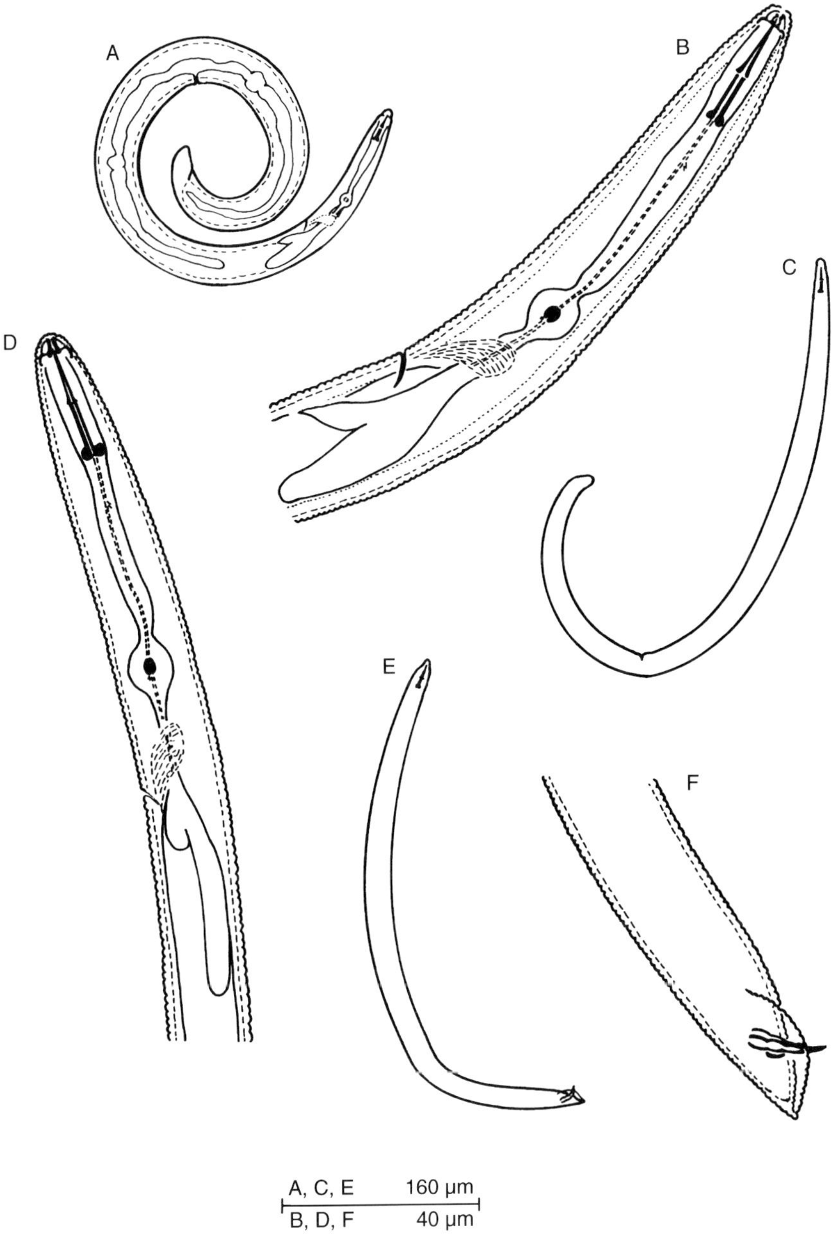

Fig. 5.1. Species of *Helicotylenchus*. *H. dihystera*: A, female; B, female anterior end. *H. multicinctus*: C, female; D, female anterior end; E, male; F, male tail.

increased by 20% by both treatments, compared with untreated cane (Jonathan *et al.*, 1991).

Helicotylenchus dihystera is adequately controlled in a number of crops by treating the soil with small amounts of effective, non-fumigant nematicides, e.g.

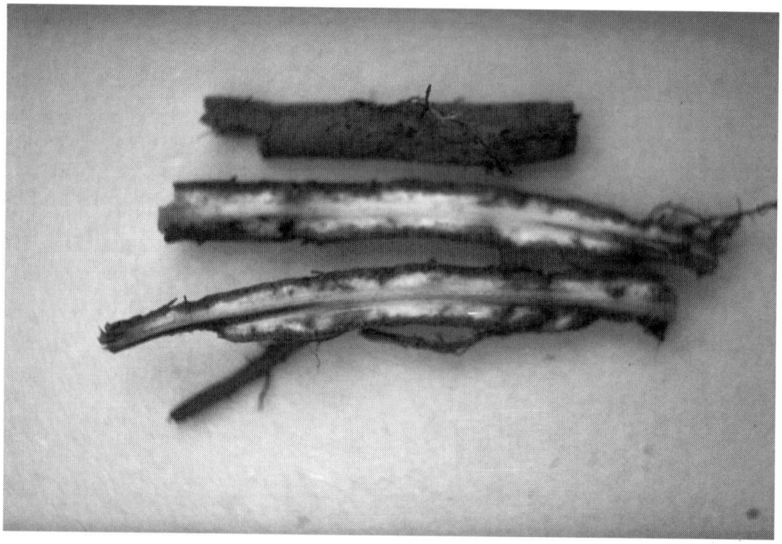

Fig. 5.2. Injury to banana roots by *Helicotylenchus multicinctus* (Photo: J. Bridge).

oxamyl for tomato (Cadet, 1990), fensulphothion, ethoprophos, aldicarb or carbofuran for sweet potato (Martin and Birchfield, 1977), aldicarb, phorate or carbofuran for 'patchouli' (Kumar and Nanjan, 1984) and aldicarb or phosphamidon for the bean *Lablab purpureus* (Sundaram and Velayutham, 1988).

Helicotylenchus multicinctus (Cobb, 1893) Golden, 1956 (Fig. 5.1)

As *H. dihystera*, *H. multicinctus* is a sedentary semi-endoparasite or endoparasite of roots that multiplies on many crop plants. Unlike *H. dihystera*, *H. multicinctus* multiplies by amphimixis. It occurs in many tropical and warm temperate countries and is especially important as a pest of bananas (Fig. 5.2). It has caused serious injury to bananas in Israel (Minz *et al.*, 1960), Australia (Colbran and Saunders, 1961) and the Windward Islands (Edmunds, 1969). The nematodes that feed on cortical cells of banana roots cause small, superficial, brown lesions, which can be very numerous and may even coalesce. The corm may also become infested. Feeding adults may be completely embedded in cortical tissue of banana roots (Blake, 1966). Banana plants infested with *H. multicinctus* may be stunted, take longer to mature, yield smaller bunches and degenerate faster than uninfested plants. The nematode often occurs together with *Radopholus similis* in banana roots in warm, tropical soils but in cooler or drier (sub-optimal) soils the latter nematode may not accompany *H. multicinctus* in banana roots.

Control

Crop plants resistant to *H. multicinctus* have not yet been bred and control of this pest by crop rotation is difficult, because it can multiply on many species of plant. In bananas and plantains, control is limited to minimizing injury through heavy mulches of organic matter and in propping up the fruiting stems, as bunch weights increase, to prevent premature toppling, or, if the costs of treatment can be borne by the grower, applying nematicides at regular intervals.

In the Jordan Valley, Israel, where the nematode can cause banana yields to decline after the first ratoon crop, DBCP at about 17 l ha^{-1} in irrigation water increased yields of affected plants (Minz *et al.*, 1960). The withdrawal of DBCP from commercial use has resulted in dependence on soil-applied, non-phytotoxic, granular organophosphate or carbamate nematicides. Treatment, usually aimed primarily at *R. similis*, is also effective against *H. multicinctus* and other nematode parasites of banana.

Treatment of banana suckers before planting them can decrease or eliminate infestations. Paring the corms and dipping them for 30 min in solutions of aldoxycarb (up to 3% concentration), oxamyl (up to 1%) or carbofuran (up to 3%) eliminated both *R. similis* and *Helicotylenchus* spp. from Giant Cavendish bananas, without phytotoxicity (Inomoto and Monteiro, 1991). Numbers of *H. multicinctus* as well as *R. similis* and *Pratylenchus* spp. were controlled best and yields of Robusta and Dwarf Cavendish bananas were increased most by dipping pared or unpared suckers in a clay slurry and then coating them with 42 g carbofuran or 17 g fensulphothion per sucker (Rajagopalan *et al.*, 1977).

In dessert bananas for export, control of *R. similis* is essential to prevent serious yield losses due to 'toppling' disease and repeated applications of organophosphate or carbamate nematicides are made, which control it and *H. multicinctus*. In Brazil with cv. Naniçao, 4 g aldicarb, carbofuran, fensulphothion or oxamyl per plant, applied to the soil at planting, followed by repeated treatment four months later, decreased nematode numbers over a 15 month monitoring period and reduced the numbers of toppled plants. Yields were earliest after carbofuran but largest after aldicarb (Zem *et al.*, 1981). Over a 24 month period both *Helicotylenchus* and *R. similis* were controlled best on cv. Giant Cavendish and yields were increased most by 3 g aldicarb per mat (Figueroa, 1980). Numbers of *H. multicinctus* were also controlled by 1.2 g fenamiphos per mat, applied five times over 20 months, or by 2.5 or 5 g aldicarb applied four times over 20 months (Jones, 1979)

Treating plantains with 3 g oxamyl or carbofuran, followed by 1.2 g per plant four months later, decreased numbers of the nematode and increased yields by 98% (oxamyl) or 69% (carbofuran) (Caveness and Badra, 1980). A total of 3 g aldicarb per plant controlled *H. multicinctus* better than *Meloidogyne javanica*, greatly increased plant growth and induced earlier plant maturity (Badra and Caveness, 1981). In five-year-old plantain cv. Agbagba at Ibadan, Nigeria, numbers of *H. multicinctus* and *M. incognita* were greatly reduced and yield was greatly increased by treating the soil with oxamyl or carbofuran. At 5 g per mat, carbofuran induced the fastest flowering (Babatola, 1988).

Rotylenchulus reniformis Linford and Oliveira, 1940 (Fig. 5.3)

The 'reniform' nematode, first described from Hawaii, is widespread in the tropics and subtropics, where it attacks and multiplies on a wide range of cultivated and uncultivated plants. Linford and Yap (1940) listed 65 plant species in 30 families that are hosts of this nematode and there are now in excess of 115 known host plant species. *Rotylenchulus reniformis* is a damaging pest of aubergine, banana, betel vine, brassicas, carrot, castor, chickpea, coffee, cowpea, cucumber, haricot bean, lettuce, maize, mango, mung bean (*Phaseolous aureus* syn. *Vigna radiata*), okra, olive, papaya, passion fruit, pea, pineapple, potato, sweet potato and tea.

Mature females of *R. reniformis* are sedentary, semi-endoparasites of roots, which feed on cortical parenchyma, the pericycle or even the phloem, depending on the host species (Fig. 5.4). The male has a poorly developed stylet and median oesophageal bulb and does not feed. Sexual reproduction is the norm but parthenogenesis has also been reported. Up to about 70 eggs (depending on host

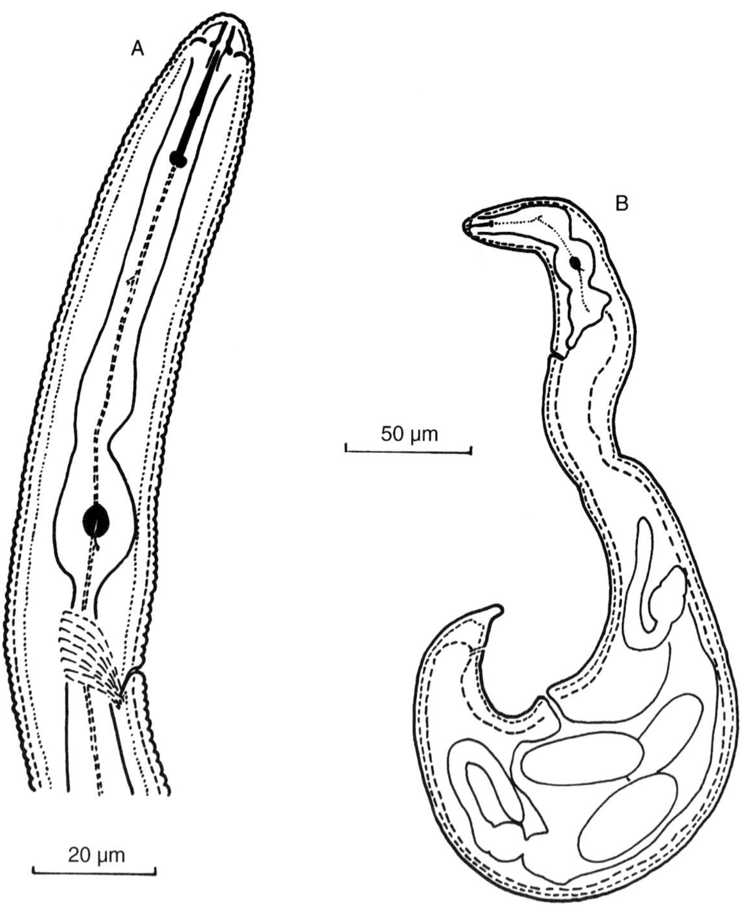

Fig. 5.3. *Rotylenchulus reniformis*: A, immature female, anterior end; B, female.

(a)

(b)

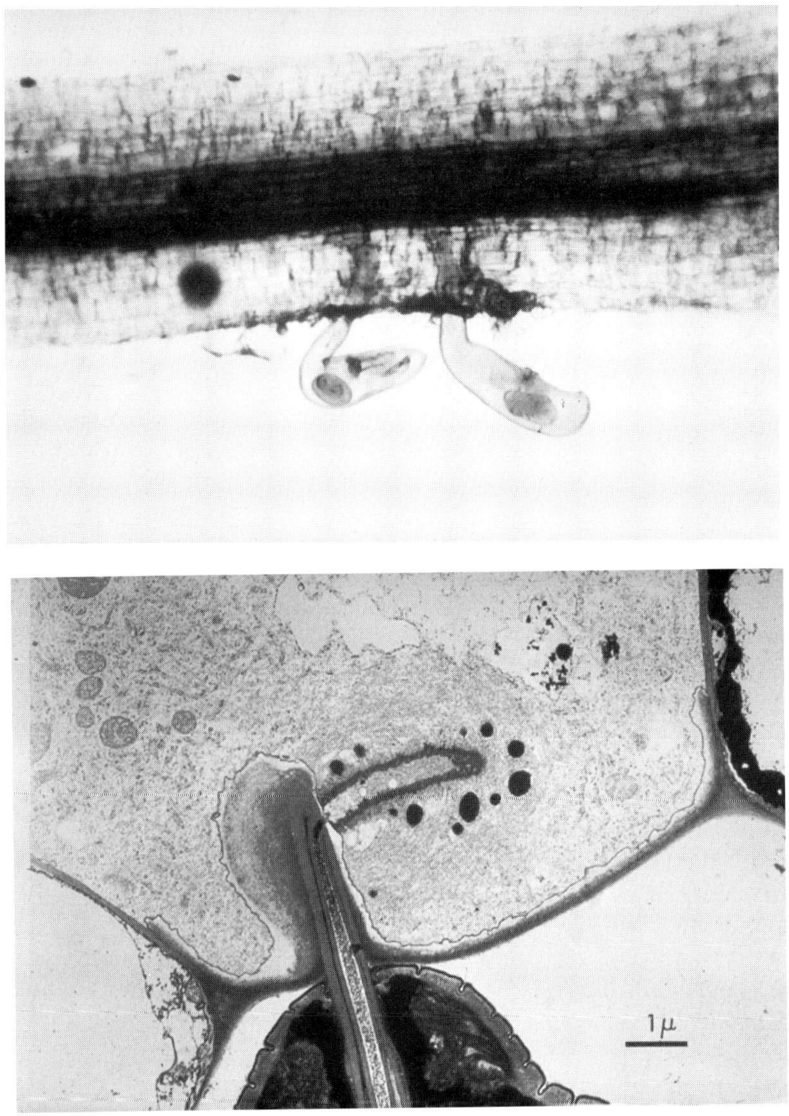

Fig. 5.4. (a) Soybean root infested with *Rotylenchulus reniformis*. (b) Stylet and feeding tube of *R. reniformis* (Photos: R.V. Rebois).

plant) are laid by the female in a gelatinous matrix (the 'egg sac'), which is often concealed by adhering soil particles. All juvenile stages are passed in the soil in as little as 10 days, the complete life cycle taking up to about 30 days, depending on host and soil conditions. In general, more than one nematode g^{-1} of soil may damage intolerant plants. In pots, black gram, cowpea, mung bean and pigeon pea were injured by 0.2 to > 14 nematodes g^{-1} of soil, depending on the tolerance of the plant. For example, 1.4 *R. reniformis* g^{-1} soil retarded the growth of intolerant pigeon pea cultivars, but ten times as many were necessary to affect tolerant cultivars in a similar way (Thakar and Yadav, 1985a). Oteifa and Osman (1974) found the growth of tomato decreased as the number of *R. reniformis* to which it was exposed was increased. The growth of cotton shoots and roots, in pots, was reduced at > 8 and > 1 g^{-1} soil, respectively (Sud *et al.*, 1984). In the field, cotton responded to nematicide use at populations of 1–2.4 nematodes g^{-1} soil (Thames and Heald, 1974; Gilman *et al.*, 1978; Palanisamy and Balasubramanian, 1983). Thames and Heald (1974) showed that cotton yield was negatively correlated with population density at planting but positively correlated with population density near harvest time. The larger the root system, the more nematodes it can support. Yield loss due to *R. reniformis* will depend on soil conditions, especially soil moisture stress, on cultivar tolerance and perhaps on the nematode population involved, as more than one race of the nematode has been recognized.

Rotylenchulus reniformis can interact with *Fusarium* and *Verticillium* wilts of cotton (Neal, 1954; Birchfield, 1962). Neal showed that 81% of cotton plants susceptible to the nematode developed wilt in soil inoculated with both the nematode and *Fusarium*, whereas only 10% of plants suffered wilt in soil inoculated with the fungus alone. In cotton plants resistant to the fungus, however, *R. reniformis* did not increase their susceptibility to wilt.

Control

The 'reniform' nematode may be controlled by a number of management techniques, including use of rotations with non-host, or poor host crop plants, growing marigolds and fallowing or solarizing the soil to kill the nematodes. Some organic soil amendments have also proved beneficial. Fumigant and non-fumigant nematicides have been generally successful in controlling this pest. Some cultivars of several crop plants are resistant to this nematode.

Management techniques

In Puerto Rico, rotating pineapple with sugarcane and pangola grass controlled *R. reniformis* (Ayala, 1971). Following infested sweet potato in Trinidad, maize reduced numbers of the nematode as effectively as fallowing the soil (Braithwaite, 1974). In Texas, following cotton, sorghum left only 10% of the nematodes found after a second cotton crop. Nematicides increased yield of cotton following four years of cotton but did not when cotton followed only one year of cotton (Thames and Heald, 1974). Two crops of maize or sorghum, or three periods of fallow controlled the nematode in the Philippines (Castillo *et al.*, 1976). In India, *Brassica juncea*, *B. campestris* and the marigold *Tagetes patula* have been found non-hosts of *R. reniformis*. In Japan, *T. patula* greatly decreased a population of the nematode, which fallowing of the soil did not (Nakasono, 1973). *Catharanthus roseus* was unaffected by the nematode and decreased it by 96–100% 45 days after inoculation (H. Patel *et al.*, 1993). Growing non-host red pepper (*Capsicum annuum*) for four months decreased numbers of *R. reniformis* in the soil by 80%. Growing it with 1 kg carbofuran, phorate, aldicarb or disulphoton ha^{-1} applied to the soil 15 days after sowing decreased the nematode by 99–100% (Sivakumar *et al.*, 1979).

Restricting irrigation to 60–110 days after planting a sweet potato crop in Brazil, instead of irrigating it throughout the vegetative growth period, minimized increase of *R. reniformis* (Rodriguez Fuentes and Caraballo Llosas, 1981). Control of the nematode in this way may, of course, be at the expense of crop yield. In the Lower Rio Grande Valley of Texas, soil solarization decreased numbers of the nematode in the top 15 cm of the soil and increased yields of lettuce and cowpea. Eggs and juveniles were killed by exposure for 1–24 h (depending on temperature) to 41–47°C and repeated exposure to lethal temperatures for sublethal times had the desired cumulative lethal effect (Heald and Robinson, 1987).

Resistant cultivars

Cultivars of several crop plants have been reported fully resistant to *R. reniformis* (Table 5.1). As several races of the nematode exist with differing host preferences, these cultivars may prove to be susceptible to some populations of the nematode. Nevertheless, resistant cultivars can lessen dependence on nematicides. In Louisiana, for example, growing the resisant soybean cv. Pickett 71 for two years obviated the need to apply a soil fumigant before growing the susceptible soybean cv. Lee 68 or susceptible cotton (Williams *et al.*, 1983). Resistance to *R. reniformis* in pigeon pea was linked to an increase in the phenolic content of the roots after inoculation with the nematode (Thakar and Yadav, 1985b, 1986). Similarly, in tomato, susceptible cultivars in which the phenolic content of roots increased most after inoculation suffered the least growth reduction and supported the least nematode multiplication (Mahmood and Saxena, 1986).

Table 5.1. Crop cultivars resistant to *Rotylenchulus reniformis.*

Crop plant	Cultivar
Castor	Bhagalpur local, HC-6 and T3.
Chickpea	PBNG 34, BGM 434, BGM 426, BG 268, BG 273, GNG-158, BGM 425, GL-83119.
Cotton	Auburn 80–180, Auburn 634.
Cowpea	CSSR-1, S-488, Pusa Phalguni, RC-48, V-16.
Crotalaria	*C . spectabilis, C. juncea, C. paulina, C. mucronata, C. grantiana, C. retusa, C. breviflora* and *C. lanceolata.*
Dolichos biflorus (Horse gram)	BGM-1, Kuruval, K-42, DP-1–1236, Kalarahang.
Papaya	Solo, Washington.
Phaseolus aureus (syn. *Vigna radiata*)	ML-62, ML-80, TT-8E, Pusa-103, PDM-14.
Pigeon pea	Pusa-33, Pusa-84, Pusa-78, Pusa-85, ICPL 87, 4658, 2387, 556–1, 4167, 3830–2, Norman.
Potato	La Rouge, Red la Soda, B 7154–8, B 7180–6, B 7642–2, Draga, Giant, Serana-28, Liset-A.
Soybean	Padre, Pickett 71, Peking, Sharkey, Coker 485, Centennial, Stonewall, Dyer, Custer and Forrest.
Tomato	Ace, Moneymaker, Napoli, Peto-108, Peto-95, Nema-1400, Ace-55, Petopride Kalyanpur Selection I and II and LA 121.
Vigna mungo	Ratnapur-1, UG-201, UG-135.

Biological control

Some recent research has shown that there are possibilities of biological control of *R. reniformis* with *Paecilomyces lilacinus*, *Streptomyces* and *Bacillus thuringensis*.

The 'reniform' nematode was effectively controlled on watermelon and *Capsicum annuum* by applying *P. lilacinus* to the soil one or two weeks before planting (Vicente *et al.*, 1991; Vicente and Acosta, 1992). In microplots, total numbers of the nematode on tomato were decreased by 59% midseason and by 36% at harvest by inoculating the soil with *P. lilacinus* (Walters and Barker, 1994). Infestation of pepper and tomato by *R. reniformis* was decreased by applying culture ID-42 of *Aspergillus niger* to the seed coats or drenching it into the soil (Zuckerman *et al.*, 1994). A species of *Streptomyces* also lessened infestation of tomato by the nematode (Dicklow *et al.*, 1993). Finally, culture CR-371 of *Bacillus thuringiensis* reduced populations of the nematode on tomato and pepper (Zuckerman *et al.*, 1993).

Organic soil amendments

Pot experiments have shown that large amounts of organic waste can control *R. reniformis*. All *R. reniformis* were killed by adding poultry or pigeon dung to soil at 0.5–2% (w/v). The droppings produced volatile substances toxic to the nematodes and the droppings remained toxic during their decomposition (Badra *et al.*, 1979). Other materials found effective in pot experiments were neem oil cake (at about 2 t ha^{-1}) (Padhi and Misra, 1987), neem sawdust plus ammonium sulphate (Siddiqui and Alam, 1990b) and water extracts of groundnut cake (Mahmood and Saxena, 1987). The drawback to all these treatments is the very large amount of material that must be applied to the soil to achieve worthwhile reductions in the numbers of nematodes – usually 1000 times as much as is needed of a non-fumigant nematicide. Some reduction in dosage was achieved by combined use of 0.5 t neem or karanj oil cake with 1 kg aldicarb plus 15 kg N ha^{-1} (Rao *et al.*, 1987). However, organic soil amendments seem only practical for treatment of small holdings and nurseries.

Fumigant nematicides

Numbers of *R. reniformis* in vegetable crop soil have been greatly reduced by preplant fumigation with 47 l 1,3-dichloropropene ha^{-1} (Heald, 1978) or 60–80 kg dazomet ha^{-1} (Chengzu *et al.*, 1987). Numbers of the nematode in cotton soil were similarly reduced by 90 l D-D or 67 l Telone ha^{-1} (Birchfield, 1968). Injected into raised beds before planting at 34 or 56 l ha^{-1}, 1,3-dichloropropene decreased numbers of *R. reniformis* on cotton 7 weeks after planting (Gazaway and Rodriguez-Kabana, 1991). Preplant row treatments with D-D, Telone or Vorlex controlled the nematode satisfactorily in sweet potatoes (Birchfield and Martin, 1968).

Rotylenchulus reniformis is a serious pest of pineapple, especially in Hawaii and the Philippines, where it is the major nematode problem and a major factor limiting cultivation of the crop in Hawaii (Caswell *et al.*, 1990). Preplant soil fumigation, especially if the soil surface is covered with plastic mulch, using 224–336 l 1,3-dichloropropene ha^{-1} can prevent injury to the plant crop

(Rohrbach and Apt, 1986), but control would need to approach 100% if the first and second ratoon crops are to be protected. This is rarely, if ever, achieved, especially as the nematodes have a short life cycle and may migrate upwards from deeper, less effectively fumigated soil. Injected 46 cm deep in pineapple fields before planting, 157 l 1,3-dichloropropene ha^{-1} killed on average about 80% of *R. reniformis*, but more close to the injection depth (Sipes *et al.*, 1994). Loss of yield in the ratoon crops can only be prevented by use of postplant (non-fumigant) nematicides. Schneider *et al.* (1992) found that 336 l 1,3-dichloropropene ha^{-1} had not prevented nematode multiplication one year after planting.

Non-fumigant nematicides

SEED TREATMENTS

Treating okra seed with carbofuran at 3 or 6% of total weight controlled both *R. reniformis* and *Meloidogyne incognita* economically and increased yield (Sivakumar *et al.*, 1973). Numbers of the nematode were decreased 85–99% up to 120 days after planting, when pineapple seed pieces were dipped in fenamiphos (600–2400 ppm) (Apt, 1979). At 700 g $(100\ kg)^{-1}$ seed, carbofuran decreased infestation of upland cotton by 82% up to two months after sowing and increased cotton yield significantly (Lordello and Brancalion, 1986).

BARE ROOT DIPS

Dipping roots of aubergine in aldicarb, carbofuran or terbufos at 500 or 1000 ppm for 15 or 30 min decreased numbers of the nematode in the roots and soil by three quarters or more (Krishnaprasad and Krishnappa, 1981).

SOIL TREATMENTS

In Egypt, aldicarb or thionazin granules applied to the soil controlled *R. reniformis* well on cotton (Oteifa *et al.*, 1970). Aldicarb also increased cotton yield by 65% in infested soil (Oteifa *et al.*, 1976). In Alabama, USA, aldicarb at 1.7 or 2.2 kg ha^{-1} over the rows of cotton controlled the nematode up to harvest (Gazaway and Rodriguez-Kabana, 1991). Applied in the furrows at sowing, 1.2 or 1.7 kg aldicarb ha^{-1} also decreased numbers of the nematode attacking cotton (Gazaway and Edmiston, 1992 a,b), though incorporation of aldicarb in a 15 cm wide band over the rows controlled the nematode better than application in the seed furrows (Gazaway, 1994). Row treatments with aldicarb, thionazin, fensulphothion or ethoprophos were also effective in controlling *R. reniformis* on sweet potatoes (Birchfield and Martin, 1968). Applied to the soil at transplanting, 2.25 kg carbofuran ha^{-1} as a 3% granule controlled the nematode on some vegetable crops (Chengzhu *et al.*, 1987). In Oahu, Hawaii, *R. reniformis* was controlled and pineapple yields were increased by preplant fumigation with 336 l 1,3-dichloropropene ha^{-1} plus 3.4 kg fenamiphos ha^{-1} applied every three months for the first year by drip irrigation, whether the crop was irrigated weekly or less frequently (Schneider *et al.*, 1992). As fenamiphos residues in soil are relatively persistent and susceptible to leaching, it has been suggested that the bi-monthly optimal 1.7 kg fenamiphos ha^{-1} for control of *R. reniformis* and *Meloidogyne javanica* should not be drip irrigated in to avoid leaching it out of the root zone (Schneider *et al.*, 1993). In soil containing 1.8–4.2 *R. reniformis*, but not in less infested soil, 6 kg carbofuran ha^{-1} improved the growth of pigeon pea (Sharma *et al.*, 1993c).

FOLIAR TREATMENTS

Oxamyl has controlled *R. reniformis*, when applied as a foliar spray to sweet potatoes (McSorley and Parrado, 1983) and to tomato (Oteifa and Osman, 1974). Foliar sprays of 2–5 kg fenamiphos ha^{-1}, applied at 1–3 month intervals to pineapples, lessened infestations of the nematode and increased pineapple yields, especially of ratoon crops. The nematicide was also effective when applied through a drip irrigation system (Apt, 1979). A 10 ml drench of 0.1% ascorbic acid lessened multiplication of the nematode on tomato (Siddiqui and Mahmood, 1993a). Al-Sayed (1991) found foliar sprays of ascorbic acid (at 1000, 5000 or 10,000 ppm) significantly decreased final nematode populations under soybeans and was more effective than 0.2 g fenamiphos per plant.

***Tylenchulus semipenetrans* Cobb, 1913** (Fig. 5.5)

'Citrus' nematode, *T. semipenetrans*, is a worldwide pest of citrus, causing 'slow decline' disease, which Thomas (1923) showed was due to the nematode (Fig. 5.6). Citrus nematode seems to have a limited host range. In the *Rutaceae*, 50 species or hybrids of *Citrus* and other rutaceous plants are hosts of *T. semipenetrans*. Nine host species belong to other plant families and include grapevine, olive, loquat, pear and persimmon.

Injury has been reported to grapevines in Australia (Seinhorst and Sauer, 1956) and the pest has been found on grapevines in California (Raski *et al.*, 1956). Very large numbers of the nematode retarded the growth of olives (Lamberti *et al.*, 1976). Persimmon is very susceptible to the nematode, large numbers of it being associated with poor growth in Californian and Israeli orchards (Nesbitt, 1956; Cohn and Minz, 1961). In Italy, losses of 20–30% were estimated in heavily infested persimmon (Di Maio, 1979). The relationship between *T. semipenetrans* and plant growth has, however, been studied most with citrus. Baines (1950), Baines and Clarke (1952) and Baines *et al.* (1962a) showed that infested young lemon and orange trees suffered considerable growth reduction and fruit-bearing trees yielded 30–50% less fruit than uninfested trees. In Israel, citrus yield decreased when there were more than 4000 female nematodes g^{-1} feeder roots (Cohn *et al.*, 1965). Although citrus fruit yield is negatively correlated with numbers of *T. semipenetrans* in the roots, the exact relationship is affected by soil and climatic factors (Reynolds and O'Bannon, 1963; Cohn *et al.*, 1965; Bindra *et al.*, 1967).

Heavy infestation of citrus feeder roots by females, which are obligate, sedentary, semi-endoparasites, leads eventually to their destruction (Fig. 5.6). Infestation and development of the nematode in citrus roots is greatest between 25 and 30°C (Baines, 1950). The anterior end of the female is deeply embedded in the root cortex, where it feeds on a number of 'nurse' cells surrounding a cavity that develops around the head of the worm (Van Gundy and Kirkpatrick, 1964). The male has a weak stylet and a poorly developed oesophagus and nei-

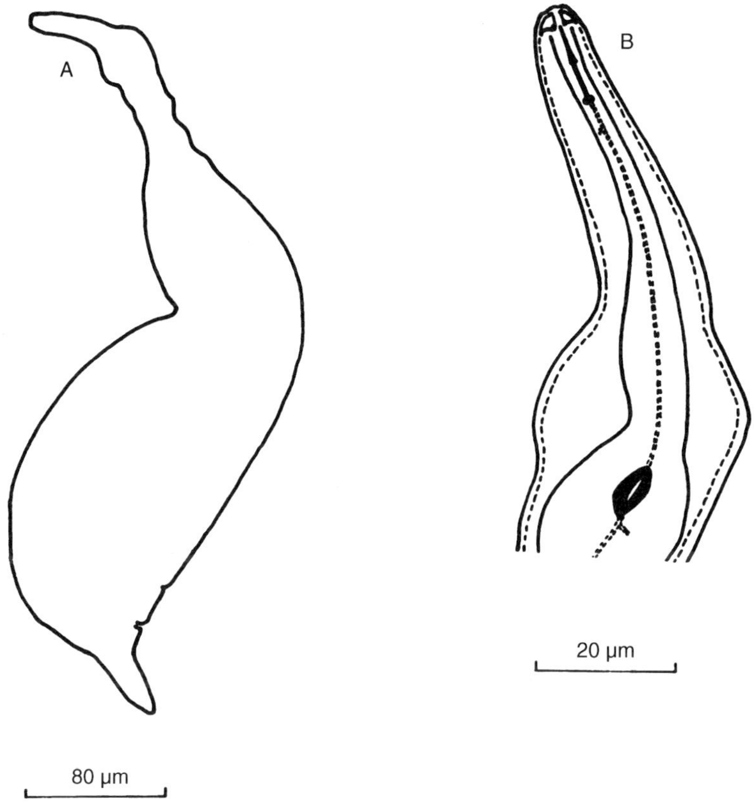

Fig. 5.5. *Tylenchulus semipenetrans*: A, female outline; B, female anterior end.

ther it nor the juvenile male feeds on the roots. In heavily infested roots there are dark lesions in the cortex, which eventually sloughs off, killing the distal part of the root.

Tylenchulus semipenetrans is amphimictic and parthenogenetic. There are six or seven generations on citrus in Sichuan, China (Deng *et al.*, 1993). The eggs are laid in a gelatinous matrix secreted, unusually, through the excretory pore, which, uniquely, is located towards the tail end of the female. Soil particles adhere to the gelatinous matrices, so that infested roots have a dirty appearance, even after they have been rinsed in water (Fig. 5.6). Citrus roots infested with the nematode are more readily infested by *Fusarium oxysporum* and *F. solani* (O'Bannon *et al.*, 1967) and other organisms, which are important in the final destruction of the feeder roots. The loss of feeder roots results in increased drought stress and decreased soil nutrient uptake, leading to retarded growth of shoots and yellowing and loss of leaves. Affected trees do not die, but have an unthrifty appearance and yield fewer, smaller fruits than uninfested trees. In citrus nurseries, citrus seedlings are usually devoid of symptoms of infestation above ground but they grow very slowly and if transplanted into soil heavily infested with *T. semipenetrans* they show symp-

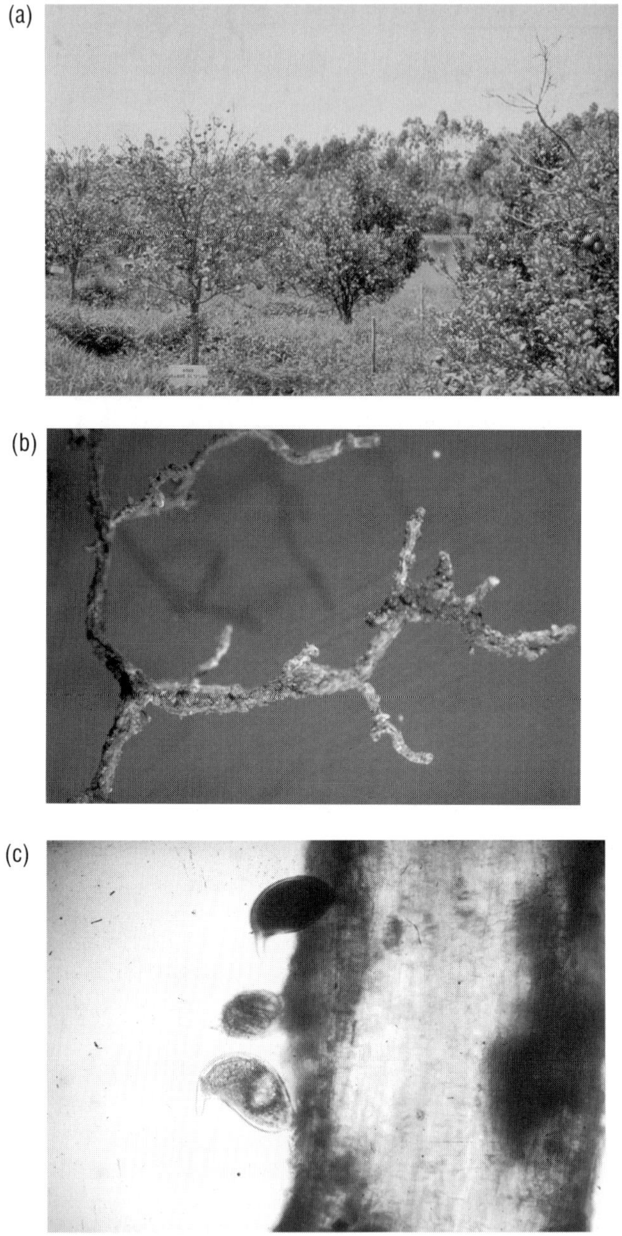

Fig. 5.6. Slow decline of citrus due to *Tylenchulus semipenetrans*. (a) affected citrus grove; (b) infested roots with 'dirty' appearance, due to stickiness of the gelatinous egg-sacs of the nematode; (c) sedentary females attached to citrus root (Photos: J. Bridge).

toms similar to those found in older trees. Despite its limited host range outside the *Rutaceae*, the nematode can survive for several years after the removal of citrus trees from the soil (Baines *et al.*, 1962b). *Tylenchulus semipenetrans* is killed if the soil dries out, or if the soil temperature much exceeds 30°C. Deep in the subsoil,

Table 5.2. Races of *Tylenchulus semipenetrans.*

Differential host	'Citrus'	'Poncirus'	'Mediterranean'
Poncirus trifoliata	+/−	+	−
Olive	+	−	−

+/− = Multiplies poorly on; − = no multiplication on; + = multiplies on.

where soil moisture and temperature are more favourable, the nematode can survive on citrus roots, which can persist for many years after citrus trees have been removed from the soil.

Three races of *T. semipenetrans* are recognized. All three multiply on *Citrus* spp. but they can be differentiated by their ability to multiply on *Poncirus trifoliata* and olive (Inserra *et al.*, 1980) (Table 5.2). A fourth race ('grass' race) has been assigned to two other species of *Tylenchulus* – *T. graminis* and *T. palustris* (Inserra *et al.*, 1988). The 'citrus' race is widespread in USA citrus-growing regions and in Italy, the 'Poncirus' race occurs in California and the 'Mediterranean' race occurs in citrus-growing countries around the Mediterranean, in South Africa and perhaps in India.

Control

The nematode is controlled by preventing its spread to uninfested soil, by crop husbandry techniques to lessen plant injury, by the use of resistant and/or tolerant rootstocks and by nematicides.

Prevention of spread

Unaided, the nematode spreads slowly from infested to uninfested soil, so losses from 'slow decline' can be decreased by taking measures which prevent rapid spread of the nematode. Uninfested soil should be planted or replanted with uninfested nursery stock. This could be obtained by planting stem or leaf cuttings rooted in virgin soil, well away from infested orchards. Alternatively, the nursery soil, if infested, could be thoroughly fumigated. However, as large amounts of some fumigants, notably methyl bromide, dazomet and metham-sodium, also kill endomycorrhizal fungi, which are important to the nutrition of citrus seedlings, the soil may need to be reinoculated with them (Milne, 1974) after all traces of fumigant have dispersed from the soil. Reinoculation of the soil with mycorrhizae might not be necessary if a less fungitoxic fumigant, like 1,3-dichloropropene, is used. To ensure the nematodes are killed near the soil surface, the latter should be covered with gas tight sheeting during fumigation. Even then, not every seedling grown in the fumigated soil may be totally free of *T. semipenetrans*.

Nursery stock has been freed of citrus nematode before transplanting it by dipping the bare roots in water at 45°C for 25 min (Baines, 1950). In Brazil, the nematode was eliminated from *Citrus limon* roots by dipping them in water at 50°C for 10–20 min (Silva *et al.*, 1987a). Neither treatment harmed the plants.

In contrast, although 50°C for 10 min eradicated the nematode from citrus roots, it killed the rootstocks and 47°C for 10 min, though non-phytotoxic, did not kill all the nematodes (Scotto la Massese, 1965). Clearly, hot water treatment of citrus stocks is not infallible. Dipping *Citrus limon* roots in 1% aqueous carbofuran for 15 min, followed by 0.2 g aldicarb applied to the soil around each plant three weeks after transplanting, eradicated the nematode from the roots without phytotoxicity (Silva *et al.*, 1987b). All means of spreading the nematode – by watcr, farm implements and machinery and by infested plants from infested to uninfested orchards or areas of orchards, should be prevented, thereby avoiding the expense of subsequent curative control measures.

Crop husbandry techniques

Ensuring that infested plants have adequate moisture and nutrients and that they are not subject to unhealthy soil conditions or attack from other pests and diseases may lessen the impact of *T. semipenetrans* on plant growth and yield. Such practices are, however, unlikely to obviate the need for direct control of the nematode.

Resistant rootstocks

Resistance to *T. semipenetrans* was found in selections of trifoliate orange (*Poncirus trifoliata*) in Florida, California, Brazil, India and southern France and in *Citrus amblycarpa* cv. Nasnaran, *C. aurantifolia* (sour orange), *C. sinensis* (sweet orange) cv. Simmon as well as in *Severinia disticha*, which is related to *Citrus*. In India, *C. aurantifolia* was reported resistant, whereas in California it was reported tolerant to the nematode. Most commercially resistant rootstocks have been derived from *P. trifoliata* hybrids and include: Troyer, Corrizo and Savage citranges (*P. trifoliata* × *C. depressa*); Scalo citrumelo and Swingle citrumelo (*P. trifoliata* × *C. paradisi* (grapefruit)); and Swingle trifoliate orange × Shekwasha (*P. trifoliata* × *C. depressa*). Swingle citrumelo has proved very resistant to all three races of citrus nematode (Duncan and Cohn, 1990) and is resistant to a feeder root rot due to *Phytophthora parasitica* and to *Tristeza* virus. A population of the nematode has recently been found to attack it in Florida (Duncan *et al.*, 1994). Troyer and Corrizo citranges and Scalo citrumelo have been reported quite resistant to the nematode, but, according to nematode population, they may vary from susceptible to resistant.

Some selections of *P. trifoliata* × Poorman orange (*Citrus* × hybrid of unknown origin) have yielded rootstocks 44/12 and 48/21 resistant to the 'Poncirus' and 'Mediterranean' races of *T. semipenetrans* and to *Phytophthora citrophthora* and *Tristeza* virus (Gottlieb *et al.*, 1986; Spiegel-Roy *et al.*, 1988).

Grapevine rootstocks resistant to *T. semipenetrans* have been obtained from *Vitis champini* or *V. riparia*. Ramsey and Dog Ridge rootstocks were selected from *V. champini*. Resistant K 51–32 and K 51–40 were derived from *V. champini* × *V. riparia* hybrids and Freedom from *V. champini* × 1613. Swartzmann was obtained from *V. riparia* × *V. rupestris* and 1616 from *V. riparia* × *V. solonis*. The breadth of resistance in these rootstocks to races and populations of *T. semipenetrans* seems not to have been studied.

How long the resistance in citrus and grapevine rootstocks will remain effective is a matter for conjecture. Sooner or later, it is likely that virulent biotypes of the nematode will be selected.

Fumigant nematicides

Large amounts of most soil fumigants are needed to rid soil of the nematode. Soil fumigation can prevent infestation of nursery plants and transplanted citrus but fungicidal compounds – methyl bromide, dazomet and metham-sodium – may eliminate endomycorrhizae essential to the establishment of young plants (Cohn *et al.*, 1968a; Milne, 1974). If these fumigants are used, the land may have to be left for a year or more to allow repopulation with the mycorrhizae before it can be replanted with citrus. Alternatively, it may be possible to inoculate the planting stock with mycorrhizae. D-D and 1,3-dichloropropene are much less fungitoxic.

Baines *et al.* (1959) studied the vertical diffusion patterns and efficacy of D-D applied to loam and silt loam soils. At 1123–1684 l ha^{-1}, D-D killed all citrus nematodes in soil to 1.2–1.5 m deep, provided the soil was rolled or, preferably, covered with polyethylene sheet immediately after injection. Deeper control required deeper injection but nematode control was then less than 100% in the upper layers of the soil. A two-depth injection programme would then be needed. Even 2246 l D-D ha^{-1} had little effect on soil fungi (Baines *et al.*, 1962a). Methyl bromide at 65 l ha^{-1}, Telone at 374 l ha^{-1} or Vorlex (20% methyl isothiocyanate : 80% dichloropropenes) at 187 l ha^{-1} killed the nematode in the top 1.2 m of the soil, without killing soil fungi (Baines *et al.*, 1962b). D-D at 374 or 561 l or Telone at 299 or 449 l ha^{-1} as broadcast or strip fumigation treatments controlled the nematode and increased citrus yields in Florida (O'Bannon and Tarjan, 1973). Excellent control of *T. semipenetrans* was obtained in many countries by treating the soil with dibromochloropropane (DBCP), a fumigant of high specific gravity, which killed nematodes as it sank through the soil profile. Smaller doses of this compound than of other fumigants were effective, especially when applied in irrigation water. For example, over 99% of citrus nematodes were killed when 37 or 56 l of 75% (w/w) e.c. ha^{-1} was applied to the soil in a 30-year-old citrus orchard and citrus yields were increased. It was five years after this treatment before the nematode had increased to damaging numbers, requiring further successful treatment (O'Bannon and Reynolds, 1967). All other recognized soil fumigants are too toxic to be applied in citrus orchards, so the withdrawal of DBCP for toxicological reasons and the contamination of water supplies with traces of other soil fumigants, often injected deeply and in large amounts, has stimulated the search for alternative nematicides. Several non-fumigant nematicides have proved effective.

Non-fumigant nematicides

Three carbamates (aldicarb, oxamyl and carbofuran) and four organophosphates (fenamiphos, fensulphothion, ethoprophos and cadusafos) have controlled *T. semipenetrans* in citrus orchards. Of these, aldicarb and fenamiphos have received most attention.

Table 5.3. Percentage control of *Tylenchulus semipenetrans* by aldicarb.

Soil depth fraction	Furrow irrigation	Sprinkler irrigation
(cm)		
0 – 30.5	3	93
30.5 – 61	49	99
61 – 91.5	90	97

CARBAMATES

In a 25-year-old Valencia orange orchard in California, 11.2 or 22.4 kg aldicarb ha^{-1} controlled *T. semipenetrans* when application to the soil was followed by 72 h sprinkler irrigation. Sprinkler irrigation was more effective than furrow irrigation and did not contaminate the drainage water (Elgindi *et al.*, 1978) (Table 5.3).

At 5.7 or 11.4 kg ha^{-1}, aldicarb controlled the nematode as effectively as 65 kg DBCP ha^{-1} and increased orange and grapefruit yields (Timmer and French, 1979). Application to the soil of 5.5 or 11.3 kg aldicarb ha^{-1} for three consecutive years controlled *T. semipenetrans* on 12-year-old orange on a sour orange rootstock in Texas and increased fruit yield in two out of three years' observation (French and Timmer, 1979). Similarly in Arizona and California, 5.6–11.2 kg aldicarb ha^{-1} decreased numbers of the nematode and increased citrus yields after the first year (Van Gundy and Garabedian, 1982). In Cyprus, annual application of 30 kg aldicarb ha^{-1} controlled the nematode and increased fruit yields (Philis, 1988). In Arizona and California, oxamyl at 16.8 kg ha^{-1} controlled the nematode and after the first year increased citrus yields (Van Gundy and Garabedian, 1982). At 20 kg ha^{-1}, oxamyl was similarly effective in Cyprus (Philis, 1988). In plots of 20-year-old orange trees on Troyer citrange rootstock in California, 60% of the feeder roots were located in the top 30 cm of the soil and populations of citrus nematode peaked in the roots in spring, summer and winter, suggesting at least three generations per annum. Applying oxamyl, at 1.1 kg ha^{-1} every 3 weeks over 183 days from April to the end of September, decreased numbers of female *T. semipenetrans* in roots in the top 30 cm of the soil by 60% and kept soil populations below the economic threshold from two months after the first application to the end of the season (Hamid *et al.*, 1989). Carbofuran applied every four months at 5 g per tree controlled the nematode on citrus in Colombia (Olmedo *et al.*, 1980).

ORGANOPHOSPHATES

Fenamiphos at 40 kg ha^{-1} decreased *T. semipenetrans* in the year of application to soil around 3-year-old Valencia orange trees and increased fruit yields. These effects did not, however, persist into a second or third year (Milne and Villiers, 1978). Nematode control was maintained when 5 g fenamiphos was applied to each tree every four or six months (Olmedo *et al.*, 1980). Application of 10 ml of 40% e.c. of fenamiphos (m^2)$^{-1}$ to flood irrigation basins around orange trees in Spain and South Africa controlled citrus nematode and increased fruit yields (Milne and Willers, 1979a,b). The same dosage of fenamiphos applied to the soil by minisprinklers controlled the nematode on orange trees for three years in Cyprus (Philis, 1988). Placing 22.3 kg fenamiphos ha^{-1}, in bands 1.5 m wide under the tree canopy to the drip-line of grapefruit trees on sour orange rootstocks in East Florida, decreased citrus nematode in the roots and by 89% in the soil and increased grapefruit yield 16 months later (Duncan, 1989). Chhabra and Bindra (1972) found that 5 kg fensulphothion ha^{-1} controlled the nematode on citrus as effectively as did 37 kg DBCP ha^{-1}. At 30 kg ha^{-1}, fensulphothion controlled the nematode on *Citrus maxima* and increased fruit yield (Chhabra *et al.*, 1978). In Colombia, 5 g fensulphothion per tree every four months also controlled the nematode (Olmedo *et al.*, 1980). Ethoprophos at 40 kg ha^{-1} in India (Chhabra *et al.*, 1978) and at 30 kg ha^{-1} annually in Cyprus (Philis, 1988) controlled the nematode and increased citrus yields. Ethoprophos at 4.7 kg and isazophos at 5 kg ha^{-1}, applied to lemon trees through sprinkler irrigation in Cyprus, decreased *T. semipenetrans* over three years by 50% and 43%, respectively, and the increased benefit : increased cost ratio was 3–4 : 1 (Philis, 1993). Finally, cadusafos at 4 kg ha^{-1} also controlled the nematode on Satsuma orange in China (Zou *et al.*, 1992).

Tarjan (1976) found that fenamiphos and oxamyl, both of which have good basipetal translocation properties in plants, can control *T. semipenetrans*, when injected into the trunks of citrus trees. The technique has the advantage of restricting the nematicide to tree tissues, thereby avoiding any possible risk of environmental contamination. Applied over three years, 5.25–30 g fenamiphos per tree controlled the nematode as effectively as 45 g per tree broadcast to the soil and it increased yields of oranges and grapefruit by 19% and 27%, respectively (Monke *et al.*, 1992).

Migratory Endoparasites of Roots and Tubers (*Hirschmanniella*, *Pratylenchus*, *Radopholus* and *Scutellonema*) 6

Roots and tubers may be seriously injured by nematodes which feed on and migrate through the tissues, causing lesions or cavities to form there, which may in turn be invaded by fungi and bacteria resulting in rotting diseases. The most important of these migratory endoparasitic nematodes are 'rice root' nematodes, species of *Hirschmanniella* Luc and Goodey, 1964, ' root-lesion' nematodes, species of *Pratylenchus* Filipjev, 1936, 'burrowing' nematodes, species of *Radopholus* Thorne, 1949 and the yam nematode, *Scutellonema bradys* (Steiner and Le Hew, 1933) Andrassy, 1958.

Hirschmanniella spp.

Seven species of *Hirschmanniella* have been found in the roots of rice, their principal host crop (Fig. 6.1). The name 'rice root' nematode is, however, misleading, as there are other nematodes that also attack rice. *Hirschmanniella* spp. are widely distributed in the major rice-growing regions of the world. They are found in some other crop plants, notably upland cotton (*Gossypium hirsutum*), tomato, maize, wheat and sugarcane, but are most numerous in rice roots, causing stunting, chlorosis and reduced tillering in the shoots, with consequent loss of grain yield (e.g. Jonathan and Velayutham, 1987). The damage is most evident in soils of poor nutrient status but is also influenced by the tolerance of the cultivar grown, among other modifying factors. The nematodes feed on the root parenchyma, moving through the roots in the air channels. Reproduction is probably sexual and there are up to three generations in a rice plant, depending on soil temperature. Maximum multiplication is relatively small (13-fold in Java according to van der Vecht and Bergmann, 1952). *Hirschmanniella* spp. parasitize many weeds of rice fields, especially *Gramineae* and *Cyperaceae*. In the absence of host plants, *H. oryzae* (van Breda de Haan, 1902) Luc and Goodey, 1964 (Fig. 6.2) can survive up to a year in saturated soil and at both high and low temperatures (8–45°C). However, 12 months fallow decreased numbers of *H. oryzae* by 88% (Muthukrishnan *et al.*, 1979) or even eradicated them (Fortuner and Merny, 1979).

(a)

(b)

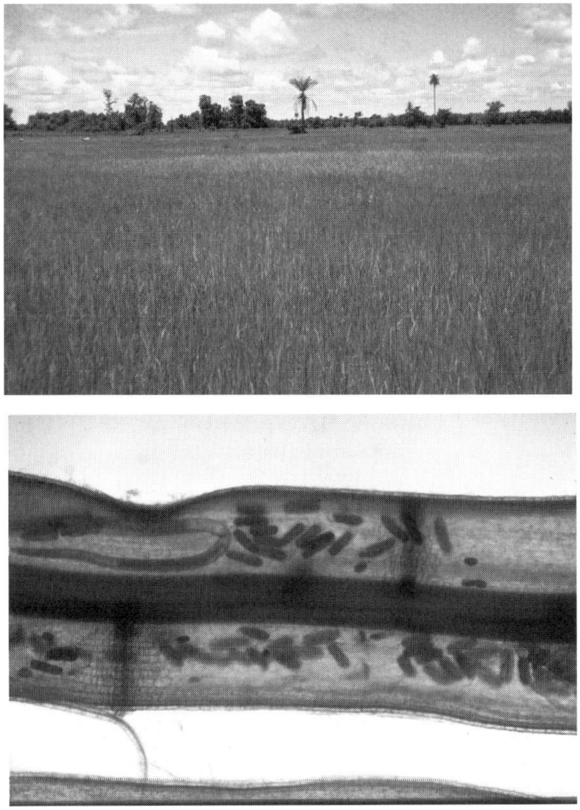

Fig. 6.1. (a) Stunting and yellowing in rice associated with infestation of the roots by *Hirschmanniella oryzae*. (b) Female and eggs of *H. oryzae* in rice root (Photos: J. Bridge).

Control

Hirschmanniella spp. have been controlled to a greater or lesser extent by crop rotation, weed control, use of less susceptible and more tolerant cultivars of rice and by nematicides. Control is complicated by the number of species that attack rice, several of which may occur in individual fields and plants.

Crop rotation

Where rice is not grown in the dry season, numbers of *Hirschmanniella* are decreased by growing non-host crops, e.g. millet, cotton or wheat against *H. oryzae* (Mathur and Prasad, 1973), but the host status of crop plants must be assessed locally against the species of *Hirschmanniella* present. As these species may vary in distribution from field to field, it may be difficult to make reliable predictions of the efficacy of specific rotational crops in decreasing the nematode populations. Crops resistant or immune to all species present in a rice-growing region would therefore be valuable. Some green manure crops have been shown to decrease numbers of *Hirschmanniella* in the soil. *Sesbania rostrata* acts as a trap

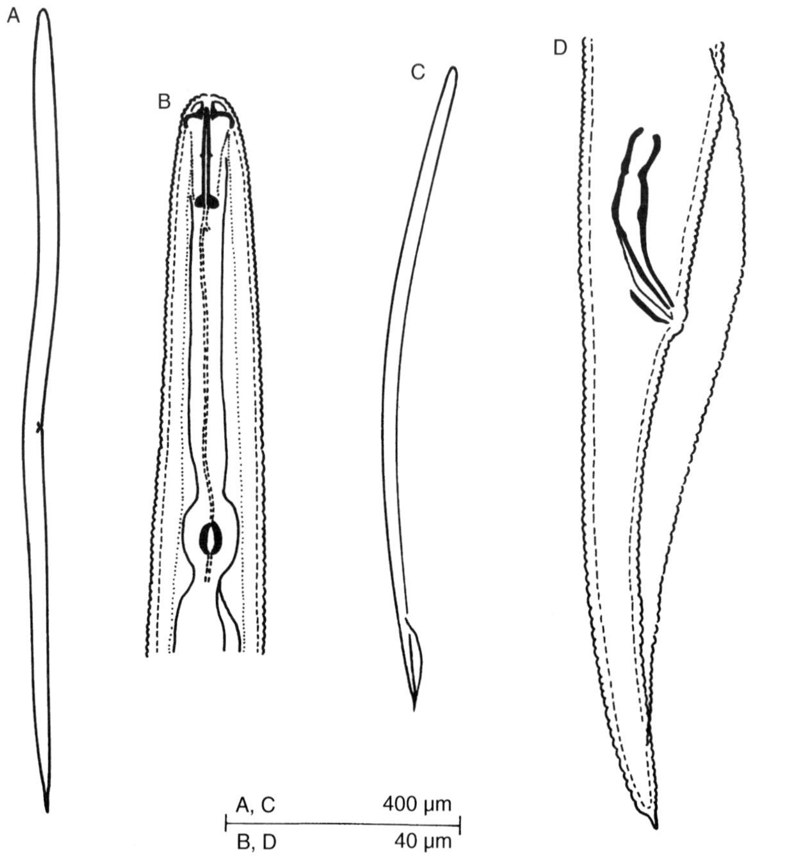

Fig. 6.2. *Hirschmanniella oryzae*: A, female; B, female anterior end; C, male; D, male tail.

crop for *H. oryzae* and *H. spinicaudata* (Schuurmans Stekhoven, 1944) Luc and Goodey, 1964, the principal species in flooded rice in West Africa and it and *Aeschynomene afrasperma* as green manure crops contributed more than 100 kg N ha^{-1} to a following rice crop (Rinaudo *et al.*, 1988). The green manurial effect of *S. rostrata* was of greater benefit to a following rice crop than the trap-crop effect (Pariselle and Rinaudo, 1988). In an earlier experiment (Germani *et al.*, 1983), *S. rostrata* greatly decreased *H. oryzae* and increased yields of flooded rice 2- to 3-fold, despite the removal of all shoot tissue of *S. rostrata* from the plots. *Sesbania rostrata* and *A. afrasperma* are fully resistant to *H. oryzae* and *H. mucronata* (Das, 1960) Luc and Goodey, 1964 (Hendro *et al.*, 1992). Although *S. rostrata*, followed by *A. afrasperma* controlled *H. oryzae* and *H. mucronata* and significantly increased yield of a following rice crop, the practice was uneconomic (Prot *et al.*, 1992). Another leguminous green manure crop, *Sphenoclea zeylanica*, which exudes toxic compounds from its roots, can decrease numbers of *Hirschmanniella* by 99% (Mohandas *et al.*, 1981). Infestation of the soil by *Hirschmanniella* spp. can be lessened, somewhat, by good weed control in rice and by fallowing the land.

Resistant and tolerant cultivars

Very few of the rice cultivars grown at present are resistant to *Hirschmanniella* spp. One cultivar, TKM 9 appears resistant to *H. oryzae* (Ramakrishnan *et al.*, 1984) and a few others appear partially resistant to *Hirschmanniella*. In view of the number of species of the genus that can infest rice and the possible existence of pathotypes of differing virulence within each species, progress in developing good resistant cultivars is likely to be slow. Local selection for tolerance is probably made unwittingly by rice growers seeking better yielding cultivars.

Nematicides

As currently available fumigant nematicides are all phytotoxic and would need to be applied to well-drained soil several weeks before rice could be planted safely, their potential use would be limited to dry season rotational cropping periods, either before or after a non-host crop. The costs of treatment would have to be borne largely by rotational crops of greater cash value than rice. Not surprisingly, therefore, recent research has concentrated on the efficacy of non-fumigant nematicides, which are non-phytotoxic and easier to apply.

Applying nematicides to rice seedbeds, dipping rice seedlings in nematicidal solutions or suspensions, or coating the roots with nematicide before transplanting them can give early protection from attack by *Hirschmanniella* spp. Applied to rice nursery seedbeds at 1.3 kg ha^{-1}, carbofuran decreased numbers of *H. oryzae* in rice transplants for 30 days after transplanting them into the field (Jonathan and Velayutham, 1984). Dipping seedlings in 0.2% carbosulphan or isofenphos significantly decreased numbers of *H. mucronata* in the roots and in the soil. The carbosulphan and isofenphos dips increased rice yields by 95% and 46%, respectively (Prasad *et al.*, 1986b). Combined treatment of 1 kg carbofuran ha^{-1} to the seedbed before sowing and dipping the seedlings in 0.2% fenamiphos or carbosulphan before planting them in the field lessened infestation of rice roots with *H. oryzae* and increased rice yields (Subramanian and Velayutham, 1983).

Numbers of *H. oryzae* in rice roots were decreased by 90% by applying 2 kg phorate, aldicarb, carbofuran or fensulphothion ha^{-1} to the soil in which rice seedlings were planted (Chhabra and Dhaliwal, 1979). Numbers of this nematode in rice roots were also decreased up to 60 days after transplanting rice into soil given 1 t neem cake or 10 t sugarcane press mud ha^{-1} (Jonathan and Pandiarajan, 1991). Repeated applications of non-fumigant nematicides to the soil were also effective. Numbers of *H. mucronata* in rice roots were decreased 90% and grain yield was almost doubled when 1 kg carbofuran ha^{-1} was applied to the soil 50 days after rice seedlings from a carbofuran-treated nursery had been planted in it (Routaray *et al.*, 1987). In a similar procedure, ethoprophos or carbofuran were effective against *H. oryzae* (Singh *et al.*, 1989). Similarly effective were 1 kg carbofuran ha^{-1} at planting followed by 1 kg ha^{-1} 15 days later against *H. mucronata* (Prasad and Rao, 1979a) and 0.5, 1 or 2 kg carbofuran ha^{-1} 3 days before and again 30 days after transplanting against *H. mucronata* and *H. oryzae* (Aung, 1990).

Pratylenchus spp. (Figs 6.3, 6.4)

These 'root-lesion' nematodes are obligate, root endoparasites, injurious to many crop plants (Figs 6.5, 6.6). They feed primarily on root cortex, through which they migrate, creating cavities and channels in which their eggs are deposited singly or in small groups. Before penetrating the roots, the nematodes sometimes browse on root surfaces and on root hairs. Small lesions, first yellow, then brown to black, develop where the nematodes enter and feed in the roots. These often enlarge and may eventually girdle the root. Cells on which the nematodes feed and the cavities and channels, through which they pass, become necrotic and

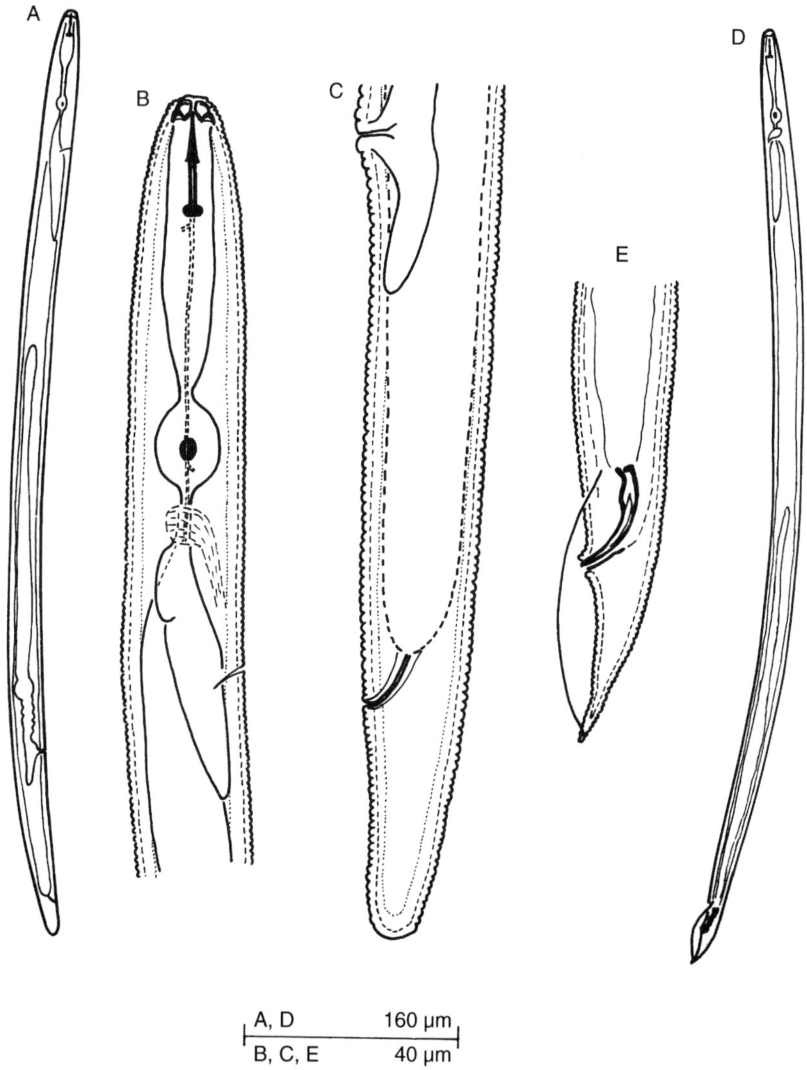

Fig. 6.3. *Pratylenchus penetrans*: A, female; B, female anterior end; C, female posterior end; D, male; E, male tail.

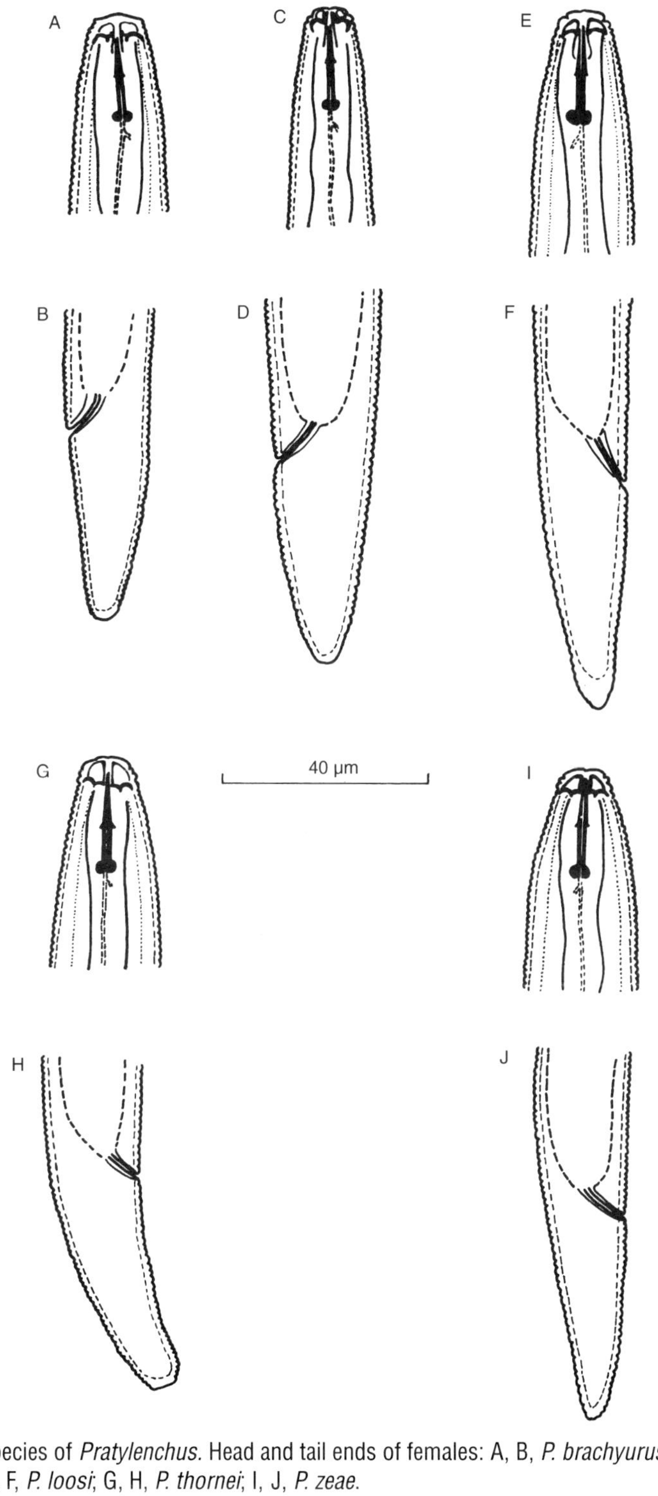

Fig. 6.4. Species of *Pratylenchus*. Head and tail ends of females: A, B, *P. brachyurus*; C, D, *P. coffeae*; E, F, *P. loosi*; G, H, *P. thornei*; I, J, *P. zeae*.

(a)

(b)

Fig. 6.5. (*and opposite*) *Pratylenchus* injury. (a) *P. brachyurus* damage to peanut roots and pods (Photo: D.W. Dickson); (b) *P. coffeae* damage to yam (Photo: J. Bridge); (c) dieback in tea due to nematode root rot, Sri Lanka (Photo: J. Bridge and C. Gnanapragasam); (d) tea root rot due to *P. loosi.*

secondary invading bacteria and fungi may rot the roots. When heavily infested plants are pulled from the ground, girdled root cortex is left in the soil, giving affected roots a bristly appearance. *Pratylenchus brachyurus* (Godfrey, 1929) Filipjev and Schuurmans Stekhoven, 1941 and *P. penetrans* (Cobb, 1917) Filipjev and Schuurmans Stekhoven, 1941 also feed to some extent on phloem parenchyma and xylem elements. Damage to the roots causes chlorosis and stunting of the plant and dieback of its branches. Yields are reduced and plants may be killed, especially if bacteria or fungi are involved in the disease. Populations of some species of less than $0.1\ g^{-1}$ soil can be very damaging to certain crop plants.

Most of the most economically important species considered here prefer warm soils, but *P. penetrans* is widespread in temperate areas of the world. *Pratylenchus brachyurus, P. coffeae* (Zimmermann, 1898) Filipjev and Schuurmans Stekhoven, 1941, *P. scribneri* Steiner, 1943 and *P. zeae* Graham, 1951 are widespread in tropical and subtropical regions. *Pratylenchus loosi* Loof, 1960 is a serious pest of tea in

(c)

(d)

Fig. 6.5. *Continued*

cool soils in Sri Lanka and Japan and in tea and citrus in north-east India, and it is also reported in Bangladesh and China. *P. goodeyi* Sher and Allen, 1953 is a widespread pest of bananas in East Africa and is also known in Australia, Canary Islands and Crete but there is little information available on its control. *Pratylenchus thornei* Sher and Allen, 1953 is an important pest of wheat and occurs in warm, temperate, subtropical and some tropical soils. *Pratylenchus vulnus* Allen and Jensen, 1951 is found in warm, temperate and subtropical soils.

There is a wide range of host plants for *P. brachyurus*, *P. coffeae*, *P. penetrans*, *P. scribneri*, *P. vulnus* and *P. zeae* but *P. goodeyi*, *P. loosi* and *P. thornei* have been reported from fewer crop plants. Males are common or abundant in populations of *P. coffeae*, *P. goodeyi*, *P. loosi*, *P. penetrans* and *P. vulnus* and reproduction is sexual. Males are very rare and reproduction is parthenogenetic in *P. brachyurus*, *P. scribneri*, *P. thornei* and *P. zeae*. The life cycle takes several weeks to complete, depending on species, host plant and soil temperatures but several generations are usually completed in a season. For example, the life cycle of *P. penetrans* takes 30 days at 30°C but 92 days at 15°C (Mai *et al.*, 1977).

Three species of *Pratylenchus* are known to interact with some bacterial and

(a)

(b)

(c)

Fig. 6.6. *Pratylenchus* injury. (a) *P. penetrans* damaged strawberry (Photo: L.R. Krusberg); (b) *P. vulnus* damage to strawberry roots (Photo: J. Bridge); (c) *P. vulnus* on boxwood: five untreated rows on right, DBCP-treated rows on left (Photo: C.M. Heald).

fungal pathogens, the resultant disease complex doing greater damage to the host crop than that due to individual elements of the complex. *Pratylenchus brachyurus* interacts with *Pseudomonas solanacearum* on tomato, the black shank fungus of tobacco and the cotton stunt fungus disease complex. Soybeans were damaged more by *P. brachyurus* plus *Rhizoctonia solani* than by either alone. *Pratylenchus penetrans* has been reported to interact with *Aphanomyces euteiches* on peas, *Cylindrocarpon radicicola* on oats and bulbs, *Fusarium oxysporum* on lucerne and peas, *Trichoderma viride* on lucerne and *Verticillium albo-atrum* and *V. dahliae* on several crop plants. Resistance in peas to *F. oxysporum* f. *pisi* was broken when the roots were inoculated with *P. penetrans*. Wheat attacked by *P. thornei* may be attacked by smut and *F. solani*.

Pratylenchus brachyurus

Control

Cultural control

With its wide host range, this species, as other *Pratylenchus* spp., is not controlled adequately by crop rotations alone. Some reduction in soil infestation may be achieved by growing a poor host, such as groundnut in Florida (Good *et al.*, 1954). In South Carolina, however, although there were few of the nematodes in the roots, groundnuts left appreciable numbers of *P. brachyurus* in the soil because the 'pegs' and nuts were infested, as subsequently demonstrated in Georgia. The presence of more than one species of *Pratylenchus* in the soil and the probable existence of races within a species, with differing host preferences, make crop rotation an unreliable control measure on its own. The elimination of reservoir weed hosts may help decrease numbers of *P. brachyurus* in the soil, e.g. *Digitaria horizontalis* in maize crops and in intervening fallow periods (Stradioto *et al.*, 1983). Intercropping pineapple in the Ivory Coast with groundnut, tomato or okra did not increase the nematode, whereas maize or *Capsicum frutescens* did and tomato and *C. frutescens* reduced pineapple yields through competition (Osseni, 1985). The application of 40, 65 or 90 kg powdered cocoa pod husk ha^{-1} to maize plots infested with *P. brachyurus* decreased numbers of the nematode by 78, 79 or 85%, respectively, and increased maize yields by 70, 78 or 127%, respectively. There were also residual benefits to a second maize crop (Egunjobi, 1977).

Fumigant nematicides

At 45 mg l^{-1} for 24 h at 25°C, methyl bromide almost eliminated *P. brachyurus* from the shells of whole pods of groundnut, without affecting seed germination (Minton and Gillenwater, 1973). Soil infestations can be dramatically decreased by applying a dichloropropene fumigant to the topsoil or to predetermined crop rows before planting (e.g. Nusbaum, 1960). Preplant fumigation with 250 l D-D ha^{-1} was as effective in controlling the nematode on pineapple as 30 l DBCP ha^{-1} (Sarah, 1980).

Non-fumigant nematicides

Non-fumigant nematicides have also been effective in controlling the nematode in maize, groundnut and citrus and, in conjunction with preplant fumigation of the soil, in pineapple. *Pratylenchus brachyurus* was well controlled on 5-year-old tangerine trees budded on rough lemon rootstocks by 34 kg fenamiphos (in granular form) or 67 kg fensulphothion or DBCP ha^{-1} applied in liquid form around the trees ('basin' treatment). Fruit size in the first crop and yield in the second crop were increased after fenamiphos or DBCP (O'Bannon *et al.*, 1974).

In pineapples, the nematode was controlled in South Africa by spraying the butt end of suckers with 2.5 kg fenamiphos ha^{-1} plus the same spray concentrate just after planting and every three months thereafter to the foliage. Dipping the planting material for 10 min in fenamiphos (600 ppm), followed by spraying the foliage at three monthly intervals, was also effective (Keetch and Webster, 1977). In the Ivory Coast, following preplant fumigation of the soil with DBCP or D-D, 0.15 g carbofuran or 0.2 g fenamiphos per plant sprayed on the foliage early in April or in October–November (beginning and end of the rainy season) kept populations of the nematode small on pineapple. At 0.2 g per plant, or when infestations of *P. brachyurus* were well established, 0.3 g per plant, sprayed on pineapples 1 and 5.5 months after planting controlled the nematode, without preplant fumigation (Sarah, 1980). In a subsequent report (Sarah, 1983), foliar sprays of carbofuran were said to depress pineapple yield, whereas fenamiphos was highly effective as a foliar spray, both in controlling the nematode and in stimulating plant growth, except when applied after flowering, when it was toxic to young plant tissue.

Carbofuran at 3.25 kg ha^{-1} in Brazil controlled both *P. brachyurus* and *P. zeae* on maize and doubled yield (Lordello *et al.*, 1983). At 2.5–4.5 kg ha^{-1} at planting and at tasselling, carbofuran significantly decreased *P. brachyurus* on maize and increased its yield in Nigeria (Atu and Duru, 1990). Granular carbofuran was also effective against the nematode on groundnut in Texas (Boswell and Grichar, 1983).

Pratylenchus coffeae

Yams

A dry rot of yams (*Dioscorea rotundata* and *D. cayenensis*) in Jamaica, caused by *P. coffeae*, increased in storage at 22–31°C, not at 12–13°C, because nematode multiplication was prevented at the lower temperature range (Thompson *et al.*, 1973). In similarly diseased *D. rotundata* tubers in Puerto Rico, the number of nematodes was greatly decreased by dipping the tubers for 15–60 min in water at 35–54°C, or in thionazin or fensulphothion (1250 ppm) for up to 60 min, without affecting their germination (Acosta and Ayala, 1976). In Jamaica, however, dipping *D. rotundata* tubers in water at 51°C injured the distal end of the tubers, whereas dipping them in aqueous oxamyl (1200 or 2400 ppm) lessened nematode infestation, without harming the tissues (Coates-Beckford, 1977). *Dioscorea cayenensis* planting material was disinfested by dipping it in water at 45°C for 45 min or in oxamyl solution (2000 ppm) for 30 min, thereby increasing tuber yields 23% or 36%, respectively (Hutton *et al.*, 1982). In Guadeloupe, the nematode can be very damaging to *D. cayenensis rotundata* tubers, sometimes even killing the plants. In winter storage of the tubers at 25–30°C, the nematode increased from 145 g^{-1} skin at harvest to 1450 g^{-1} skin four months later. The pest has been controlled by rotation of yams with non-host legumes and graminaceous crops, use of nematicides, planting uninfested tubers or tubers dipped for up to 35 min in water at 51°C and control of weeds, especially *Rottboelia exaltata* and *Setaria barbata*, which are very good hosts of *P. coffeae* (Kermarrec *et al.*, 1988). Castagnone-Sereno (1988) recommends immersion of yams in water at 50°C for 40 min before planting, to control both *P. coffeae* and *Scutellonema bradys*, another yam-rotting nematode.

Other Crop Plants

Satisfactory plantain plants were obtained from planting material, even when it was heavily infested with *P. coffeae* and *Helicotylenchus multicinctus*, by deeply paring away infested corm tissue, dipping it for 25 min in water at 55°C and applying 3 g carbofuran per plant to the soil in which it was planted (Sampedro *et al.*, 1987). Growing intercrops of *Tagetes*, lucerne, sunn hemp (*Crotalaria juncea*) or coriander (*Coriandrum sativum*) for four months in a banana grove lessened infestation of banana roots with *P. coffeae* and *Radopholus similis* (Naganathan *et al.*, 1988).

Seedling roots of citrange were freed of *P. coffeae* by immersing them for 10 min in water at 49°C (Kaplan, 1982). Applied to the bark of rough lemon (*Citrus jambhiri*) infested with *P. coffeae*, a paste containing 4.5–18 mg technical fenamiphos per plant had a systemic effect, reducing numbers of the nematodes in the roots and increasing plant weight (Tarjan, 1972).

In coffee nurseries, injecting the soil with D-D or mixing dazomet into it controlled *P. coffeae* (Abrego and Holdemann, 1961) but the efficacy of D-D was increased when the soil was covered with a plastic sheet (Thorne and Scheiber, 1962). Large doses of oxamyl, carbofuran, fenamiphos or aldicarb applied to the soil controlled *P. coffeae* in coffee nurseries in El Salvador. Application of the nematicide ten days after sowing was best (Abrego, 1974; Abrego and Cardona, 1976). Applied to the soil in which young coffee trees were growing, 25 or 50 mg fenamiphos (in 5% granules) per tree controlled *P. coffeae* for 90 days but repeated treatment was then needed to maintain control (Kumar, 1982). Grafting coffee cultivars susceptible to the nematode on resistant rootstocks, such as *Coffea conuga* (de Fluiter, 1947) and *C. canephora* (Pattabhiraman, 1949) has also been successful in the control of *P. coffeae*. As *C. canephora* var. *robusta* tolerated attack by the nematode, it may also be a useful rootstock for the intolerant *C. arabica* or *C. excelsa* (Palanichamy, 1973; Anon., 1974).

Aldicarb or fensulphothion at 4 kg ha^{-1}, applied as granules to the soil, controlled *P. coffeae* on groundnut in the Punjab, India (Chhabra and Mahajan, 1976). At 2–4 kg ha^{-1}, cadusafos (as 10% granules ('Rugby')) controlled the nematode satisfactorily on ramie (*Boehmeria nivea*) in Hunan, China (Anon., 1990).

Pratylenchus loosi

This nematode is a particularly important pest of tea grown between altitudes of 900 and 1800 m in Sri Lanka and from 0 to 300 m in Japan. It attacks both the feeder roots and the main roots of tea plants, restricting the uptake of nutrients and water from the soil by the diseased root system. As a result, the bushes become unthrifty and chlorotic, yield badly and may eventually die. Several years elapse before newly planted tea produces its first crop and, if the bushes are already infested with *P. loosi*, the loss of production may result in serious financial losses over many years.

Control

Control measures aim to ensure that uninfested tea seedlings or rooted cuttings, preferably resistant or tolerant of this pest, are planted in uninfested or chemically disinfested soils and that, in established tea, cultural practices are adopted which favour the plant and not the nematode.

Soil fumigation

Kerr (1963) stressed the importance of replanting tea estates, in the higher parts of Sri Lanka, with cuttings free from *P. loosi*. Despite widespread fumigation in nurseries, 36 of 60 above 900 m were infested. Where both the bags of soil into which the cuttings were transplanted and the nursery beds had been fumigated, infestations, if present, were up to 5 $(100\ g)^{-1}$ soil and up to 40 g^{-1} root, but where only the bags of soil were fumigated, the estates were as heavily infested as those where the nurseries were not fumigated at all. Fumigating the beds and the bagged soil together, using 997 kg methyl bromide ha^{-1} was more effective than fumigating them separately with D-D (Kerr and Vythilingham, 1967). However, infestations of transplants can be minimized by 280–560 l D-D ha^{-1}, correctly applied to the nursery soil, especially in conjunction with the use of gas-tight plastic sheeting. Sivapalan (1969a) found 499 kg methyl bromide ha^{-1} gave excellent control of the nematode in nursery beds but 997 kg ha^{-1} was needed for comparable control in pre-bagged soil. At 200 kg ha^{-1} (or 135.7 g $(m^3)^{-1}$), 98% dazomet (Basamid) incorporated in the soil also gave excellent control, and with less risk to operatives, so that roots of tea raised in it were free of *P. loosi* one year later (Sivapalan *et al.*, 1980a). Basamid or methyl bromide at 176.7 g $(m^3)^{-1}$ of soil killed most *P. loosi* but tea plants grew better in dazomet-treated than in methyl bromide-treated soil (Gnanapragasam, 1982).

In infested soil in which nine-month-old tea plants were subsequently planted, the nematode was controlled satisfactorily over a 15 month observation period by 'Di-Trapex' (methyl isothiocyanate plus D-D) at 135 or 269 kg ha^{-1} and better than by D-D at 269 or 538 kg ha^{-1} (Sivapalan, 1969b). In Japan, however, the nematode was controlled, in soil in which tea was to be planted, by fumigating it with 200–300 l D-D ha^{-1}, injected 20–30 cm deep (Takaji, 1969).

Non-fumigant nematicides

In Sri Lanka, the application of 0.7 or 1.4 g aldicarb, 0.1 or 0.2 g carbofuran or 0.175 g fenamiphos per planting hole to transplants of susceptible tea clone TRI 2024 lessened nematode infestation and promoted better plant growth than in untreated soil (Sivapalan *et al.*, 1980b). Following pruning of tea bushes with a high yield potential, fenamiphos, at 0.5 g per plant mixed with fertilizer, can lessen or prevent infestation of new feeder roots, which must remain healthy if yield is to be maintained (Gnanapragasam, 1987).

Cultural control

Several cultural practices lessen the abundance of *P. loosi* in the soil or the impact it has on tea bushes. These were reviewed by Sivapalan and Gnanapragasam (1990). Addition of large amounts of organic materials, such as dung, waste tea, copra waste or neem cake, to the soil lessens the numbers of the nematodes it contains. Forking the soil disrupts compaction, which otherwise limits feeder root replenishment. Appropriate fertilizer application can lessen the impact of *P. loosi* on tea and nitrogen applied as urea may suppress populations of the nematode. Of the grasses used to recondition old tea soils, prior to replanting, Guatemala grass (*Tripsacum laxum*), weeping love grass (*Eragrostis curvula*) and Mana (*Cymbopogon confertifolius*) suppress the nematode. Weeping love grass has also suppressed the 'burrowing' nematode, *Radopholus similis*, which can also parasitize tea. Allowing tea bushes six to eight weeks in which to recover from pruning, before plucking is restarted, is essential if they are suffering root injury by *P. loosi*. Finally, when old tea land is being cleared, it is important to remove as many of the old roots as possible, because the nematodes can survive two or three years in pieces of large roots left behind and these can result in failure of replanted tea.

Resistance and tolerance

As tea plantations can be productive for 50 or 60 years, the most practical means of controlling *P. loosi* would seem to be the use of resistant or tolerant transplants. Of the many tea clones screened for resistance and tolerance in Sri Lanka, three are resistant or tolerant to both *P. loosi* and *R. similis*, i.e. Balangoda DG 7, Diyagama DN and Tangakelle CY 9, and many more are resistant or tolerant to *P. loosi* alone. There would, therefore, seem hope of using one or more of the former as rootstocks for the best tea clones. This assumes, of course, that the rootstocks will not select, in time, virulent forms of *P. loosi* and/or *R. similis*. At present, there is no evidence of races within *P. loosi*, though there are races of *R. similis* and the widespread use of resistant clones or rootstocks could perhaps select virulent forms of both nematodes from field populations. This is unlikely if the rootstocks are tolerant and susceptible but more likely if they are tolerant *and* resistant.

Pratylenchus penetrans

Many crop plants are attacked by *P. penetrans* in temperate soils. This species of *Pratylenchus* has been more closely studied than any other in this genus. It is particularly troublesome in perennials, notably fruit trees, raspberry, strawberry and roses (Fig. 6.6). Several annual crop plants have also suffered notable injury and narcissus, lilies, potato, tomato and tobacco have needed control measures to prevent serious crop losses.

Root Rotting Disease

Fruit trees

Pratylenchus penetrans has been associated with retarded growth of fruit trees, especially in the Netherlands (Hoestra and Oostenbrink, 1962) and in replanted orchards in New York State, USA and in Ontario, Canada. Mai and Parker (1956) found the nematode in the roots of cherry showing poor growth and witches-broom type roots. Treating the soil with 504 kg dichloropropene ha^{-1} before planting killed nearly all the nematodes in the topsoil, and 21 months later there were still only 3% of the numbers found in untreated soil. In south-west Ontario, Mountain and Boyce (1957, 1958) found *P. penetrans* the only nematode clearly associated with replanting disease of peach. Fumigating the soil with D-D prevented injury to peaches in the first two years after planting them in old peach orchards. Fumigating infested soil with D-D before replanting with sour cherry allowed good establishment and prevented death of trees at the end of the third year, as occurred in unfumigated soil. However, numbers of the nematode had greatly increased over three years and growth of trees on mazzard cherry rootstocks decreased sooner than those on mahaleb cherry rootstocks (Mai and Parker, 1967). Similarly, 468 l Telone ha^{-1} before replanting greatly improved the growth and early yields of cherries and apple (Mai and Parker, 1970, 1972). In another experiment, preplant application of 299 l Telone ha^{-1} reduced *P. penetrans* numbers from the second to the sixth year after treatment, but nematode numbers increased to pretreatment numbers after nine years. Cumulative cherry yields were almost doubled by fumigation plus PKMgCa fertilizer. However, in a healthy orchard, preplant fumigation with 327 l Telone ha^{-1}, or 234 l Vorlex ha^{-1} greatly increased yields in the first two harvests, i.e. by 88% or 199%, respectively. Vorlex killed soil fungi as well as the nematodes and soil fumigation appeared to increase tree resistance to cold injury (Mai, 1972). Response to soil fumigation cannot, therefore, be proof of the pathogenicity of *P. penetrans*.

In the Niagara Peninsula of Canada, 72 l Telone or 112 l Vorlex ha^{-1} injected before planting in the tree rows controlled *P. penetrans* for at least two years in peach and apple, but only 112 l Vorlex ha^{-1} significantly increased tree growth. Post-plant applications of DBCP, despite controlling *P. penetrans*, failed to improve tree growth (Marks and Davidson, 1973). After pre-plant soil fumigation, especially with Vorlex, yields of apples were increased for six successive crops and this was thought due to control of *P. penetrans*, allowing a healthy root system to develop. Net returns were twice those of trees in untreated soil (Arneson and Mai, 1976). In Queensland, Australia, where *P. penetrans* was shown to stunt Granny Smith apple seedlings, soil fumigation, bare fallowing, application of sawdust or black polyethylene mulch, all of which decreased nematode numbers greatly, improved the growth of apple seedlings. Cover crops of cowpea, rye or lupin, which increased *P. penetrans*, were injurious to apple seedlings (Colbran, 1979).

Some success in controlling *P. penetrans* in established fruit trees has been obtained with non-fumigant nematicides. Numbers of the nematode were greatly decreased in apple trees sprayed four times at fortnightly intervals with 0.9 or 1.8 kg oxamyl ha^{-1} in 379 l water containing a spreader-sticker. In apple or peach seedlings, the nematode was best controlled by spraying at or before inoculation (Abawi and Mai, 1971). Up to 43 days after spraying apple or peach seedlings with oxamyl, the roots were toxic to *P. penetrans* and fewer nematodes were found in them than in unsprayed seedlings. The effect was greatest when the seedlings had been sprayed at or before inoculation with the nematodes (Abawi and Mai, 1975). In pots, sprays equivalent to 1.12 kg oxamyl $(450\ l)^{-1}$ water ha^{-1} reduced numbers of the nematode in the roots of fruit tree rootstocks (Allen and Marks, 1977). Numbers of *P. penetrans* in the roots were decreased by over 95%, when peach and pear seedlings were sprayed twice with 4.8 g oxamyl l^{-1}, though total elimination of the nematode was not achieved (Abawi and Mai, 1978). Numbers of the nematode were also decreased over 95% in the feeder roots by applying twice per season 5.6 or 8.4 g aldicarb (as 10% granules) cm^{-1} trunk diameter in a 46 cm wide band around each newly planted apple tree. Such treatment over a three year period increased shoot growth and trunk diameter (Herne *et al.*, 1973). *Pratylenchus penetrans* was well controlled and yields of sour cherries on *Prunus mahaleb* rootstocks were increased during the first four years of cropping, when, to soil around trees planted in autumn 1979, 5, 10 or 15 kg aldicarb or 10 or 15 kg oxamyl ha^{-1} was applied twice in August 1980 and again in May 1981, May 1982 and in August 1983 after the first crop. The largest doses were, however, somewhat phytotoxic (Szczygiel and Rebandel, 1990). Combining treatment of the soil with aldicarb and treatment of the foliage with oxamyl decreased soil and root populations of *P. penetrans* and increased the girth of apple tree trunks (P.W. Johnson *et al.*, 1982).

Mai (1972) stressed the importance of combined use of nematicides and appropriate cultural practices to control *P. penetrans*-induced replant problems of fruit trees. Nematode resistant cover crops grown for two or three years prior to replanting, soil fumigation in late summer or in early autumn before replanting, use of disease-free nursery stock and maintenance of recommended soil nutrient status and pH, used together, can minimize the replant problem. Apple rootstocks may differ widely in their tolerance of *P. penetrans*, as responses to D-D in infested soils differed greatly (Parker and Mai, 1974). Equating tolerance levels with levels of response to D-D soil fumigation is, however, debatable.

Nursery stocks

Pratylenchus penetrans is also a serious pest in tree nurseries. In the Netherlands, the nematode has been held responsible for 'soil tiredness' in tree nurseries. This was satisfactorily controlled by the use of nematicides, especially D-D at 600–1200 l ha^{-1}, which was economic, particularly for annually lifted crops (Meijnecke, 1959). In Sweden, a 'soil sickness' of apple nurseries, which appeared to be due to *P. penetrans*, was eliminated by green manuring, which stimulated nematophagous fungi, or by the addition of milled peat to the soil, which increased numbers of nematode-predatory tardigrades (Kauri-Pääsuke, 1973a). Ploughing in large amounts of green matter, whether of *P. penetrans*-

susceptible or -resistant plants, decreased nematode density, apparently by enhanced activity of nematophagous fungi (Kauri-Pääsuke, 1973b). There is also evidence that the nematodes may be harmed by toxic actinomycetes or bacteria (Walker *et al.*, 1966) or toxic chemicals resulting from the decomposition of plant tissue, e.g. butyric acid (Sayre *et al.*, 1965).

The nematode has also been troublesome in conifer seedlings. In seedlings of Douglas fir (*Pseudotsuga menziesii*) the nematode was controlled best by applying 6.7 kg fenamiphos ha^{-1} to the soil, which decreased the nematodes in the roots and in the soil by 88 and 90%, respectively, and increased seedling height and root weight by 23 and 45%, respectively (McElroy, 1985). In Victoria, Australia, fenamiphos applied to infested nursery soil eliminated the nematode from *Pinus radiata* roots and greatly improved seedling growth (Marks *et al.*, 1987).

Root Lesion Disease

Strawberry

Injury to strawberries by *P. penetrans* has been recognized for a long time. Seinhorst (1957) reported small numbers of the nematode damaging to strawberries and other crop plants. Surface-sterilized nematodes were able to incite a black root-rot in strawberries (Chen and Rich, 1962; Townshend, 1963). The nematode has been controlled by fumigating the soil before planting. As strawberries occupy the ground for two or three years, a large percentage of the nematodes must be killed, if long-term disease control is to be achieved.

Injected into infested soil in September, 650 l D-D ha^{-1} almost eliminated the nematode from the topsoil but the treatment was phytotoxic to strawberries planted in April of the following year (Klinkenberg, 1955). Applied in summer or autumn in New Brunswick, Canada, before strawberries were planted, 140 l D-D ha^{-1} controlled *P. penetrans* and root-rot and very significantly increased fruit yields (Morgan, 1964). Dazomet (methyl isothiocyanate precursor) and mixtures of methyl isothiocyanate and D-D (Vorlex, Di-Trapex) have also proved effective. At 468 l ha^{-1}, Vorlex (20% methyl isothiocyanate in a D-D-type mixture) controlled the nematode and increased strawberry yields (Keefer *et al.*, 1978). In Poland, the nematode was well controlled by preplant fumigation of the soil with 300 or 600 l Di-Trapex ha^{-1} and strawberry yields were greatly increased over three years (Szczygiel and Rebandel, 1982). More recently, 500 l Di-Trapex or 500 kg Basamid (98% dazomet) ha^{-1} applied to the topsoil controlled *P. penetrans* and *Verticillium dahliae* well, allowing strawberries to follow strawberries, without the need for intervening rotational crops. Yields were increased by 9–15% for cv. Senga Sengana and by 2–5% for cv. Dukat over three years (Szczygiel and Rebandel, 1991).

Raspberry

Raspberries may also be damaged severely by *P. penetrans*. McElroy (1975) found *P. penetrans* and *Xiphinema bakeri* associated with decline of red raspberry (*Rubus idaeus*) in loamy soils in British Columbia, Canada. Dormant red raspberry rooted cuttings of five cultivars were freed of *P. penetrans* by dipping them for 15 min in water at 47.8–48.3°C, without harming them. A sixth cultivar, Haida, was injured but not at 46.6–47.2°C. for 15 min, which left less than ten nematodes g^{-1} root (McElroy, 1973). Numbers of *P. penetrans* and *Xiphinema bakeri* were decreased by spraying the foliage four times from full leaf in May and at two week intervals thereafter with oxamyl at approx. 1000 or 2000 ppm. Raspberry yields in both the treatment and following year were increased (McElroy, 1975). According to McElroy (1977), *P. penetrans* killed 24% of new raspberry plantings and reduced yields by 47%. In eastern Scotland, Trudgill and Brown (1978) found the nematode in raspberry roots from 34 of the 52 farms surveyed. Raspberries grew better in infested soils, which had been air-dried, frozen to −20°C, to which aldicarb at 6.7 kg ha^{-1} had been applied, or which had been treated with dazomet.This suggested that *P. penetrans* could adversely affect the growth of raspberries. However, in the field one or more fungi also seemed to be involved in raspberry replant disease (Trudgill, 1983). Over three years, 220 or 330 kg dazomet ha^{-1} applied to the soil before planting controlled the nematode and increased raspberry yields (Trudgill, 1979). Preplant fumigation of the soil profitably increased the yields of raspberries in land heavily infested with *P. penetrans* in Scotland. Dazomet, dichloropropene mixture or metham-sodium were effective, dazomet controlling the nematode better in the top 20 cm of the soil than below 20 cm, whereas dichloropropene or metham-sodium controlled the nematode deeper in the soil than dazomet but less effectively nearer the soil surface (Trudgill, 1986). In Poland, numbers of *P. penetrans* in Malling Jewel raspberries were decreased in soil treated before planting with dazomet (300 or 500 kg Basamid ha^{-1}), D-D (500 l ha^{-1}) or Di-Trapex at 300 or 500 l ha^{-1}. The nematode was almost eliminated from the soil in the first year after planting by Di-Trapex and raspberry yields were greatly increased over three years by all three fumigants. The largest dose of dazomet, and to a lesser extent that of Di-Trapex, were initially phytotoxic (Rebandel and Szczygiel, 1989).

Roses

Aldicarb applied at 12.2 kg ha^{-1} to beds of roses under glass controlled *P. penetrans* and increased flower production by 16% (Johnson and McClanahan, 1974). Repeated small doses of aldicarb or oxamyl controlled *P. penetrans* and *P. vulnus* on five-year-old roses better than one large dose, the benefits of such treatment lasting from six months to one year (Jakobsen and Rasmussen, 1976).

Flower bulbs

To ensure lily bulbs are uninfested before they are planted they can be dipped in water at 46°C for 20 min (D'Herde *et al.*, 1960) or 47°C for 45 min (Saigusa and Aihara, 1977). At 400 l ha^{-1}, D-D controlled *P. penetrans* in soil and doubled the yield of lily bulbs (D'Herde *et al.*, 1960). Dichloropropene mixtures or chloropicrin injected into the soil killed 98% of *P. penetrans* and increased yields of top grade King Alfred narcissus (Apt and Gould, 1961). Telone at 374 l ha^{-1} effectively controlled *P. penetrans* on subsequently planted Easter lilies (Maggenti *et al.*, 1967). In fumigated soil, in-furrow application of 3 kg aldicarb ha^{-1} (as Temik 10G) during planting of *Lilium regale* bulbs, whose roots were infested with the nematode, lessened infestation of new roots and increased bulb yields. Retreatment with aldicarb in the second year improved rooting but not bulb yield. Incorporating aldicarb in the ridge soil before planting was less effective than in-furrow application (Boontjes *et al.*, 1978). To obviate the risk of 1,2-dichloropropane getting into ground water from soil fumigation with 1,3-dichloropropene–1,2-dichloropropane mixtures, *P. penetrans* can be decreased in soil and in the roots of Oriental lilies by applying ethoprophos, fenamiphos, oxamyl or sodium tetrathiocarbonate to the soil by drip irrigation, or by applying oxamyl to the foliage. Combined use of one of these non-fumigant nematicides drip irrigated into the soil and oxamyl applied to the foliage was especially effective (Westerdahl *et al.*, 1994).

Potato

Oostenbrink (1958, 1961) demonstrated that *P. penetrans* could be very damaging to potatoes in sandy soils in the Netherlands. The nematode is also important in parts of Canada, the USA and Europe. The yield response of a potato cultivar, to nematicide applied to soil infested with the nematode, was inversely related to its tolerance of attack (Bird and Vitosh, 1978). The nematode causes a root-rot resulting in yield losses of up to 50% and a necrotic 'scabbiness' of the tuber surface, making the tubers unattractive and difficult to market. In Ohio, USA, *P. penetrans* increased the severity of early senescence of potatoes in the presence of one pathotype of *Verticillium dahliae* but not in the presence of another pathotype (Botseas and Rowe, 1994). The nematode is controlled by soil fumigation and by applying non-fumigant nematicides to the soil, to the seed tubers or to the foliage.

FUMIGANT NEMATICIDES

Autumn soil fumigation with 468 l Vapam (33% metham-sodium), 234 l Vorlex or 468 l Trapex (20% methyl isothiocyanate) ha^{-1} controlled *P. penetrans* in potatoes planted the following spring and increased tuber yield and quality (Cetas and Harrison, 1963). Depending on the nature and amount of soil fumigant applied, the method of application and the soil conditions in the growing season, the nematode was controlled in up to three successive potato crops by Vorlex, Telone, D-D or D-D-chloropicrin mixture (Miller and Hawkins, 1969). The nematode was controlled, and infection of potatoes by *Verticillium albo-atrum* was delayed, by fumigating the soil with 234 or 374 l of 1,3-dichloropropene ha^{-1} or 187, 281 or 374 l of 1,3-dichloropropene–chloropicrin (85% : 15%) mixture ha^{-1} (Biehn *et al.*, 1971). Olthof (1987) found Vorlex at 55 l ha^{-1} and Telone II b at 75 l ha^{-1} more effective in suppressing *P. penetrans* than aldicarb at 2.2 kg ha^{-1}, carbofuran at 3.3 kg ha^{-1} or oxamyl liquid at 9 or 18 l ha^{-1}. However, Vorlex plus aldicarb was the most effective treatment and at harvest left the smallest numbers of *P. penetrans* in the soil after Russet Burbank potatoes. Metham-sodium at 3.1 or 4.5 g a.i. m^{-1} row or 1,3-dichloropropene at 8.2 g a.i. m^{-1} row plus 0.21 g aldicarb m^{-1} row decreased numbers of the nematode at planting and throughout the growing season and increased tuber yields 23–46% (Olthof, 1989a). Applied broadcast in irrigation water, metham-sodium at 76 or 152 l a.i. ha^{-1} decreased *P. penetrans* in the soil by 100 and 99% or 90 and 88%, respectively, in two years. Despite the almost complete initial control with 152 l ha^{-1}, at harvest the nematode had increased, though less than in plots given smaller doses of the fumigant (Olthof, 1989b).

NON-FUMIGANT NEMATICIDES

Granular carbofuran, oxamyl, ethoprophos or fenamiphos controlled *P. penetrans* best when applied in-furrow at planting. At 3.4 kg ha^{-1}, aldicarb (as a 15% granule) controlled the nematode equally well, whether applied in a band before planting or in wide or narrow in-furrow bands at planting and increased tuber yields most consistently of the five compounds (Schultz and Cetas, 1977). Noling and Bird (1979) similarly found 3.3 kg aldicarb ha^{-1} effective in decreasing both soil and root populations of the nematode. Tuber yields were largest in plots that received both aldicarb and the largest dose of nitrogen fertilizer. Applied in-furrow at planting, aldicarb at 1.1 or 2.2 kg ha^{-1} reduced numbers of *Pratylenchus*, mainly *P. penetrans*, in potato roots more than 2.2 kg oxamyl ha^{-1}. Both compounds increased yields of cvs Superior and Russet Burbank by 15.5% and 9.3%, respectively, but did not increase the yield of cv. Shepody, which is presumed tolerant of attack by the nematode (Kimpinski and Sanderson, 1989).

Tomato and tobacco

Both fumigant and non-fumigant nematicides have controlled *P. penetrans* adequately on tomato and tobacco. With a polyethylene cover over the soil, D-D at 327 l ha^{-1} applied two weeks before transplanting, or oxamyl at 75 l of 10% solution ha^{-1}, significantly decreased numbers of the nematode in the soil and in the roots of tomato. Despite the nematode control achieved, *Verticillium* wilt was not decreased (Farley and Riedel, 1971). Of 15 nematicide treatments assessed, *P. penetrans* was controlled best and tobacco leaf yields were increased most by 90 l Telone C (1,3-dichloropropene plus chloropicrin) or 8.6 kg triazophos ha^{-1} (Marks and Elliot, 1976a). Telone at 90 l ha^{-1} controlled the nematode better in soil to 20 cm deep when it was injected 15 cm deep, than when it was injected more deeply (Marks *et al.*, 1972). Metham-sodium at 209 l a.i. ha^{-1} or mulches of grass or hay suppressed multiplication of *P. penetrans* in tomato roots. Control was increased when metham-sodium and a soil mulch were applied to the soil (Miller, 1977). Applied to the foliage two and five weeks after transplanting, 1.12 kg oxamyl in 500 l water ha^{-1} controlled the nematode on tobacco and increased leaf yields (Marks and Elliot, 1976b). The penetration of *P. penetrans* into tomato roots was decreased by 80–90% for six weeks or more by dipping them in activated charcoal, papain or gelatin powder. Activated charcoal was less effective when it was incorporated in infested soil before planting (Miller and McIntyre, 1976).

Control of *P. penetrans* by Seed Dressings

Soaking lucerne seeds for 17 h in oxamyl (32,000 ppm) before sowing them in infested soil prevented *P. penetrans* entering the roots of the seedlings (Townshend and Potter, 1979). More recently, Townshend (1989) has found that coating lucerne seeds with a mixture of 160 mg oxamyl ml^{-1} of 1:5 polymer sticker : water lessened infestation of lucerne roots by 90% initially and by 60% at 11 weeks. Although the sticker delayed emergence by four days, the seed dressing subsequently improved the growth of seedlings in infested soil compared with seedlings grown from untreated seeds.

Management of *P. penetrans* Populations

The numbers of *P. penetrans* in the soil may be decreased by bare fallowing the land for several months (provided this does not harm soil structure or encourage soil erosion), by the efficient control of weed hosts in perennial crops and by cultivation of *Tagetes patula*.

The nematode population in the soil was decreased 80–90% by growing *T. patula* for three to four months in infested soil (Oostenbrink *et al.*, 1957). Caubel and his co-workers dramatically improved the growth and yield of globe artichokes in Brittany, France by controlling *P. penetrans* with *T. patula* grown for two years. In subsequent studies, *T. patula* grown amongst artichokes controlled the nematode and increased yields as effectively as did the application of 400 l D-D ha^{-1} applied one month before planting the artichokes. D-D killed 93% of the nematodes and increased yields from 1 t ha^{-1} in untreated soil to 5–6 t ha^{-1}, whereas *T. patula* cv. Legion d'Honneur sown mid-May at 0.6 × 0.6 m spacing and hoed in three months later or sown at 6 kg ha^{-1} in interrows of 0.25 m and hoed in four and a half months later reduced soil numbers by 82% and increased yields two- to fivefold (Caubel *et al.*, 1978, 1981).

Pratylenchus thornei

This is a widespread species with a wide host range. It is an important pest of wheat in Utah, USA, Australia, India, Mexico, former Yugoslavia and Italy, causing stunting, reduced tillering, chlorosis and small or sterile ears. It retards growth of barley and apple in the Netherlands, maize and oats in Utah and can also be troublesome in potatoes and chickpeas. *Pratylenchus thornei* has been controlled by fumigant and non-fumigant nematicides and by soil solarization but for wheat and other relatively low value crops these are too expensive or impractical.

Wheat

In Mexico, where the nematode is a widespread pest of wheat, stunting the plants and reducing plant stands, nematicide treatments, though effective, were uneconomic and no cultivar resistant to the nematode has been found. The practical solution of the problem was to sow wheat in cool soil (15°C), which does not favour *P. thornei*, growing cultivars tolerant of the nematode and ensuring that the plants received adequate nitrogen fertilizer (Van Gundy *et al.*, 1974). In Italy, the numbers of *P. thornei* in wheat roots were decreased by amending the soil with nitrogen plus phosphorus and growing the crop in rotation, rather than continuously (Tacconi *et al.*, 1988). In New South Wales, Australia, the nematode has been found to 120 cm deep in the soil. Although 4 kg aldicarb ha^{-1} almost eliminated it from the top 10 cm of the soil under wheat, it had no effect on the nematode below 15 cm deep. Methyl bromide killed it to 30–90 cm deep and also controlled root-rot and crown rot, giving yields 9% greater than achieved by the use of aldicarb. Grain yield responses to the aldicarb treatment correlated well with the tolerances to *P. thornei* of a range of wheat genotypes, so the nematode seems at least partly responsible for wheat yield decline in north-western parts of New South Wales (Doyle *et al.*, 1987).

Potato

In Israel, covering the soil for 31 days with clear polyethylene sheeting 0.04 mm thick raised soil temperature sufficiently to kill 80–100% of *P. thornei*, comparable with a 90% reduction from soil fumigation with a mixture of EDB and chloropicrin. Weeds and the microsclerotia of *Verticillium dahliae* were fully controlled by soil solarization, not by EDB–chloropicrin fumigation, although the latter reduced *Verticillium* wilt of potatoes by 55%, due to the reduction in *P. thornei*, a predisposing factor for the disease (Grinstein *et al.*, 1979). Early dying of potatoes in Israel, due to *V. dahliae* was also controlled by applying metham-sodium in overhead sprinkler irrigation water, which killed *P. thornei*. The use of metham-sodium and cultivars tolerant of the disease allowed yields of 40–60 t of tubers ha^{-1} to be obtained (Nachmias and Krikun, 1985). Oxamyl delayed infection of potatoes by *V. dahliae* by controlling *Pratylenchus*, mainly *P. thornei* (Scholte, 1989).

Chickpeas

In Syria, all chickpeas tested were susceptible to *P. thornei*. Seed treatments were ineffective, whereas aldicarb at 5–10 kg ha^{-1} applied to the soil suppressed invasion of roots by the nematode and increased chickpea yields. More practically, turnip, parsley, *Cannabis sativa* and *Nasturtium fantarum* are non-hosts and carrot, radish, coriander, spinach and sugarbeet are poor hosts of *P. thornei*, offering the possibility of control of this pest through selected crop rotations (Greco *et al.*, 1988a). Earlier work by Walia and Seshadri (1985) had found seed dressings of aldicarb, carbofuran, fensulphothion or phorate at 1–2% of chickpea seed weight decreased populations of the nematode left in the soil after the treated chickpeas had been grown. In southern Italy, summer soil solarization for six or eight weeks in 1988 lessened infestation of chickpea roots in 1989 by about 50% and increased chickpea yields (Greco *et al.*, 1991b).

Pratylenchus vulnus

Although it has a wide host range, *P. vulnus* is notable as a serious pest of nut and fruit trees – cherry, fig, apple, walnut, peach, citrus, olive, avocado, of roses and of grapevines. The nematode does not occur out-of-doors in the cool soils of northern Europe but it is found there in glasshouses, especially on roses. The nematode prefers warm temperate and subtropical soils and has a worldwide distribution. As other species of *Pratylenchus*, *P. vulnus* is normally a parasite of the root cortex but it may also harm phloem and sometimes xylem elements. Reproduction is sexual. Although it is not possible to give precise threshold populations above which yield losses occur, 5–10 nematodes $(100\ g)^{-1}$ of soil may be considered potentially damaging. Infested roots show typical lesions, which may eventually girdle the root. The loss of functional feeder and sometimes main roots results in leaf chlorosis, twig die-back and yield loss.

Fumigant nematicides

Large doses of fumigant nematicides applied to the soil before planting can prevent serious injury to woody perennials for up to four years, provided nearly all the nematodes are killed. In California, USA, *P. vulnus* is the chief pest of avocado trees. Good control has been achieved with 1123 l D-D ha^{-1} and tree growth was improved (Sher *et al.*, 1959). Improved growth of roses in the field was related to reduction in numbers of *P. vulnus* by D-D, EDB, chloropicrin or metham-sodium. Nematode numbers remained small for at least nine months after planting (Sher and Munnecke, 1958). Numbers of *P. vulnus* in sandy silt and clay loam soils in California were greatly decreased by soil fumigation with D-D or EDB, which increased the growth of newly planted walnut trees for three seasons. After two seasons, however, numbers of the nematode had increased to pre-treatment levels (Lownsbery and Sher, 1957, 1958). *Pratylenchus vulnus* is the most important nematode pest of walnuts in California. Its control relies heavily on careful selection of nursery and orchard sites, rootstocks and in preplant fumigation of the soil with dichloropropene (Lownsbery *et al.*, 1969). The Chinese wingnut, *Pteracarya stenoptera* is tolerant of attack by the nematode and is often used as a rootstock (Lownsbery *et al.*, 1974a). The loss of DBCP for control of *P. vulnus* and other nematodes in established woody perennials has drawn attention to the possible use of non-fumigant nematicides for this purpose in the short term. In the longer term, however, there will be need for resistant or tolerant rootstocks.

Treatments for roses and grapevines

Young rose plants were freed of *P. vulnus*, without phytotoxicity, by dipping their roots in water at 48.3°C for 35 min, after the plants had been cold hardened for three weeks at 2.2°C, then heat hardened for 24 h at 37.8°C, prior to dipping (Towson and Lear, 1982). Complete control, without phytotoxicity, was also achieved by dipping rose seedlings in fenamiphos at 50 ppm for 30 min or 100 or 200 ppm for 10 min (Veld *et al.*, 1988).

Several non-fumigant nematicides control *P. vulnus* in roses. The nematode was damaging to roses at population densities above 25 $(500\ g)^{-1}$ soil and, in soils of differing infestation levels, bloom yields were increased more by treating the soil with oxamyl than by treating it with aldicarb (den Ouden and Beuzenberg, 1972). Bergé and Cuany (1972a) found *P. vulnus* was controlled best with aldicarb in eight-year-old *Rosa indica* rootstocks when applied at 10 kg ha^{-1} in May and again in August. Bloom size, not number, was also increased. Rose numbers and weight were increased and the number of *P. vulnus* was greatly decreased following application to infested soil of 20 kg fenamiphos (as 5% granules) ha^{-1} before planting. Repeated treatments at nine week intervals of

20 kg fenamiphos (as 10% granules) or of 20 l fenamiphos a.i. (as 400 E.C.) ha^{-1} decreased populations of the nematode better than aldicarb or oxamyl (Amsing, 1988). In roses grown hydroponically, oxamyl applied to the rock wool medium in the tank gave 95–99% control of the nematode (Amsing, 1990). Injecting non-fumigant nematicides under pressure into the trunks of grapevines and walnut trees also controlled *P. vulnus*. Injected into the trunks of two-year-old walnuts, carbofuran, fenamiphos, oxamyl or sulphocarb significantly decreased root infestations of the nematode (Viglierchio *et al.*, 1977).

Pratylenchus zeae

Although of worldwide occurrence and with a wide host range, *P. zeae* is important as a pest of maize, sugarcane and rice in warm temperate, subtropical and tropical soils.

Control

As other species of *Pratylenchus*, *P. zeae* has been controlled with soil fumigants (D-D, EDB) and with non-fumigant nematicides (aldicarb, carbofuran, fensulphothion, fenamiphos or ethoprophos at 6.7 kg ha^{-1}) (Johnson and Chalfont, 1972). The cost of such treatment is, however, too great for maize or rice. In Egypt, the nematode was controlled with 40, 70 or 100 gal Vapam (metham-sodium) per feddan, which increased maize yields but at high cost. Some cultivars were, however, more tolerant of attack by *P. zeae* than others (Oteifa and Taha, 1964). At 15 kg ha^{-1}, carbofuran controlled the nematode on sugarcane in north-east Argentina (Costilla, 1973). At 3 kg ha^{-1}, carbofuran reduced numbers of *P. zeae* and *Meloidogyne javanica* up to six months after planting, whereas 30 t filter cake ha^{-1} had no effect on numbers of the nematodes. Both treatments increased sugarcane yields by 23% 17 months after planting. Used together, these two treatments increased cane yields by 41% (Novaretti and Nelli, 1985). As with maize, there is evidence of tolerance in sugarcane to the nematode. After application of 3 kg aldicarb or carbofuran ha^{-1} to the soil, cv. NA 56–79 yielded 9.5 t more cane ha^{-1} than untreated sugarcane, cv. SP 71–799 yielded 19.2 t more cane ha^{-1} and SP 71–1406 yielded 21.9 t more ha^{-1} than untreated sugarcane (Novaretti *et al.*, 1988). In Brazil, the sugarcane cultivar SP 70–1143 is resistant to *P. zeae* and *M. javanica* (Novaretti, 1992). At Coimbatore, India, the sugarcane cultivars Co 88020, Co 89009 and Co 89034 are resistant to *P. zeae* (Mehta *et al.*, 1994).

In pots of soil containing 6.3 or 30 *P. zeae* ml^{-1} the growth of upland rice cv. IR 36 was greatly decreased and the nematode increased to 11,705 or 13,219 g^{-1} root, respectively. In the field, the rice cultivar UPL Ri 5 yielded 29% more grain in plots given 4 kg carbofuran ha^{-1} than in untreated plots, despite the very small numbers of the nematode at sowing (0–1.1 ml^{-1}). In contrast, the equally susceptible cultivar Kinandong Patong did not respond to carbofuran treatment of the soil, which may indicate that it is tolerant of attack by *P. zeae*.

The infestation of the roots of cv. UPL Ri 5 at harvest (94–1720 g^{-1} root) was negatively correlated with rice yield ($r = -0.59$) (Plowright *et al.*, 1988).

Compared with maize and rice, which are good hosts of *P. zeae*, cowpea and mung bean (*Phaseolus aureus*) are poor hosts of the nematode and two successive crops of either of them can decrease numbers of the nematode dramatically. Rice following cowpea yielded 37% more grain than rice following rice but the nematode multiplied rapidly on rice and there was no residual benefit from either legume in the second rice crop (Aung and Prot, 1990). At least part of the benefit of the legumes to the rice crops may be an increase in residual soil nitrogen. Sunn hemp (*Crotalaria juncea*) is also a poor host of *P. zeae* (Sundararaj and Mehta, 1990). As Gaur and Khan (1994) found that 8% of *P. zeae* survived and a quarter of these were able to invade maize roots after five and a half years in slowly air dried soil, dry season bare fallowing may decrease but will not eradicate the nematode.

Radopholus spp.

Two species of *Radopholus* ('burrowing nematodes'), *R. similis* (Cobb, 1893) Thorne, 1949 and *R. citrophilus* Huettel, Dickson and Kaplan, 1984, are of economic importance, mainly as pests of perennial crop plants.

Radopholus similis

First described from diseased bananas in Fiji, *R. similis* (Fig. 6.7) is a major pest of bananas worldwide. It is, however, unknown in Crete, Cyprus, Mauritius, Taiwan and the Canary Islands. It was formerly unknown in East Africa but, probably as the result of importations of infested breeding stock, it is now a serious pest in Tanzania and Uganda. In bananas and plantains, this migratory endoparasite of root cortex causes rotting of the fibrous root system, resulting in the felling of the stems, usually just before harvest, when leverage on the enfeebled roots is greatest (Plate 3). Severe yield losses occur. The relationship between growth of the banana plant (Y) and infestation of the roots by the nematode (P = no. $(100\ g)^{-1}$ root) was first demonstrated by Guérout (1972). The relationship fitted the Seinhorst equation

$$Y = cz^{P} = z^{P-T} \tag{6.1}$$

where c and z are constants and T is the tolerance level of infestation of the roots (Seinhorst, 1965). In infertile soil this was 10 *R. similis* g^{-1} root. A close relationship between infestation level and plant growth and yield was found in experiments in Cameroon, but the exact relationship was affected by climate, soil conditions and cumulative effects from one cropping cycle to the next (Melin and Vilardebó, 1973a). The nematode is also damaging to sweet potato and abacá (*Musa textilis*), to lowland tea in Sri Lanka, causes 'yellows' disease of black pepper, which devastated the crop on the island of Bangka, Indonesia, and damages coconut and areca nut (*Areca catechu*) in south India. Cocoa, coffee (several

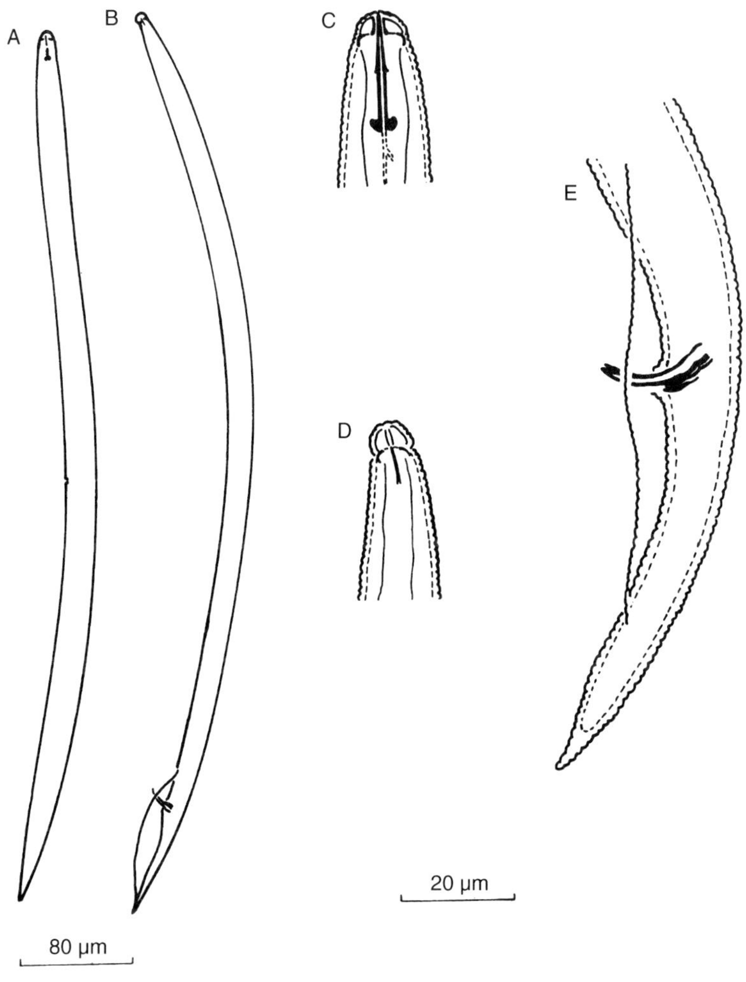

Fig. 6.7. *Radopholus similis*: A, female; B, male; C, female head end; D, male head end; E, male tail.

species), cowpea, pigeon pea, sugarcane and pineapple have also been reported infested by *R. similis*.

The nematode moves intercellularly in the cortex of banana roots, feeding on parenchyma cells, causing extensive cavitation and characteristic reddish-brown lesions, which extend to the root surface as longitudinal fissures. The cortex of the roots may be attacked to the bases of the roots, some 10 cm deep in the rhizome. The rhizome cortex may also be attacked, though tunnelling by the banana weevil, *Cosmopolites sordidus*, is the greater problem in this tissue. *R. similis* usually reproduces sexually, sometimes parthenogenetically. The life cycle, which takes about three weeks at 25°C, may occur wholly within the root, nematodes emerging from roots that are decaying to invade healthy roots. All juvenile stages as well as adults are infective.

There is evidence that *R. similis* occurs as biotypes, which differ in host preference. A Panamanian isolate differed from a Honduran isolate in host range (Edwards and Wehunt, 1971). Venkitesan (1976) found that isolates of the nematode from the roots of black pepper, coconut, areca nut and banana in south India differed in host specificity. More recently, Koshy and Jay (1991) have identified ten races of *R. similis* among 28 populations in India, using 17 differential host plants, of which citrus, cloves, pineapple and sugarcane cv. Co 997 were immune to all 28 populations. In populations of the nematode from Central and South America, biotypes were found which differed in their speed of multiplication on banana, in their host preferences and in the percentages of pointed : rounded tails in females (Pinochet, 1979; Tarté *et al.*, 1981). All this complicates control by biological methods.

Bananas and Plantains

Cultural control

Ideally, uninfested banana suckers should be transplanted to uninfested land. In areas affected by the nematode, it may not be possible to find land completely free of the pest. However, Loos (1961) found that *R. similis* survived no more than six months without host plant roots or live rhizome tissue. Bare fallowing and/or the growing of fully resistant cover crops has, more recently, been shown to decrease heavy infestations of the soil to trace populations. All old banana rhizomes and weed hosts must be removed. In Panama, where rice and weeds favoured nematode survival, the application of ethoprophos at the start of a 12–18 month period of bare fallowing or weed control in the fallow period greatly decreased *R. similis* in soil previously heavily infested (Salas *et al.*, 1976). In peaty soil in the Ivory Coast, 11 months bare fallow almost eliminated the nematode from the soil, so that populations remained less than 25 g^{-1} root for 14 months after replanting. A similar dramatic decrease in numbers of *R. similis* in the soil was observed, following 3–7 weeks of flooding (Sarah *et al.*, 1983). Bare fallows may decrease soil fertility and encourage soil erosion. Flooding the soil, which is only feasible on very flat land, may harm soil structure. In Martinique, following bananas, a one year fallow under the grass *Brachiaria decumbens* or sorghum plus siratro (*Macroptilium atropurpureum*) grew and yielded well, improved soil fertility and satisfactorily decreased *R. similis* and other soil pests (Ternisien and Melin, 1989). In Queensland, Australia, finger grass (*Digitaria milanjiana* cv. Jarra), which is resistant to the nematode, provides good ground cover and has been released for use as a pasture ley, and, in north Queensland, as a rotation crop for bananas to control *R. similis* (Hall *et al.*, 1993).

Treatment of planting material

Disinfested land should be planted with disinfested banana transplants. This can be achieved by hot water treatment of infested suckers and/or thorough paring away of infested rhizome tissue (Plate 3), or perhaps in the future by the use of micropropagated ex-plants grown on in nematode-free nurseries. Loos and Loos (1960) disinfested banana suckers by peeling the rhizomes and cutting away all discoloured root and rhizome tissue and then dipping the rhizomes in an aqueous mixture of 0.2% DBCP, 5% copper sulphate and 5% lime, to prevent reinvasion by nematodes and fungi, then drying the suckers for at least 24 h before planting them. In New South Wales, Blake (1961) found that there was less risk of the suckers succumbing to thorough paring if they were more than 10 cm in diameter but if they were less than 13 cm in diameter they could be freed of *R. similis* by immersing them in water at 55°C for 20 min, so preventing transfer of the nematodes to uninfested or disinfested soil. Colbran and Saunders (1961) also found immersion of suckers in water at 53–55°C for 20 min freed them of the nematode. In Cuba, hot water treatment of banana suckers at 55°C for 20 min was also very effective (Casamayor *et al.*, 1967; Decker *et al.*, 1970). More recently in Queensland, Broadley (1979b) found heavy pruning of rhizomes followed by dipping them for 20 min in water at 53–55°C, or light paring then dipping in aqueous ethoprophos (600 ppm, 20 min) or fenamiphos (250 ppm, 10 min) controlled *R. similis* well.

Chemical soil treatment

Cultural control of *R. similis*, though desirable, is often impractical. Availability of suitable land for profitable banana production may make fallowing impractical. Replanting bananas is also expensive. Hot water treatment of rhizomes is costly, as suckers are bulky; temperature control is important, requiring an expensive, thermostatically controlled water bath and nematode control without phytotoxicity is not always achieved, particularly as the nematode may be deep in the rhizomes. In the Mungo region of Cameroon, the control of *R. similis* with nematicides was found to be more cost-effective than fallowing and hot water treatment of suckers for transplanting (Melin and Vilardebó, 1973b). Luc and Vilardebó (1961) found the nematode could be controlled in the Ivory Coast by fumigating the soil at planting with 300 l D-D or 150 l EDB ha^{-1} and using disinfested suckers. Phytotoxicity of these fumigants was the drawback, especially in applications to established bananas. DBCP, however, at 40 l ha^{-1} at planting in May or June and at 25 l ha^{-1} in October and 15 l ha^{-1} in March thereafter, controlled the nematodes well and increased banana yields profitably. Its widespread use in the Ivory Coast and Cameroon gave way to granular non-fumigant nematicides, which were found to be more effective and easier to apply than DBCP (Guérout, 1970; Vilardebó *et al.*, 1972; Vilardebó and Guérout, 1976). As DBCP has been withdrawn from general use, growers are now dependent on granular non-fumigant nematicides for control of banana nematodes. Those found effective against *R. similis* were organophosphates (fenamiphos, fensulphothion, ethoprophos and isazophos) and carbamates (carbofuran and aldicarb).

Based on successful applications in the Ivory Coast (Guérout, 1972, 1974), Cameroon (Melin and Vilardebó, 1973a, 1976) and Madagascar (Beugnon and Vilardebó, 1974), Vilardebó and Guérout (1976) recommended the application of 3 g fenamiphos per banana plant to 1 m^2 of the soil around the stool three times a year, ahead of expected rainfall. Such treatment controlled *R. similis* and maintained or increased yields of banana plantations. Similar treatment was also effective in Surinam (Hasselbach, 1973), while in South Africa 4.5 kg fenamiphos ha^{-1} controlled the nematode in heavily infested bananas but did not increase yields (Keetch *et al.*, 1976). In Queensland, Australia, 12 kg fenamiphos ha^{-1} at planting prevented early increase of *R. similis*, *Helicotylenchus multicinctus* and *Meloidogyne javanica* on bananas. In established bananas, 25 kg ha^{-1} rapidly decreased *R. similis* and populations of it were kept small by applying 12 kg ha^{-1} every four months, especially when treatments were made to 1 m^2 soil around the base of each plant (Burnett *et al.*, 1974). Broadley (1979a) found 2 g fenamiphos per plant, applied to the soil six times a year, more effective in controlling the nematode than the same dose three times a year. The best treatment increased yield by up to 60%. Evaluated in a five year experiment in north Queensland, the application of 375 g fenamiphos (100 m $row)^{-1}$ per year lessened root rot effectively, whether the nematicide was applied in granular form or by irrigation. Application of 31 g (100 m $row)^{-1}$ per month (April to November) plus 125 g (100 m $row)^{-1}$ in December by drip or mini-sprinkler irrigation was simpler, safer and more cost-effective than granular applications. Mini-sprinkler application increased bunch weight and plant girth more than other treatments (Schipke and Ramsey, 1994).

Three other organophosphate nematicides have been assessed against *R. similis* of banana – ethoprophos, fensulphothion and isazophos. Though not as reliable as fenamiphos, ethoprophos controlled the nematode satisfactorily in several experiments when 5 g was applied to 1 m^2 around the base of each plant three times a year in the Ivory Coast (Guérout and Pinon, 1973; Guérout, 1974). For the Ivory Coast, Madagascar and Cameroon, Vilardebó and Guérout (1976) recommended the application of 4.5 g ethoprophos per plant concentrated in 1 m^2 around the stool three times a year. Applying 10 g initially, followed by 5 g doses every four months, rapidly decreased *R. similis* on bananas (Guérout, 1974). At 5 g $(m^2)^{-1}$, fensulphothion virtually freed bananas of *R. similis* (Guérout, 1972). In Queensland, 125 g fensulphothion applied in a 1 m wide band to 100 m row, twice a year, controlled *R. similis* well to the end of the fourth cropping cycle, which the same amount of fenamiphos did not. Fensulphothion also increased yields by up to 35% over this period (Broadley, 1976a). In a subsequent report (Broadley, 1979a), 2 g fensulphothion per stool at planting in July and in December and June each year thereafter controlled *R. similis* better and increased banana yields more than the same dosage of ethoprophos, fenamiphos or oxamyl. Speich (1982) reported good control of the nematode in banana trials in the Ivory Coast, Cameroon and the Caribbean with 2.5–3 g isazophos per plant applied three times a year. In Cuba, Perez *et al.* (1987) found isazophos, applied in an 0.5 m band around each Giant Cavendish banana or plantain plant at 3 g per plant at the onset of the rains and again five months later, controlled the nematode and increased yields by 41 and 55%, respectively.

Vilardebó and Guérout (1976) recommended application of 3 g carbofuran to 1 m^2 around the base of each banana plant three times a year in the Ivory Coast and Madagascar. In Cuba, 3 g per plant at planting followed by the same dose five months later controlled *R. similis* effectively and increased banana and plantain yields (Perez *et al.*, 1987). Aldicarb, however, would seem to be the most powerful of the non-fumigant nematicides against *R. similis*. In an experiment in Dominica, West Indies in 1968–1969, aldicarb applied around bananas badly affected with *R. similis*-induced 'toppling disease' revitalized the grove, greatly increasing its productivity (A.G. Whitehead and J.E. Edmunds, unpublished). Since then, the efficacy of aldicarb has been widely confirmed. In Peru, two applications of 3.75 g aldicarb per plant, applied seven months apart, decreased both *R. similis* and banana weevil for six to seven months after the second application and increased banana yields by 12.8 t ha^{-1} (Ricavarren-Herrera *et al.*, 1976). Similar treatment in Ecuador gave similar results (Hasing-Lama *et al.*, 1976). In Puerto Rico, aldicarb most effectively controlled both the nematode and the banana weevil when 3 or 4.5 g per plant was applied to the soil around each stool every four months or when 3 g was similarly applied every six months (Román *et al.*, 1983). In Cameroon and Martinique, 1.5, 2 or 3 g aldicarb per plant three times a year increased banana yields better than other nematicides in soil infested with both *R. similis* and the banana weevil (Vilardebó *et al.*, 1988). In the Ivory Coast, 1.5 g per plant three times a year gave very good control of both nematode and weevil and increased bunch weights and shortened the length of the cropping cycles (Sarah *et al.*, 1988). Aldicarb has the advantage over organophosphates of being active in a wide range of soils, even those with large organic matter fractions. As with organophosphates, it is less effective in very alkaline soils (pH > 8). Fenamiphos and ethoprophos, for example, worked well in kaolin-type clay soil in the Ivory Coast but poorly in peaty soil (Guérout *et al.*, 1976). Where the organic matter fraction of the soil is small, a wide range of nematicides is effective against *R. similis*. In Puerto Rico, for example, fenamiphos, fensulphothion, ethoprophos, aldicarb or carbofuran granules at recommended dosages and frequencies prevented yield loss in plantains, due mostly to *R. similis*, allowing, for the first time in Puerto Rico, three successive crops to be harvested without the need to replant (Román *et al.*, 1976).

Resistance

As yet there are no commercial banana or plantain cultivars fully resistant to *R. similis*. In the Philippines, 24 cultivars of 78 screened were only lightly or moderately attacked by *R. similis*. None of them was resistant to *Meloidogyne incognita*. In Karnataka, South India, cv. Yelakkibale is considered resistant. The cultivars Padali Moongil, Kunnan and Ayiramka Poovay were the least infested of those cultivars assessed in Bangalore, South India. Diploid clones of *Musa AA* vary from slightly resistant to immune to *R. similis* and the Pisang Jari Buaya clones offer the best prospect as parental breeding material for bananas and plantains (Pinochet, 1988). Hybrid SH-3142, derived from the Pisang Jari Buaya clones, is highly resistant to Black Sigatoka disease, race 4 of *Fusarium* wilt and *R. similis*. It is, however, very susceptible to *Pratylenchus coffeae* and *M. incognita*. The resistance appears dominant and controlled by a few genes (Pinochet,1992). However, the genetics of *Musa* spp. are complex, breeding is difficult and at least three pathogenic forms of *R. similis* occur in Central America and the Caribbean alone, so progress towards useful resistant cultivars is likely to be slow. According to Ploetz (1994), breeding programmes have developed dessert and cooking bananas resistant to *R. similis*. Their commercial success is yet to be established.

Black Pepper (*Piper nigrum*)

Radopholus similis has been shown to cause 'yellows' disease and the very similar 'slow wilt' disease of black pepper (van der Vecht, 1950 and Venkitesan, 1976, respectively), though secondary invasion of the roots by fungi, e.g. *Fusarium solani*, may hasten the demise of infested plants. More than 250 *R. similis* g^{-1} root was very frequently associated with 'slow wilt' affected vines in Kerala (Anon., 1986). In another report (Ramana, 1986), 250 *R. similis* g^{-1} root was constantly associated with 'slow wilt' in Kerala. Affected vines have chlorotic foliage, especially in drought conditions, shed their inflorescences and die over three to five years. Chemical control of the nematode in black pepper is difficult, as the crop is perennial and repeated applications of nematicides may not be economic. Furthermore, in India, black pepper is grown alongside trees susceptible to *R. similis* and may be associated with other susceptible crop plants grown in mixed perennial multicrop systems. At 3 g per vine, phorate (10% granules) applied in May–June and again in September–October controlled the nematode over a period of three years, leading to a remission of the symptoms of 'slow wilt' (Mohandas and Ramana, 1987). To be fully successful, such treatments would have to begin at planting, using uninfested seedlings, and continue from year to year. In Sri Lanka, GK 49, KW 31, KW 33, MW 21, Panniyur and TG 7 black peppers are reported resistant to *R. similis*.

Coconut

Coconut suffers foliar chlorosis and yield loss when attacked by *R. similis*, due to rotting of the outer cortex of the younger, unsuberized roots, initiated by nematodes invading them just behind the root tips (Koshy and Sosamma, 1987, 1988). The infestation of coconut by this nematode has been noted in Jamaica, Florida, India and Sri Lanka. The south India coconut population has a very wide host range, including several important crop plants. An inoculum of one nematode $(3.5 \text{ ml})^{-1}$ soil halved the height and girth of coconut plants five years later and as few as one nematode $(3.5 \text{ l})^{-1}$ soil reduced the height and girth by 13% and 24%, respectively, five years later (Koshy and Sosamma, 1988). The nematode is widespread in Kerala, Karnataka and Tamil Nadu, India.

The nematode was completely controlled in the nursery by applying 25 kg fenamiphos or phorate ha^{-1} three times to the soil (September, December and May) (Koshy *et al.*, 1986). Nematicide applications to established palms is impractical in terms of cost and safety. There could be residues in harvested nuts and in other crops often grown in close association with coconuts. Some coconut cultivars are less susceptible or more tolerant of *R. similis*. Planting seedlings of these cultivars with *R. similis*-free intercrops, liberally mulched and fertilized to offset damage and retard nematode multiplication would seem the best prospect for control at present. The complexity of the multicropping system, however, makes control of *R. similis* in coconut a difficult goal.

Areca Nut

Areca or betel nut (*Areca catechu*) is injured by *R. similis* in much the same way as black pepper and coconut. Although repeated applications of fensulphothion and aldicarb to the soil around the vines controlled the nematode and increased nut yields (Sundararaju and Koshy, 1986), the practicality of such treatment is questionable. Some cultivars are tolerant of *R. similis*, i.e. Indonesia 6, Mahuva 8 and Andaman-5. The hybrid Indonesia 6 × Singapore (VTL-17) is highly resistant to the nematode (Sundararaju and Koshy, 1982; Griffith and Koshy, 1990). *Areca triandra* cultivars Sri Lanka 3 and Indica are also resistant to *R. similis*. Choice of cultivar would therefore seem a better prospect for control than use of nematicides.

Radopholus citrophilus

Differences between the 'citrus' and 'banana' races of *R. similis* were considered sufficient to justify their recognition as two sibling species, *R. citrophilus* (citrus race) attacking both citrus and banana and *R. similis* (banana race) attacking banana but not citrus. *R. citrophilus* has n = 5 chromosomes, whereas *R. similis* from India (Kerala) and Indonesia has n = 4. However, neither a population of *R. citrophilus* from Hawaii with five chromosomes (Huettel *et al.*, 1986) nor a population of *R. similis* from Puerto Rico with five chromosomes attacked citrus. Although small differences in the morphology of the two species may occur (Baldwin *et al.*, 1978), it is clearly difficult to distinguish these two species confidently.

Radopholus citrophilus is a much less important pest, worldwide, than *R. similis*, but in Florida it has been a major problem in citrus grown in the sandy soils of the central ridge area, where it has been demonstrated primarily responsible for 'spreading decline' of citrus (Fig. 6.8). In 1956, over 1000 citrus groves were known to be infested with the nematode, as well as 130 citrus and 179 ornamental plant nurseries (Suit and DuCharme, 1957). Losses of yield caused by spreading decline were 50–80% for grapefruit and 40–70% for oranges in affected groves and the smaller crops were unprofitable. Affected trees grow slowly, become chlorotic and produce small and fewer fruits. The trees only die if they are neglected, but they cannot be fully rehabilitated by improved cultural techniques. The disease is due primarily to the rapid destruction of the tree's feeder roots. Ford (1952, 1953) found that, whereas 30% of the feeder roots were lost in soil 25–75 cm deep, 90% of them were lost below 75 cm deep in trees affected by spreading decline and diseased trees had only half as many feeder roots as healthy trees. This short life of feeder root flushes deprives the trees of nutrients and water, leading to drought stress and mineral deficiencies. Suit and DuCharme (1953) found *Radopholus citrophilus* in feeder roots in diseased areas, not in healthy areas of 28 out of 29 groves sampled and found none of the nematodes in the feeder roots of 26 healthy groves studied. They showed that at 23.3–24.4°C sour orange seedlings grew very slowly and had a reduced root system in decline subsoil from four citrus groves, compared with seedlings of the

Fig. 6.8. Citrus affected by spreading decline due to *Radopholus citrophilus* (Photo: D.W. Dickson).

same age grown in subsoil from healthy areas of the same four groves. In contrast, there was no difference in growth of seedlings in topsoil from decline and healthy areas of the four groves. No other nematode was associated with spreading decline. *R. citrophilus* was shown to invade the fine feeder roots from behind the root cap to the zone of root elongation and up to the zone of root suberization. Their penetration, feeding and multiplication in the roots leads to the formation of brown to black lesions on the roots and cavitation of the cortex below. Unlike *R. similis*, *R. citrophilus* also penetrates and feeds in the phloem-cambium of the roots, which may be completely destroyed (DuCharme, 1968).

The host range of *R. citrophilus* is probably as wide or wider than that of *R. similis* but, as the two species may occur together in Florida and elsewhere, their separate host ranges have yet to be determined. Furthermore, two biotypes of *R. citrophilus* have been identified in Florida, which differ in their ability to reproduce on four rootstocks. Biotype 1 reproduces poorly on Milam lemon and moderately on Ridge Pineapple, Albritton sweet orange and Carrizo citrange, whereas biotype 2 reproduces well on all four rootstocks (Kaplan and O'Bannon, 1985). These two biotypes may also differ in their host ranges, as may biotypes of *R. similis*.

The life cycle of *R. citrophilus* at the optimum temperature of about 24°C is short (18–20 days), and can be completed wholly within the roots, so infestation can increase rapidly (DuCharme and Price, 1966). The nematode can reproduce parthenogenetically and sexually. Destruction of roots seems to be hastened by invasions of soil fungi (*Fusarium*, *Sclerotium*), mites and oligochaetes (DuCharme, 1968). Feldmesser *et al.* (1959) had earlier shown in glasshouse experiments that the application of captan at a dosage of 56 kg ha^{-1} to sour orange seedlings infested with *R. citrophilus* dramatically improved their health in six months. Leaves were greener, roots twice as heavy and shoots three times as heavy as in untreated seedlings. There were twice as many nematodes in the treated as in the untreated root systems, no doubt because the increased roots could support the extra worms. This showed that the secondary invaders of nematode-damaged roots were responsible for the collapse of the roots, leading to the severe dieback of spreading decline. Harder to explain is why they also found that citrus seedlings planted in spreading decline soil from which *R. citrophilus* had been removed suffered from decline symptoms. Perhaps the soil still contained viable eggs of the nematode.

Control

Treatment of nursery stock

Birchfield (1954) found that immersing citrus roots for 10 min in water at 50°C killed all *R. citrophilus* in them. To prevent lasting injury to the plants, they had to be transferred immediately after heat treatment to cold water for 10 min and given extra water and 'care' after planting. More conveniently, dipping the roots in thionazin, fensulphothion or ethoprophos at 1000 ppm for 30–60 min or in fenamiphos at 250–600 ppm for 30 min also eradicated the nematodes but without phytotoxicity (O'Bannon and Taylor, 1967).

Treatment of established citrus

Because spreading decline could not be cured in Florida citrus, drastic action was taken to rehabilitate affected groves and to prevent its spread within and between groves. Studies on the persistence of *R. citrophilus* in soil and roots suggested that the nematode might be eradicated from infested soil if it was kept free of host roots for up to one or two years. Birchfield (1957) found the nematode in soil or roots died in four months, when buried about 90 cm deep in the field. Tarjan (1961) found the nematode did not survive in roots buried 31–152 cm deep in sandy soil in central Florida for more than six months, whereas *Hoplolaimus coronatus* and *Pratylenchus brachyurus* did. In contrast, Hannon (1963) found citrus feeder roots infested with *R. citrophilus* nine months and in one case 14 months after removal of citrus trees from the soil. Living root fragments with feeder roots were also found below 152 cm 24–30 months after soil had been fumigated with D-D at 560 l ha^{-1} and these could enable *R. citrophilus* to survive deep in the sandy subsoil for longer than was previously thought possible (Collins and Feldman, 1965). Despite these reservations, both spreading decline and *R. citrophilus* were well controlled by the 'pull-and-treat' method, first suggested by Suit and Ford (1950) and detailed by Suit *et al.* (1955). Essentially, this consisted of first mapping accurately the area in a citrus grove affected by spreading decline, confirmed by detailed soil sampling for *R. citrophilus*, then removing all trees in this area plus a further two trees beyond the margins, which might also have been infested. The trees and all recoverable roots were burnt on site and the land was ploughed, disked and levelled. The soil was injected 30 cm deep with 560 l D-D ha^{-1}, preferably six months later to allow most remaining roots to disintegrate and improve the efficacy of fumigation. The fumigated land was then kept in bare fallow for two years, or under the non-host *Crotalaria spectabilis* for part of that period. This proceedure, though drastic and very costly, prevented infestation of neighbouring healthy trees and kept replanted trees free of infestation for at least five years. To ensure their isolation, bare fallowed, frequently fumigated zones 5–30 m wide were established around the treated decline areas

(Suit and Feldman, 1965). Collins and Feldman (1965) found that D-D at 1870 l ha^{-1} eliminated the nematode from all roots 3.65 m deep and at 4680 l ha^{-1} killed all citrus seedling roots buried 4.57 m deep. Despite their efficacy, large doses of D-D, 1,3-dichloropropene or EDB contaminated wells used for drinking water and their use ceased in 1983. An alternative proceedure would involve deep trenching, to prevent citrus roots from neighbouring trees crossing the buffer zones, and the use of herbicides to ensure the maintenance of bare fallow. The difficulty of controlling *R. citrophilus* below 45 cm and the success of citrus grown in Florida in shallow soils suggests that the problem of spreading decline in the deep sands of central Florida might be resolved through frequent gentle irrigation and other practices to encourage the development of extensive shallow root systems, where conditions are more hostile to *R. citrophilus* (Tarjan and O'Bannon, 1977). Of course, this might well exacerbate other problems.

McBeth (1960) reported excellent control of *R. citrophilus* from 37.4 l DBCP ha^{-1} metered into flood irrigation water three times, at monthly intervals. Unfortunately this option is no longer available for established citrus trees. O'Bannon and Tomerlin (1977) found that oxamyl applied to the foliage in six sprays per annum decreased soil populations of the nematode and increased citrus yields significantly in the following two years. In subsequent work, oxamyl applied to the foliage three to five times per annum significantly increased tree vigour, drought resistance and fruit yield but control of *R. citrophilus* was not so clear cut (O'Bannon and Selhime, 1981). Split applications of fenamiphos decreased the nematode and increased citrus yield (O'Bannon and Tarjan, 1979).

Resistance and tolerance

Very little resistance or tolerance to *R. citrophilus* has been detected among the large number of species, cultivars and hybrids of *Citrus* that have been assessed. Milam lemon, Estes rough lemon, Ridge Pineapple sweet orange and Corrizo citrange were released as rootstocks resistant or tolerant of attack in 1958. Unfortunately, they all proved susceptible to some populations of the nematode (Poucher *et al.*, 1967). Kaplan and O'Bannon (1985) found a population (biotype 2), which reproduced on Milam lemon, Ridge Pineapple and Albritton sweet oranges and Corrizo citrange distinct from another population (biotype 1), which reproduced poorly on Milam lemon and moderately on the other rootstocks. At present, therefore, there is no citrus rootstock reliably resistant to *R. citrophilus* in Florida, which, coupled with the difficulties and cost of control with nematicides, leaves the problem still only partially solved.

Scutellonema bradys (Fig. 6.9)

In 1933, Steiner and Le Hew described this nematode from tubers of a yam (*Dioscorea* sp.) grown in Jamaica, which had suffered heavy infestation and exhibited extensive brown discoloration of the outer tissues. Goodey (1935) reported a similar 'dry rot' disease of yams in Nigeria, associated with the same nematode (Fig. 6.10).

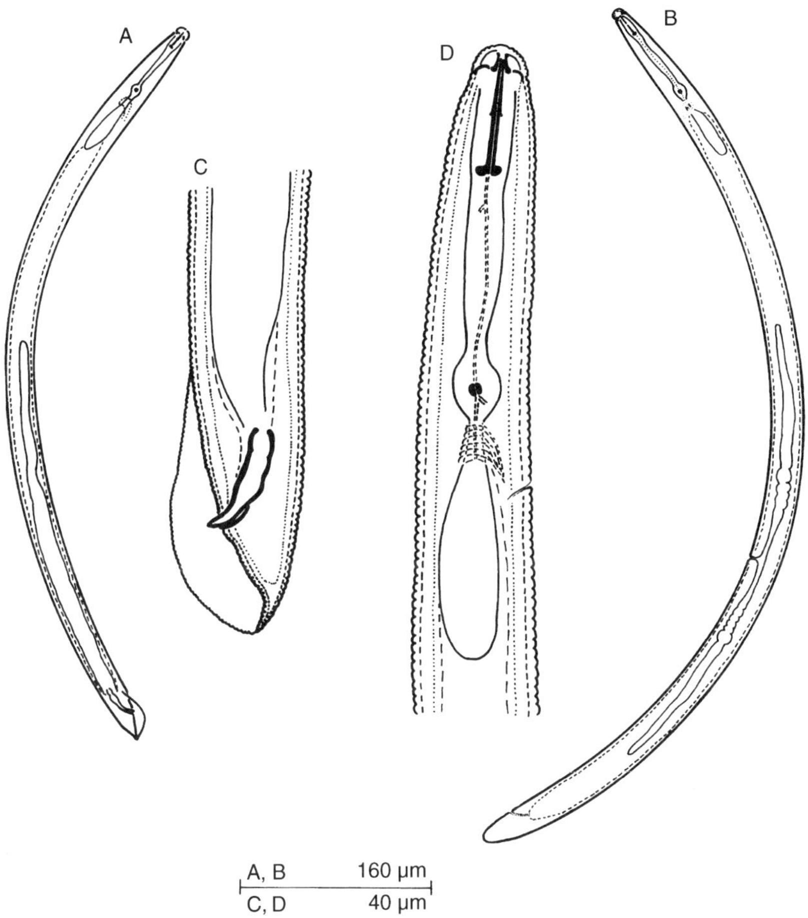

Fig. 6.9. *Scutellonema bradys:* A, male; C, male tail; B, female; D, female anterior end.

Scutellonema bradys has quite a wide host range and has been found in the Caribbean, Central America, Florida, West Africa, Brazil and India. It is, however, of importance as a pest of food yams (*Dioscorea* spp.), causing 'dry rot' disease, which makes the tubers unsaleable and may make them prone to fungal and bacterial rots in storage. In the field, yield losses may reach 30% and, if wet rot follows in storage, losses may exceed 80%. The nematode is a migratory endoparasite of roots and tubers, which feeds intracellularly in the outer 1–2 cm of the tuber. Males are abundant and multiplication would appear to be sexual. Populations can reach over 6000 g^{-1} of tuber tissue. The 'dry rot' may be due to the nematode alone or in combination with a bacterium, *Corynebacterium* sp. In *D. alata*, infestation in the seed tuber transfers first to the roots, then to the new tuber. Multiplication of the nematode, which is abundant, coincides with the onset of dormancy but is greatest in storage (Cadet and Quénéhervé, 1994).

Fig. 6.10. *Scutellonema bradys* dry rot of yam (Photo: J. Bridge).

Control

Resistance to *S. bradys* is unknown in *Dioscorea* spp. and the nematode is controlled by cultural methods, disinfesting tubers or tuber pieces before planting and the application of nematicides to infested soil.

Cultural methods

When it is practical, rotation of yams with non-hosts, such as groundnut, chillies, tobacco, cotton, maize or sorghum will limit damage by *S. bradys*. Rotation or intercropping with cowpeas, sesame, green gram (mung bean), tomato, okra, melon or pigeon pea, which are good hosts of the nematode, will exacerbate the problem (Adesiyan, 1976). Uninfested tubers or portions of tubers should be used as planting material or infested material should be disinfested before planting it (see below). Applying cattle manure to yam mounds at about 1.5 kg per mound can lessen infestation and increase yam yields, as can the use of fertilizer containing phosphorus (Adesiyan and Adeniji, 1976).

Disinfesting planting material

Although it is possible to raise uninfested nursery stock from seed, or by propagation from leaves or stems in a sterile medium, yam tubers or tuber pieces can be disinfested by dipping them in hot water or in nematicidal solutions or suspensions.

Dipping diseased yam tubers in water at 50°C for 30–40 min (Smit, 1967), or 15 min (Ayala and Acosta, 1971), was effective in eliminating *S. bradys* from them. The nematode was eliminated, without phytotoxicity, from *D. alata* and *D. cayenensis* tubers by immersing them in water at 50–55°C for 40 min (Adesiyan and Adeniji, 1976). However, phytotoxicity can occur, depending on the exact treatment adopted, the species of yam and its physiological state. The nematode was also controlled, without phytotoxicity, by dipping infested tubers in fensulphothion (1250 ppm) for 15 or 30 min, or in thionazin (625 ppm) for 60 min (Ayala and Acosta, 1971). A 30 min dip in aqueous D-D, carbofuran, oxamyl, calcium hypochlorite or formalin (all at 1000 ppm of a.i.) reduced infestation of *D. alata* tuber pieces but only D-D, carbofuran or formalin treatment increased tuber yields (Badra and Caveness, 1979). After five weeks of storage, *S. bradys* was completely suppressed by dipping tubers of *D. alata* in thionazin (1000 or 2500 ppm) without leaving significant thionazin residues in them (Adesiyan, 1981).

Numbers of *S. bradys* in yam tubers were decreased by up to 80% by γ irradiation from Cobalt 60. Although treatments completely suppressed sprouting and yam deterioration in storage, complete elimination of the nematode required a dosage of radiation harmful to the tubers (Adesiyan, 1977). The use of γ irradiation on yam tubers, however, seems impractical.

Nematicides applied to the soil

In Nigeria, infestation of *D. rotundata* by *S. bradys*, 24 weeks after planting, was lessened and tuber yields were increased by 85–137%, when 2 kg aldicarb, carbofuran, oxamyl or isazophos ha^{-1} was applied to the soil in which yams were planted. Negligible residues of aldicarb, carbofuran or isazophos were found in harvested tubers and no residues of oxamyl were detected in tubers (Adesiyan and Badra, 1982).

Sedentary Endoparasites of Roots and Tubers (I. *Globodera* and *Heterodera*)

7

Cyst-nematodes (Fig. 7.1) are generally the most important nematode pests in temperate soils. Several cyst-nematodes are also troublesome in the tropics, notably potato cyst-nematodes (*Globodera rostochiensis* (Wollenweber, 1923) Skarbilovich, 1959 and *G. pallida* Stone, 1973) on potatoes grown at high altitudes. In the tropics as a whole, root-knot nematodes (*Meloidogyne* spp.) are the most important plant nematode pests, dealt with in Chapter 8.

Fig. 7.1. Cyst-nematode shape. Lemon-shaped cyst characteristic of *Heterodera* spp. Orange-shaped cyst characteristic of *Globodera* spp. (Photo: P.R. Burrows).

Globodera rostochiensis and *G. pallida* (Figs 7.2, 7.3)

The two species of potato cyst-nematode currently recognized almost certainly evolved, along with their principal host, potato, on high plateaux of the Andes in South America. From there they have been widely spread as cysts on the tubers, or in soil around them, or other agricultural produce. Evans and Stone (1977) listed 48 countries in which one or both species had been reported and 58 or more countries are now known to have one or both species (EPPO, 1992).

In South America, *G. rostochiensis* occurs in Chile, Bolivia, Venezuela and

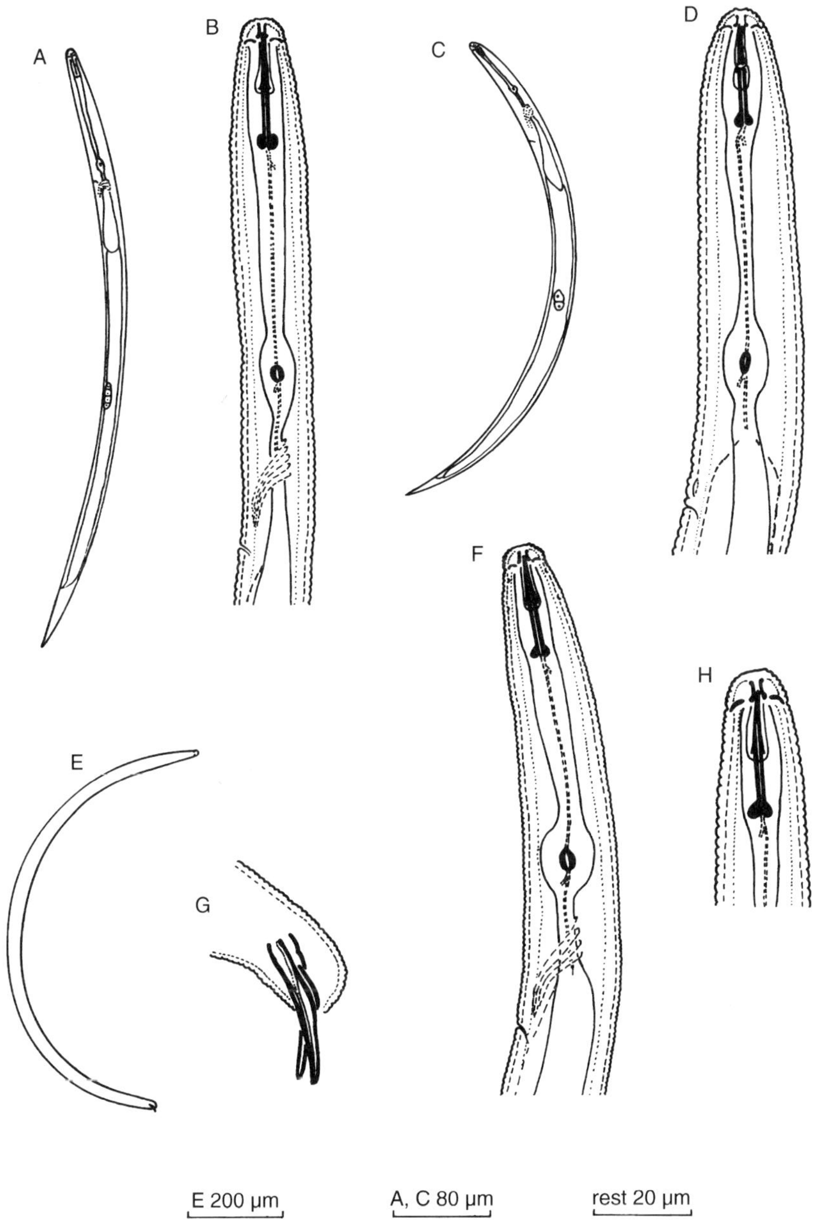

Fig. 7.2. *G. pallida*: A, second stage juvenile; B, anterior end of second stage juvenile; H, anterior end of male. *G. rostochiensis*: C, second stage juvenile; D, anterior end of second stage juvenile; E, male; F, male anterior end; G, male tail end.

south Peru and *G. pallida* is found in Bolivia, Venezuela, Colombia, Ecuador and in most of Peru (Evans *et al.*, 1975). The nematodes occur widely in Europe and Russia, though *G. pallida* seems less common or is absent in southern Europe. Both species can occur together in the same field and in the same one hectare

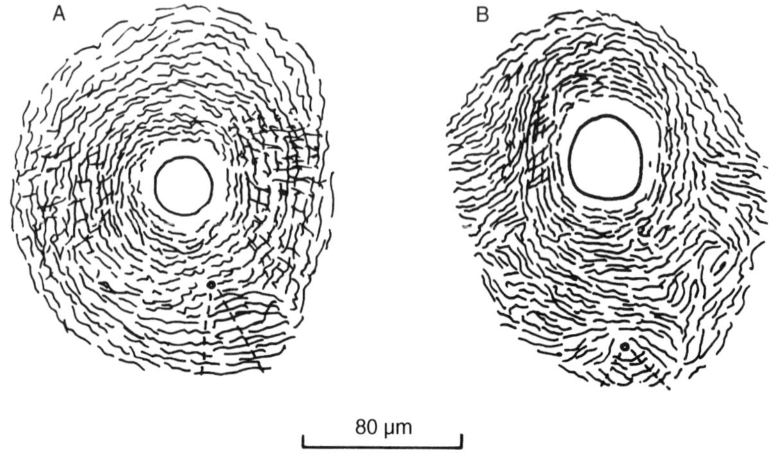

Fig. 7.3. Cyst cones of potato cyst-nematodes. A, *G. pallida*; B, *G. rostochiensis*.

blocks within a field, their relative abundance being determined by the resistance/susceptibility of previous potato crops (Whitehead and Westerdijk, 1987). *Globodera pallida* and *G. rostochiensis* do not interbreed successfully (Parrott, 1972) but each species has a number of host races, identification of which is, at present, unsatisfactory.

Host Range and Distribution in Soil

Potato cyst-nematodes multiply on *Solanaceae*, especially potato, tomato and aubergine (*Solanum melongena*), and on wild species of *Solanum* and *Lycopersicon*. The nematodes are most important as pests of potato (Plate 4), although they may also injure tomato and aubergine. On potato, all underground tissues (roots, stolons and tubers) are prone to infestation and, although cysts may be found on the deepest roots, heavy infestations are usually limited to the top 20 to top 50 cm of the soil profile (Whitehead, 1977a). The depth distribution of the cysts in the soil is affected by the ability of thc roots to penetrate the soil profile, previous potato cropping history, depth of ploughing and susceptibility of the potato grown.

Host Efficiency

Potato cultivars susceptible to potato cyst-nematodes can differ in host efficiency. For example, in England *G. rostochiensis* in the top 20 cm of the soil multiplied, on average, 31.5-fold on cv. King Edward grown in 20 soils with 6–194 eggs g^{-1} soil but only 12.1- and 10.8-fold, respectively, on cvs Pentland Crown and Desirée in

24 and 14 soils, respectively, with 6–180 eggs g^{-1} soil. In a number of experiments, *G. pallida* increased much less on cvs Record and Romano than on cv. Desirée (Whitehead, ex published experiments). In general, the larger the crop of susceptible potatoes, the larger the infestation left in the soil after harvest.

Life Cycle

Potato cyst-nematodes multiply sexually (den Ouden, 1960a). All the fertilized eggs are retained in the body of the female, the body wall of which becomes tanned, as she dies, to form a tough, protective cyst housing the embryonated eggs. The females of *G. rostochiensis* become first golden-yellow to orange then brown cysts, whereas those of *G. pallida* remain a creamy white or become a pale lemon yellow before turning into brown cysts (Plate 4). The eggs hatch very largely in response to solanaceous root diffusate. In potato root diffusate *in vitro*, juveniles of *G. pallida* emerged more slowly from the cysts than juveniles of *G. rostochiensis* (Whitehead, 1992b). As for many other cyst-nematodes, the first moult occurs in the egg, the hatched juvenile being second stage. Second stage juveniles penetrate roots, often just behind the root cap and become sedentary, the head close to the endodermis. They also penetrate stolons and developing tubers, coming to rest just beneath the periderm. Enlarged, multinucleate (syncytial) cells form around the head of the sedentary juveniles in response to nematode stylet probing and feeding. After three further moults, pear shaped, sedentary females and vermiform, active males are formed. The males leave the roots and are attracted, chemically, to young females, most of whose bodies protrude from the roots and the surfaces of the stolons and tubers later on. Up to 765 eggs may be contained in a cyst (Hesling, 1959), though more typically there may be 200–500 per cyst. The cyst can remain intact in the soil for 15 years or more. The number of eggs remaining in the cyst declines exponentially over many years but the rate of decline can vary (see below). Usually there is only one full generation per potato crop, because the new eggs must first pass through diapause and a period of low soil temperatures (or perhaps a long drought) is needed to break such dormancy. Some juveniles, however, may hatch before the female becomes a cyst, i.e. before the onset of diapause and, if there is adequate soil moisture, such juveniles may invade underground host tissue and complete a (smaller) second generation (e.g. Whitehead, 1993).

On potato, multiplication rates vary inversely with the initial population density. At very low density, at or below the tolerance threshold, it may be as much as 150-fold, at high (equilibrium) density it is 1-fold and at even greater densities there are fewer nematodes after harvest in the soil than there were before planting. Populations can exceed 1500 eggs g^{-1} soil.

Damage and Yield Losses in Potatoes

In heavily infested soil, intolerant cultivars of potato grow slowly, because the roots are stunted, which limits the uptake of water and nutrients from the soil (e.g. Trudgill, 1987). Heavily infested roots have a 'bottle brush' appearance and may have an unhealthy brown colour. As a result, the haulms emerge slowly from the soil, remain small, may wilt in dry weather and show symptoms of nutrient deficiencies, despite a soil nutrient status adequate for healthy plants. Nitrogen and magnesium deficiency symptoms are commonly observed. Affected plants may senesce prematurely, though this may be due to increased susceptibility to *Verticillium* wilt disease. Tubers remain small and are often covered with females and cysts mid-season. The skin of the tuber may then have a russetted appearance and later bears many small pits, previously occupied by the adults. Tuber yield and tuber quality are decreased. The severity of symptoms and yield loss are related to the number of viable eggs, or more precisely the number of hatched juveniles that emerge into the soil. The relationship, over a wide range of initial population densities, can be described as a sigmoid curve, relating total plant yield with log nematode density (Seinhorst, 1965). The curve has two inflexions, the first at a low density, at greater than which yield is significantly decreased and the second at high population density, above which yield is at a minimum. The curve equation (Seinhorst, 1965) is:

$$Y = Y_{min} + (1 - Y_{min})\, cz^{P_i} \tag{7.1}$$

where Y = yield as a fraction of maximum yield (1.0), Y_{min} = minimum yield as a fraction of maximum yield, P_i = number of nematodes at planting, c is a constant (1.05–1.15), representing 5–15% compensation by the plant for nematode injury and z is a constant (usually *c.* 0.995), representing the fraction of maximum yield left after one nematode has fed (Jones, 1978). Rarely, if ever, can this theoretical curve, derived from pot experiments under controlled conditions, be reproduced from field data. A sufficiently wide range of P_i values is very difficult to contrive in the field without imposing other confounding effects, small numbers of nematodes cannot be estimated reliably and Y_{min} can be zero. In most situations, however, the relationship can be fitted to the middle part of the sigmoid curve, i.e. to a virtually straight line relationship. The shape of the curve and hence P_t, the tolerance level for damage, the slope of the middle portion of the curve and the size of Y_{min} can all vary, according to the cultivar of potato, the cultural conditions, particularly soil moisture stress, and the involvement of other pathogens. Yield loss in intolerant potatoes can often occur when there are more than 2 or 3 eggs g^{-1} soil, whereas for tolerant potatoes yield loss only occurs at much greater population densities.

Based on linear regressions of tuber yield on eggs g^{-1} soil, Brown (1969) in eastern England found losses of tubers (mostly cvs Majestic and King Edward) over 3.8 cm diam. averaged 2.3 t ha^{-1} (0.4–4.0) per 20 eggs g^{-1} soil, with a maximum yield loss of 17.6 t ha^{-1} More recently, Brown and Sykes (1983), using data from trials in the West Midlands of England in addition to the earler data (Brown, 1969), found that yield loss was greater, over the range 0–40 eggs g^{-1} soil, than over the range 40–160 eggs g^{-1} soil, and that the average yield loss over the range 0–160 eggs g^{-1} soil was 2.75 t ha^{-1} for tubers over 3.2 or 3.8 cm diam. per 20 eggs g^{-1} soil, with a maximum loss of 22 t ha^{-1}.

In unirrigated, sandy loam infested with *G. rostochiensis* at populations ranging from about 5–300 eggs g^{-1} soil, maximum loss of tubers over 3.8 cm diam.

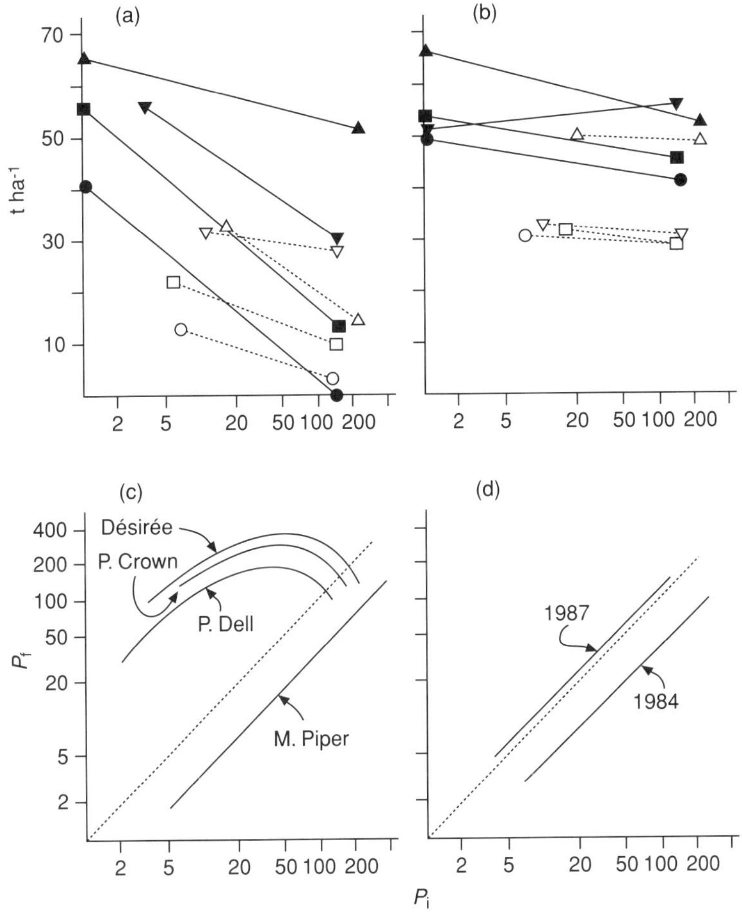

Fig. 7.4. Effect of number of *G. rostochiensis* eggs g^{-1} soil at planting (P_i) on yield of potato tubers over 38 mm (1984) or 40 mm (1987) and on eggs g^{-1} soil after harvest (P_f) in a field of sandy loam soil. cv. Pentland Crown 1984 (△), 1987 (▲); cv. Desirée 1984 (□), 1987 (■); cv. Pentland Dell 1984 (○), 1987 (●); cv. Maris Piper 1984 (▽), 1987 (▼); (a) and (c) soil untreated, (b) and (d) 5.6 kg oxamyl ha^{-1} incorporated in the seedbed just before planting. In (d) for all cultivars P_f/P_i was less than 1.0 in 1984 and somewhat greater than 1.0 in 1987, following oxamyl treatment of the seedbed.

was around 30 t ha^{-1} in two dry years and about 13.5 t ha^{-1} in a third, wet year (Whitehead *et al.*, 1980a). In a similar soil, over the range 2–95 *G. rostochiensis* eggs g^{-1} soil, loss of total Desirée tubers was about 8.2 t ha^{-1} per 20 eggs g^{-1} soil, with a maximum yield loss of 38 t ha^{-1} (Whitehead *et al.*, 1984).Yield loss can also vary between cultivars. For example, in an unirrigated sandy loam, in two years yield loss associated with *G. rostochiensis* in cv. Maris Piper was less than that in three other cultivars (Whitehead, *et al.*, 1987b), but in the same field in another year the slopes of all four regression lines were parallel (Whitehead, unpublished data) (Fig. 7.4). Yield losses due to *G. pallida*, in soil in which a range of initial infestations had been obtained by previously growing resistant and susceptible potatoes, were shown by Elston *et al.* (1991) to fit an inverse linear relationship with nematode numbers, equally as well as an exponential (Seinhorst, 1965) relationship but with one less parameter and better than linear (Brown, 1969) or log linear (Oostenbrink, 1966) relationships. In the five partially resistant and one susceptible genotypes used, yield loss was not related to their level of resistance to the nematode.

Clearly, yield loss depends on the numbers of hatchable nematodes in the soil at planting, the cultivar grown, soil moisture stress, the virulence of the nematode population and on the presence of interacting organisms, e.g. *Verticillium dahliae* (Whitehead, 1983). Although Jones and Uncles (1976) found nematicide use economic at fairly high population densities (35–53 eggs g^{-1} soil), it was uneconomic at high (70–128 eggs g^{-1} soil) or lower (16–24 eggs g^{-1} soil) densities. Wright *et al.* (1986) found the application of 5 kg oxamyl or 3 kg aldicarb ha^{-1} to seedbeds, mostly for intolerant cv. Record, was economic for sites with 1–5 eggs g^{-1} soil as well as for sites with 6–24 eggs g^{-1} soil and the cost of treatment was repaid when one or more eggs g^{-1} soil was present. No other pest or disease was found, which might have explained the responses to the two nematicides.

Control

Potato cyst-nematodes have been controlled, more or less successfully, by crop rotations, use of resistant potato cultivars, trap cropping, soil solarization, use of nematicides and, most successfully, by the combined use of two or more of these measures (integrated control).

Crop rotation

As potato cyst-nematodes only multiply on *Solanaceae*, particularly potato, tomato and aubergine, growing them infrequently decreases soil infestations and thereby lessens injury to the next solanaceous crop. Growing potatoes continuously on infested land leads to heavy infestations and uneconomic yields. In a monoculture experiment started in 1933 in Germany, the nematode increased rapidly and after 44 years yields were 4.9 t ha^{-1} and 9.1 t ha^{-1}, respectively, on mineral fertilized and FYM fertilized plots (Dowe and Decker, 1981). Apart from other pests and diseases, exacerbated by potato monoculture, infestation of the tubers by potato cyst-nematodes would probably have made them unsaleable. At least partial control is achieved in rotations in which non-solanaceous crops predominate. The number of non-host crops that must be grown to prevent injury to a host crop will depend on the abundance of the nematodes in the soil and their rate of decline in the absence of host plants. The annual decline rate has been estimated at about 20% (Cole and Howard, 1962; den Ouden, 1970), 33% (Cooper, 1953; Huijsman, 1961; Winfield, 1965a), 30–40% (Duggan, 1978) and 49% (den Ouden, 1974). Faster rates of decline have been reported in Sweden (Andersson, 1987). Although speed of decline is generally held to be independent of population density, there is evidence that it may be faster for large than for small populations in some soils (Whitehead *et al.*, 1980a; Andersson,

Table 7.1. Years rest from solanaceous crops for decline of potato cyst-nematode to 5 eggs g^{-1} soil.

Annual decline	Eggs g^{-1} soil		
(%)	50	200	800
50	3.3	5.3	7.3
30	6.5	10.3	14.2
20	10.3	16.5	22.7

1987). To prevent damage to potatoes, when soil infestations are large and decline rates are slow, very long rotations are needed (Table 7.1). Ten or more years between host crops would usually be unacceptable, commercially.

Because the rate of decline may differ between fields, infestations should be monitored at intervals to determine when the next potato crop can be grown without significant loss of tuber yield and quality. If the interval in years is too long, additional control measures must be applied. In microplots under continuous barley, three populations of *G. rostochiensis* and five of *G. pallida* declined at 13–25% per annum, while a fourth population of *G. rostochiensis* declined at 41% per annum (Whitehead, 1995). The decline in egg numbers in the cyst seems to be the result of spontaneous hatching in the spring (den Ouden, 1960b), though destruction of eggs by predation may also occur. In England, in the light sandy, silty or peaty soils in which most ware potatoes are grown, satisfactory crops have been obtained in six-, or seven-year rotations (Cooper, 1953; Winfield, 1965a). Some damage may still occur in patches, where the nematode infestation was heavy in the previous crop.

Solanaceous weeds are controlled with herbicides in cereal crops, so they are unlikely to affect the rate of decline of potato cyst-nematodes in the soil. Volunteer potatoes, if numerous, however, can influence the speed of decline between potato crops. Population densities smaller than the equilibrium density were significantly increased under oats by four volunteer potatoes m^{-2} (den Ouden, 1967). It is therefore important to keep the number of volunteers in the soil to a minimum. Although tractor-hoeing in widely spaced row crops may destroy many volunteers from potato seed and restrict the growth of volunteers from tubers and tuber pieces left in the soil from a previous potato crop, it cannot eliminate them between the rows, nor have much effect on those in the crop rows. Furthermore, the nematode can complete its life cycle on the roots of deeply buried tubers, whose shoots may not reach the soil surface. After a potato crop, the soil should be lightly cultivated to allow frost to kill unharvested tubers, most of which are left near the soil surface.When volunteers are numerous, killing them before the life cycle of the nematode is completed may hasten nematode decline. Spraying volunteers in the first three weeks of June in the Netherlands with glyphosate reduced the numbers of viable eggs of both species of *Globodera* (Jellema and Brinkman, 1986).

Resistant, susceptible and tolerant cultivars

As South American populations of potato cyst-nematodes are, genetically, more diverse than European populations and are able to overcome resistance conferred

Table 7.2. Some cultivars resistant to potato cyst-nematodes.

Nematode	Cultivar
Globodera rostochiensis	Aula, Ariadna, Aguti, Anosta, Atica, Atlantic, Alleghany, Alhamra, Aistes, Alcmaria, Amigo, Aminca, Astarte, Aurora, Ausonia, Accent, Adora, Agria, Belchip, Belorusskil 3, Berber, Brodie, Buchan, Bereginya, Cardinal, Ceres, Christa, Coastal Chip, Castile, Concorde, Cara, Costella, Cultra, Dramant, Donna, Diamant, Dundrod, Darwina, Element, Elvira, Eden, Esta, Fala, Foka, Franzi, Fresco, Fianna, Fronika, Genesee, Gitte, Granola, Gigant, Gloria, Hampton, Hudson, Highlat Russet, Heidrun, Hydra, Hertha, Islander, Irga, Ilse, Junior, Jewel. Kristall, Kanona, Karaka, Krostar, Kingston, Kestrel (p.r.), Lada, Lady Rosetta, Liseta, Maris Anchor, Maris Court, Maris Piper, Meta Prigozhit 2, Multa, Marijke, Morene, Minerva, Miranda, Mondial, Morag (p.r.), Naroch', Nema Rus, Nida, Nicola, Nadine, Navan, Omega, Ovina, Obelix, Prevalent, Procura, Pentland Meteor, Prigozhit 2, Prigozhiiz, Pansta, Prominent, Peconic, Première, Prior, Provita, Pentland Javelin, Penta, Provan, Picasso, Rosa, Rocket, Rathlin, Stina, Saturna, Spekula, Simcoe, Skutella, Scala, Steuben, Stemster, Skirza, Shurminskii 2, Santé, Senator, Sherine, Stroma (p.r.), Saxon, Tarpan, Tunika, Taiga, Torva, Ukama, Valor, Vil'nya, Vikhola, Wachtel, Xenia.
G. pallida	Atrela, Benol, Berlina, *Cromwell*, *Darwina*, Elles, *Karaka*, Kaimai, Kestrel, *Multa*, Maria Huanca, *Nadine*, Pansta, *Prevalent*, Producent, *Santé*, Sovereign, Senator, Sierra, Stroma, Valor.

p.r. = Partially resistant to *G. rostochiensis*; cultivars listed under *G. pallida* are usually only partially resistant to it; those in italic type are also resistant or partially resistant to *G. rostochiensis.*

by genes, which have been incorporated in resistant cultivars in Europe (Franco and Evans, 1978) and elsewhere, it is important to prevent any further dissemination of cysts from South America. Strict quarantine regulations should at least delay, if not prevent, long-term, further introductions of potentially virulent populations.

As a fully resistant cultivar, by definition, prevents multiplication of invading nematodes, it can decrease infestation of the soil by a potato cyst-nematode by 80% in a season. Three consecutive years of growing it should therefore decrease soil infestation by 99%, although such a practice could exacerbate *Rhizoctonia solani* damage. Four consecutive years of a fully resistant potato decreased *G. rostochiensis* by 99% (Zawiślak *et al.*, 1981).

Resistance to potato cyst-nematodes was reported by Ellenby (1954) in some cultivated lines of *Solanum tuberosum* ssp. *andigena* (tetraploid) and in wild *S. vernei* (diploid). In the late 1960s, potatoes resistant to *G. rostochiensis* became available to growers in England. These had been bred from *S. tuberosum andigena* × *S. tuberosum tuberosum* hybrids (Howard, 1969) and have single, dominant gene (H_1) resistance to pathotypes Ro 1 and Ro 4. One of these cultivars, Maris Piper, is now the most widely grown potato in England. Many other cultivars fully resistant to these pathotypes of *G. rostochiensis* have now been bred. Some, with resistance from *S. vernei*, are only partially resistant and a few have resistance to several other pathotypes (Table 7.2). Esta and Heidrun are resistant to Ro 1 and Ro 5, Pansta to Ro 1–Ro 4 and Miranda and Franzi to Ro 1–Ro 5. Resistance to *G. pallida* has been incorporated into new cultivars, mainly using resistant lines of *S. tuberosum andigena* and *S. vernei*, sometimes with other wild *Solanum* spp. included in the pedigrees. The resultant resistance is partial, lessening multiplication of *G. pallida*. Some of these cultivars are listed in Table 7.2. Some of them are also fully or partially resistant to *G. rostochiensis*. Some 16 species of *Solanum* from Argentina, Bolivia and Peru are resistant to European populations of both species of *Globodera* (Turner, 1989).

Resistance to either species of *Globodera* could not be related to the solanidine glycoside content of leaves or tubers of seven wild species of *Solanum* (Hoogendoorn *et al.*, 1992).

Pathotypes within both species of *Globodera* have been identified by differential host reactions (Canto-Saenz and de Scurrah, 1977; Kort *et al.*, 1977). The host differentials used were common to both species of *Globodera*, except that *S. tuberosum* ssp. *andigena* (H_1) for *G. rostochiensis* was replaced by *S. multidissectum* (H_2) for *G. pallida* (Table 7.3). Such a scheme may distinguish pure populations of each pathotype but is unsuitable with unknown mixed populations, as is often the case with field populations. Furthermore, there may be differing host reactions to different populations of the same pathotype. The unsatisfactory nature of the pathotyping scheme has prompted the assertion that only three pathotypes of *G. rostochiensis* can be recognized reliably, i.e. Ro1 (= Ro1 + Ro4), Ro2 (= Ro2+ Ro3) and Ro5 and that there are more or less virulent populations of *G. pallida* (Pa) (Nijboer and Parleviet, 1990). Earlier work by Bakker (1988) had shown no relationship between two-dimensional electrophoresis of general proteins of pathotype populations and their identification by differential host reactions.

The subsistence potato grower and the starch potato grower can select their resistant cultivars on the basis of yields, whereas the grower of potatoes for

Table 7.3. Identification of pathotypes of potato cyst-nematode by host differentials.

Host differential	Pathotypes designated					
Europe	Ro1	Ro4	Ro2	Ro3	Ro5	
South America	R_1A	R_1B	R_2A	R_3A	–[a]	
Solanum tuberosum tuberosum	+	+	+	+	+	
S. tuberosum andigena (H_1)	–	–	+	+	+	
S. kurtzianum KTT/60.21.19	–	+	–	+	+	
S. vernei GLKS.58.1642.4	–	+	–	-	+	
S. vernei $(VT^n)^2$62.33.3	–	–	–	–	+	
Europe	Pa1	–[a]	–[a]	–[a]	Pa2	Pa3
South America	P_1A	P_1B	P_2A	P_3A	P_4A	P_5A
S. tuberosum tuberosum	+	+	+	+	+	+
S. multidissectum (H_2)	–	–	+	+	+	+
S. kurtzianum KTT/60.21.19	+	+	–	+	+	+
S. vernei GLKS.58.1642.4	+	+	+	–	+	+
S. vernei $(VT^n)^2$62.33.3	–	+	–	–	–	+

a = Comparable pathotype not reported.

pre-packing or processing must choose only those cultivars which meet the specialized quality standards acceptable to the market. As a result, the choice of cultivar is much more restricted for the latter than for the former group of growers. Five first early and nine second early and main crop potatoes resistant to *G. rostochiensis* (Ro1/Ro4) are recommended or provisionally recommended (for UK use) in the 1997 British NIAB list of potato cultivars (Anon.,1996). Of these, the most widely grown is the main crop cv. Maris Piper, but other cultivars resistant to *G. rostochiensis* are less popular than susceptible cultivars (Pentland Dell, Desirée, Record, Estima and Maris Bard). Only the first early cv. Rocket and the main crop cvs Santé, Nadine and Valor, which are partially resistant to *G. pallida* (Pa2/Pa3), are recommended by NIAB (Valor provisionally recommended) at present. All four cultivars are fully resistant to *G. rostochiensis* (Ro1/Ro4). On cv. Santé and two other cultivars partially resistant to *G. pallida*, 25 English populations of this species multiplied little, compared with the susceptible cv. Desirée but three other populations of the nematode were rather more virulent. Nevertheless, the results suggested such cultivars could be generally useful in England in lessening increase of *G. pallida* (Whitehead, 1991a).

The widespread and often frequent cultivation of Maris Piper, which is susceptible to *G. pallida*, has greatly increased the latter in southern England, so that *G. pallida* is now the predominant species in England, having long been the predominant of the two species in northern England (Whitehead, 1991a). A similar situation has arisen in the Lower Rhine area of Germany, where repeated cultivation of Ro1 (Ro1/Ro4)-resistant cultivars has replaced this pathotype with Ro2, Ro3 (=Ro2) and Ro5 and with Pa2 and Pa3 (=Pa2/Pa3) (Thomas, 1994). There is, therefore, need for more agronomically acceptable cultivars resistant to both species. Continued, widespread use of cv. Maris Piper and other cultivars with the same narrow resistance will intensify selection of *G. pallida* and may reveal hitherto undetected Ro pathotypes, as occurred in an experiment by Øydvin (1978). Similarly, cultivars partially resistant to *G. pallida* may select virulent strains from field populations (Turner, 1990; Whitehead, 1991b). Cultivars with only *S. vernei* in their pedigrees increased *G. rostochiensis*, other than Ro1, and *G. pallida* Pa3 (Olsson, 1985). In north-east Netherlands, populations of *G. pallida* virulent on cvs Producent, Elles and Darwina have been identified, highlighting the need for new resistant starch cvs for this area, if the intensive potato growing practised there is to be maintained (Burgt *et al.*, 1993, 1995). A population of *G. pallida* very virulent in some partially resistant potatoes in Scotland may have been a separate introduction from South America (Philipps and Trudgill, 1993).

Nematode multiplication varies between susceptible potato cultivars. Jones (1970a) reported differences between seven susceptible cultivars grown continuously for six years in microplots and suggested that the size of the root system may determine such differences. This is supported by data from many subsequent field experiments by Whitehead and his co-workers, which showed that, in general, multiplication of either species in untreated plots increased with tuber yield (and presumably with increase in root system size). On cv. King Edward, *G. rostochiensis* increased twice as much as on five other cultivars (Whitehead *et al.*, 1980b) and, on average, twice as much on King Edward as on Desirée or Pentland Crown in numerous other experiments. Similarly, the variation noted in multiplication of *G. pallida* on susceptible cultivars of potatoes in pots by Dale and Philipps (1985) was also encountered in field experiments, in which this nematode multiplied much more on cvs Cara and Desirée than on cvs Romano and Record (Whitehead *et al.*, 1987b; Whitehead, 1993).

Whether they are resistant or susceptible to one or both species of *Globodera*, potato cultivars differ in their tolerance of nematode attack. This may be expressed in a larger value of P_t and/or a smaller regression of yield against nematode number in a tolerant cultivar (Evans and Haydock, 1990). *In vitro* studies have also shown that the speed at which a potato genotype induces hatching, as well as the growth of its invaded roots, is part of the tolerance/intolerance mechanism (Arntzen *et al.*, 1994). Tolerance in a susceptible cultivar will increase nematode multiplication, so it is more usefully linked with full resistance, preventing significant yield losses at larger population densities. If the resistance base is narrow, however, tolerance may promote increase of other pathotypes to which the potato is susceptible. Dale (1989) found that vigorous and tolerant, partially resistant genotypes increased potato cyst-nematodes more in heavily infested soil than did intolerant, susceptible genotypes of potato. Tolerance to the two species of potato cyst-nematode may differ in a cultivar (Trudgill and Cotes, 1983). Maris Piper, generally tolerant of attack by *G. rostochiensis*, has shown intolerance of this species under certain conditions. Susceptibility to *Verticillium dahliae* may affect tolerance to potato cyst-nematode attack (Evans, 1987). These findings show that tolerance is a variable phenomenon and precise classification of cultivars for tolerance/intolerance is not, at present, feasible. The effect of the environment on the expression of tolerance/intolerance needs further clarification.

Trap cropping

The numbers of potato cyst-nematodes in the soil may be diminished by growing potatoes long enough to allow most juveniles to hatch and invade the roots, then uprooting and destroying them, so no new eggs are produced. First early potatoes, if harvested early, have such a trap cropping effect. Harvested a little later, they may still permit little multiplication (Webley and Jones, 1981). Improved growth of potatoes had been observed over 40 years earlier in infested soil in which two 'trap crops' had been grown the previous year (Carroll and McMahon, 1939). More recently, the numbers of potato cyst-nematodes in the soil were decreased by 45–73% by growing a potato trap crop for about one month before ploughing it in (Carlsson and Videgård, 1971). As seed tubers can sprout several times from remaining dormant 'eyes' after the shoots have been destroyed, ploughing in the trap crop could generate a serious volunteer potato problem. Removal of the trap crop, including the seed tubers, from the soil at the appropriate time is essential to the success of the technique. Whitehead (1977b) showed that small tubers of cv. Pentland Crown planted at 15 cm spacing decreased numbers of *G. rostochiensis* in the top 20 cm of a sandy clay soil by 45% more than did bare fallow, six weeks after planting them. Four weeks trap cropping decreased the nematode less and eight weeks trap cropping increased the number of cysts in the soil, so that there were as many eggs g^{-1} soil as in the bare fallowed plots.

Using the very tolerant cv. Cara as a trap crop in spring or early summer and planting the tubers 20 cm apart in ridges 91.4 cm apart or in single ridges in microplots, heavy infestations of *G. pallida* (40–465 eggs g^{-1} soil) were decreased by 75% or more in about six weeks. The haulms with seed tubers attached were then forked out of the ground or lifted by a tractor-drawn harvester, the crop residue being carted off the field. To lessen the disposal problem, the haulms can be mown to about 10 cm above the ridge tops, though in loose peaty soils this might separate the seed tubers from the haulms. The trap crop could be followed by a late-planted crop such as peas, carrots or a brassica (Whitehead, 1992b; Whitehead *et al.*, 1994). In France, *G. pallida* was reduced by 80% per annum by trap cropping. Trap crop plus use of ethoprophos gave a 98.5% reduction of *G. pallida* in the soil (Mugniery and Balandras, 1984). The percentage reduction in soil infestation, using cv. Cara, is largest with heavy infestations, probably because heavy infestations stimulate proliferation of roots of this very tolerant cultivar and thereby increase hatching, through enhanced secretion of root diffusate into the soil. For success, it is most important that the trap crop is removed before the females are fertilized, otherwise the soil infestation may be increased. Ideally the trap crop should be removed as soon as the first females can be seen, just emerging on the surface of the oldest roots.

Biological control

There have been several reports of biological control of potato cyst-nematodes, *in vitro* or in pots. As yet there is little evidence of effective control in the field. *Cladosporium herbarum* parasitizes eggs of *G. rostochiensis* early in their development and is said to be more effective in this than *Verticillium chlamydosporium* or *Paecilomyces lilacinus* (Roessner, 1987). In pots, *P. lilacinus* applied to seed potato tubers lessened the number of cysts of *G. rostochiensis* in the soil, after the plants had been grown, and increased tuber yield six-fold (Gul *et al.*, 1988). *Verticillium chlamydosporium* was found to infest young females of *G. rostochiensis* but not enough to lessen nematode population density after potatoes had been grown in the pots (Crump and Irving, 1992).

Soil solarization

Covering the soil with one or two layers of polyethylene film and allowing the sun to raise soil temperatures sufficiently to kill the nematodes can be as effective as fumigating the topsoil in a hot climate, or (in summer) in glasshouses in a temperate climate. In Oman, for example, 62 days of solarization of plots, covered with 80 gauge black polyethylene, reduced *G. rostochiensis* on potatoes by 95%, comparable with a 100% reduction with 4 kg oxamyl ha^{-1} (Mani *et al.*, 1993). Out-of-doors in temperate climates, however, lethal temperatures may only be achieved in the top few cm of the soil. In New York State, USA, solarization killed about 97% of *G. rostochiensis* eggs in the top 10 cm of the soil in 1981 but far fewer in the cooler summer of 1982 (LaMondia and Brodie, 1984). Control of potato cyst-nematodes and other nematodes in glasshouse soils might be enhanced in temperate climates, when the soil is fumigated under polyethylene sheeting, by leaving the sheets undisturbed for a period in hot weather to allow adequate solar heating of the soil to increase nematode kill in the top 10–15 cm of the soil. Soil solarization would be most appropriate for small fields and for glasshouse use. The costs of the sheeting and its installation and removal make soil solarization a rather expensive technique.

Nematicides

Many fumigant and non-fumigant nematicides have been assessed against potato cyst-nematodes. Relatively few have been effective and of these even fewer have been used in commercial practice, because of their cost, toxicology or other drawbacks.

FUMIGANT NEMATICIDES

Two halogenated, aliphatic hydrocarbons – methyl bromide and 1,3-dichloropropene (in D-D, Vidden D and Telone II) are currently available for control of potato cyst-nematodes in glasshouses and fields, respectively. Two methyl isothiocyanate precursor compounds (metham-sodium and dazomet) as well as methyl isothiocyanate mixtures are also available.

Under gas-tight polyethylene sheeting, methyl bromide at 488–1464 kg ha^{-1} has prevented increase of potato cyst-nematodes on tomato or potato (Hague and Sood, 1963; Peachey *et al.*, 1963a; James and Hague, 1967; Whitehead *et al.*, 1973b). At 122 kg ha^{-1}, the gas penetrated soil to 46 cm deep, so for control of potato cyst-nematodes on tomato, standard dosages of 488 or 977 kg ha^{-1} could be reduced, because the more important grey sterile fungus was also controlled (James and Hague, 1967).

At 156–1153 kg ha^{-1}, D-D mixture has lessened multiplication of potato cyst-nematodes on potato in some soils (Lear *et al.*, 1952; Mai and Lear, 1952; Whitehead *et al.*, 1972b, 1973c). At 504 kg ha^{-1}, D-D applied 10 cm deep at 25 cm centres decreased the number of viable nematodes by 99% in a heavily infested field. This treatment, repeated yearly, kept populations small, despite growing susceptible potatoes annually (Mai and Lear, 1952). Even greater kill was achieved by 252 kg ha^{-1}, followed after ploughing by 504 kg ha^{-1} 11 days later (Lear *et al.*, 1952). Such effective control is rarely, if ever, achieved in commercial practice, because of less efficient sealing of the soil surface, allowing more rapid escape of the gas and because of adsoption on soil organic matter in soils with large organic matter fractions. Different samples of D-D mixture were also found to differ considerably in the dosage required to kill in excess of 90% of potato cyst-nematode eggs (Fenwick, 1951). Perhaps some samples contained much more 1,3-dichloropropene, subsequently shown to be the major nematicidal constituent of D-D, than other samples. However, 1,3-dichloropropene mixture (Telone) at rates up to 448 kg ha^{-1} only lessened nematode increase on potato or had no significant effect on nematode increase (Whitehead *et al.*, 1973b,c). Kills of 98–100% of *G. pallida* were obtained with Telone at 448 kg plus dazomet at 434 kg ha^{-1} under polyethylene sheeting in a glasshouse soil (Whitehead *et al.*, 1979b). Repeated application of 1,3-dichloropropene to soil can result in poor control of *G. rostochiensis*. Sterilizing the soil then restores the efficacy of a following application of 1,3-dichloropropene. A *Pseudomonas* sp. is thought responsible for this accelerated degradation of dichloropropene (Lebbink *et al.*, 1989). Although 622 kg chloropicrin ha^{-1} applied to sandy loam in autumn prevented multiplication of *G. rostochiensis* on susceptible potatoes grown the following year (K. Evans, personal communication), it is such a powerful lachrymatory compound that its widespread commercial use seems undesirable.

Methyl isothiocyanate and compounds that release it on decomposition in the soil are generally more effective, dose for dose, when mixed with the soil, than are D-D or Telone injected into the soil (Peachey, 1962; Peachey *et al.*, 1963b). However, in a silt loam infested with *G. pallida*, Telone II (about 94% 1,3-dichloropropene) was more effective than metham-sodium, when both were injected into the soil by a Rumpstad Combiject applicator in autumn (Whitehead, unpublished data). At 150–312 kg ha^{-1}, methyl isothiocyanate lessened or prevented multiplication of potato cyst-nematodes on tomatoes in glasshouses and potatoes out-of-doors (Peachey *et al.*, 1963a, 1965; Lewis and Stone, 1964). Raschke and Pieroh (1962) found 98% reduction of viable potato cyst-nematodes in soil treated with 1500 l ha^{-1} 20% methyl isothiocyanate and that winter treatment, though it took longer, was as effective as in autumn, provided that the soil was well drained. There is, of course, less risk of the soil pores becoming blocked with moisture in the autumn than in winter, so efficient soil fumigation has a greater chance of success in autumn than in winter, except on very well drained soils. At 520 kg ha^{-1}, metham-sodium lessened or prevented increase of potato cyst-nematodes on tomato or potato (Atanasov and Collingwood, 1962). Dazomet at 165–628 kg ha^{-1} in dust or fine granule ('prill') formulation mixed in the soil has also lessened or prevented nematode increase on tomato and potato (Peachey *et al.*, 1963a,b; Jones, 1973; May and Osborne, 1973; Whitehead *et al.*, 1973 b,c,d). However, although in one organic soil dazomet lessened nematode increase on potatoes (Whitehead *et al.*, 1975b), in others it did not (J.M.Jones, 1973; Smith and Whitehead, 1975). Dazomet is usually too expensive for use on potatoes but its cost can be borne by tomatoes grown under glass. In soil infested with *G. pallida*, yields of tomatoes were greatly increased by 441 kg dazomet ha^{-1} mixed with the top 15 cm of the soil. The treatment did not, however, prevent the nematode increasing in the top 10 cm of the soil, whether a dust or a prill formulation was used and whether the soil surface was sealed with polyethylene sheeting, water or left uncovered (Whitehead *et al.*, 1975a). Combined use of dazomet with Telone or Di-Trapex CP (20% methyl isothiocyanate : 65% D-D : 15% chloropicrin) prevented nematode increase on tomatoes and inhibited 'corky root' damage by the grey sterile fungus (*Pyrenochaeta lycopersici*) (Whitehead *et al.*, 1975a, 1979b). The alternation of steam sterilization and D-D in successive years is justified, both in lower costs and in effective control of glasshouse pests and diseases of tomato, including potato cyst-nematodes (Murdoch and Murdoch, 1963).

The efficacy of a soil fumigant in controlling potato cyst-nematodes is affected by the method of application. Methyl bromide was distributed more uniformly under gas-tight sheeting, when it was fed in via a perforated pipe, than when it was distributed from a single entry port (Hague *et al.*, 1964). Applying D-D in two equal doses, one before and one after ploughing, killed more *G. rostochiensis* than a single, equivalent dose (Lear *et al.*, 1952), which kills relatively few nematodes near the soil surface. Applying dazomet in two doses, one before and one after ploughing was also more effective than a single equivalent dose applied after ploughing in a silt loam and was as effective in a peaty loam, but less effective in a sandy loam compared with a single equivalent dose applied after ploughing. In silt loam, when 213 kg dazomet ha^{-1} was rotavated into the top 15 cm of the soil and 224 kg Telone ha^{-1} was injected 25 cm deep into the soil, multiplication of the nematode on cv. King Edward potatoes was prevented. Although ridging or rolling sandy or silt loam soils after application of dazomet or Telone did not affect the number of potato cyst-nematodes killed (Whitehead *et al.*, 1973c), other research has shown that rapid escape of gas from the soil surface must be prevented, if the greatest number of nematodes is to be killed by a soil fumigant (Peters, 1955). Slits left by injection tines must be immediately filled in and the soil surface, which should be free of organic debris, rolled flat. The Rumptstad Combiject sprays liquid fumigants 20 cm deep in the soil behind 75 cm wide A-blade coulters, rotary cultivates the top 7 cm of the soil to a fine tilth and with a powered, flat roller, running faster than the forward speed of the tractor, leaves a very flat soil surface (Fig. 7.5). Whether or not there is rapid escape of gas from such an initially smooth, compacted surface will depend on

Fig. 7.5. Rumptstad Combiject for liquid fumigant application to soil (upper). Horstine Farmery type applicator on spike rotavator for incorporating granular nematicides in the top 15 cm of the soil (lower).

the nature of the soil and whether or not the soil surface dries out. If the surface dries rapidly and the soil has an appreciable clay or silt fraction, it may fissure, facilitating escape of fumigant gas. Speed of gaseous diffusion will also depend on soil temperature and, most importantly, on soil moisture. If the soil is moist but well drained, the gas can reach the cysts, but if it is at or near field capacity, diffusion will be greatly reduced and few nematodes will be killed. In glasshouses, the nematodes can be controlled just as well by fumigating the soil with 448 kg D-D ha^{-1} and then wetting the top 5–7.5 cm with 5.5 or 11 l m^{-2} of micellar solution* of chlorphenol or *p-m*-cresol (Staniland and Stone, 1953; Stone, 1957) as by fumigating the soil under polyethylene sheeting with methyl bromide. More conveniently, such a level of control has been obtained by combined use of Telone and dazomet under polyethylene sheeting, as already mentioned. Stone (1961) found that new cysts formed on tomato roots were mostly killed by trickle irrigating 0.5% *p-m*-cresol, solubilized with an alkyl sulphate-type detergent, into the soil at the end of the growing season.

Because cultivation after ploughing some clay and heavy silt loams in autumn may make seedbed preparation difficult the following spring, the soil should be fumigated in autumn as soon as there is adequate soil moisture, so that the land can be reploughed before winter, to maximize 'frost mould' production.

The condition of the cysts may also affect the percentage of eggs within them that are killed by a soil fumigant. Harrison (1956) found young (i.e. full) cysts more susceptible to D-D than cysts a year or more old, whereas Hague (1959) found that eggs in full cysts were less susceptible to fumigation than those in partially emptied cysts.

NON-FUMIGANT NEMATICIDES

Some non-fumigant nematicides have been more or less effective in controlling potato cyst-nematodes at dosages much smaller than needed for control by fumigant nematicides. Most of these are organophosphates and carbamates, which are powerful anticholinesterase inhibitors, but being non-phytotoxic can be applied at planting (Whitehead *et al.*,1973f).

Organophosphates. A number of organophosphates has been shown to control *G. rostochiensis* partially or fully, i.e. fensulphothion, thionazin, fenamiphos, disulphoton, diazinon, terbufos, ethoprophos and fosthietan (Fraser and Lindley, 1962; Dern, 1968; Whitehead *et al.*, 1972a, 1985; Homeyer, 1973; Gladkaya and Ponin, 1976; Homeyer *et al.*, 1987). Organophosphates, generally, are lipophilic and adsorbed by soil organic matter, so they require intimate mixing with the soil, if they are to be effective. They are usually ineffective in very organic soil. For example, at 11.2 or 13.5 kg ha^{-1}, ethoprophos and isazophos (CGA 12223) lessened increase of *G. rostochiensis* and mixed populations of *G. rostochiensis* and *G. pallida* in sandy and silty loams, but were ineffective against *G. rostochiensis* in very organic soil (Moss *et al.*, 1975). At about 5 kg ha^{-1}, ethoprophos, fenamiphos,and fosthietan lessened *G. rostochiensis* increase in heavily

* Relatively insoluble nematicides are 'solubilized' to 'micellar' solutions by emulsification with a detergent of alkyl sulphate type and then adding more detergent until the emulsion becomes clear.

infested sandy loam by more than 75% but were less effective in peaty loam (Whitehead *et al.*, 1985). In sandy, silty and peaty loams, ethoprophos at 5.6 kg ha^{-1} did not lessen multiplication of *G. pallida* on susceptible potatoes in plots also treated with oxamyl. At 11 kg ha^{-1}, ethoprophos lessened *G. pallida* on potatoes in silty loam but not in peaty loam (Whitehead *et al.*, 1994). At present, ethoprophos is the only organophosphate used commercially (in The Netherlands and UK) for control of potato cyst-nematodes. Most of the organophosphates effective against *G. rostochiensis* in pots were *O,O*-diethyl phosphorothioates or phosphorodithioates, whereas none of the 14 *O,O*-dimethyl phosphorothioates or phosphorodithioates was very effective, probably because the latter decompose more rapidly in the soil (Whitehead *et al.*, 1985). An earlier field experiment in a sandy clay soil had shown that six diethyl phosphorothioates (fensulphothion, disulphoton, thionazin, diazinon, chlorfenvinphos and chlorpyrifos) were more effective against *G. rostochiensis* than six dimethyl phosphates (phosphamidon, demeton-S-methyl, Schradan, Folimat, endothion and Bidrin) (Whitehead *et al.*, 1972a). Ethoprophos, fenamiphos and fosthietan belong to none of these groups of compound.

Carbamates. Field experiments in The Netherlands and in England showed that potato cyst-nematodes might be controlled by aldicarb broadcast on the soil after planting (den Ouden, 1968) or incorporated in the seedbed before planting potatoes (Whitehead and Tite, 1969). In another experiment on the same sandy clay soil, Whitehead *et al.* (1972a) showed that 2 kg aldicarb or 103 kg Isolan ha^{-1} incorporated in the seedbed before planting prevented multiplication of *G. rostochiensis* on cv. Majestic potatoes, whereas 7 kg methomyl ha^{-1} did not. For safety and efficacy, carbamate nematicides are best applied in granular form to the seedbed just before planting.

Small amounts of aldicarb (as 10% granules) were more effective in controlling *G. rostochiensis* than fumigating soil with large amounts of D-D or dazomet (Whitehead *et al.*, 1972a, 1973d). Bunt (1972) showed in pots that oxamyl prevented multiplication of potato cyst-nematodes when applied at 2.5 ppm to the soil or at 1.25 ppm to the soil plus a foliar spray of 500 ppm, the foliar spray being much less effective than soil treatment. In field experiments on sandy loam infested with *G. rostochiensis*, 4.8 kg oxamyl ha^{-1} incorporated in the seedbed prevented nematode increase. Dipping shoots of infested young potato plants in oxamyl solution (2000 ppm) did not affect nematode developement (Whitehead *et al.*, 1973a). In contrast, Harrison (1974) found spraying foliage of susceptible potatoes in infested soil with 1.1, 3.4 or 6.7 kg oxamyl ha^{-1} or 2.2 or 9.0 kg carbofuran ha^{-1} gave over 99% control. However, Brodie (1983) found five foliar applications of 1.12 kg oxamyl or carbofuran ha^{-1}, beginning at 90% emergence of potato haulms from the ridges, suppressed increase of *G. rostochiensis* as did 5.6 kg ha^{-1} in the rows at planting but combined soil and foliar treatments were, as already found by Whitehead *et al.* (1973a), no more effective than soil application alone. In a further experiment (Whitehead *et al.*, 1985), up to six sprays of up to 4 kg oxamyl ha^{-1} in each spray lessened multiplication of *G. rostochiensis* but results were variable and not as successful as those obtained with 5 kg oxamyl ha^{-1} incorporated in the seedbed before planting. At four sites in

Scotland, numbers of juvenile *G. rostochiensis* in susceptible Pentland Crown roots eight weeks after planting were decreased 80% by 3.4 kg aldicarb ha^{-1} and 92% by 5.6 kg oxamyl ha^{-1} (Trudgill *et al.*, 1978).

Application of aldicarb or oxamyl to the seed furrow during planting was less effective or ineffective compared with mixing the nematicide into the seedbed soil before planting (Whitehead *et al.*, 1975b; Duggan, 1978) because the treatment was too localized and roots growing away from the treated soil were not protected systemically from infestation (Table 7.4).

The oxime *N*-methylcarbamates aldicarb, oxamyl and Tirpate were shown effective against potato cyst-nematodes in a wide range of soils. At 11.2 kg ha^{-1}, all three compounds fully controlled potato cyst-nematodes in a silty and in a peaty loam, when incorporated in the top 15 cm of the soil before it was ridged and planted with susceptible potatoes (Whitehead *et al.*, 1973e). Similarly applied to a sandy, a silty and a peat loam soil, aldicarb or oxamyl at 5.6 kg ha^{-1} lessened or prevented nematode increase on susceptible potatoes. Less reliable results were obtained with 3.4 kg ha^{-1}, or when the nematicide was incorporated in the top 7.5 rather than the top 15 cm of the soil (Moss *et al.*, 1975, 1976). In Sweden, *G. rostochiensis* was controlled in a sandy clay soil by 3.36 kg aldicarb or 3.6–5.0 kg oxamyl ha^{-1} (Tülkikkala, 1981). In pots, only carbofuran of seven *N*-methylcarbamates of phenols was reliably effective against *G. rostochiensis*, whereas five of eight *N*-methylcarbamates of oximes (aldicarb, oxamyl, Tirpate and two undisclosed compounds, SDS 47187 and SDS 46995) were effective against *G. rostochiensis*. In the field, at about 5 kg ha^{-1}, all five compounds plus cloethocarb were effective in sandy loam but carbofuran and cloethocarb were less effective than aldicarb or oxamyl in peaty loam (Whitehead *et al.*, 1985).

Granules are not mixed uniformly with the top 15 cm of the soil by harrowing them in – typically 70% or more are left in the top 5 cm of the soil (Fig. 7.6). Thorough mixing of the granules with the upper layers of the soil is obtained by use of a rotary cultivator with spike or L-bladed tines (Fig. 7.5). Provided this is operated at slow forward tractor speed, fast rotor speed and with the rear shield lowered to break up any clods, the granules are distributed evenly to the working depth, preferably 15 cm for potato cyst-nematode control. Thorough mixing is important because success is not then dependent on redistribution of the nematicide by rainfall, which may be lacking after planting, because the lipophilic

Table 7.4. Control of *Globodera rostochiensis* on potatoes by aldicarb rotary cultivated into the top 15 cm of the seedbed or applied in the seed furrows at planting.

Treatment	kg ha^{-1}	Tubers over 3.8 cm (t ha^{-1})	*P_f/P_i
Untreated	0	47.7	51.3
Aldicarb rotary cultivated in	3.3	53.7	5.6
	6.5	48.2	1.0
Aldicarb in seed furrows	3.4	48.2	63.1
	5.6	47.8	49.0

*P_f/P_i = eggs g^{-1} soil at harvest / eggs g^{-1} soil at planting.

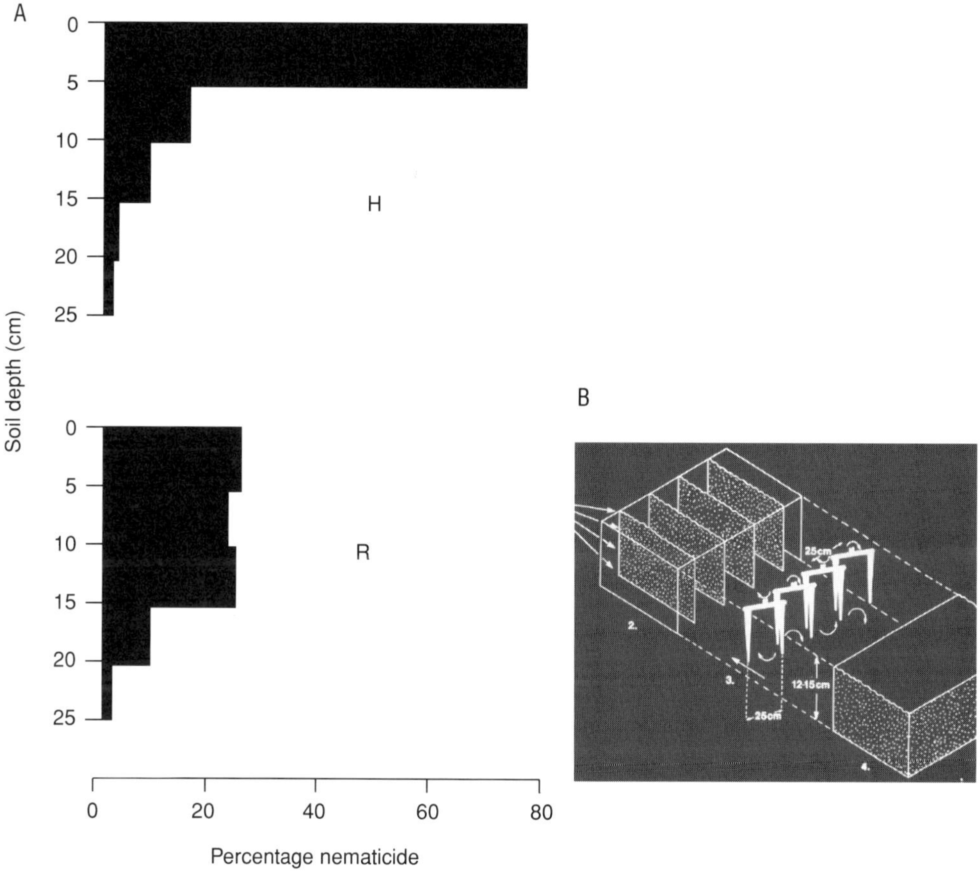

Fig. 7.6. A, depth distribution of nematicide granules by harrows (H) and by rotary cultivators (R). B, vertical band–rotary harrow granule application principle.

organophosphate nematicides are not redistributed by rainfall and because there is little or no water flux in organic soils in spring, at least in north-west Europe (Whitehead, 1988a). In some soils, notably poorly structured silty loams, rotary cultivation may result in poor growth of potato plants due to soil compaction, leading to waterlogged, anaerobic seedbeds. Faster, effective incorporation, with little or no harm to the seedbed was achieved by the Vertical Band–rotary harrowing technique (Bromilow and Lord, 1979; Whitehead *et al.*, 1981) (Figs 7.6, 7.7) and the Vertical Band–Dutch harrow technique (Whitehead, 1988b). Currently, granular nematicides are often applied to the seedbed in conjunction with clod and stone separation from the soil. Although a limited study of this technique was encouraging (Spaull and Tones, 1986a,b), distribution of the granules in the soil can be very uneven down the soil profile, resulting in poor nematode control. Application of aldicarb or oxamyl granules in front of the stone separator machine can result in better mixing of the granules with the soil than applying them over the first separator web. If the granules are distributed

(a)

(b)

Fig. 7.7. (a) Vertical band granule applicator. (b) Cutaway of a vertical band applicator tine, showing air-borne distribution of granules in soil.

through the 25–30 cm tilth usually required for bed formation after stone separation, the greater dilution, while it may not affect tuber yield, is likely to lessen control of nematode multiplication. Further study of this method of applying granular nematicides to soil is needed to establish the optimum procedures for satisfactory nematode control. Granules can also be incorporated to 24 or 33 cm deep with a rotary spader (Sprong, 1989) but this too would overdilute permitted dosages of nematicides.

In the field experiments of Moss *et al.* (1975,1976), increases of *G. rostochiensis*, *G. pallida* and mixed populations of both species were well controlled by aldicarb or oxamyl, except for one population of each species. Greater variation in control of both species on cvs Pentland Crown and Pentland Dell by 5.5 kg oxamyl ha^{-1} was noted in seven field experiments by Trudgill *et al.* (1983). At three sites in Scotland, Alphey *et al.* (1988) found aldicarb significantly lessened or prevented multiplication of *G. pallida* or *G. rostochiensis* on susceptible potatoes at both 3.36 and 1.68 kg ha^{-1}. Data in Brown and Sykes (1983) shows that aldicarb or oxamyl (presumably at 3.4 or 5.6 kg ha^{-1}, respectively) at six sites infested with *G. pallida* and eight infested with *G. rostochiensis* controlled the two species more or less equally, i.e. multiplication on various susceptible cultivars of

Table 7.5. Percentage reduction in P_f/P_i by 5.6 kg oxamyl ha^{-1} in 138 experiments and observation plot assessments.

	No. expts/ obs.	*Globodera rostochiensis*		No. expts/ obs.	*G. pallida*	
		Average	(Range)		Average	(Range)
Pentland Crown	26	85	(57–97)	3	39	(21–56)
King Edward	25	81	(52–96)	9	39	(15–69)
Desirée	16	87	(75–99)	34	46	(+198–96)
Record	8	74	(37–98)	9	37	(+184–91)
Romano	1	74	–	7	49	(+122–91)
Average		80	(52–99)		42	(+198–96)

potato decreased, on average, by 65.5% (21–81) for *G. pallida* and 71.4% (35–95) for *G. rostochiensis*. Evidence of differential control of the two species became apparent when data from many other field experiments, on sites with pure, or nearly pure, populations of *G. pallida* or *G. rostochiensis*, were analysed (Whitehead *et al.*, 1984 and Table 7.5). Although *G. pallida* was sometimes well controlled by oxamyl (Fig. 7.8), in other experiments it was not and, on average, oxamyl lessened multiplication of *G. rostochiensis* twice as much as multiplication of *G. pallida*.

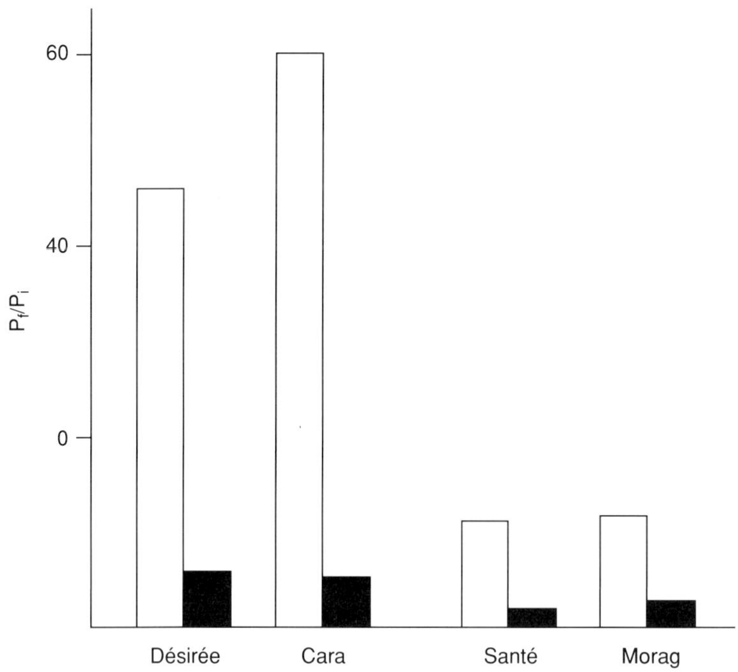

Fig. 7.8. Control of *G. pallida* increase (P_f/P_i) in two susceptible cultivars, Desirée and Cara, and in two partially resistant cultivars, Santé and Morag. Increase in soil treated with 5.6 kg oxamyl ha^{-1} shown by solid columns and increase in untreated soil shown by open columns.

In more recent field experiments in a range of soils in England, control of *G. pallida* multiplication by 5.6 kg oxamyl ha^{-1} varied greatly and supplementary treatments with formalin to kill microorganisms that might speed degradation of oxamyl in the soil, carbofuran or ethoprophos to extend control, or sodium metavanadate to speed nematode hatching were of no significant benefit (Whitehead *et al.*, 1991a,1994). No physical or chemical explanation could be found for the poor control of *G. pallida* in a number of soils. The rate of breakdown of oxamyl in the soil was as in soils in which control was satisfactory, the granules held 10% of active ingredient and they were well incorporated in the top 15 cm of the soil by rotary cultivation. Furthermore, oxamyl in pots was equally effective in soil with or without a history of oxamyl usage against *G. rostochiensis* and *G. pallida*. The explanation may lie in the generally slower hatching and longer persistence of second stage juveniles of *G. pallida* in the soil compared with *G. rostochiensis*, so that a small percentage of the former may still be hatching after oxamyl has degraded in the soil. Some juveniles of *G. pallida* that hatched earlier may still be capable of invading potato roots late in the season, when they are no longer inactivated by oxamyl. Upward migration of persistent juveniles from below the treated layer of soil may also occur mid-season, if there is adequate soil moisture, and may give rise to a second generation. However, in root observation boxes in a cool glasshouse, oxamyl applied to the top 15 cm of two peaty loam soils and one sandy loam soil, at a dosage equivalent to 6 kg oxamyl ha^{-1} in the field, largely prevented multiplication of *G.pallida* migrating upward from heavily infested soil 15–30 cm deep over more than three and one half months. Speed of hatching does, however, vary amongst *G. pallida* populations and one population, hatching as fast as *G. rostochiensis*, was equally as well controlled by oxamyl as was *G. rostochiensis* in sandy loam at Woburn, England (Whitehead, 1992b). Further, cv. Cara can hatch 80% of a heavy infestation of *G. pallida* in about six weeks in late spring or early summer in England and Evans (1983) showed that although cultivars differed little in their ability to hatch *G. rostochiensis*, they differed considerably in their ability to hatch *G. pallida*.

Faster degradation of aldicarb sulphoxide, and to a lesser extent aldoxycarb, was shown to occur in soils previously treated with aldicarb. Dosage and frequency of treatment were important factors in the reduction of the half-life of aldicarb at 15°C. from 25–50 days in previously untreated soils, to 5–17 days in previously treated soils (Suett and Jukes, 1988). Microbial breakdown of carbofuran probably explains the earlier report of poor control of *Pratylenchus scribneri* on maize in previously treated soils (Smolik, 1978). Enhanced degradation of oxamyl could perhaps explain some instances of poorer control of both species of *Globodera*, though it is more likely with *G. pallida* than with *G. rostochiensis*. Multiplication of *G. pallida* on tomatoes grown in organic sandy clay loam was prevented by incorporating 11.2 kg oxamyl ha^{-1} in the top 15 cm of the soil in both a heated and an unheated glasshouse (Whitehead *et al.*, 1975a), though tomato is a less efficient host of the nematode than potato and the dosage of oxamyl was twice the maximum permitted dose for potatoes. Further research into the reasons for site-to site-variation in control of potato cyst-nematodes might lead to improved control with non-fumigant nematicides.

Mode of action of organophosphates and *N*-methyl carbamates. Hatching of potato cyst-nematodes was inhibited or retarded by applying aldicarb at 5 ppm, but not 1 ppm, to the soil (Hague and Pain, 1970) and by diffusate from the roots of potato plants sprayed with oxamyl (Harrison, 1971). Osborne (1973) found that, except at much higher concentrations than approved for commercial use, the inhibition of hatch of potato cyst-nematodes by aldicarb was reversible, i.e. when the nematodes were transferred to plain water they regained motility. At 2 ppm in soil over a period of 8 weeks, oxamyl but not carbofuran retarded hatching for 4 weeks, whereas isazophos or fenamiphos inhibited hatching. These effects were more pronounced at 8 ppm (Hague, 1975). Storey (1984) found that at 11.2 kg ha^{-1} (*c.* 5 ppm) oxamyl delayed hatch of *G. rostochiensis* in a peaty loam soil but did not in a sandy loam soil. As oxamyl at 2.5 ppm or less, lessens or prevents multiplication of *G. rostochiensis* on susceptible potatoes in sandy loam soils, it therefore exerts its main effect after hatching. Being cholinesterase inhibitors, organophosphate and *N*-methyl carbamate nematicides in soil water around the infective juveniles inhibit or prevent their locomotion. *In vitro*, the effects of small concentrations of aldicarb were reversible, when the nematodes were transferred to water but 10 ppm was lethal (Nelmes, 1970). Aldicarb at 1 ppm prevented juveniles invading potato roots (Hague and Pain, 1970). Second stage juveniles are prevented from invading the roots of young potato plants as a result of inhibited locomotion and because roots that have absorbed aldicarb, thionazin or fenamiphos are toxic and repel the nematodes before or soon after penetration. Dosing the roots 11 days after the nematodes had invaded did not affect their developement (den Ouden, 1971).

Benzimidazoles. Hoestra (1976) found benomyl strongly inhibited hatching of potato cyst-nematodes and in pots suppressed new cyst formation on potatoes in soils with low, but not in soils with high, organic matter contents. Similar hatching results were obtained by Whitehead (1992b), who also found that benomyl at 0.1 ppm (3.4×10^{-7}M) increased hatching of two populations of *G. pallida* in potato root diffusate. Although, in pots, benomyl, carbendazim, thiabendazole, thiophanate ethyl and thiophanate methyl, all of which, except thiabendazole, have an *N*-methyl carbamate group, controlled *G. rostochiensis* more or less as effectively as oximecarbamates, in the field they were less effective, especially in peaty loam soil (Whitehead *et al.*, 1985).

Sodium azide. In Poland, Giebel (1967) found that encysted eggs of potato cyst-nematode were killed by sodium azide at 1 ppm in 10 h and Giebel *et al.* (1982) and Kornobis (1989) found 100 kg as liquid Nemazyd 20 (20% sodium azide plus stabilizers) ha^{-1} watered on reduced numbers of *G. rostochiensis* in the top 30 cm of the soil by 80–90%. Sodium azide is non-persistent, was not detected in ground water and no azide ions were found in potatoes. It is, however, a powerful, hypotensive drug, so its commercial use is questionable.

Isothiocyanates. Apart from methyl isothiocyanate, already discussed in the section on soil fumigants, a number of other isothiocyanates has been assessed as potential nematicides. A 0.001% solution (10 ppm) of phenyl isothiocyanate killed cysts after 24 h and mixed in soil at a rate equivalent to 251 kg ha^{-1} completely controlled potato cyst-nematode (Smedley, 1939). Ellenby (1945a) found that allyl isothiocyanate (the mustard oil of black mustard seed) totally inhibited hatching of potato cyst-nematodes at 500 ppm and permanently decreased it at 50 ppm. In field trials (Ellenby, 1951), allyl isothiocyanate soaked on peat and applied to the potato drills increased tuber yields in infested soil, without increasing the number of cysts in the soil. At about 12.5 kg ha^{-1} applied in granular peat in the drills, allyl isothiocyanate doubled tuber yield in infested soil (Ellenby, 1945b). In pots, allyl isothiocyanate, at a dosage equivalent to 35 kg ha^{-1} in the top 15 cm of the soil, lessened numbers of *G. rostochiensis* females on potato roots but was phytotoxic above 70 kg ha^{-1}. At about 14 kg ha^{-1}, ethyl and 4-chlorophenyl isothiocyanates were similarly effective and not phytotoxic. At 700 kg ha^{-1} (equivalent dosage in pots), the percentages of encysted eggs killed were 99% (4-chlorophenyl-), 98% (4-methylphenyl- and 2-methylphenyl-), 97% (4-fluorophenyl-) and 94% (2-ethylpyridyl-). At 140 kg ha^{-1}, allyl isothiocyanate killed 95% (Whitehead *et al.*, 1985).

Hypochlorites. Cysts may also be destroyed if they are immersed in hypochlorite solutions (Smedley, 1936). More recently, immersing tubers contaminated with cysts in sodium hypochlorite (1% available chlorine) completely destroyed the cysts in 24 h (Wood and Foot, 1975) and in 30–45 min (Wood and Foot, 1977), though destruction of cysts was temperature dependent but not dependent on pH above 7.0. To prevent damage to the tubers, the skins should be mature and thoroughly washed after treatment. In Northern Ireland, however, dipping tubers in the same concentration of sodium hypochlorite did not kill all cyst contents (Anon., 1978). Sodium hypochlorite with 1% available chlorine was more effective than calcium hypochlorite, but disintegration of the cysts was not hastened by using a stronger solution (Sarakoski, 1977). However, a solution with 9% available chlorine dissolved cysts in 30 min, without harming the tubers (Manoharan *et al.*, 1978).

Picrolonic acid. Numbers of *G. rostochiensis* were decreased by 84–88% following treatment of the soil with 94.5 kg ha^{-1} of the artificial hatching agent, picrolonic acid, and growing timothy grass, or *Festuca pratensis*, compared with 57–60% reduction by growing the grasses alone (Timofeev and Ponin, 1973). Picrolonic acid at up to 34.4 kg ha^{-1} in a sandy clay soil did not increase hatch of *G. rostochiensis* (Whitehead, 1977b). Another artificial hatching agent, sodium metavanadate, applied to bare sandy loam at 36.7 kg ha^{-1} and rotary cultivated into the top 15 cm of the soil stimulated hatching of *G. rostochiensis* from mid-March to mid-September (Whitehead, 1992b).The same amount applied to plots infested with *G. pallida* and also treated with 5.6 kg oxamyl ha^{-1} did not improve control by oxamyl. It seems likely that most of the compound was absorbed by the roots of the developing potato crop before it could significantly increase early hatch (Whitehead *et al.*, 1994). The use of this hatching agent would therefore seem limited to fields that can be bare fallowed for six months. It is doubtful if the cost of the chemical plus the cost of maintaining land in bare fallow for this period would be economic.

Integrated control. Preventing or delaying infestation of the soil by potato cyst-nematodes is highly desirable. In the USA, expensive quarantine of all possible means of introducing the nematodes from widespread infestations on Long Island, New York and elsewhere, together with the detection and quarantine of foci of infestation on the mainland, have spared the mainland potato crop the ravages of these pests (Mai, 1977). Although *G. rostochiensis* is widespread in south and central Sweden, *G. pallida* was known in only one locality. Strict quarantine measures were proposed to prevent spread of *G. pallida*, or at least retard its spread until cultivars resistant to it were available (Andersson, 1975). However, in a survey of nearly 6000 potato fields in Sweden from 1979 to 1984, 6% of the 44–66% with potato cyst-nematodes were infested with *G. pallida* (Olsson, 1985). For many countries, internal quarantine measures are difficult and expensive to implement but keeping out species or pathotyopes not already in the country is well worthwhile.

Existing infestations are best controlled by a combination of fully or partially effective control measures, i.e. by integrated control. Repeated use of a single control measure may select resistant or virulent nematodes from a field population so, in the long term, successful control is more likely from an integrated control programme. Integrated control is not a new concept. For example, Spears *et al.* (1956) reported that in a field, in which potatoes had been grown in 1944, 26% of the cysts recovered from the soil, before it was fumigated with D-D in 1946, were viable, but only 7% were viable after fumigation. After a total of 11 years rest from potatoes, following fumigation, no viable cysts were detected. More recently, Gus'kova and Makovskaya (1982) also found a combination of D-D at 600 l ha^{-1} plus three years of non-host crops fully effective in controlling *G. rostochiensis*. Soil fumigation, where effective, and cultivation of non-host crops was the only integrated control available until the mid-1960s, when resistant cultivars and more widely effective non-fumigant nematicides became available. Stelter and Semmeler (1967) found that growing one resistant potato crop in a six or seven year rotation decreased the nematodes (presumably *G. rostochiensis*) to barely detectable numbers. Stelter (1970) found *G. rostochiensis* (=Ro1) declined at rates varying from 40 to 45% in the first year after potatoes to 20 to 25% in the fifth year and that susceptible potatoes grown every sixth year increased numbers of the nematodes, whereas combined use of resistant potatoes and non-host crops decreased them dramatically. In a three year rotation of lupin, winter wheat and resistant potatoes, numbers of *G. rostochiensis* in the soil were decreased 90–92% (as predicted in the model of Jones, 1970a) (Timofeev and Ponin, 1975). Finally, in Finland, resistant (H_1) and susceptible potatoes alternating in three-, four- or five-course rotations decreased *G. rostochiensis* below the economic threshold or even below detection level (Tülkikkala, 1991). Successful control by a resistant potato depends on the absence of species or pathotypes to which it is susceptible. In fields around St Petersberg, Kaliningrad, Novgorod and Pskov, Russia, resistant potatoes (with gene H_1) work well because the nematode populations there lack aggressive pathotypes (Gus'kova and Makovskaya, 1982).

Although repeated cultivation of potatoes fully resistant to a species or pathotype of the nematode may rapidly decrease it, or prevent small populations increasing (e.g. Evans and Brodie, 1980), such a practice is undesirable in terms of crop health and tuber quality and may fail through selection of species or virulent pathotypes. For example, when potatoes resistant to *G. rostochiensis* were grown continuously for six years the nematodes were decreased 94% in the first year but increased in the fifth and sixth years by selection of aggressive biotype B (probably Pa1), thereby negating the earlier population reductions (Kiel, 1969).

Combined use of resistant cultivars and nematicides can be advantageous. Control of *G. rostochiensis* was just as effective from the combined use of a small dose of thiazone (dazomet) and the resistant cv. Sagitta as from the application of 1000 kg thiazone ha^{-1} to the soil (Gus'kova, 1967). In heavily infested soil, yields of resistant potatoes were small but, where the soil was fumigated with thiazone, *G. rostochiensis* was decreased by 83–84% and the yields of resistant potatoes were greatly increased (Gus'kova *et al.*, 1969). Based on the assumptions that a resistant cultivar of potato decreased the nematode population by 80%, that a non-host crop decreased it by 33% and that a nematicide killed 70 or 95% of the nematodes, Jones (1970a) showed how the nematodes were decreased by up to 99.9% by these control measures used singly or combined over a number of years. Efremenko and Klimakova (1971) showed that *G. rostochiensis* could be fully controlled in Lithuania by growing resistant potatoes in the first and fourth years of a four-course rotation, coupled with the use of a nematicide in the first year, if there were more than 250 viable cysts [? = full cysts] m^{-2} (equivalent to 150 l soil to 15 cm deep), i.e. more than a trace infestation. The resistant cv. Sagitta decreased the percentage of viable cysts in the soil from 37–38% to 0–1.1%. Irrespective of the initial infestation of the soil, *G. rostochiensis* was also fully controlled by 270 kg tiazon (dazomet) ha^{-1} applied to the soil in autumn, followed by the resistant potato 61–8/1 grown in the next two years (Gladkaya, 1973). Combined use of 270 kg thiazone (dazomet) or 80 kg heterophos ha^{-1}, resistant potato cultivars and non-host crops 'eradicated' *G. rostochiensis* in two to three years (Ponin *et al.*, 1978)

True eradication of potato cyst-nematodes is not practical or even possible in commercial practice, but full control of *G. rostochiensis* and perhaps *G. pallida* is feasible. Nollen and Mulder (1970) showed how, in the starch potato area of north-east Netherlands, *G. rostochiensis* could be controlled fully in a two-course rotation by a combination of autumn soil fumigation with D-D, coupled with growing resistant and susceptible potatoes alternately in successive rotations and non-host crops in years between potato crops. In a long term three-course rotation experiment with potatoes, sugarbeet and barley, *G. rostochiensis* was fully controlled by growing resistant Maris Piper potatoes or by fumigating the soil in autumn with 336 (or more) kg dazomet ha^{-1} or treating it with 5.6 kg oxamyl ha^{-1} in spring before planting susceptible Pentland Crown potatoes. The nematicides greatly increased the yields of both cultivars in moderately to fairly heavily infested soil (Whitehead *et al.*, 1980a). In two-, four-, six-, seven- or eight-course rotations, oxamyl greatly decreased or prevented multiplication of *G. rostochiensis* and greatly increased yields of cvs Desirée and Maris Piper in heavily infested soil. Infestations in the two and four-course rotations, in which oxamyl was applied to the soil before planting potatoes, stayed at the levels found in six-, seven- or eight-course rotations (Whitehead *et al.*, 1991b). In short rotations with susceptible potatoes, the residual benefit of population control by nematicide may or may not be large enough to permit restriction of nematicide usage to alternate rotations (Whitehead and Nichols, 1992a). Partial disruption of a compacted layer in the topsoil may increase tuber yield response to treating the seedbed in spring with oxamyl (Whitehead and Nichols, 1992b).

Jones (1970b) quantified integrated control in the general equation:

$$P_f = P_i\,(F_O)^a\,(F_R)^b\,(F_N)^c\,......\,(F_x)^y \qquad (7.2)$$

where P_f = nematode population achieved by integrated control, P_i = initial population, F_O = fraction of population left after one non-host crop, F_R = fraction left after one resistant cultivar grown, F_N = fraction left after nematicide used once and a,b,c,y = number of times used. This equation assumes the individual measures operate independently of each other. As a simple example:

$$P_f = 100\,(0.67)^4(0.25)\,(0.2) = 1.0 \qquad (7.3)$$

i.e. a combination of four years of non-host crops, one resistant potato crop and use of a soil fumigant (0.2). If after then growing one susceptible potato crop P_f/P_i is less than 100, full control of the pest has been achieved. While such integrated control is quite adequate for the control of *G. rostochiensis*, it is not appropriate to the control of *G. pallida*, because the nematode may decline more slowly in the soil between potato crops than *G. rostochiensis*, there are no cultivars fully resistant to *G. pallida* and non-fumigant nematicides may not control it as effectively as *G. rostochiensis*. If a trap crop of cv. Cara potatoes decreases *G. pallida* by 80%, a soil fumigant decreases it also by 80% and the nematode declines at 25% per annum, the equation becomes:

$$P_f = 100\,(0.75)^4\,(0.2)\,(0.2) = 1.3 \qquad (7.4)$$

In organic soils, however, soil fumigation is ineffective and, as partially resistant potatoes increase populations, it may be necessary to increase the length of the rotation, i.e.

$$P_f = 100\,(0.75)^9\,\,(0.2) = 1.5 \qquad (7.5)$$

A nine year break between potato crops is, however, unacceptable to most potato growers. Although long rotations may lessen the need for, or the amount of, nematicide required to prevent injury to potatoes, the economics of long intervals between potato crops may be less attractive than those of shorter rotations with conventional nematicide use (Wijnands *et al.*, 1992).

Against *G. pallida*, Gurr (1987, 1992) found that in heavily infested soils multiplication was lessened more by cultivars partially resistant to it than by growing susceptible potatoes with 3.36 kg aldicarb ha^{-1}. Resistant cultivars control multiplication less when P_i is small (Jones, 1970b; Philipps, 1984; Whitehead, 1993). Because most potatoes partially resistant to *G. pallida* are intolerant of attack, they must be grown, in significantly infested soil, with a nematicide. In numerous experiments in England, although damage to susceptible Desirée potatoes by *G. pallida* was usually prevented when the seedbed was treated with 5.6 kg oxamyl ha^{-1}, multiplication of the nematode was lessened at some sites and not at others. Generally, partially resistant potatoes and oxamyl together controlled the nematode better and increased tuber yields more than did either alone. Results suggested that trap cropping and/or soil fumigation in addition to non-fumigant nematicide use could greatly improve control of *G. pallida* in conventional rotations, especially if partially resistant potatoes were also employed (Whitehead *et al.*, 1991a,1994) (Fig. 7.9). Recent observations in commercial fields showed that *G. pallida* multiplication could be satisfactorily controlled in silt loam soils by combined use of Telone II soil fumigation in autumn and aldicarb or ethoprophos applied to the seedbed in spring, not from either alone (Whitehead, unpublished data).

Heterodera Schmidt, 1871

Species of *Heterodera* have lemon-shaped cysts, as distinct from species of *Globodera*, which have orange or pear-shaped cysts (Fig. 7.1). Of the numerous *Heterodera* spp. that have been described, nine damage crop plants to an extent that warrants their consideration here. Several of these have long been recognized as serious crop pests, others have only reccntly been described and so their control needs further study. Three species, *H. avenae* (cereal cyst-nematode), *H. glycines* (soybean cyst-nematode) and *H. schachtii* (sugarbeet cyst-nematode) are particularly important pests and, not surprisingly, their biology and control have been given most attention. *Heterodera* spp. are sedentary endoparasites of roots.

Heterodera avenae Wollenweber, 1924 (Fig. 7.10)

The cereal cyst-nematode parasitizes cereals and grasses and is of worldwide distribution (7.11). The brown to black cysts may contain 200 or more eggs, which must pass a period of two or more months in a dormant state before they are able to hatch. Hatching can occur at 4–25°C, the optimum being around 10°C. Although spontaneous hatching occurs, it may be increased by the exudates of graminaceous roots. The hatched juveniles are second stage and invade roots just behind their growing points. They settle in the cortex with the head end close to the endodermis, where they feed, stimulating the formation of multinucleate syncytia ('transfer' or 'giant' cells). As occurs in cyst-nematodes generally, these juveniles moult three times to become swollen females or vermiform males. The life cycle takes some 9–14 weeks in a spring-sown cereal, longer in an autumn-sown cereal, when temperatures in winter may be low enough to delay development. There is only one generation per crop, irrespective of climate. Eggs of *H. avenae* in dry climates are adapted to withstand drought, whereas those in cool temperate soils are not.

Yield loss in cereals varies considerably, according to soil infestation level, the cereal grown and environmental factors. Andersen (1960) showed that oat yields decreased over a range of one to 20 juveniles g^{-1} soil by 21 to 85%, whereas spring barley losses over the same range of infestations increased from 16 to 55%. Damage thresholds differ between differing climatic regions. In the Mallee of Victoria, Australia, seedling dry weight and wheat grain yield were

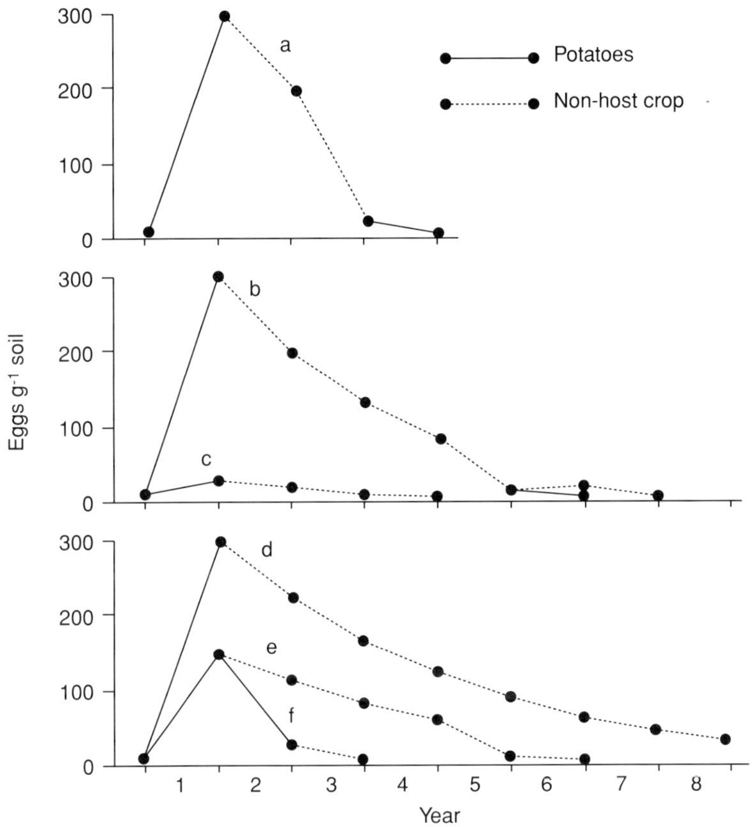

Fig. 7.9. Integrated control programmes for *Globodera* spp. on potatoes: a–c for *G. rostochiensis* and d–f for *G. pallida*. a, Susceptible potatoes in year 1, soil fumigation in year 3 and resistant potatoes in year 4; b, susceptible potatoes year 1, non-host crops years 2–7, or years 2–5 plus resistant potatoes or soil fumigation in year 6; c, susceptible potatoes grown with effective nematicide (aldicarb/oxamyl) year 1, non-host crops years 2–4; d, susceptible potatoes year 1, non-host crops years 2–7, may not give full control; e, control improved by use of aldicarb or oxamyl in year 1; f, as e but soil fumigation or use of a Cara trap crop in year 2 and a non-host crop in year 3 may result in rapid control of *G. pallida*.

negatively correlated with the severity of damage caused by the nematode to the roots ($r = -0.79$ and -0.66, respectively) (Boer *et al.*, 1991). Damage to the seminal roots is especially important, but the effects of this damage on grain yield is mitigated to some extent by the subsequent formation of prop roots. Yield loss due to *H. avenae* differs between cereals. It is greatest in spring oats, followed by spring wheat, spring barley and least in rye. Winter cereals suffer less injury than spring cereals. Damage may also be affected by the time of sowing, being greater in June or July sowings than in April or May sowings in Australia (Brown, 1982). In pots, damage to wheat was increased in the presence of *Rhizoctonia solani* (Meagher *et al.*, 1978), so the damage threshold population of *H. avenae* was smaller. In general, damage to cereals is increased when the spring is cool and

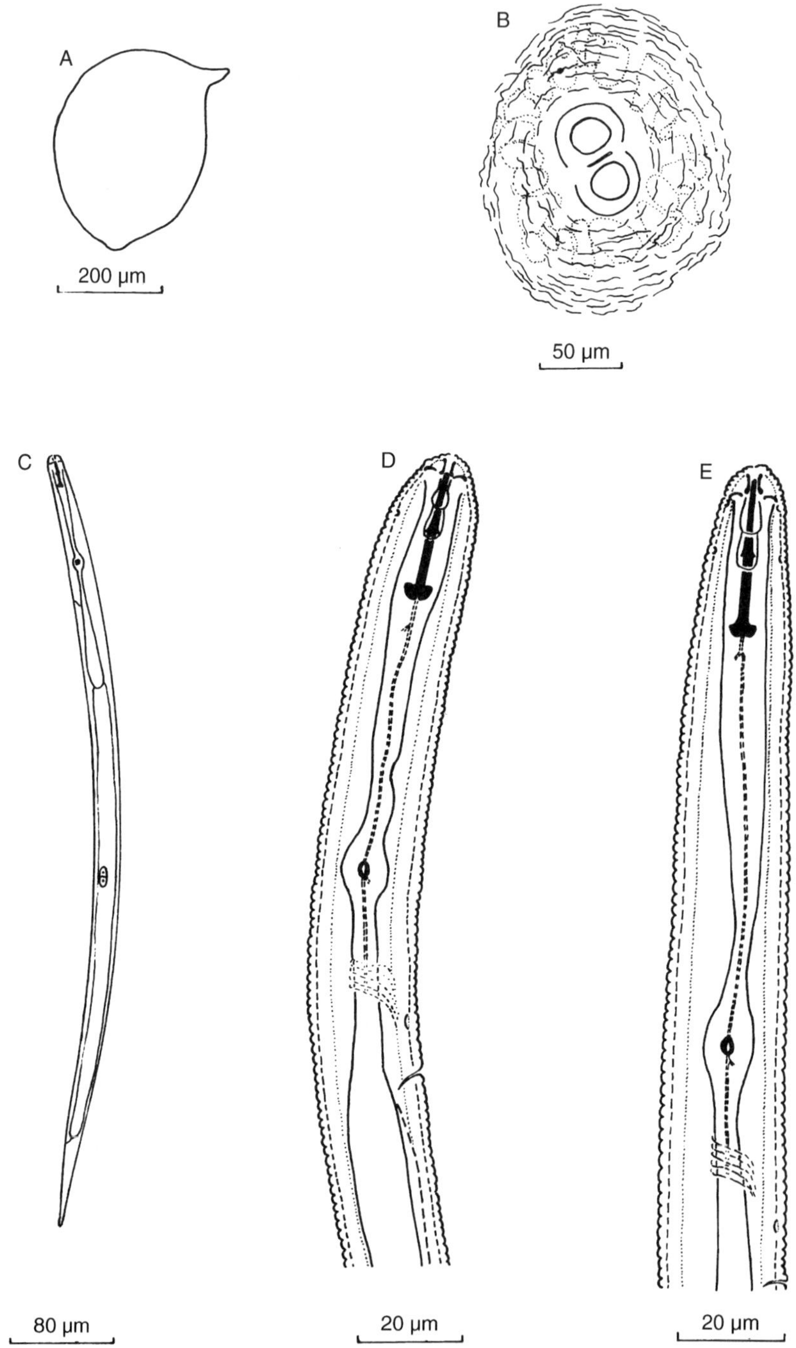

Fig. 7.10. *Heterodera avenae*: A, mature female; B, cyst cone; C, second stage juvenile; D, anterior end of second stage juvenile; E, male, anterior end.

wet and when this is followed by a hot, dry summer. Multiplication of *H. avenae* is greatest on oat, followed in descending order by spring wheat, winter wheat, spring barley and winter rye (Ehrenpfordt *et al.*, 1985). In the UK, spring oats are the most susceptible cereals, followed by autumn oats, and spring sown cereals are more heavily parasitized than autumn sown cereals (Kerry *et al.*, 1982a).

Control

Cereal cyst-nematode may be controlled, more or less, by tillage, crop rotation, use of resistant cereals, antagonistic organisms and by nematicides.

Tillage effects

Damage to wheat is affected by the tillage regime adopted. At two sites in Australia, tillage increased injury due to *H. avenae* (Neate, 1988). Reduced tillage in South Australia reduced nematode infestation but increased take-all (*Gaumannomyces graminis*) and *Rhizoctonia solani* (Roget, 1988). In Victoria, damage to wheat by cereal cyst-nematode and take-all was reduced by direct drilling and weed control, but in north-east Victoria, this increased *Pseudocercosporella herpotrichoides* and *Pyrenospora tritici-repentis*. Soil cultivation before sowing lessened root-rot due to *Rhizoctonia solani* (Boer and Kollmorgen, 1988). Tillage effects that lessen cereal cyst-nematode may therefore exacerbate fungal pathogens of cereals, so they may be counterproductive.

Crop rotation

Rotation of cereals with non-graminaceous crops prevents the developement of heavy infestations, whereas in cereal monocultures populations may increase greatly in a few years. In Bavaria, growing maize (resistant to *H. avenae*) and other cereals (susceptible to *H. avenae*) in alternate years decreased soil infestations (Bachthaler and Hien, 1976). In Poland, rotations with 60% cereals usually leave less than one egg g^{-1} soil, whereas, in rotations with more than 66% cereals, populations increase. However, if rye and maize are included in such a rotation, nematode population growth is restricted, because both these cereals are poor host plants (Glaba and Kus, 1990). Whereas wheat monocultures gave peak populations of 30 and 44 eggs g^{-1} soil, three years of vetch reduced populations by 98% to 0.5 eggs g^{-1} soil. In the rotations, yield loss was less than 10% but reached 50–60% in wheat monocultures (Romero *et al.*, 1991). When economics dictate that cereals be grown in monocultures, the grower is dependent on the use of resistant cultivars, nematicides or natural biological control.

Resistance

Resistance to *H. avenae* has been bred into a number of wheat, barley and oat cultivars. Table 7.6 lists some of these. Their use is complicated by the existence of at least 15 races of the nematode and by the superior agronomic features of more numerous susceptible cultivars. Cereals resistant to *H. avenae* have low cytokinin contents and cereals susceptible to it have high cytokinin contents. Low cytokinin and high auxin contents favour male nematode development, high

(a)

Fig. 7.11. (a) Poor growth of oats attacked by cereal cyst-nematode, *Heterodera avenae* (especially in plots on left) (Photo: J. Beane). (b) Wheat roots without (left) and with *H. avenae* (Photo: T.D. Williams).

cytokinin favours female development (Lung, 1987). Resistance in oats is based on that in the Swedish cultivar Nelson and that in line 1376 (*Avena sterilis*). In Nelson, resistance is controlled by one major dominant gene and one other dominant gene, that in 1376 by three dominant genes (Clamot, 1985). Resistance in barley is based on genes Ha1 (in cv. Drost), Ha2 (in KVL 191), Ha3 (in cv.

(b)

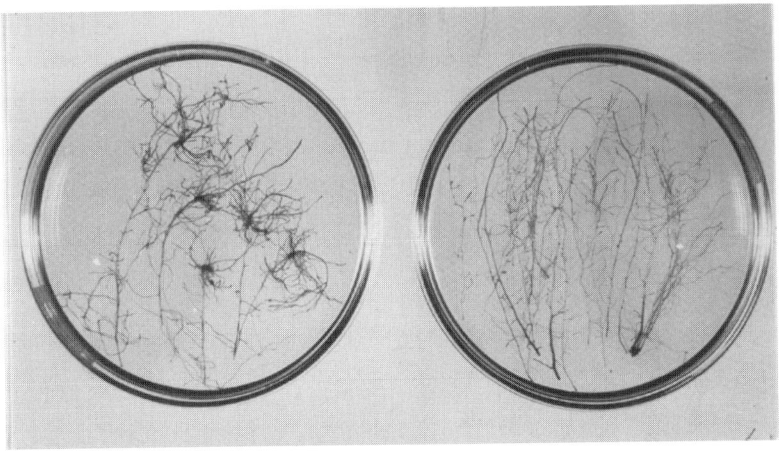

Fig. 7.11. *Continued*

Morocco) and one other gene, Has (in cv. Sabarlis). Resistance in cv. Loros is due to a single dominant gene. The genetics of resistance to *H. avenae*, however, is incompletely understood.

Resistance operates, as in cyst-nematodes generally, by the degeneration of the syncytia (e.g. Grymaszewska and Golinowski, 1987). As a result, few, if any, egg-laying females develop, though many males may emerge from resistant cereal roots. The soil infestation level declines in consequence. The presence of small numbers of mature females on resistant cereal roots does not necessarily indicate a virulent pathotype of *H. avenae* has been selected, for O'Brien and Fisher (1978) showed that at low P_i, on single, inoculated roots per seedling, as

Table 7.6. Cereals resistant to *Heterodera avenae* (resistance gene and races to which resistant, when known).

Cereal	Cultivar
Wheat (*Triticum aestivum*)	Katyl (tolerant), RE 607 (tolerant), Molineux, Festiguay, Dirk, Loros, AUS 10894, ex Loros (Aus 11577) (races Fr1–4), AUS 10894 (races Fr1–4), V640/74 (races 1,2 and Vaxtorp), Red River, Adler, ws-1812.
(*T. durum*)	Russe, Dwarf 457.
Barley (*Hordeum vulgare*)	Hulda (Ha2, several races), Morocco (Ha3), CY 3902 (Ha3), Siri (Ha2), Drost (Ha1, race 11), Bendicte (Ha2), Kara (Ha2, races 1&2), Nemex (Ha2), Sabarlis (Has), Bajo-Aragon (races 1,2 and Vaxtorp), Zita (race 12), P 31.3221 (races Fr2–4), Welana (races 1&2), Prisca (races 1&2), K. 6808, Dalmatische, Rika, P 31322, Martin, DL nos. 349, 350, 375, 376, 379, Rajkiran, Athenai, c 18147, Marocaine 079, Nile, Stange, Regatta, Decor.
Oat (*Avena fatua*)	Hedvig, Selma, Tam 0–301, Tam 0–312, Panema, Trafalgar, Krupnozernyi 67, Keeper, Rollo.
x *Triticosecale*	Tahara.

many females developed on resistant wheats cvs Loros (AUS 90248) and AUS 10894 as on susceptible Halberd wheat. The resistant reaction was induced by subsequent inoculation of juveniles, but it did not spread to other roots of the same seedling. The resistant reaction therefore appears to be localized and initiated when P_i is fairly large, not when it is small. The effect of root invasion on root growth varies according to the susceptibility and tolerance of the genotype. In intolerant, resistant wheat lines, root elongation is strongly inhibited in proportion to the number of juveniles in the roots, but in tolerant lines it is less affected and in susceptible lines numerous new rootlets form at infection points (Virville and Person-Dedryver, 1989). Similarly in oats, root tip growth ceased when resistant roots were invaded by *H. avenae*, in contrast to susceptible roots, in which invasion had less effect on root tip growth (Lung and Moltmann, 1984). However, in both susceptible Sun II oats and in near-isogenic resistant Sun II × *Avena sterilis* oats, root extension and lateral root production were inhibited by *H. avenae* invasion. Fewer juveniles (J2) were found in resistant than in susceptible roots (Rawsthorne and Hague, 1986). The greater the infestation of roots, the greater the percentage of juveniles that become males (e.g. in barley (Gotke and Swarup, 1985a)). In barley, the percentage of juveniles invading resistant and susceptible seedlings differed at 100 or 1000 per pot but did not differ at 10,000 per pot (Gotke and Swarup, 1985b).

Biological control

In cereals grown intensively, *H. avenae* may be controlled naturally by predatory fungi, especially *Verticillium chlamydosporium* Goddard and an *Entomophthora*-like fungus (*Nematophthora gynophila* Kerry and Crump) (Kerry, 1978). *Nematophthora gynophila* destroys females (Kerry and Crump, 1980) and *V. chlamydosporium* destroys females and eggs. In wheat monoculture, populations of cereal cyst-nematode declined from 30 or 44 eggs g^{-1} soil to 4.5 g^{-1} soil after five consecutive crops. Drenching soil, heavily infested with *H. avenae*, with formalin, increased nematode numbers five-fold on a subsequent cereal crop by destroying soil fungi, especially *V. chlamydosporium* and *N. gynophila* (Kerry *et al.*, 1982b). These authors estimated that 60% of females were killed by the two fungi in moist soil. The nematode multiplied less and more females were parasitized by *V. chlamydosporium* in moist than in dry soils in pots (Graham, 1980). In fields with soils prone to drying out rapidly, e.g. thin soils over chalk, or in droughty seasons, little control may be exerted by such fungi. In Victoria and South Australia, fungal parasitism of *H. avenae* was rare, occurring only in the small percentage of moister fields that were surveyed (Stirling and Kerry, 1983). In India, multiplication of *H. avenae* on barley was lessened by *Fusarium moniliforme* and *Helminthosporium gramineum* (Gill and Swarup, 1977). Increasing the moisture status of field soils, where feasible, should enhance fungal control of *H. avenae*. However, the deliberate use of these fungi as biological control agents is not yet practical, as they are subject to antagonism from other soil organisms.

Nematicides

Treating all the topsoil with fumigant or non-fumigant nematicides is too expensive for cereal crops but placing small amounts of nematicide in the cereal rows at sowing may be economically justified. Brown (1973) found that 2.2 kg aldicarb ha^{-1} in wheat drill rows significantly increased grain yield and controlled the nematode well, so that no white females could be found on the roots. Gurner *et al.* (1980) described a low-volume applicator, developed in Australia, for liquid soil fumigants, which, using capillary tubing of 1 mm internal diameter, placed as little as 2.5 l EDB ha^{-1} in the seed furrows during sowing. At 7.5 kg ha^{-1} applied in this way, EDB controlled *H. avenae* and gave the best benefit : cost ratio in wheat, the increased grain yield from 2 ha more than repaying the cost of chemical and applicator. Similarly, Brown (1981) in Australia reported wheat yields increased by 0.5–1.0 t ha^{-1} and the number of cysts in the soil decreased by the use of 1.9–11 l EDB or 3.7 or 7.4 l DBCP ha^{-1}. Brown *et al.* (1982) reported field experiments in Victoria, Australia (1974–1979), which showed significant reduction in number of *H. avenae* cysts on wheat and increased grain yields from 6 and 12 l DBCP ha^{-1}, while best yield and nematode control from an experiment that included EDB were given by 7.4 l EDB ha^{-1}. In a more recent publication, however, the number of female *H. avenae* developing on wheat roots was not decreased by EDB, in contrast to their decreased number on resistant wheat roots (Georg *et al.*, 1989). Clearly soil fumigation may fail if soil conditions do not favour rapid diffusion of the gas through the soil. Nevertheless, nematicides which reduced numbers of *H. avenae* in the soil, increased wheat yields in Victoria and South Australia (Brown, 1983a,b,c; Georg *et al.*, 1983). At Woburn, England, there was no yield benefit to susceptible Maris Tabard oats, in sandy loam infested with 7 *H. avenae* eggs g^{-1} soil, from 2.4, 4.6 or 9.4 l of a mixture of 80% EDB : 20% dichloromethane, nor from 3.7, 4.7 or 9.9 l Telone II ha^{-1} applied to the seed furrows during spring sowing. The EDB–dichloromethane mixture was clearly phytotoxic and Telone II may also have been. The dichloromethane was used to lower the freezing point of EDB from 9°C to prevent it solidifying in the capillary tubes (Whitehead *et al.*, unpublished data).

Granular, non-fumigant nematicides can be applied to cereal seedbeds at sowing time with less risk of phytotoxicity than from fumigant nematicides. Grain yield responses to treatment will depend on soil infestation, the moisture and nutrient status of the soil and the cereal grown. Gosse (1975) obtained much larger yield responses to aldicarb or methyl bromide from oats than from barley and larger responses at 20 than at 4 eggs g^{-1} soil. The granules can be broadcast onto or into the seedbed or applied in or around the seed furrows during sowing. The active compounds may also be applied directly to the seeds. At 4 or 5 kg ha^{-1}, aldicarb spread through the seedbed has given excellent control of *H. avenae*, allowing few, if any, females to develop in the roots of oats or wheat (Steudel and Rumpenhorst, 1978; Williams and Beane, 1980; Brown and Pye, 1981). In South Australia, yield of wheat cv. Condor was not increased by aldicarb (or DBCP) in the absence of the nematode, whereas at seven infested sites, yield increases of up to 0.9 t ha^{-1} were obtained and 64% of the yield response to aldicarb was shown by covariate analysis to be explainable by the decrease in numbers of *H. avenae* (Rovira *et al.*, 1981). In infested sandy loams at Woburn, England, oxamyl applied to, or under, the rows at sowing increased yields of spring wheat and spring oats, but aldicarb, similarly applied, did not

Table 7.7. Effect of granular nematicide placement on yield of three cereals (t ha^{-1}) in soils infested with *Heterodera avenae*, Woburn, England.

	Spring barley (1974)		Spring wheat (1975)		Spring oats (1981)	
P_i (eggs/g soil):		32		18		7
Untreated		4.8		2.2		3.7
Aldicarb	0.7 kg ha^{-1}	4.9	Oxamyl 1.1 kg ha^{-1}	2.6	Oxamyl 0.7 kg ha^{-1}	4.0
	1.5 "	4.8	2.2 "	2.3	1.4 "	5.4**
(s.s.)	3.0 "	4.8	(s.s.) 2.4 "	3.4**	2.7 "	5.5***

,* greater than untreated at $P<0.01$, 0.001, respectively; (s.s.) granules applied 5 cm deep in seedbed soil.

increase the yield of spring barley (Table 7.7). Despite the increased yields, there were fewer *H. avenae* after harvest than before sowing.

Broadcast applications of 4 or 5 kg aldicarb ha^{-1} are uneconomic for cereals but seed dressings and some seed furrow treatments can be. In Rajasthan, India, drilling nematicides with the seed was more effective than broadcast treatments, 1.5 kg aldicarb ha^{-1} being the most economic of the row treatments (Mathur *et al.*, 1986). As in Table 7.7, row application of a granular nematicide does not guarantee increased grain yield in infested soil, for the cereal grown and the soil environment are important factors. Spring wheat responded to oxamyl applied 5 cm deep (s.s.) in the seedbed but not to a similar amount applied in the seed furrows, which may have been phytotoxic. Spring oats, on the other hand, responded well to oxamyl in the seed furrows in lightly infested soil.

Non-fumigant nematicides are most economic when applied as seed dressings. Kaushal and Seshadri (1986) found that the number of *H. avenae* cysts formed on wheat roots was significantly decreased by dressing seed with 2.5 or 5% of its weight of aldicarb, aldicarb sulphone or carbofuran, while invasion of the roots by juveniles was significantly decreased by 1–5% seed dressings of isofenphos. Seed dressings of oxamyl have also been beneficial in Australia (Brown, 1984). In soil heavily infested with *H. avenae* and *Pratylenchus mediterraneus*, coating seeds of wheat with 10 g furathiocarb or carbofuran or 3.6 g oxamyl kg^{-1} seed decreased *H. avenae* in the roots by 80–90% two months after germination and increased grain yields by 38–48% (Orion and Shlevin, 1989). The health risks associated with cereal seed dressed with these compounds need to be fully appreciated and may restrict their use to fully automated loading and drilling machinery.

Heterodera cajani Koshy, 1967

The pigeon pea cyst-nematode, *H. cajani*, occurs in several States in India and has been found on cowpea in Egypt. The nematode has stunted pigeon pea and black gram plants in pots, according to the number of nematodes in the soil. In the field, a tolerance level of 2.6 eggs or juveniles ml^{-1} of soil has been found for pod yield (Sharma *et al.*, 1993a). The nematode multiplies on many *Leguminosae* and

Pedaliaceae, including the leguminous crop plants chickpea, horse gram (*Dolichos uniflorus*), hyacinth bean (*Lablab niger*), soybean, tepary bean (*Phaseolus acutifolius* var.*latifolius*), moth bean (*P. aconitifolius*) and species of *Vicia* and *Phaseolus*. The nematode is adapted to high soil temperatures, optimum emergence from the cysts occurring at 28°C (Sharma and Swarup, 1984). One generation is completed in 16 days at a soil temperature of 29°C (Koshy and Swarup, 1971). Female developement and multiplication were greater on pigeon pea roots at 25°C than at 10, 15, 20 or 30°C (Singh and Sharma, 1994). On cowpea, the nematode suppresses *Rhizobium* nodule formation (Sharma and Sethi, 1975). Added to soil ten days before or at the time of inoculation with *Fusarium udum*, *H. cajani* exacerbated wilting and growth reduction of pigeon peas (Singh *et al.*, 1993).

Control

As yet, control of this nematode is not fully understood. Crop rotations, avoiding frequent leguminous hosts, should limit yield losses in legumes. Although cultivars resistant to the nematode have not been bred and no evidence of it was found in 58 cultivars of pigeon pea (Sharma *et al.*, 1993b), several distinct races of the nematode have been detected (Walia and Bajaj, 1986, 1988; Siddiqui and Mahmood, 1993b). Some wild relatives of pigeon pea are resistant to *H. cajani* (Sharma *et al.*, 1993a).

Temporary control of the nematode has been obtained by soaking pigeon pea seeds in nematicidal solutions. Early penetration of the roots by juveniles was prevented and development of subsequently invading juveniles was retarded, when the seeds had been soaked in 0.01 or 0.001% aqueous triazophos (Prasad, 1985). Prolonged soaking of seeds in high concentrations of carbofuran, fensulphothion, phorate, or, especially, aldicarb, were effective in lessening infestation of cowpeas (Zaki and Bhatti, 1986). Although treating infested soil with 6 kg carbofuran ha^{-1} increased yield of pigeon peas and gave good early protection from invading juveniles (Sharma *et al.*, 1993a), such treatment is probably uneconomic.

***Heterodera carotae* Jones, 1950** (Fig. 7.12)

Carrot cyst-nematode, *H. carotae*, can be a serious pest of carrots, especially where they are grown frequently in sandy soils in Europe. In parts of France, Italy, Switzerland and Germany, crop losses can be severe. The damage threshold is as small as 0.2 eggs or juveniles g^{-1} or ml^{-1} of soil (Ambrogioni and Marinari-Palmisano, 1976; Bossis and Mugniery, 1989a). In the field, damage appears as irregular stunting of carrot tops, with yellowing, reddening or, in severely infested soil, drying out of the leaves. There is an abnormal proliferation of secondary roots around the nematode invasion sites, as well as stunting and forking of the tap root, which is not infested. Numerous tiny females and cysts are just visible, to the naked eye, attached to the rootlets, from 260–360 day-degrees

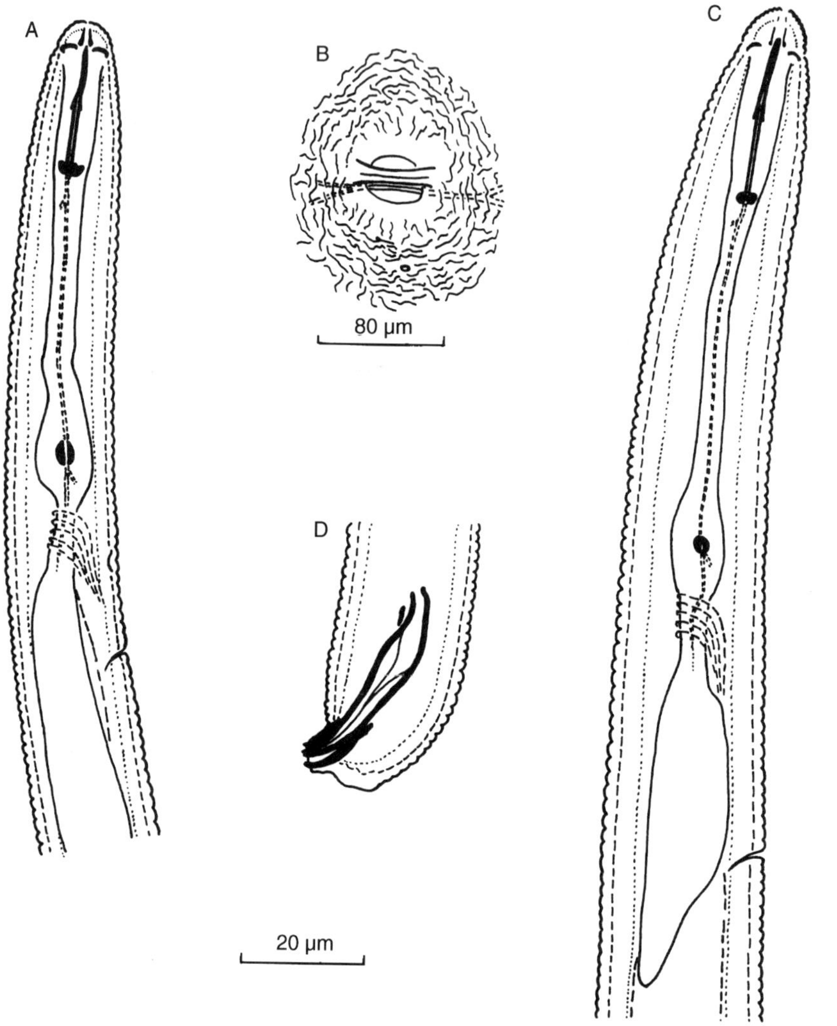

Fig. 7.12. *Heterodera carotae*: A, anterior end of second stage juvenile; B, cyst cone; C, male anterior end; D, male tail end.

after sowing. In a three year carrot monoculture on infested land, yield declined from 50 t to 25 t ha^{-1} in the third year (Harranger, 1971). In 1982, in a trial in Normandy, France, total yield fell from over 60 t ha^{-1} at 3 J2 g^{-1} soil to about 25 t ha^{-1} at 70 J2 g^{-1} soil (Bossis *et al.*, 1989). In Italy, yield losses at 16 and 32 eggs ml^{-1} soil were 40 and 70%, respectively (Greco and Brandonisio, 1980).

Heterodera carotae has a narrow host range. It is only known to multiply on wild and cultivated carrot (*Daucus carota*), *Daucus pulcherrimus* and three wild *Umbelliferae* (*Torilis arvensis*, *T. japonica* and *T. leptophylla*). Eggs hatch faster from the comparatively large egg-sac than from the cyst in response to diffusates from the roots of wild and cultivated carrots, *T. arvensis*, *T. japonica* and especially from a non-host, *Chaerophyllum temulum* (Aubert, 1986). The optimum temperature for hatching is 15–20°C. Hatching ceases at 25°C. In early harvested carrots there may be one generation of *H. carotae*, in late harvested carrots two, or three, if the crop is not harvested until the following spring (Bossis, 1986). As many as four generations may occur when two crops are grown in one year (Vallotton, 1983). In Italy and France, the equilibrium population density ($P_f = P_i$) is about 50 eggs ml^{-1} soil or 50–70 J2 g^{-1} soil, respectively (Greco and Brandonisio, 1980; Bossis, 1986) but in the Valais, Switzerland, it may be more than 300, despite a 70% yield loss (Aubert, 1986). In the absence of a host crop, the soil population of *H. carotae* may decline in one year by 33% (Bossis, 1986) or 49% (Ambrogioni and Marinari-Palmisano, 1976).

Control

Crop rotation

The very narrow host range of *H. carotae* suggests that it could be readily controlled by growing other crop plants between carrot crops. However, in the poor, sandy soils in which many carrot crops are grown, frequent carrot crops may be economically essential. The efficacy of short rotations is doubtful, because damaging infestations may decline very slowly. Although the decline may be up to 50% in the year following a carrot crop, the loss may be almost entirely from juveniles hatching from the egg-sacs, which may hold twice as many eggs as the cysts. Egg hatch in subsequent years may be very slow because eggs in the cysts hatch only in response to leachates from carrots (Bossis, 1988) and the other umbellifers already mentioned. As the damage threshold population is very small, it may take many years of non-host crops to bring the field population to such small numbers.

Resistance

There are no carrot cultivars resistant to *H. carotae.*

Soil solarization

Mulching the soil with polyethylene sheets for one to eight weeks in the period June to August in Italy killed 35–76% of *H. carotae* eggs but there was no effect on eggs below 25 cm soil depth (Greco *et al.*, 1985). Similar soil solarization for four, six or eight weeks killed 23–49% of the eggs but when this was done with 100 or 250 l 1,3-dichloropropene ha^{-1}, it almost eradicated *H. carotae* (Greco *et al.*, 1990).

Nematicides

Serious damage to carrot crops in sandy soils can be prevented by fumigating the soil with EDB, D-D. 1,3-dichloropropene, 1,3-dichloropropene plus methyl isothiocyanate or by dazomet. In southern Italy, the nematode was controlled and carrot yield increased when the soil was fumigated with 300 or 400 l Di-Trapex ha^{-1}, 400 or 800 l D-D ha^{-1} or 500 kg dazomet ha^{-1} (Greco *et al.*, 1976; Greco and Lamberti, 1977a). Unless the soil is covered with gas-tight sheets, damage may not be fully prevented in very heavily infested soils by D-D, 1,3-D or 1,3-D plus methyl isothiocyanate. Control of *H. carotae* in organic soils is not achieved by soil fumigants, because the gases are adsorbed by the organic matter. Kills in excess of 80% are hard to achieve, even in sandy soil, without gas tight soil covers. Although carrot yields may be dramatically increased, in part due to a stimulant effect (Bossis and Mugniery, 1989b), kill may be insufficient to prevent subsequent soil population increase, especially when the crop is grown long enough to permit two or more generations (Bossis *et al.*, 1989). Kill was not increased by using more than 250 l of 80% D-D : 20% methyl isothiocyanate (Greco and Brandonisio, 1980). The non-fumigant nematicides aldicarb, fenamiphos and oxamyl were less effective in controlling *H. carotae* in Italy than soil fumigants (Greco and Lamberti 1977a,b) and in France, Bossis *et al.* (1989) found carbofuran much less effective than D-D or EDB.

Biological control

In trials in Switzerland and western France, the biological control agent *Paecilomyces lilacinus* did not exert significant control in the short term on *H. carotae* (Vallotton, personal communication; Bossis, 1988).

Heterodera ciceri Vovlas, Greco and Di Vito, 1985

This cyst-nematode is a serious pest of chick peas (Cicer arietinum) in northern Syria. The damage threshold was found to be one egg ml^{-1} soil and yield was reduced to zero at about 60 eggs ml^{-1} soil (Greco *et al.*, 1988b). The nematode multiplies well on chickpea, lentil, pea and grass pea (*Lathyrus sativus*) but little on lucerne, lupin, *Vicia* spp. or haricot bean. There are very few host plants that are not legumes.

Control

Crop rotation and soil solarization are known to control this nematode. After one year of non-hosts, barley or wheat, nematode numbers in the soil declined by 37–57% and after two to three years by 84–88%. After one, two or three consecutive years of these crops, chickpea yields were increased 2.1, 4.5 or 7.9-fold, respectively. The maximum number of nematodes maintained on chickpea or grass pea was 63 eggs g^{-1} soil (Saxena *et al.*, 1992), at which population density five years of non-host crops would be needed for numbers to decline to 2 g^{-1} soil. Soil solarization in summer for six or eight weeks greatly reduced numbers of the nematode in the topsoil (Di Vito *et al.*, 1991). Although some chickpea lines are resistant to *H. ciceri*, resistant cultivars are not yet available (Di Vito *et al.*, 1988).

Heterodera cruciferae Franklin,1945 (Fig. 7.13)

Brassica cyst-nematode, *H. cruciferae*, is not a major pest but is widespread in Europe, and western and central regions of the former USSR. It also occurs in California and Australia. The nematode multiplies on a range of cruciferous

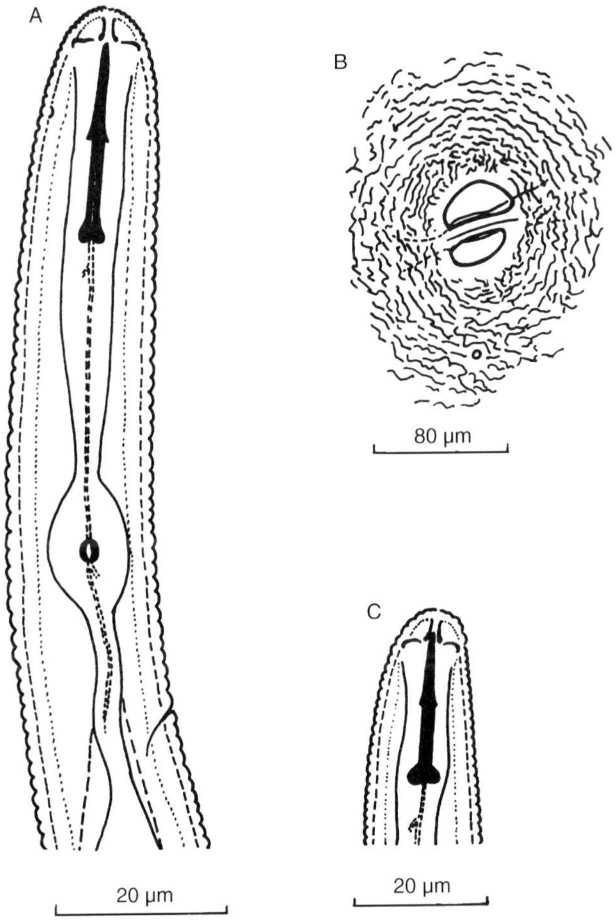

Fig. 7.13. *Heterodera cruciferae*: A, anterior end of second stage juvenile; B, cyst cone; C, male head end.

crops and weeds. Some *Labiatae* are also hosts but brassicas are particularly good hosts. Although Doncaster (1952) found no damage to cabbage seedlings in Lincolnshire, England, in soil containing 125 juveniles g^{-1}, transplanted cauliflowers were injured at 25 eggs g^{-1} soil and severely damaged when the soil contained more than 75 eggs g^{-1} (Sykes and Winfield, 1967). Transplanting brassicas into heavily infested soil can result in crop failure. Symptoms of attack by *H. cruciferae* are stunting, reddening of the leaves or interveinal chlorosis and marked proliferation of lateral roots. One to three generations have been reported on brassica crops, depending on the duration of vegetative growth and on soil temperatures. A synergistic effect of *H. cruciferae* with 'club root' (*Plasmodiophora brassicae*) has been reported (Lear *et al.*, 1966a).

Control

Crop rotation

As *H. cruciferae* is much more damaging to brassica seedlings than to established plants, seedlings should be raised in soil free of the nematode and should not be transplanted into heavily infested soil. Because the nematode has such a narrow host range, soil populations can be decreased by growing crops other than brassicas for several years. However, as the nematode is stimulated to hatch principally by *Brassicae*, its speed of decline in their absence is slow.

Nematicides

In infested soil, yields of Brussels sprouts were greatly increased by soil fumigation with D-D or by applying DBCP in water to the soil at transplanting (Lear *et al.*, 1965, 1966a). Infestation of cabbage seedlings was prevented by sowing seed in nursery soil fumigated with 500 kg methyl bromide or 400 kg Basamid (98% dazomet) ha^{-1} (Korunić, 1975). Total and marketable yields of cabbage were greatly increased by applying 225 l D-D ha^{-1} to infested soil in South Australia (Stirling and Wicks, 1975). Incorporating oxamyl at 5 kg ha^{-1} in the top 15 cm of the soil controlled the nematode better than applying oxamyl to the soil surface or spraying it on the foliage and it increased the percentage of class I cauliflower heads (de Moel, 1986). Applying aldicarb at sowing to winter oilseed rape in heavily infested soil increased yield by 0.66 t ha^{-1} but did not lessen invasion of the roots by *H. cruciferae* or nematode multiplication (Harris and Evans, 1988). The relatively short persistence of oximecarbamate nematicides in soil, coupled with leaching by winter rainfall make autumn treatment unlikely to affect nematode multiplication in late spring and summer of the following year.

Resistance

Brassicae resistant to *H. cruciferae* have not been bred.

Heterodera glycines Ichinohe, 1952 (Fig. 7.14)

The soybean cyst-nematode, *H. glycines*, causes 'yellow dwarf' disease of soybean (*Glycine max*), which can result in serious yield losses (Plate 5). The nematode is known in Japan, Korea, People's Rebublic of China, USA, the Amur region of former USSR, Colombia, Egypt, Indonesia, Argentina and Brazil. In the USA, it is an increasing problem, despite control measures, occurring in 24 States and all major soybean areas, except for the northern States of Nebraska, North and South Dakota and Pennsylvania (Miller, 1986). Miller put yield losses due to *H. glycines* in the USA in 1984 at $248 million.

The nematode multiplies mainly on *Leguminosae*. Riggs and Hamblen (1962) reported 64 species of this family as hosts. Of most economic importance among these are soybean, snap bean (*Phaseolus vulgaris*), adzuki bean (*P. angularis*), some species of *Lespedeza* and *Melilotus*, but pea (*Pisum sativum*) and most clovers (*Trifolium* spp.) are non-hosts. Some species of *Scrophulariaceae*, *Caryophyllaceae* and *Labiatae* are also hosts of *H.glycines*. In China, mung bean or green gram

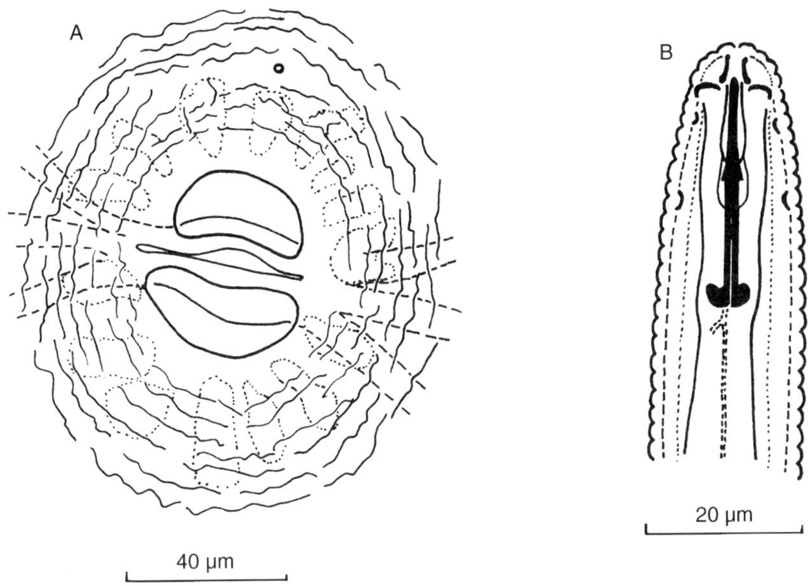

Fig. 7.14. *Heterodera glycines*: A, cyst cone; B, anterior end of second stage juvenile.

(*Phaseolus aureus*), cowpea (*Vigna unguiculata*), pea and sesame (*Sesamum indicum*) are hosts of race 4 of the nematode (Dong, 1988).

When *H. glycines* parasitizes soybean roots, it may suppress binding of *Rhizobium japonicum* to the root hairs, so preventing nodule formation. Nodules that do form may be smaller and contain less leghaemoglobin. The severity of this effect varies with the race of soybean cyst-nematode, race 1 being more harmful than races 2 and 4. The result of nodule loss is stunting and nitrogen starvation of the plant, which results in yellowing of the foliage. Not surprisingly, yield is negatively correlated with the number of eggs and juveniles in the soil at sowing or mid-season (Phipps and Elliott, 1983; Schmitt *et al.*, 1983). Those soybean cultivars little affected, i.e. tolerant, have a deep rooting habit and enhanced root growth. Their tolerance is race independent. Schmitt *et al.* (1987) found that the tolerance of a cultivar may vary between different soil types. Johnson *et al.* (1993) found that the syncytia in the roots of a susceptible soybean cultivar are located mainly in the stele in a dry silt loam but mainly in the cortex in the same but wetter soil, where they may interfere less with water transport, so making the plants more tolerant of infestation. 'Sudden death' syndrome of soybean, due to the blue strain of *Fusarium solani*, was exacerbated when the roots were inoculated with both *H. glycines* and the fungus (McLean and Lawrence, 1994).

Juveniles of *H. glycines* emerge from cysts at 16–36°C with an optimum of 24°C (Slack and Hamblen, 1961). A hatching factor from root exudates of kidney beans (*Phaseolus vulgaris*) was isolated and purified by Masamune *et al.* (1982). Called glycinoeclepin A, it is a dibasic acid of molecular weight 446 and empirical formula $C_{25}H_{34}O_7$ and its structure has been determined (Fukuzawa *et al.*, 1985). Glycinoeclepin A stimulates hatching of *H. glycines in vitro* at 10^{-11}–10^{-12} g ml^{-1}. In the USA, there may be three or four generations in the cooler soybean areas but six or seven in the hotter regions (Noel, 1986). However, Lawn and Noel (1986) found that most of the population increase on soybean came from the first generation. In Heilongjiang Province of China, where race 3 is dominant, there may be three generations on soybean per annum (Liu *et al.*, 1988). In Shanxi Province, there may be four generations per annum on soybean (Dong, 1988).

Control

Soybean cyst-nematode may be controlled by rotating soybeans with non-host crops, by growing resistant cultivars and to a greater or lesser extent by judicious use of a nematicide. Combined use of control measures is more likely to be successful than dependence on a single measure.

Crop rotation

Soil infestations of *H. glycines* may decline quite rapidly under non-host crops, so that short rotations may be effective in reducing populations to levels not harmful to soybeans (Plate 5). For example in Missouri, USA, after three years of maize, cotton (*G. hirsutum*) or sorghum (*Sorghum bicolor*), *H. glycines* was barely detectable in the soil and soybean yields were increased, compared with soybeans in monoculture. Decline of the nematode population was, however, dependent on the initial infestation level (Francl and Dropkin, 1986). In Alabama, maize is not a host of races 3 and 4 and *S. bicolor* cv. Pioneer 8222 and two potentially useful forage or green manure crops, *Aeschynomene americana* (American joint vetch) and *Indigofera hirsuta* (hairy indigo) are poor or non-hosts of race 4 (Weaver *et al.*, 1988b; Rodriguez-Kábana *et al.*, 1988b). Numbers of race 4 were greatly decreased by growing Pioneer 8222 sorghum for two years, which greatly benefitted the following soybean crop (Rodriguez-Kábana *et al.*, 1990a). Two years of non-host crops reduced *H. glycines* to very small or undetected numbers in North Carolina. Planting early-maturing soybeans mid- to late June and rotating them with a non-host crop kept populations of the nematode small (Schmitt, 1991). Similarly, *H. glycines* was barely detectable in the soil after two years of non-host crops and early maturing (group V) soybeans left smaller numbers of the nematode in the soil than late maturing (group VII) soybeans (Koenning *et al.*, 1993). Choice of early soybean maturity group (III, IV, V) did not affect the size of the nematode population left in the soil after harvest (Todd, 1994).

In Japan, the nematode was controlled best by a resistant cultivar or a four year rotation, which included non-host crops and soil amendments (Shimizu *et al.*, 1989). The speed of decline of *H. glycines* can vary considerably, so that the number of non-host crops needed to decrease soil infestations to a non-damaging level (around 2 eggs g^{-1} soil) may also vary considerably. In Heilongjiang Province of China, decline under non-hosts is about 22%, whereas in Alabama it declined 95–100% under American joint vetch or hairy indigo grown for one year (Rodriguez-Kábana *et al.*, 1990b).

The major drawback to control of soybean cyst-nematode by crop rotation is economic. It may not be possible to market some or all of the non-host crops, or they may be less profitable than soybeans.

Resistant cultivars

Heterodera glycines is most economically controlled by growing soybeans which are resistant to it, so that nematode multiplication is much reduced or prevented (Plate 5). The use of cultivars with rather narrowly based resistance has revealed a number of races of the nematode. Golden *et al.* (1970) using four differential genotypes recognized four races (1–4) and Inagaki (1979) using the same differentials recognized five (Table 7.8).

The number of races recognizable by the use of differential soybean genotypes is the number of infestation categories to the power x, where x = the number of differential genotypes. So, with four differentials and two categories (+ and −) 16 races are recognizable, with five differentials 32, six 64, and so on. Sixteen races of *H. glycines* have, indeed, been recognized, using the four differentials of Golden *et al.* (1970) (Riggs and Schmitt, 1988). In 1981, Riggs *et al.*, using 13 differential soybean genotypes, recognized 25 'races' in 33 populations of *H. glycines* from the USA and Japan. Thirteen differentials could, in theory, distinguish 8192 races, i.e.2^{13}.

As parasitism of a resistant soybean can change greatly within seven generations (Triantaphyllou, 1975), different races can occur in the same field (Miller, 1966) and they can interbreed, it is not surprising that populations able to parasitize a resistant cultivar may appear fairly quickly, especially if the resistant soybean is grown repeatedly in the same soil. It must not be forgotten that a cultivar allowing almost 10% of the multiplication on a susceptible cultivar is a partial resister, i.e. the resistance is quantitative, not qualitative, as in a fully resistant cultivar. The differential scheme is, therefore, rather arbitrary in that no distinction is made between full and partial resistance to a particular population. Liu *et al.* (1993) have found that race 3 selected continuously for nine to ten generations on resistant cv. Harbin behaved as race 2 on resistant cv. Changlihei, behaved as race 4 on resistant cv. Peking, behaved as race 14 on resistant cv. Franklin and on susceptible cv. Lee 68 behaved as race 3. Selection on the resistant soybeans also increased the pathogenicity of the populations.

Table 7.8. Identification of races 1–5, using soyabean genotype differentials.

	Race				
Soybean genotype	1	2	3	4	5
cv. Lee 68	+	+	+	+	+
cv. Pickett	–	+	–	+	+
cv. Peking	–	+	–	+	–
PI 88788	+	+	–	–	+
PI 90763	–	–	–	+	–

+ = 10% or more of the number of white females on Lee 68;
– = <10% of the number on Lee 68.

In the USA, the first resistant soybeans released, cvs Pickett, Dyer and Custer, followed by Pickett 71, Forrest, Mack, Centennial and others, were resistant to race 3 (Table 7.9), but race 4 was recognized as highly pathogenic and this led to release of cvs Bedford, Fayette, Bradley and others resistant to races 3 and 4. Resistance to race 3 in the USA was derived from Peking and to race 4 from PI 88788, so the genetic base for resistance is quite narrow (Anand, 1986). Only one cultivar, Hartwig, is resistant to races 1–6 and 14, while Peking is resistant to races 1,3,5 and X (Table 7.9).

One drawback to the use of cultivars resistant to *H. glycines* is that they may yield less grain than newer, high-yielding but susceptible cultivars. Another problem is the selection of virulent (i.e. resistance-breaking) biotypes of the nematode. Selection of such forms from a race 3 population was shown to occur in two generations on Pickett 71 and PI 88788 and in four generations on Peking and PI 90763 (Zirakparvar and Norton, 1981). Populations selected on Pickett 71 reproduced as much on it as did unselected populations on susceptible cv. Williams (McCann *et al.*, 1982). After seven and 11 generations of selection on cv. Bedford and PI 88788, the selected nematodes multiplied greatly on them but little on PI 89772 or PI 90763, whereas selection on the latter gave nematodes that multiplied little on Bedford or PI 88788 (Young, 1982). The resistance of Bedford and Peking broke down to populations selected by continuous cultivation of them over three to six years in Tennessee (Young, 1984a).

Selection for virulence is not necessarily inevitable, for Hartwig *et al.* (1982) did not detect any significant increase in *H. glycines* in the field as a result of growing race 3-resistant cv. Centennial continuously for six years in Mississippi. In other experiments in Mississippi, selection for virulence did occur (Slack *et al.*, 1981). Anand *et al.* (1994) found 83% of 58 populations of *H. glycines* from ten

Table 7.9. Soybean cultivars resistant to races of *Heterodera glycines* (maturity group).

Race	Cultivar
1	Thomas (VII), Twiggs, Centennial, Bryan (IV), Gordon, Jeff, Bedford, Chibuliuheidon, Xingxian Huipizhi Heidon (XHH), Hartwig (V), Karikei 222, Karikei 225, Gaozuo No.1, Forrest, Qihei no. 2.
2	Hartwig (V).
3	Hartwig (V), Thomas (VII), Twiggs, Centennial, Bryan (IV), Gordon, Jeff, Bedford, Chibuliuheidon, XHH, Saline (III), Peking, Karikei 222, Karikei 225, Cordell, Bell, Jack, Fayette(III), Bradley (VI), Epps (V), Alpha (I), Avery, Egyptian (IV), Pyramid (IV), Pickett, Narow M, Walters (V), Gaozuo No.1, Yu Don 8, Newton (II), Hagood (VII), Stonewall, Carter, Maxey (VIII), Lamar, McNair 770, Gregg (VII), Custer, Pickett 71, Forrest, Mack, Avery (IV), Doles (VI), Lyon (VI), Qihei No.2
4	Jeff, Bedford (IV), Chibuliuheidon, XHH, Hartwig (V), Cordell, Bell, Jack, Fayette (III), Bradley (VI), Epps (V), Avery, Egyptian (IV), Pyramid (IV), Nathan, Rhodes (V).
5	Bedford (IV), Chibuliuheidon, XHH, Hartwig (V), Peking, Karikei 222, Karikei 225, Cordell, Gaozuo No.1.
6	Hartwig (V).
9	Hartwig (V)
14	Hartwig (V), Saline (III).
X	Peking, Ilsoy.

major soybean-producing States were able to multiply to some degree on Pickett, although race 3 (to which Pickett is resistant) occurred in 45% of the locations. Reliance on the resistance in cultivars derived from Pickett and Peking, especially in southern soybean States, is therefore unsound and cultivars derived from PI 90763, PI 88788 and PI 437654 should be relied on instead, assuming they are satisfactory agronomically. Selection of a race 5 population for up to 14 generations on cv. Bedford did not change the race, whereas on cv. Cordell the population changed to race 14, to which Bedford is moderately resistant. Bedford and Cordell are both resistant to races 3, 4 and 5 (Young, 1994).

Primary selection for virulence does not fix the genes for reproductive ability. However, a population selected on PI 89772 was able to multiply on cv. Cloud after four cycles of selection but was then less able to multiply on PI 89772. The reverse also occurred (Luedders and Dropkin, 1983). Similarly, following primary selection on Bedford and PI 88788 and secondary selection on D 72–8927 and J 74–85, the nematodes were less able to multiply on the primary selector and *vice versa* (Young, 1984b). Similar findings were reported by Anand and Shumway (1985). In 1982, Riggs and Robbins reported that populations had been selected in the glasshouse capable of multiplying on all resistant soybean cultivars then available. In view of this, the monoculture of a resistant cultivar should be avoided. To prevent, or at least delay, the formation of new races of *H. glycines*, they recommended a rotation of a non-host crop, a resistant cv. and a susceptible cultivar. Anand *et al.* (1995) have shown that a 1:1 blend of a resistant and a susceptible soybean cultivar may minimize selection of virulent races, while yield is maintained at an acceptable level. Used in rotations, a tolerant, susceptible cultivar relieves selection pressure that would result from use of a resistant cultivar, without serious yield loss (Hussey and Boerma, 1992). However, it would greatly increase the number of non-virulent nematodes left in the soil after harvest.

The cytological basis of resistance would seem to be similar in soybeans to that in other crops resistant to cyst-nematodes, i.e. degeneration of the syncytia. In resistant cv. Forrest, syncytia form, as in a susceptible cultivar, but they become surrounded by necrotic cells and subsequently degenerate (Kim, 1986). As a result, females are not produced or die before laying eggs.

Biological control

An unidentified fungus (Arkansas fungus 18, probably an ascomycete) was shown capable of decreasing *H. glycines* in pots by 90%. The fungal hyphae were able to penetrate the cyst wall directly and eggs and first stage juveniles within them were invaded (Kim *et al.*, 1992).

Nematicides

Although damage to soybeans can often be prevented by use of an effective nematicide appropriately applied to the soil, limiting or preventing multiplication of soybean cyst-nematode is much more difficult, because several generations are completed on the roots. The non-phytotoxic soil fumigant DBCP applied at sowing was economic, easy to apply and effective but it can no longer be used (Noel, 1986). Other soil fumigants are phytotoxic and not usually as effective. Up to

37 l EDB ha^{-1} at sowing, however, was not phytotoxic to two soybean cultivars in Alabama and controlled the nematode as much as did DBCP (Rodriguez-Kábana *et al.*, 1979c). Also applied at sowing, D-D or 1,3-dichloropropene (Telone II) at 18.7–84.2 l ha^{-1} was not phytotoxic but controlled *H. glycines* less than 9.4 l DBCP ha^{-1} (Rodriguez-Kábana *et al.*, 1979d). The nematode was controlled better and soybean yields were increased more by EDB (4.7–18.6 l ha^{-1}) plus anhydrous ammonia (56 or 112 kg ha^{-1}) than by either compound alone (Rodriguez-Kábana *et al.*, 1981a). However, in a series of experiments in Alabama with resistant and susceptible soybeans, EDB increased grain yields but also increased *H. glycines* (Weaver *et al.*, 1988a).

Control of the nematode with non-fumigant nematicides has also been variable. Although active in a wide range of soils, aldicarb has been effective in some parts of the USA and not in others. In Delaware, aldicarb at 3.3 or 5.6 kg ha^{-1} decreased the number of cysts on the roots and in the soil around the roots of susceptible cv. Essex and increased the grain yield (Mulrooney, 1990). In Illinois, aldicarb gave the most consistent control of *H. glycines* of non-fumigant nematicides tested (Edwards and Noel, 1981). In Virginia, ethoprophos at 14.6 kg ha^{-1} controlled multiplication of the nematode better than aldicarb, fenamiphos or carbofuran, although all four nematicides increased grain yields (Phipps, 1987a). Furthermore, although aldicarb at 3.9, 7.8 or 11.2 kg ha^{-1} significantly increased the grain yields of eight soybean cultivars, it had no effect on nematode increase (Phipps, 1987b).

The efficacy of a non-fumigant nematicide is affected by its rate of degradation in the soil, the reproductive potential of the nematode and the method of applying the nematicide to the soil. In the hotter soils, nematicide breakdown is likely to be faster and there may be more generations of the nematode than in the cooler soils. Narrow banded or seed furrow applications may be too localized and banded treatments to the soil surface may not be dispersed through the soil by rain or irrigation. Sowing soybean seeds in vertical bands of soil treated with a non-fumigant nematicide could improve control of *H. glycines* (Whitehead, 1988a). It is unlikely, however, that any currently available nematicide could control the nematode as effectively as a resistant cultivar, which prevents or minimizes nematode increase throughout the soil explored by its roots. Nematicides would seem best used to limit damage by soybean cyst-nematode in the early stages of growth. Nematode populations would seem best controlled by the use of non-host crops in rotation with soybeans and by the judicious use of resistant cultivars of soybean.

***Heterodera goettingiana* Liebscher, 1892** (Fig. 7.15)

Since its discovery in 1890 by Liebscher, the pea cyst-nematode, *H. goettingiana*, a very damaging pest of pea crops, has been reported in much of Europe (Germany, the Netherlands, UK, Belgium, Spain, Portugal, France, Italy) and in Israel, Algeria, Malta, the former USSR and the USA. Damaging attacks have long been recognized in the UK, France, the Netherlands and Italy. Heavy infestations can also damage crops of broad and field beans.

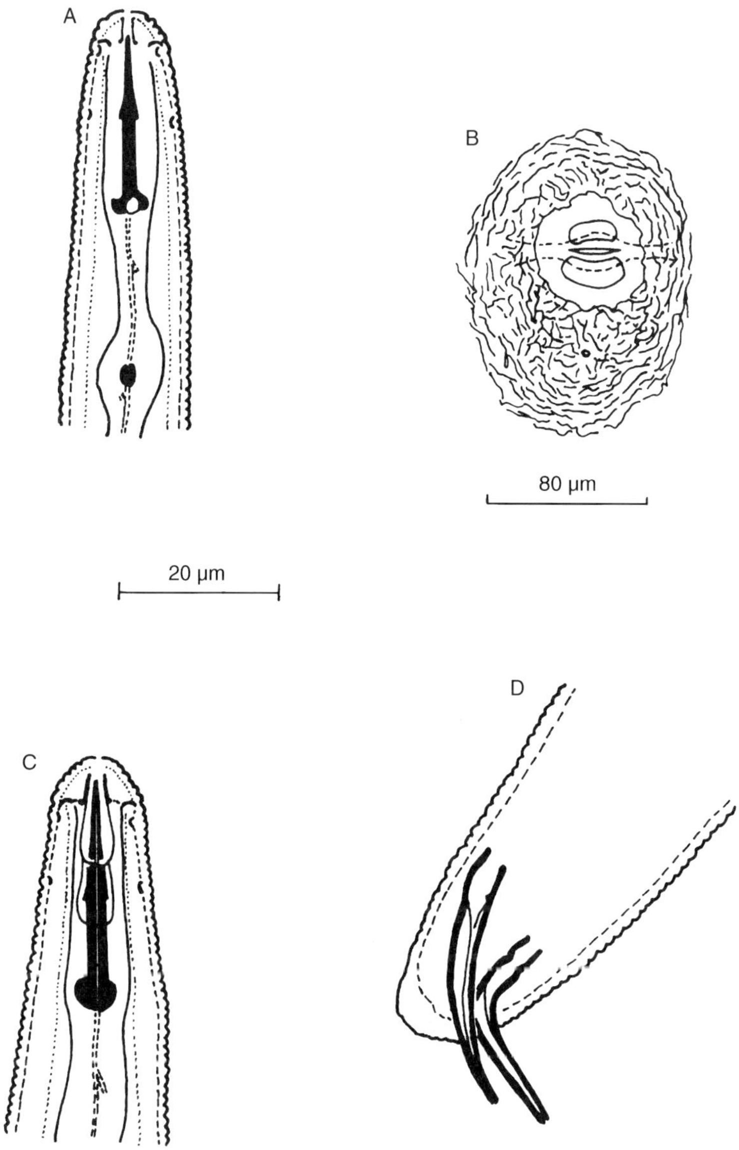

Fig. 7.15. *Heterodera goettingiana*: A, anterior end of second stage juvenile; B, cyst cone; C, male head end; D, male tail end.

Pea cyst-nematode multiplies on field and garden peas (*Pisum arvense* and *P. sativum*, respectively), broad and field beans (*Vicia faba*), vetches (*Vicia* spp.) and grass pea (*Lathyrus sativus*), as well as on weed species of *Vicia* and *Lathyrus*. Lentil (*Lens culinaris*) was reported a host of the nematode in the Netherlands (Oostenbrink, 1951) but was resistant to six populations of the nematode in Italy (Di Vito and Greco, 1986). Earlier reports of the susceptibility of sweet pea (*Lathyrus odoratus*), chickpea, soybean and species of lupin were contradicted by

(a)

(b)

Fig. 7.16. (a) Peas grown in plots infested with pea cyst-nematode, *H. goettingiana*, with granular aldicarb or oxamyl mixed into the seedbed at sowing (dark plots) or left untreated (light plots). (b) Seed drill equipped to apply granular nematicide in the seed furrows at sowing.

several studies of the host range of the nematode (Jones, 1950; Oostenbrink, 1951; Winslow, 1954; Di Vito and Greco, 1986). Such conflicting results may be explained by the presence of *Heterodera trifolii* Goffart, 1932 in the inoculum used, use of different cultivars differing in susceptibility to the nematode, or perhaps to differences in virulence between the populations of *H. goettingiana*.

In the field, affected pea plants are stunted and the foliage turns yellowish-green, then yellow and finally brown as the plants senesce prematurely. Pods are few and small and pea seed yield greatly reduced (Fig. 7.16). The root system is sparse with few or no *Rhizobium* nodules and foot rots are commonly exacerbated by pea cyst-nematode infestations. Large numbers of females and cysts are visible on the surface of the roots. Although broad and field beans are injured less than peas, they too suffer similarly in heavily infested soils.

Using a cultivar of peas resistant to *Fusarium oxysporum* and *F. solani*, Oostenbrink (1955) was the first to demonstrate clearly, in pots, that *H. goettingiana* was responsible for the field symptoms of attack, including the absence of

Rhizobium nodules on the roots and for great yield loss. In the field, Winfield (1965b) was able to demonstrate increasing yield loss with increase in soil infestation of the nematode and ultimately total crop loss. At the same time, Jones *et al.* (1965) demonstrated dramatic reductions in the weight of pea haulms with increasing numbers of the nematode in the soil. Against a wide range of initial population densities (4–359 eggs g^{-1} soil), there appeared to be a sigmoid relationship between haulm weight and soil infestation, with a minimum yield about a quarter of the maximum yield and a tolerance threshold of less than 5 eggs g^{-1} soil. In a pot experiment, Di Vito and Greco (1986) demonstrated a tolerance threshold for haulm weight of 3.5–5.3 eggs g^{-1} soil and a minimum yield also one quarter of the maximum. In the field in Italy, the tolerance threshold may be less than one egg g^{-1} soil (Greco *et al.*, 1991a; Ferris and Greco, 1992). In microplots, Moriarty (1963a) demonstrated a significant negative regression of broad bean yield on log (initial eggs g^{-1} soil).

In the UK, there is only one generation of the nematode on peas but two may occur on field beans (Jones, 1950). In southern Italy, three generations can occur (Greco *et al.*, 1986). Although intolerant, peas can increase small populations of the nematode fifty-fold or more (Jones *et al.*, 1965). In contrast, broad bean, which is more tolerant of attack, is a much less efficient host, barely maintaining populations at low initial densities (Moriarty, 1963a). Field beans are probably similar to broad beans in this respect. Over a wide range of initial densities, populations on peas increased to a ceiling level (Jones *et al.*, 1965).

Control

At present, pea cyst-nematode can only be controlled by crop rotations and by nematicides. Although crossing susceptible pea cv. Progress 9 with resistant line MG 101791 of *Pisum abyssinicum* yielded hybrids, which in the F_2 generation had 22% of plants resistant to the nematode, resistant pea cultivars have not yet been produced (Di Vito and Greco, 1986). Earlier work had failed to reveal any resistance among pea cultivars (Jones, 1950).

Crop rotation

Pea cyst-nematodes do not hatch in the absence of root diffusates from host plants (Shepherd, 1963). Hatching at 15°C is much greater in diffusate from pea plants four to six weeks old than two, eight or ten weeks old (Perry *et al.*, 1980). Hatching in southern Italy is suppressed in summer, when soil temperatures exceed 25°C but resumes in autumn, as temperatures decrease (Di Vito and Greco, 1986). In the absence of host plants, *H. goettingiana* eggs have been reported to decline at about 25% per annum in fallow soil (Winslow, 1955), 44% per annum (Moriarty, 1963b) and 39% per annum (Brown, 1958) but significant year to year differences in decline rate were noted. In southern Italy, the nematode declined by about 68% in the first year under fallow, wheat or oats but more slowly thereafter, so that after four years there was still 18% of the initial population (Di Vito and Greco, 1986). At a decline rate of 50% per annum, five years or more would be needed without host plants before a heavy infestation of

100 eggs g^{-1} soil had declined to a tolerance threshold of 3 eggs g^{-1} soil. As infestations may exceed 100 eggs g^{-1} soil, at least in patches of affected fields, it is not surprising that damaging populations may still occur in soil rested from host crops for ten years or more, as reported by Brown (1958), if the decline rate is much less than 50% per annum.

Assuming a constant decline rate (d), the population density (P_f) after x years, given an initial density of P_i, is given by the equation:

$$P_f = P_i (1 - d)^x \quad (7.6)$$

Clearly, the interval between host crops must be long, if injury to pea crops is to be avoided. Five or six years between pea crops should suffice, provided field beans or another host crop is not grown in the rotation and provided weed hosts are well controlled. Growing peas frequently on the same land risks serious injury from pea cyst-nematode and from fungal diseases, which it may then take many years of rotation to overcome, or necessitate expensive chemical control measures.

Nematicides

Pea cyst-nematode can be controlled by applying large amounts of soil fumigants or small amounts of non-fumigant nematicides to infested soil.

FUMIGANT NEMATICIDES

At 300 l ha^{-1}, D-D, Telone or Di-Trapex has given adequate control of *H. goettingiana* and increased pea yields. Dazomet at 300 kg ha^{-1} has been quite effective but all these fumigants must be applied to the soil several weeks or months before sowing to avoid phytotoxicity (Di Vito *et al.*, 1973; Di Vito and Lamberti, 1976a). Combined use of Telone and dazomet was very phytotoxic to semi-leafless peas in Norfolk, England, probably as a result of the destruction of mycorrhizae (D. Hayman, personal communication). It may also be necessary to reinoculate fumigated soil with an appropriate strain of *Rhizobium*, especially if a methyl isothiocyanate precursor fumigant (metham-sodium, Di-Trapex or dazomet) has been applied.

NON-FUMIGANT NEMATICIDES

Incorporated into the seedbed at sowing time at 2.8 or 5.6 kg ha^{-1}, aldicarb or oxamyl greatly increased pea yields in infested sandy clay and clay loam soils in England and lessened nematode multiplication (Whitehead *et al.*, 1974). The nematode was controlled in southern Italy with 12 kg carbofuran or 30 kg fenamiphos ha^{-1} incorporated in the seedbed one week or one month, respectively, before sowing. Pea yields were increased more and the nematode was controlled better by fenamiphos or carbofuran than by Di-Trapex, D-D or Telone (Di Vito and Lamberti, 1976b). Experiments in England showed that *H. goettingiana* was more reliably controlled by aldicarb or oxamyl when the granules were rotary cultivated into the seedbed, just before sowing, than when they were applied to the seed furrow during sowing or broadcast at the sowing depth of 5 cm, because the latter treatments were dependent on heavy rainfall to leach them down the soil profile (Whitehead *et al.*, 1979a) (Fig. 7.16).

Unfortunately, rotary cultivation may harm soil structure and negate the benefit of frosting in heavy soil, in which peas are often sown. Further experiments in England showed that pea yields were greatly increased by oxamyl at 1.4–5.6 kg ha^{-1} broadcast on the soil at sowing or at 6, 8.6 or 16 g $(100 \text{ m row})^{-1}$, in-furrow application. Although numbers of second stage juveniles in the pea roots were decreased 58 days after sowing by these treatments, nematode multiplication was not lessened. Drenching oxamyl over the rows at 8.6 g $(100 \text{ m row})^{-1}$, 31 and 41 days after sowing, lessened multiplication of the nematode but did not increase pea yields (Green *et al.*, 1981), presumably because the damage had already been done to the roots before treatment with oxamyl. For peas grown in widely spaced rows, control of *H. goettingiana* with a granular nematicide could be improved and nematicide dosage decreased by placing the granules in narrow vertical bands of soil, in the top of which the seeds are sown (Whitehead, 1988a). This would lessen the need for redistribution of the nematicide by leaching and protect the tap root and developing lateral roots from early attack. Small doses of non-fumigant nematicide, appropriately placed in the soil could be cost-effective in protecting pea crops from injury by pea cyst-nematode in shortened rotations.

Heterodera schachtii **Schmidt, 1871** (Fig. 7.17)

Beet cyst-nematode, *H. schachtii*, an important pest of sugarbeet (Plate 6), was discovered by Schacht (1859a,b) on sugarbeet near Halle, Germany and shown responsible for 'beet weariness' (Rubenmüdigkeit) of intensively cropped land (Liebscher, 1878; Kühn, 1881). The practice of growing sugarbeet continuously in fields close to sugar factories in Germany led to wholesale crop failures and the closure of 24 factories in 1874. The pathogenicity of *H. schachtii* to sugarbeet seedlings was clearly demonstrated in monoxenic culture by Polychronopoulos and Lownsbery (1968). The pest is now known in 38 countries (Cooke, 1987) and especially in Europe, the former USSR, Canada and the USA.

The nematode multiplies on a wide range of plants in 23 families, especially *Chenopodiaceae* and *Cruciferae*. Host crops include beets and mangolds (*Beta vulgaris*) and spinach (*Spinacea oleracea*) (*Chenopodiaceae*), brassicas, garden cress (*Lepidium sativum*), radish (*Raphanus sativus*) and white mustard (*Sinapis alba*) (*Cruciferae*). The relative abundance of *H. schachtii* down the soil profile can vary considerably, following a sugarbeet crop. Thorne and Giddings (1922) found only 2% of the nematodes in soil 35–76 cm deep in three sandy loam soils, whereas in England in three peaty loam soils there were as many nematodes 20–40 cm deep as 0–20 cm, or more 20–60 cm deep than 0–20 cm deep (Whitehead, 1977a). The spatial distribution of the cysts in the soil is assumed to reflect the distribution of host roots. In deep, uncompacted soils, a large network of fine roots can develop beneath the sugarbeet storage root. The widespread growing of oilseed rape cultivars with low erucic acid and glucosinolate contents is likely to increase *H. schachtii*, as these were shown, in pots, to be better and more tolerant hosts of the nematode than low erucic acid or conventional cultivars (Decker and Dowe, 1991). Less efficient host crops include soybean, cowpea, tomato, annual lupin

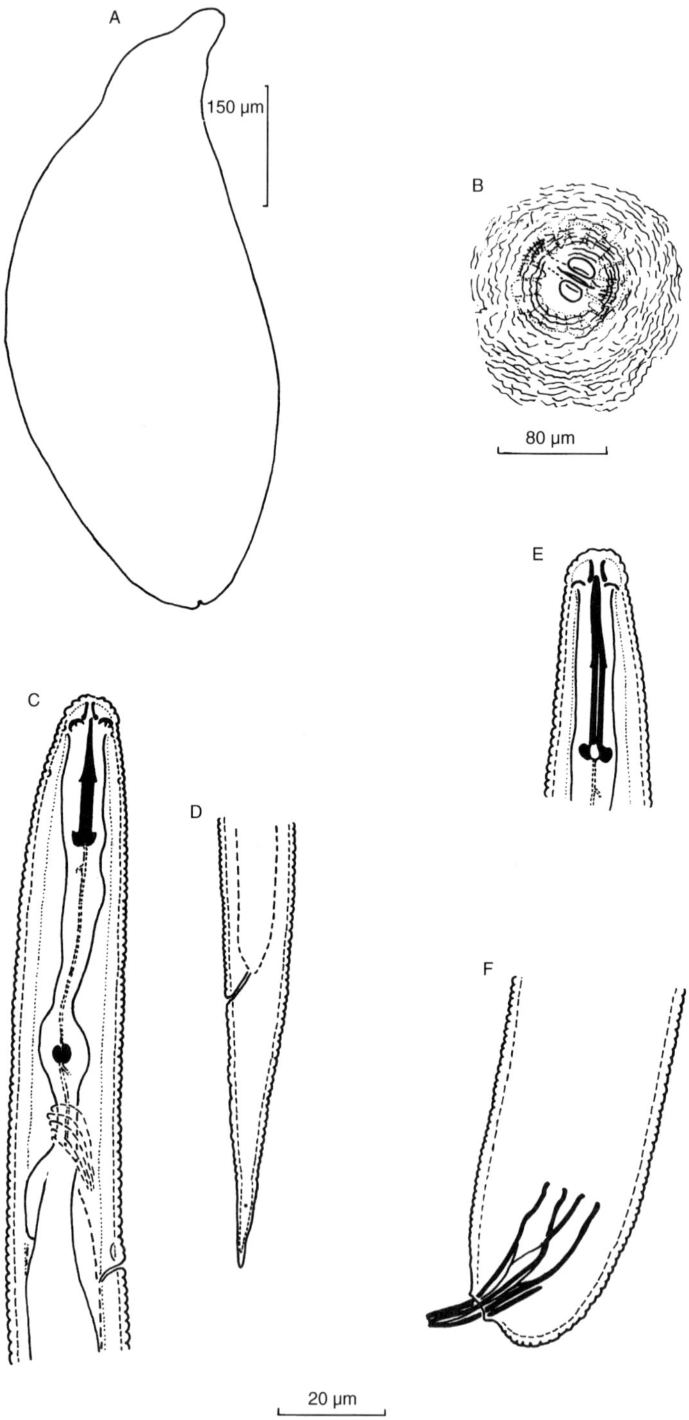

Fig. 7.17. *Heterodera schachtii*: A, mature female; B, cyst cone; C, second stage juvenile anterior end; D, second stage juvenile tail end; E, male anterior end; F, male tail end.

and rhubarb. Many common weeds, especially those belonging to the *Chenopodiacae* and *Cruciferae* are also hosts.

The number of generations of *H. schachtii* on a host plant depends on how heavily infested the soil is, the length of the plant growth phase and the soil temperature and moisture regime. In cool, temperate soils there are one or two generations on sugarbeet. In Ireland, Duggan (1959) demonstrated two generations on mangolds. In the warmer, irrigated soils of the Imperial Valley, California, however, there may be three to five generations on a beet crop (Thomason and Fife, 1962). A generation takes about 300 day-degrees above 10°C (Greco *et al.*, 1982a), suggesting two generations are possible in northern Europe and three in southern Europe. Multiplication of *H. schachtii* is inversely related to its population density, because there is competition between invading nematodes for limited root space. At very low nematode density (0.25 eggs g^{-1} soil, or less), increase may exceed 300-fold (Cooke *et al.*,1979; Greco *et al.*, 1982b), whereas at high population density roots may be so damaged that the nematode infestation of the soil is smaller after harvest than it was before sowing. Jones (1956) showed that numbers of beet cyst-nematode increased to a 'ceiling' population, which was well below what would be expected, theoretically, from more than one generation on the host plant and this was independent of the initial population density of the soil. Steudel (1972) reported a 'ceiling' of only 20–30 eggs g^{-1} soil in beet grown in monoculture for eight years in Germany. In soils moderately infested with the nematode in England, population increase was modest and in heavily infested soils post-harvest population densities were smaller than the pre-sowing densities. In both cases, biological control of the nematode seems to have been partly responsible (Whitehead *et al.*, 1979c, 1987c). Similarly, Cooke (1984b) obtained only a six-fold increase of *H. schachtii* in soil infested with 0.3 eggs g^{-1} soil and only a two-fold increase in soil with 24 eggs g^{-1} soil. Drought and other factors may also limit nematode increase.

Secondary roots of sugarbeet, invaded by second stage juveniles, cease to grow, causing a proliferation of tertiary roots, giving a bushy or bearded appearance to the root system. The tap root, which becomes the storage root, is stunted and the foliage is reduced, chlorotic and may wilt in hot weather, because the shallow root system is unable to supply adequate moisture to the leaves. This crop injury reduces root yield increasingly, as the initial population density exceeds the yield loss threshold. Actual yield loss is determined by several factors. Griffin (1977a) was able to show that populations of *H. schachtii* may differ in their pathogenicity to sugarbeet. Yield loss was also greater at higher than at lower soil temperatures, but decreased, the older the plants were when they were invaded by the nematode (Griffin, 1981a). Cooke (1987) plotted yield against initial nematode number (P_i), for data from California, England and Italy, according to Seinhorst's (1965) equation:

$$\text{Yield } (Y) = Y_{min} + (Y_{max} - Y_{min})\, z^{P_i - T} \qquad (7.7)$$

in which Y_{max} = yield in absence of the nematode, Y_{min} = minimum yield, z is a constant less than 1 and T is the tolerance limit for P_i. The curves differed significantly in shape and the data was rather dispersed about the fitted curves. Values of z differed somewhat. T (eggs $(100\ g)^{-1}$ soil) was about 100 (California), 180

(Italy) and 200 (England; combined data fitted T=0). He concluded that it was not possible to predict (globally) yield loss at a particular value of P_i. According to the data, yield losses at P_i = 10 eggs g^{-1} soil were 64% (Imperial Valley, California (Cooke and Thomason, 1979)), 37% (mineral soil, England (Cooke, 1984b)), 19% (Italy (Greco *et al.*, 1982a)) and 1–14% (ten organic soils in England (Cooke, 1991)). In Kirghiz SSR, crop losses at 2 and 13.2 juveniles ml^{-1} soil were, respectively, 10% and 50% (Gus'kova *et al.*, 1982). In Kazakh SSR, yield losses at 2–4.9, 7.5–12 and 15–35 eggs and juveniles ml^{-1} soil were, respectively, 8%, 35% and 55% (Sagitov and Tulengutova, 1985). Yield losses may be increased as a result of secondary invading organisms. For example, sugarbeet seedlings infested with *H. schachtii* are more susceptible to *Rhizoctonia solani* than uninfested seedlings (Polychronopoulos *et al.*, 1969).

Cruciferous crops may also be damaged by *H. schachtii*. In heavily infested soil, cabbage, cauliflower, broccoli, mustard and Swiss chard were severely injured and yields were very small (Lear *et al.*, 1966b). Olthof *et al.* (1974) found tolerance thresholds for cabbage, cauliflower and rutabaga, respectively, 6, 18 and 1 g^{-1} or ml^{-1} of soil. At populations from 9 to 64 eggs and juveniles g^{-1} soil, yield losses increased correspondingly from 21% to 54% (Abawi and Mai, 1980) and similar results were obtained by Olthof *et al.* (1974).

In soil under non-host plants, the number of viable eggs declined at 50% per annum in microplots in England (Jones, 1956), 49–63% under lucerne in the Imperial Valley, California and, under annual non-hosts or fallow, at 56% (0–30 cm deep) and 80% (30–60 cm deep) (Roberts *et al.*, 1981). All these comparatively fast rates of decline are partly due to the readiness of the nematodes to hatch in water and partly to death and predation of eggs.

Control

At present, *H. schachtii* is controlled by rotation of host crops with non-host crops, use of non-host or resistant catch or green manure crops and by nematicides.

Crop rotation

As the nematode population declines fairly rapidly in the soil in the absence of host plants (about 50% per annum), growing non-host crops for several years in succession decreases infestation below the damage threshold. When the soil is heavily infested this may take five years or more, and trap cropping or nematicides may be necessary to shorten the interval between sugarbeet crops to keep the rotation economically viable. In temperate soils, a four-course rotation of sugarbeet and three years of non-host crops prevents injury to sugarbeet and nematode population increase. For example, Fichtner *et al.* (1984) reported that, over 80 years, the nematode had not increased in a plot following a four-course rotation with sugarbeet, whereas, in plots following a two- or three-course rotation since 1973, populations exceeded 5 eggs ml^{-1} soil after six to nine years and sugarbeet yield decreased in the tenth year, compared with the four-course plot. Non-hosts of *H. schachtii* include potato, cereals, onion, maize, peas and beans.

However, there is evidence of host races of the nematode, differing in their ability to parasitize cultivated plants (Steele, 1964; Graney and Miller, 1980), so the host status of crop plants in the rotation must be checked against local nematode populations. Rigorous weed control is also essential, if the full potential decline of the nematode under non-host crops is to be achieved. In Italy, soybeans and cereals, grown for at least three years, decreased nematode numbers below the economic threshold* and increased sugarbeet yields (Tacconi and Santi, 1991) and, despite soybeans's partial susceptibility to *H. schachtii* in the USA, the nematode declined as much under it as under cereals. However, in heavily infested soil, three years of non-hosts without resistant oil radish (cv. Pegletta) did not decrease *H. schachtii* below the economic threshold (Tacconi *et al.*, 1990).

Although they are non-hosts of *H. schachtii*, lucerne and navy beans (*Phaseolus vulgaris*) stimulate the nematode to hatch, so hastening decline of soil populations (Golden and Shafer, 1959; Steele and Price, 1965). Similarly, buckwheat (*Fagopyrum esculentum*) is a non-host, which stimulates hatching of *H. schachtii* (Schlang, 1985). In the hotter soils of California, where the damage threshold is only 1 egg or less g^{-1} soil, and where there may be more than three generations on a sugarbeet crop, rotations used as the sole control measure would need to be longer than four-course. In contrast, in northern Europe there are usually no more than two generations and the threshold of damage in the cooler soils is larger (3–8 eggs g^{-1} soil in the Netherlands). For this reason, nematicides have been applied in the USA to prevent injury to sugarbeet and brassica crops in short rotations. In several countries in Europe, resistant oil radish (*Raphanus sativus oleiferus*) and resistant mustard (*Sinapis alba*) cultivars are grown as catch crop or green manure crops, which can speed decline of soil populations of *H. schachtii*.

Trap cropping

Rape seed sown five or six times per season between sugarbeet rows, beginning before sugarbeet emergence, stimulated hatching of *H. schachtii*, which invaded and developed in the rape roots. Provided the plants were uprooted and destroyed before the females had laid eggs, a portion of the soil population was destroyed and sugarbeet yield was increased (Kühn, 1881). Such a procedure was impractical, involving much labour and the risk in bad weather of not destroying the trap plants before completion of the nematode life cycle. Sown in late August or early in September in the Ukraine, cabbage, susceptible oil radish or mustard reduced populations of *H. schachtii* and, as the soil temperature decreased, the nematodes that had invaded the roots did not complete their life cycle (Babich *et al.*, 1991).

A more attractive technique is the use of resistant catch or green manure crops sown after an early harvested non-host crop. Research in Germany and the Netherlands has resulted in the widespread use in Europe of resistant oil-radish and white mustard cultivars, which stimulate hatch but allow little or no nematode multiplication on the roots. The most resistant of these can decrease popula-

* Economic threshold is that population level at and above which the cost of chemical control is justified by the increased yield (Cooke and Thomason, 1979). It therefore varies according to the cost of chemical treatment and the value of the crop.

tions by 70–90%, i.e. oil radish cvs Adagio, Nemex, Pegletta, Trick and Ultimo and white mustard cvs Asta, Condor, Emergo, Maxi, Oscar, Salvo and Serval (Müller, 1991). In heavily infested soil, one or two crops of oil radish cv. Pegletta can decrease *H. schachtii* by 75% (Anon., 1981a).

Resistant white mustard cv. Maxi and oil radish cv. Pegletta decreased *H. schachtii* in soil to at least 45 cm deep, whereas susceptible oil radish cv. Siletina increased the nematode to 45 cm deep (Steudel *et al.*, 1985). While the equilibrium density of susceptible oil radish Siletina was 20 eggs g^{-1}soil, that of resistant Nemex was only 0.5 eggs g^{-1} soil (Müller, 1987). In the Netherlands and in England, resistant green manures had little or no greater effect on nematode numbers than fallow (Heijbroek, 1982; Cooke, 1985). For success, the seedbeds need to be warm and moist to promote rapid plant growth and vigorous nematode hatch. Caubel and Muchembled (1991) found the effect of cruciferous trap crops to vary with years and fields. Yields of sugarbeet were increased up to 10–15% under optimum conditions, the cost of such green manuring being equivalent to about 3–4 t sugar beet ha^{-1}, compared with about 10 t ha^{-1} for dichloropropene fumigation, which increased sugar yields 20–40%. The latter treatment was economic at 6–7 eggs g^{-1} soil. In the Po valley in Italy, oil radish cvs Pegletta and Nemex sown in July after barley or wheat decreased *H. schachtii* in the soil and increased the yield of sugar by 50% and 40%, respectively, compared with plots without catch crops. This technique was considered economically viable at 10 eggs g^{-1} soil (Tacconi and Venturi, 1991; Tacconi *et al.*, 1991).

Crop rotation plus trap cropping

In Italy, at 10 eggs g^{-1} soil, sugarbeet could be grown economically in two or four year rotations with resistant oil radish cv. Pegletta grown after wheat (Tacconi and Regazzi, 1990). Without Pegletta, three years of non-host crops did not decrease *H. schachtii* below the economic threshold from an initial infestation of 7 eggs g^{-1} soil (Tacconi *et al.*, 1990).

Biological control

Evidence of natural biological control of *H. schachtii* by soil fungi was obtained by Crump and Kerry (1987), who demonstrated increased multiplication of the nematode on sugarbeet in soil treated with the fungicide captafol at 60 kg ha^{-1}. Most of this control operated on the second generation and *Nectria radicicola* (syn. *Cylindrocarpon radicicola*) was the fungus most frequently isolated from females, though *Verticillium chlamydosporium* and species of *Fusarium* were among a range of fungi less frequently encountered in the females. In soils infested with *H. schachtii*, up to 77% of females were killed by naturally occurring fungi. Of these, *N. radicicola* was the commonest, attacking second generation nematodes, whereas *V. chlamydosporium* mostly attacked first generation nematodes. In pots, *N. radicicola* killed 40% of females and the most effective isolates of *V. chlamydosporium* killed 75% of first generation nematodes. Fungal control of *H. schachtii* was also shown in field trials (Crump, 1991; Crump and Irving, 1992).

Early root infestation of sugarbeet seedlings was decreased in the field by eight isolates of Gram-negative, motile rod rhizobacteria applied as seed dressings. One isolate decreased invasion of the young roots by 75% (Oostendorp and

Sikora, 1989). Infestation of cabbage roots by *H. schachtii* was decreased 50–77% by the nematophagous fungus *Hirsutella rhossiliensis* (Jaffee and Muldoon, 1989).

Fumigant nematicides

SUGAR BEET

Numbers of *H. schachtii* in the soil can be decreased by 80% or more by soil fumigation. Before the advent of current soil fumigants, Thorne (1939) was stressing the importance of soil conditions to successful soil fumigation for nematode control. He showed the importance of soil structure, soil moisture and the distribution of the nematodes in the soil. He found that 900 kg calcium cyanide ha^{-1} ploughed in 30–35 cm deep, followed by harrowing and levelling the soil, killed 95.5% of *H. schachtii* in soil from 5 to 35 cm deep. Below 35 cm few nematodes were killed and in the top 5 cm of the soil 5% of the cysts survived, because the cyanide gas escaped too rapidly from this layer into the atmosphere to kill all the nematodes. Allen and Raski (1950) noted that D-D at 234 l ha^{-1} controlled the nematode much better in Utah and Idaho than in California and that this was probably due to soil structure differences, for, at 75% of moisture equivalent, treatment was 100 times more effective in sandy loam than in clay loam or peat soil, due to poor diffusion in the clay and peat. In peat soils, generally, adsorption onto the organic fraction is a major obstacle to fumigation. Even on a light soil, ploughed and harrowed before treatment, 280 kg D-D ha^{-1} injected or ploughed into the soil killed only 44 and 52% of the nematodes, respectively, and very few were killed in the top 5 cm of the soil (Raski *et al.*, 1951). However, in clay loam soils with up to 15% organic matter, 140 or 187 l Telone ha^{-1}, applied in 44 cm irrigation water, decreased the number of cysts that developed on the roots and increased sugar yields by 200–300%. Similarly, 94–234 l Telone ha^{-1} injected into a soil in optimum fumigation condition gave comparable benefits (Warren, 1963). In the Imperial Valley, California, beet yields were greatly increased by D-D or 1,3-dichloropropene alone or with aldicarb or carbofuran applied at sowing (Plate 7). Yields were increased up to 163 t ha^{-1} from the best treatment, but although early nematode control was achieved, by harvest untreated and treated plots were equally infested (Kontaxis and Thomason, 1978).

Applying D-D or 1,3-dichloropropene at about 136 kg ha^{-1} to the rows in which sugarbeet seeds were then sown in peaty loam soils, increased sugar yield, but only 1,3-dichloropropene decreased the number of females on the roots midseason (Whitehead *et al.*, 1979c). Similarly in Utah, 168 kg 1,3-dichloropropene ha^{-1}, as 9.4 g m^{-1} row, controlled *H. schachtii* and significantly increased sugarbeet yield at planting temperatures of 8, 12 or 16°C in soils with 5, 10 or 18 eggs g^{-1} soil (Griffin, 1987b).

Soil fumigation is expensive and may not be justified economically. Optimum application methods to soil in seedbed condition and with low organic matter contents may allow effective control to be obtained with smaller dosages. In contrast, very large amounts of Soviet D-D are apparently economic. Injected 20 cm deep in soil in the Ukraine, 500, 700 and 1000 l ha^{-1} killed 57, 77 and 92%, respectively, of *H. schachtii* in the soil. In Kirghiz SSR, autumn application of

2000 l ha^{-1} reduced losses from 67% to 22% and was cost effective at 7 or more nematodes ml^{-1} soil (Gus'kova *et al.*, 1982).

BRASSICA CROPS

Brassicas have responded dramatically to the fumigation of soil infested with beet cyst-nematode. Radewald *et al.* (1971a,b) reported yields of cabbage in California increased 10- and 12-fold, respectively, by 187 and 281 l D-D ha^{-1} and 11-fold by 281 l 1,3-dichloropropene ha^{-1}. A more uniform crop and earlier harvesting are additional benefits of soil fumigation but, even if infestations were decreased mid-season, infestation of the soil at harvest was not decreased. In Poland, cabbage yield was increased by up to 69% by 100 l D-D ha^{-1} in infested soil (Brzeski *et al.*, 1973). Part of these yield increases may be due to increase in ammonium nitrogen in the soil as a result of the inhibition of soil nitrifying bacteria. Brussels sprouts, cauliflower, broccoli, Swiss chard and mustard have also responded well to row or broadcast applications of D-D or Telone in *H. schachtii*-infested soil but again infestations were only controlled early in the growing season (Lear *et al.*, 1961, 1966a,b). The most cost-effective treatment for Brussels sprouts, in soil infested with *H. schachtii*, *H. cruciferae* and club root, was preplant row application of Telone, followed by addition of pentachloronitrobenzene to the transplant water (Lear *et al.*, 1966a).

Non-fumigant nematicides

Recent research on control of beet cyst-nematode has concentrated largely on aldicarb, because this systemic insecticide–nematicide can also protect the sugarbeet crop for several months from insect pests of root and shoot. Of a number of fumigant and non-fumigant nematicides, aldicarb controlled the nematode best (Jorgenson, 1969). Applied just after sowing as a row treatment, 5 kg aldicarb (as 10% Temik granules) ha^{-1} reduced damage to the plants by *H. schachtii* by 90%, retarded the appearance of white females on the roots but did not significantly affect the final nematode density in the soil (Steudel and Thielemann, 1970; Thielemann and Steudel, 1973). Experiments in England by Dunning and Winder (1974) with doses of up to 2.2 kg aldicarb ha^{-1} in the seed furrows at sowing, in soils with 14–68% organic matter, showed significant but inadequate increases in sugar yield and no reduction in final nematode population density. Similarly in the Netherlands, aldicarb at 2.5 kg ha^{-1}, applied to the rows at sowing or singling, gave good early control of the nematode and increased yields but had no effect on the final population density (Heijbroek, 1975).

In organic soils in England, aldicarb or oxamyl, thoroughly incorporated in the top 15 cm of the soil just before sowing, minimized root invasion and hence damage to the seedlings by juveniles, greatly increased sugar yields and reduced nematode increase in four years out of five (Whitehead *et al.*, 1979c). Grujičič (1974a) had also shown that at 5–10 kg ha^{-1} aldicarb protected sugarbeet seedlings from infestation by *H. schachtii* for three months. Such doses are uneconomic. However, experiments in England have shown that the nematode can be well controlled and sugar yields greatly increased by applying smaller amounts of aldicarb than 5 kg ha^{-1} in bands of soil 15 cm wide × 15 cm deep into which sugarbeet is sown, using a Vertical Band applicator (Whitehead *et al.*, 1987c)

(Plate 7). In California, seedling infestation is inhibited if the juveniles are exposed to 1μg aldicarb ml^{-1} and migration of males to attractive females was prevented at a concentration of 0.01 μg ml^{-1}. However, in the field the percentages of females fertilized at 0, 2.5 and 5 kg aldicarb ha^{-1} were, respectively, 69, 49 and 45 (Hough and Thomason, 1975), so an adequate concentration of aldicarb and its nematicidal breakdown products is not maintained long enough in the soil to affect final nematode population density. Hartwig and Sikora (1991) have shown that the motility of second stage juveniles is affected by 0.1 ppm of aldicarb, whereas their orientation towards and invasion of roots were unaffected at concentrations up to 0.1 ppm. The speed of breakdown of aldicarb is, of course, more rapid in warm than in cool soils. The efficacy of aldicarb is also affected by soil moisture content. At 80–100% of field capacity, control was significantly better than at 20–30% of field capacity (Griffin, 1977b). At dosages of 2.8–12.8 kg ha^{-1}, aldicarb was more effective against *H. schachtii* when it was applied to the soil in a single dose at sowing, than when it was applied as a series of treatments through the growing season (Batterby *et al.*, 1980). Repeated application of aldicarb or carbofuran to field soils in which sugarbeet is grown frequently or continuously can result in failure to affect the nematode, through accelerated microbial degradation (Steudel, 1985; Deproft, 1989)

Cocoa-shell powder around seeds of sugarbeet can reduce hatching, root penetration and multiplication of *H. schachtii* and, in the field, may increase sugarbeet yield (Schlang, 1991).

Heterodera zeae Koshy, Swarup and Sethi, 1971

Maize cyst-nematode, *H. zeae*, is widespread in India. It also occurs in Pakistan, Egypt and the USA. It can multiply on maize, wheat, barley, rice, rye, sorghum and *Zea mexicana* but wheat and oats are poor hosts. In the USA, barley has been reported a good host. The host status of sorghum also differs in different reports. Such differences may be due to differences in cultivar and/or nematode populations. Evidence of three different biotypes or races of *H. zeae* has been presented (Kheir *et al.*, 1989; Bajaj and Gupta, 1994). Wheat, usually a poor host of the nematode, was susceptible at 32°C (Bajaj *et al.*, 1987).

The nematode is adapted to high soil temperatures, peak emergence of juveniles from the cysts occurring at 25°C (Srivastava, 1980) and the life cycle is completed in 15–17 days at 27–39°C. The growth of maize is inversely related to the numbers of *H. zeae* in the soil (Srivastava and Sethi, 1984). In the field, infested maize plants are stunted and chlorotic, with narrow leaves.

Control

As yet there is little information on the control of this pest. Maize cultivars differ considerably in susceptibility to *H. zeae* (Srivastava and Swarup, 1975) and a high level of resistance to the nematode has been detected in some precommercial lines of maize (Hashmi *et al.*, 1994). Although damage to maize could be lessened or prevented by growing non-host crops for several years on infested land, further research is needed to clarify the host status of economically practical rotational crops.

Dressing maize seed with aldicarb at 0.75 or 1.5% (w/w) or fensulphothion at 0.25 or 0.5% (w/w) suppressed invasion of the roots by juveniles (Sethi and Srivastava, 1986). Applied to the soil eight days before sowing, fenamiphos at 1.5 or 3.0 kg ha^{-1} or aldicarb or carbofuran at 2.4 kg ha^{-1} controlled *H. zeae* well enough to lessen the number of cysts in the soil after the maize crop, fenamiphos proving the most effective compound. Similar treatments applied eight days after sowing were less effective (Kaul and Sethi, 1988).

Sedentary Endoparasites of Roots and Tubers (II. *Meloidogyne* and *Nacobbus*)

8

Meloidogyne spp.

Root-knot nematodes are very important pests of a wide range of crop plants in warm soils. The commonest species are *M. arenaria* (Neal, 1889) Chitwood, 1949, *M. hapla* Chitwood, 1949, *M. incognita* (Kofoid and White, 1919) Chitwood, 1949 and *M. javanica* (Treub, 1885) Chitwood, 1949 which have very wide host ranges, though races differing in host preferences have been demonstrated within them. These are of worldwide distribution. *Meloidogyne hapla* is limited to temperate soils and to cool soils in the tropics, *M. arenaria* is found in tropical, subtropical and warm temperate soils and *M. incognita* and *M. javanica* are typically tropical and subtropical. In the higher altitude tropics, *M. javanica* may be more abundant than *M. incognita*. Three other species are also very important pests but with a more restricted distribution – *M. chitwoodi* Golden, O'Bannon, Santo and Finley, 1980, especially in the Pacific north-west of North America, *M. exigua* Goeldi, 1887, notably as a serious pest of coffee in humid areas of Central and South America, Martinique, Trinidad and southern India and *M. graminicola* Golden and Birchfield, 1965 on grasses in Louisiana, USA and on rice, principally in Thailand, Bangladesh, India and Laos.

All soft, undergound plant tissues, but especially the roots, are subject to attack by root-knot nematodes. The second stage juveniles, emerging from the eggs, migrate through the soil to reach young roots, which are usually invaded just behind the tips and along the sides of the roots. They penetrate to the pericycle, become sedentary and moult three times to become vermiform males or saccate or pear-shaped females. Large, multinucleate syncytia (the 'giant' or 'transfer' cells) form around the head ends of developing nematodes in response to nematode feeding on cell cytoplasm (Fig. 8.1). Cortical cells hypertrophy and the affected root becomes galled. The galls may be small and discrete or large and coalescent (Fig. 8.2), depending on nematode species, degree of infestation and host plant.

The host ranges of *Meloidogyne* spp. are usually wide or very wide, including many plants from many different plant families, though *M. exigua*, *M. chitwoodi* and *M. graminicola* may have narrower host ranges. *Meloidogyne arenaria* is very

(a)

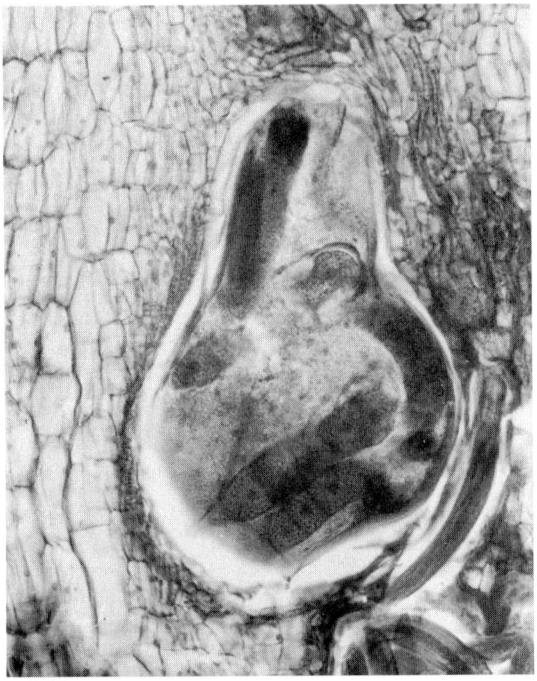

(b)

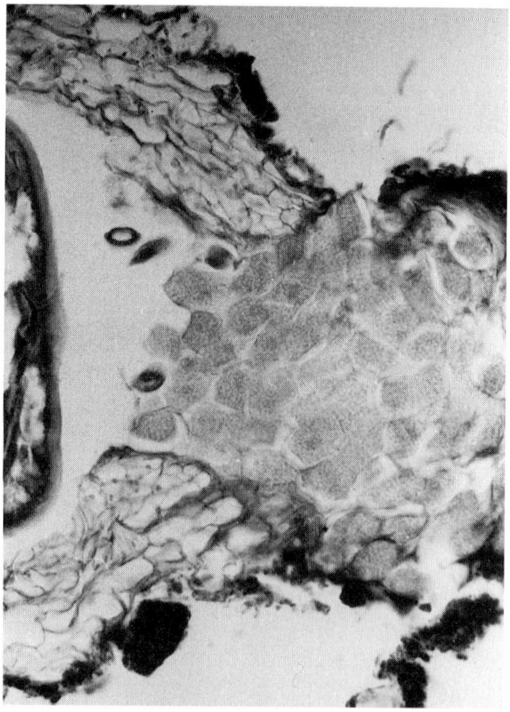

Fig. 8.1. (*and opposite*) *Meloidogyne* in tomato roots. (a) *M. incognita* and associated males. (b) *M. incognita* posterior end of female and egg-sac protruding from root surface; (c) *M. ethiopica* female head surrounded by multinucleate transfer cells; (d) multinucleate transfer cells around head end of female *Meloidogyne* sp.

(c)

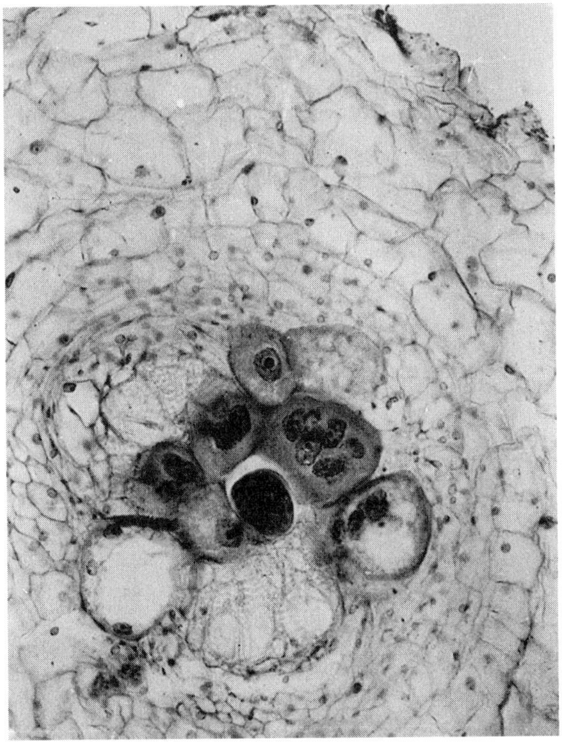

(d)

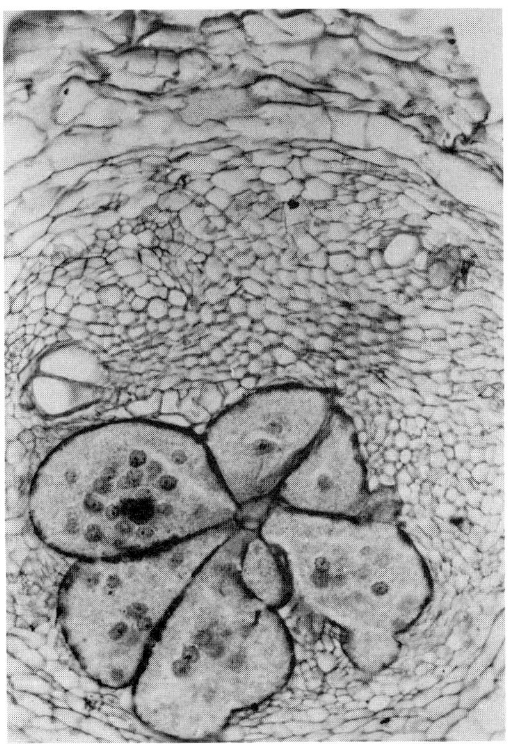

Fig. 8.1. *Continued*

Fig. 8.2. Cucumber roots heavily galled by *Meloidogyne javanica* (Photo: D.J. Hooper).

damaging to groundnuts in south-east USA and to tomatoes. *Meloidogyne chitwoodi* is a serious pest of potatoes, mainly as a result of its attacks on the tubers, which become unmarketable. *Meloidogyne exigua* is a serious pest of coffee and *M. graminicola* can damage rice. The other three species referred to are harmful to many crop plants. Heavy attacks of root-knot nematodes can cause total crop loss (Plate 8). Heavy infestations developing on annual crop plants from light soil infestations are less damaging, unless a bacterial or fungal disease is thereby exacerbated. Depending on the host plant, nematode species, soil temperatures and length of the growing season, there may be as many as ten generations per annum. Reproduction is often parthenogenetic, though sexual reproduction occurs in several species, especially when infestations become large and many juveniles, crowded together in limited plant space, develop as males. The presence of males in a population, however, is not proof of sexual reproduction. *Meloidogyne arenaria* multiplies mainly by meiotic parthenogenesis, occasionally sexually, *M. hapla* by meiotic parthenogenesis and sexually. *Meloidogyne incognita* is parthenogenetic, though males are often common and may aggregate at the posterior end of attractive females. *Meloidogyne javanica* is also parthenogenetic.

Meloidogyne spp. form disease complexes with bacteria and fungi. The nematodes are obligate parasites, which may facilitate the entry and establishment of other disease organisms or weaken the plant, making it prone to otherwise weak pathogens. Bacterial wilt disease due to *Pseudomonas solanacearum* is exacerbated by infestation of tomato with *M. hapla* and of potato or tobacco with *M. incognita*. Similarly, *P. caryophylli* wilt of carnation is increased when the plant is infested with *M. hapla*, *M. incognita* or *M. javanica*. In contrast, nodulation of soybeans by *Rhizobium japonicum* has been suppressed by infestation of the roots with *M. hapla*. Crown gall of raspberries was promoted by *M. hapla* infestation. Amongst the fungi, the various forms of *Fusarium oxysporum* wilt of many crop plants are increased, when they are infested with *M. arenaria*, *M. hapla*, *M. incognita* or *M. javanica*. Important crop plants affected include cotton, cowpea, lucerne, tomato, *Phaseolus* beans and tobacco. Interactions have been demonstrated between *Rhizoctonia solani* and *M. hapla* on soybeans, *M. incognita* on cotton and tobacco,

M. arenaria on cotton and *M. javanica* on soybeans. Black shank disease of tobacco, due to *Phytophthora parasitica* var. *nicotianae* is exacerbated by *M. incognita* or *M. javanica* and *P. sojae* of soybeans by *M. hapla*. *Meloidogyne incognita* interacts with *Pythium debaryanum* on cotton, *P. graminicola* on sugarcane and *P. ultimum* on tobacco. Other fungus–*M. incognita* interactions include *Glomerella gossypii* on cotton, *Sclerotium rolfsii* on aubergine, *Alternaria solani* on tomato and *A. tenuis* on cotton and tobacco. *Meloidogyne javanica* predisposes sugarcane to *Curvularia lunata* root-rot and kenaf (*Hibiscus cannabinus*) to *Macrophomina phaseoli* root-rot.

Fungi interacting synergistically with *M. incognita* on cotton include *Fusarium oxysporum* f. *vasinfectum*, *Rhizoctonia solani*, *Pythium debaryanum*, *Alternaria tenuis*, *Thielaviopsis basicola* and *Glomerella gossypii*. Cultivars of cotton highly resistant to *M. incognita* are therefore recommended for control of both nematode and fungi (Shepherd, 1988). In pots, 2000 juvenile *M. incognita* injured chickpeas more than a 1 g suspension of *Macrophomina phaseolina* per pot. Both together caused more damage than the total of the damage caused by the two separately (Siddiqui and Husain, 1990). In pots, *Verticillium albo-atrum* infected potato petioles two weeks earlier in *M. hapla*-infested than in uninfested plants. In the field too, the severity of the fungus disease in potatoes was increased by infestation with the nematode (Jacobsen *et al.*, 1979).

Control of *Meloidogyne* spp.

Species of *Meloidogyne* have been controlled with varying success by a number of physical, biological and chemical measures. The principal control measures are fallowing or flooding infested land, use of non-host or resistant crop plants in planned cropping systems, disinfestation or protection of planting material from root-knot nematodes, application of nematicides to soil and/or foliage and the use of organic soil amendments. Some control of root-knot nematodes has been reported with antagonistic microflora. Combined or integrated control of *Meloidogyne* spp. is desirable and practical for arable crops, though it may not be feasible for some perennial crops.

Fallowing or flooding infested land

Flooding or fallowing peat soil for 12 weeks greatly reduced the numbers of *M. incognita* and *M. javanica* it contained (Rhoades, 1964). Galling of cotton by root-knot nematodes in Arizona, USA was decreased as much by a preceding summer fallow as by the use of an effective nematicide (Reynolds and O'Bannon, 1966). Flooding soil for 15 days at 30°C killed all *M. incognita* in it, whereas flooding for 50 days at 25°C did not (Kobayashi, 1973). The longer the period of flooding, the fewer the galls of *M. graminicola* on rice roots in the Mekong Delta of Vietnam (Dang-ngoc Kinh *et al.*, 1982). Prot and Matias (1995) have shown that although this nematode can attack upland, rain watered lowland and irrigated rice, it is more prevalent in flooded than in upland fields. Although shallow flooding does not stop it reproducing, it damages fewer roots and does not affect yield compared with upland rice conditions in which it does reduce rice yields. Of 31 cropping systems assessed, bare fallowing coupled with the application of fensulphothion to the seedbed controlled *M. incognita* and *M. javanica* best on tomato (Johnson and Campbell, 1977).

Flooding and fallowing are control measures with restricted applicability. In many situations, e.g. in semi-arid areas of the world, water supplies are too limited to allow land to be flooded. Even where water is plentiful, flooding land may harm soil structure, to the detriment of some crop plants and may entail remedial cultivations and correction of soil nutrient deficiences. Bare fallowing risks soil erosion by wind or, especially on sloping land, by rain. The efficacy of flooding in controlling root-knot nematodes is due partly to death of juveniles unable to find host roots, especially at higher soil temperatures, when nematode metabolic activity is greater, and partly due to anaerobic soil conditions with toxic sulphides, notably hydrogen sulphide, actively killing the nematodes. Similarly, bare fallowing deprives the nematodes of host roots, so juveniles in the soil die in a few weeks or months. However, even if the soil is very dry, at least some of the unhatched juveniles may survive, especially inside fragments of roots.

Soil solarization

Solar heating of the topsoil has decreased numbers of viable *M. arenaria*, *M. incognita* and *M. javanica* infesting it.

In Uzbekistan glasshouses, four ten-day periods of full summer sun raised soil temperatures to 52–55°C, reducing the incidence of *M. arenaria* and *M. incognita* (Rizaeva, 1983). Covering such glasshouses with polyethylene film raised soil temperatures 7–14°C and minimized *M. arenaria*, *M. 'acrita'*, *M. incognita* and *M. javanica* in the root zone of tomatoes and increased tomato yields as much as did treating the soil with aldicarb (Khurramov, 1990). In Turkmenia, where July air temperature can reach 49°C, all *M. arenaria*, *M. incognita* and *M. javanica* were killed in uncovered soil 15 cm deep, resting on concrete, by a total of 110 h sunlight (July 12–August 12, 1987) (Aruntyunov, 1985).

The heating effect of the sun decreases with depth in the soil. In Italy, under PVC or polyethylene sheet, soil temperatures in summer were raised in open fields and glasshouses, respectively, to 38–40°C and 50–55°C at 5 cm deep and 31–32°C and 41–50°C at 20 cm deep (Garibaldi, 1987). In Sicily, temperatures of 45–50°C were achieved at 15 cm deep in glasshouse soil by watering the soil and covering it with 50 μm polyethylene sheet for 45 days from mid-July. As a result, *M. incognita* and *M. javanica* were decreased and the yield of peppers was very greatly increased (Cartia and Greco, 1987). Similarly in India, six weeks solarization with clear, thin polyethylene decreased *M. incognita*, especially in the top 10 cm of the soil, temperature frequently exceeding 40°C (Gaur and Dhingra, 1991). In Florida, solarization of a fine, sandy soil from July to September controlled *M. incognita* on the first tomato crop planted in the treated soil, increasing its yield by 25% (Overman and Jones, 1987). In the Philippines, *M. incognita* was controlled equally as well by soil solarization for 30 or 60 days as by 300 kg dazomet ha^{-1} (Raymundo *et al.*, 1986a). Solarization with double layered plastic sheeting raised soil temperatures to 60.3°C, 12°C more than with a single plastic sheet. Yields of potato tubers were increased more by soil solarization than by soil fumigation (Raymundo *et al.*, 1986b). In Korea, in vinyl-clad greenhouses, covering the soil with two layers of polyethylene, each 30 μm thick, raised the ambient 30°C to a maximum of 48.7°C at 15 cm deep and 45.2°C at 30 cm deep. Lethal temperature for the root-knot nematodes present was 40°C and this was attained at 5 cm deep in August for 17 days with 7 h sun per day. With double polyethylene sheeting, suppression of root-knot nematodes at 5, 15 and 30 cm deep was, respectively, 100, 99 and 98%. The corresponding figures for single sheeting were 96, 74 and 54% (Kim and Han, 1988). Solarization in Pakistan with double clear polyethylene sheeting reduced *M. incognita* by 84%, rather better than with single, clear or single or double black sheets (Javed *et al.*, 1994). In southern Italy, soil solarization for 6 or 8 weeks from late June to mid-August together with soil fumigation, using 300 l D-D ha^{-1}, decreased *M. incognita* and *M. javanica* by over 99% (Greco *et al.*, 1992).

There was little difference in control of *M. javanica*, or in the large yield increases of tomatoes or aubergines obtained from ten weeks soil solarization in the Jordan Valley, between black and transparent polyethylene sheet (Abu-Gharbieh *et al.*, 1991a). Soil solarization for 11 weeks raised soil temperature 3–6°C at 10 cm depth. Addition of solar heated water at 75–90°C at the beginning or end of the solarization period raised soil temperature a further 8–14°C and the maximum temperature reached was 56°C. *Meloidogyne javanica* was decreased and plant growth was greatly improved, especially from the combined treatment (Abu-Gharbieh *et al.*, 1991b).

With two model equations and the known mortality of juvenile *M. javanica* over a typical range of 30–48°C achieved with solarization, percentage mortality can be computed at hourly intervals from temperature data (Cenis and Argemi, 1988). Simulated soil temperatures attained in soil solarization increased the concentrations of ammonium N, nitrate N, P, K, Ca and Mg in non-amended soil (Stapleton *et al.*, 1991).

Non-host plants and crop rotations

A number of grasses, cereals, *Compositae* and other plants are non-hosts of *Meloidogyne* spp., which can dramatically lessen soil infestations. Ten isolates of *M. javanica* and six of *M. incognita*, all from tobacco, south Tanzania, which multiplied freely on the roots of field pea, did not multiply on the roots of cotton (Uk. 51), groundnut (Natal Common), *Crotalaria juncea*, *Tagetes minuta* or soybean (R. 184) (Whitehead, 1965). Three grasses, *Eragrostis curvula* (Ermelo), *Chloris gayana* (Katambora) and *Panicum maximum* (Sabi), grown for four years controlled *M. javanica* on tobacco (Daulton, 1963). and on potato (Mitchell *et al.*, 1971). Grown for three seasons, *Eragrostis curvula* practically eradicated root-knot nematodes from sugarcane land (Dick and Harris, 1971). Pangola grass (*Digitaria decumbens*) is completely resistant to *M. incognita* (Ayala *et al.*, 1967) and growing it for a year on pineapple land 'eliminated' the nematode from the soil (Ayala, 1971). Rye was found to be a non-host of *M. incognita* (Bergeson, 1971) and maize and wheat appeared to be non-hosts of *M. javanica* (Birat, 1970). Barley virtually eliminated *M. incognita* from infested land in Arizona, USA (Carter and Nieto, 1975) and wheat greatly decreased mixed infestations of *M. incognita* and *M. thamesi* Chitwood in Chitwood, Specht and Havis, 1952 from infested soil (Di Muro, 1975). Murphy *et al.* (1974) found *M. incognita* decreased below detectable levels in soil after four, five or more years of Bahia grass

(*Paspalum notatum*) or Bermuda grass (*Cynodon dactylon*) but other important nematodes were increased, i.e. *Paratrichodorus minor* after either grass and *Pratylenchus brachyurus, Xiphinema americanum* and *Belonolaimus longicaudatus* after Bermuda grass.

Similar reductions in the numbers of *M. incognita* in the soil have been achieved by bare fallowing the land or growing marigolds or *Crotalaria* on it for one to three years. It is doubtful if bare fallowing or growing non-hosts completely eliminates root-knot nematodes from infested soil, though numbers are often decreased so much that their detection would require the processing of large quantities of soil.

Among *Compositae*, species of *Crotalaria* (notably *C. spectabilis* and *C. juncea*) and *Tagetes* (especially *T. erecta, T. patula* and *T. minuta*) are resistant or immune to root-knot nematodes. These plants react hypersensitively to invasion by juvenile root-knot nematodes, producing nematotoxins such as α-terthienyl, which prevent the nematodes feeding in the roots. As a result, the nematodes die and the soil infestation is greatly decreased. In this way, *M. incognita* was controlled by *T. patula* (Belcher and Hussey, 1977; Motsinger *et al.*, 1977). Galling of tomatoes by *M. incognita* was greatly reduced where *T. erecta* had been grown for several months (Ducusin and Davide, 1972).

Although groundnut may be susceptible to *M. arenaria* and *M. hapla*, it may also be very resistant or immune to *M. incognita* and *M. javanica*. Daulton (1963) found groundnuts to be poor or non-hosts of *M. javanica* in Zimbabwe. In Senegal, groundnuts are non-hosts of *M. incognita, M. javanica* and *M. 'arenaria'* (Netscher, 1974) but because they are injured by heavy soil infestations, they are only suitable for decreasing light infestations. Among other non- or poor hosts of *Meloidogyne* spp. are upland cotton (*M. javanica*) (Daulton, 1963), chillies (*M. javanica*) (Birat, 1970) and castor (*M. incognita*) (Hackney and Dickerson, 1975). Johnson *et al.* (1975) found four-course rotations of maize, groundnut, cotton and soybean controlled *M. incognita*, because groundnut and soybean were poor or non-hosts of this species.

The efficacy of any of the plants mentioned may vary with the cultivar or wild plant used, the nematode isolate or isolates to which it is exposed and to the soil environment. An effective non-host in one region may therefore be less effective or ineffective in another region. The efficacy, in more recent research, of non- or poor hosts, crop rotations that include them and some management practices are now considered under each species of *Meloidogyne.*

MELOIDOGYNE ARENARIA (Fig. 8.3)

In Alabama and elsewhere in south-eastern USA, this species multiplies on groundnut and is troublesome in crops grown repeatedly on infested land, necessitating repeated use of nematicides. Several crop plants are poor or non-hosts of this species. In infested field soil, in which castor, upland cotton, American jointvetch (*Aeschynomene americana*), partridge pea (*Cassia fasciculata*) or sesame (*Sesamum indicum*) had been grown, the nematode was not detected in the soil after a following crop of groundnuts (Rodriguez-Kábana *et al.*, 1988a,b). Grown for a year, American jointvetch or hairy indigo (*Indigofera hirsuta*) decreased numbers of *M. arenaria* in the soil by 95–100% (Rodriguez-Kábana *et al.*, 1990b).

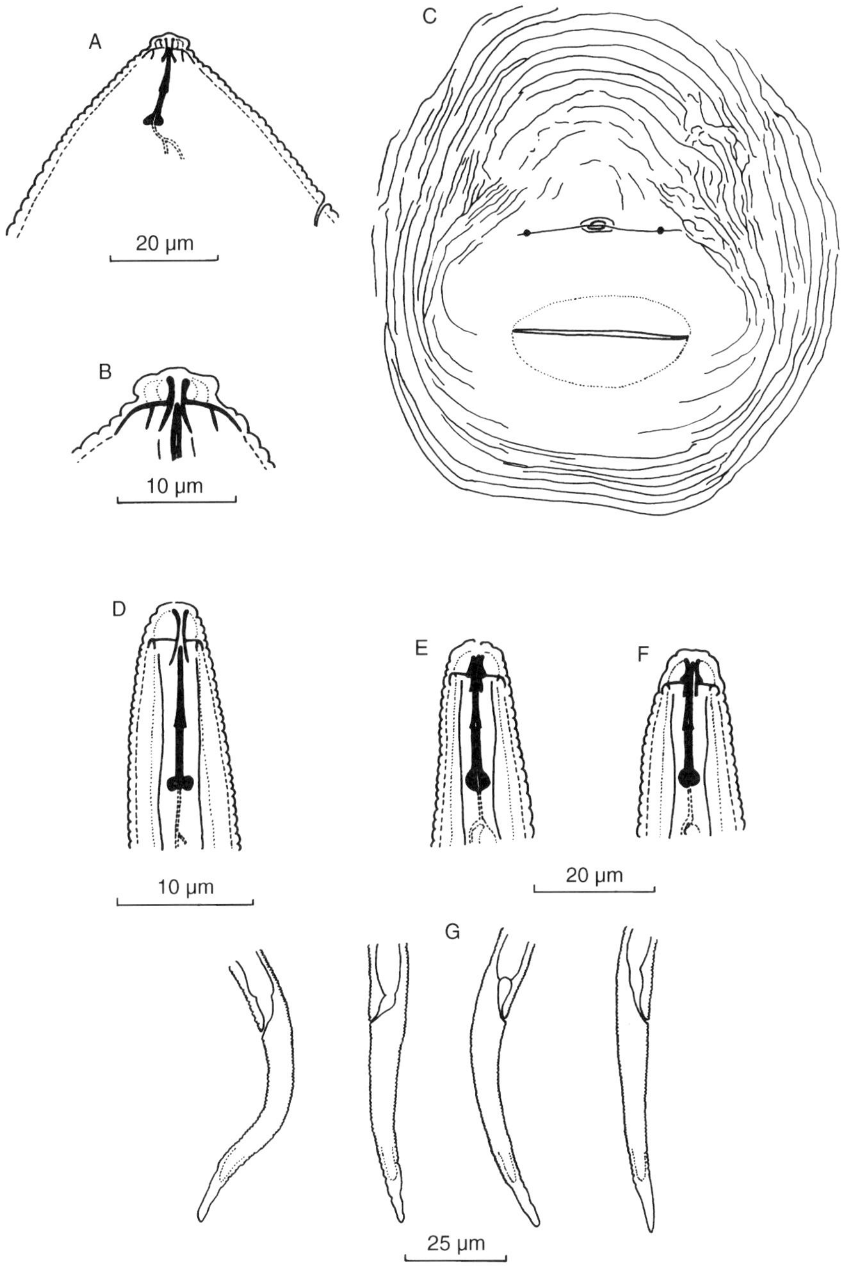

Fig. 8.3. *Meloidogyne arenaria*: A, B, female head end; C, female posterior cuticular pattern; D, anterior end of second stage juvenile; E, F, males anterior end; G, four juvenile tail ends.

Grown for two years, partridge pea suppressed the nematode better than American jointvetch. Groundnuts grown after one or two crops of partridge pea yielded as well as groundnuts grown continuously with aldicarb (Rodriguez-Kábana *et al.*, 1991a). Numbers of the nematode were also minimized by growing maize or sorghum for one or two years or growing soybean in one year and maize in another (Rodriguez-Kábana and Ivey, 1986; Rodriguez-Kábana *et al.*, 1991b). Upland cotton cv. Deltapine 90 and Bahia grass cv. Pensacola are non-hosts of the nematode in Alabama, decreasing both the nematode and southern blight (*Sclerotium rolfsii*) in affected soil, greatly increasing groundnut yields (Rodriguez-Kábana *et al.*, 1994). In Florida, upland cotton, velvet bean (*Mucuna pruriens*), *Crotalaria spectabilis*, hairy indigo, American jointvetch, sorghum–Sudan grass (*Sorghum bicolor* × *S. sudanense*) and fallow decreased race 1 of the nematode in the soil (McSorley *et al.*, 1994).

In China, cotton, groundnut, wheat, sesame and ramie (*Boehmeria nivea*) were immune to *M. arenaria* race 2 as well as to races 1 and 2 of *M. incognita* and to *M. javanica*, all of which injure jute (kenaf) (Xue *et al.*, 1992).

MELOIDOGYNE CHITWOODI (Fig. 8.4)

This is a blemishing pest of developing potato tubers in the Pacific north-west of North America and two races of it have been distinguished from ability or inability to multiply on lucerne (P_f/P_i > or < 1.0) (Pinkerton *et al.*, 1987). Viglierchio (1987) suggested that lucerne, wheat, oats and barley, being poor hosts of the nematode might be useful as rotational crops in infested potato land in North California. Neither race of the nematode increased on two Sudan grass hybrids, nor on three sorghum–Sudan grass hybrids. Used as a green manure, the Sudan grass hybrid Trudan 8 prevented upward migration of juveniles in the soil for six weeks (Mojtahedi *et al.*, 1993a). Damage to potato tubers was lessened in infested soil in which four-month-old rape plants (cv. Jupiter) had been ploughed in, commercially acceptable control being achieved in conjunction with application of ethoprophos to the soil to lessen further the infestation of tubers (Mojtahedi *et al.*, 1993b). A third race of the nematode able to reproduce on *Solanum bulbocastanum* has been recognized in California (Mojtahedi *et al.*, 1994).

MELOIDOGYNE EXIGUA (Fig. 8.5)

Compared with coffee cv. Mundo Novo and tomato which were good hosts of this species, cocoa, rice, sorghum and *Panicum maximum* were non-hosts (Almeida and Campos, 1991a). The nematode was virtually eliminated from infested soil following a rotation of nine crops over a period of 23 months, the coffee seedlings planted after it being uninfested. The nematode was greatly decreased by maize grown for nine months and by rice, sorghum, soybean, *Panicum maximum*, *Crotalaria spectabilis* or *Mucuna aterrimum* (Almeida and Campos, 1991b).

MELOIDOGYNE GRAMINICOLA (Fig. 8.6)

Jute, okra, sweet pepper (*Capsicum annuum*), millet (*Eleusine indica*) and *Digitaria adscendens* are non-hosts or poor hosts of this species, which is a pest of rice, and when planted in infested soil they decrease numbers of the nematode (Roy, 1979, 1987), as do mustard or a mustard–guzitil (*Guizotia abyssinica*) mixture

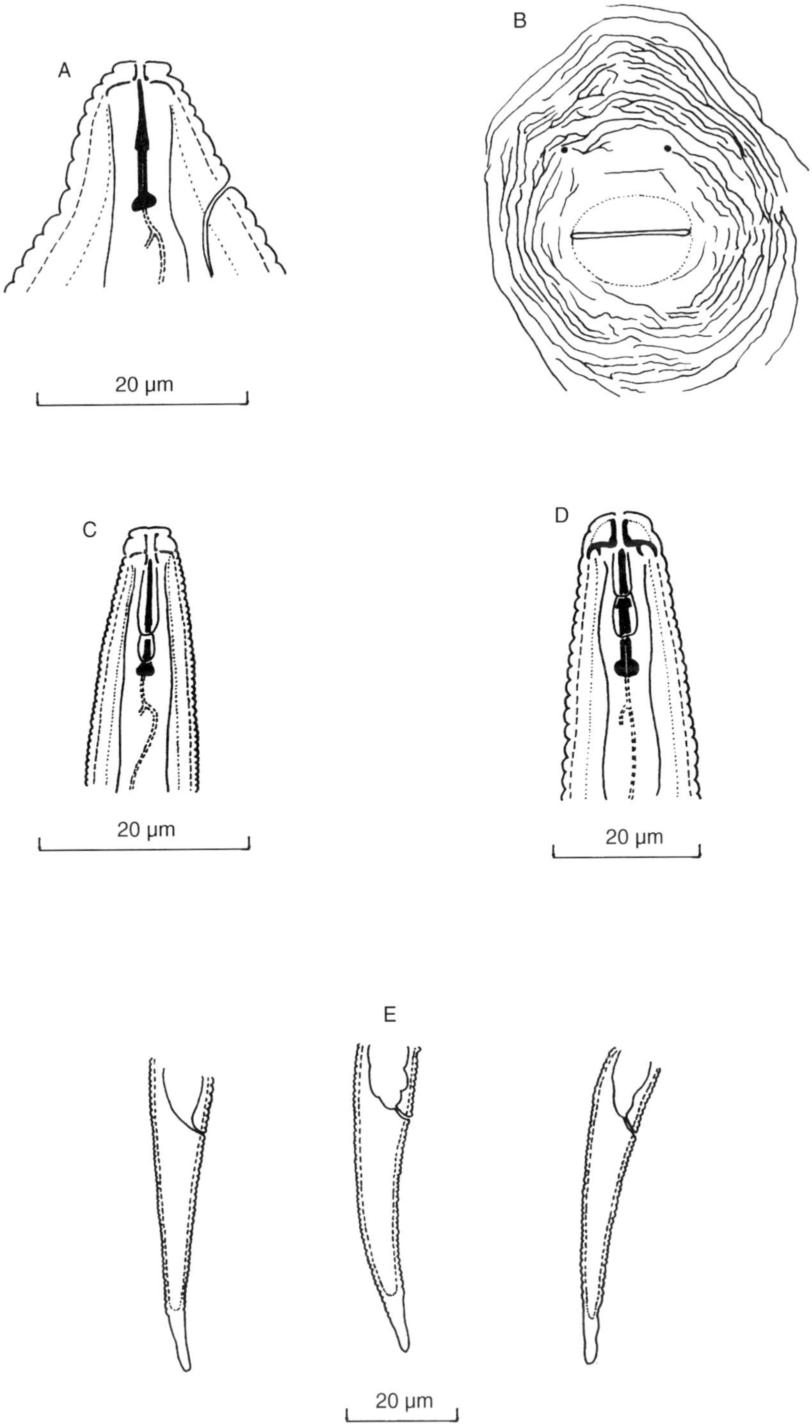

Fig. 8.4. *Meloidogyne chitwoodi*: A, female head end; B, female posterior cuticular pattern; C, anterior end of second stage juvenile; D, male anterior end; E, three juvenile tail ends.

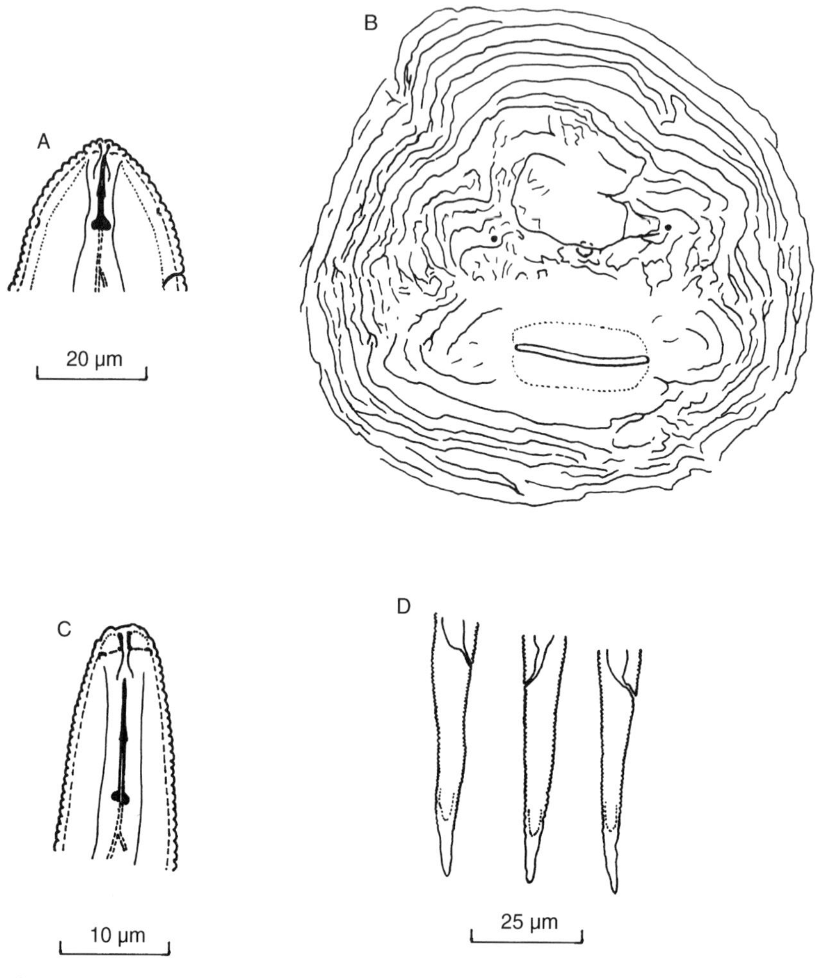

Fig. 8.5. *Meloidogyne exigua*: A, female anterior end; B, female posterior cuticular pattern; C, anterior end of second stage juvenile; D, three juvenile tail ends.

(Rahman, 1990). The nematode can survive in moist soil at 20–30°C, so short-term fallowing is unlikely to free infested soil of this pest (Soomro, 1994).

MELOIDOGYNE HAPLA (Fig. 8.7)
Maize (Ngundo, 1979), spinach, radish, oat, barley and wheat are poor or non-hosts of this nematode. Compared with carrot monoculture, a carrot–onion–oat–carrot rotation decreased *M. hapla* and increased marketable carrot yield almost three-fold, even though onion is also susceptible to the nematode (Belair, 1992). Green manures of *Tagetes minuta* cv. Nemanon, *Hesperis matronalis*, rye or *Lolium multiflorum* sown in early May in 'set-aside' land decreased *M. hapla* by 80–100% at Lelystad, the Netherlands (Lamers and Roosjen, 1993).

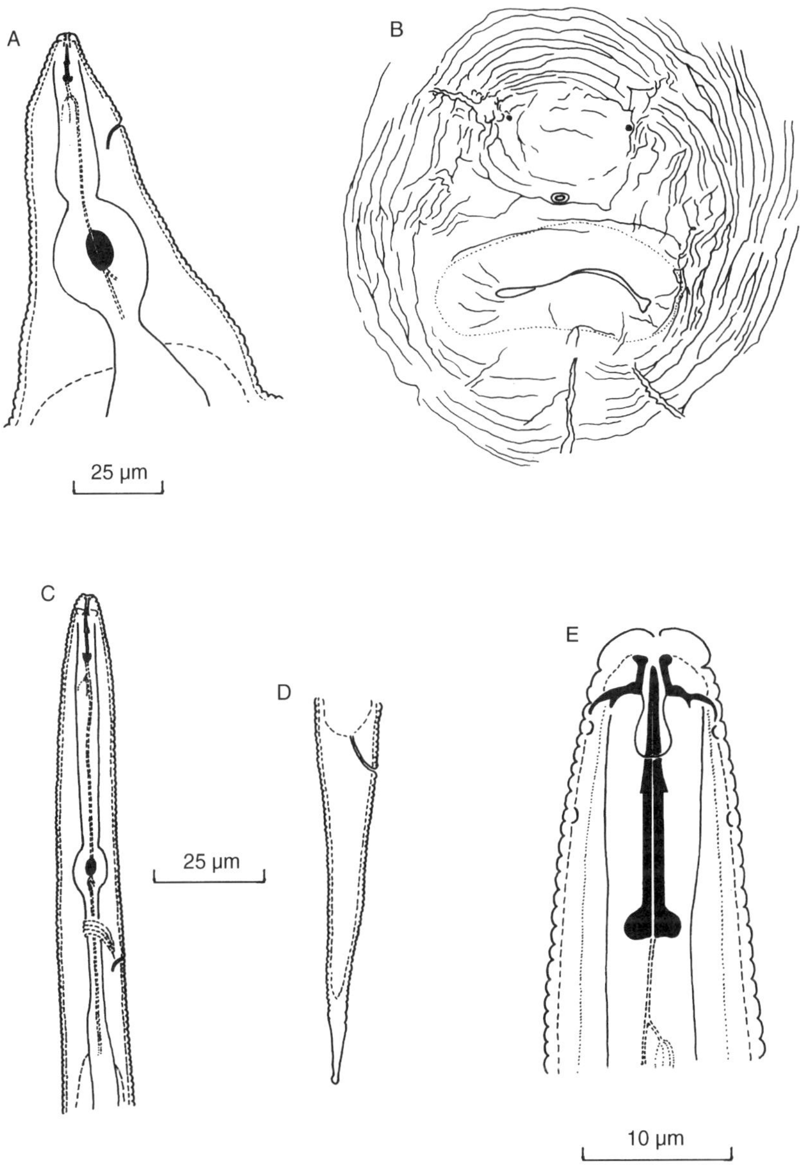

Fig. 8.6. *Meloidogyne graminicola*: A, female anterior end; B, female posterior cuticular pattern; C, anterior end of second stage juvenile; D, tail end of second stage juvenile; E, male anterior end.

MELOIDOGYNE INCOGNITA (Fig. 8.8)

A number of crop plants and cover crops are non-hosts or poor hosts of *M. incognita* but in most reports the nematode races are not identified. The nematode was greatly decreased by growing lucerne for three years and its subsequent increase on cotton in the next three years was slow (Shagalina, 1979). In India, wheat, groundnut, mustard and sesame suppressed the nematode and, of 16 rotations

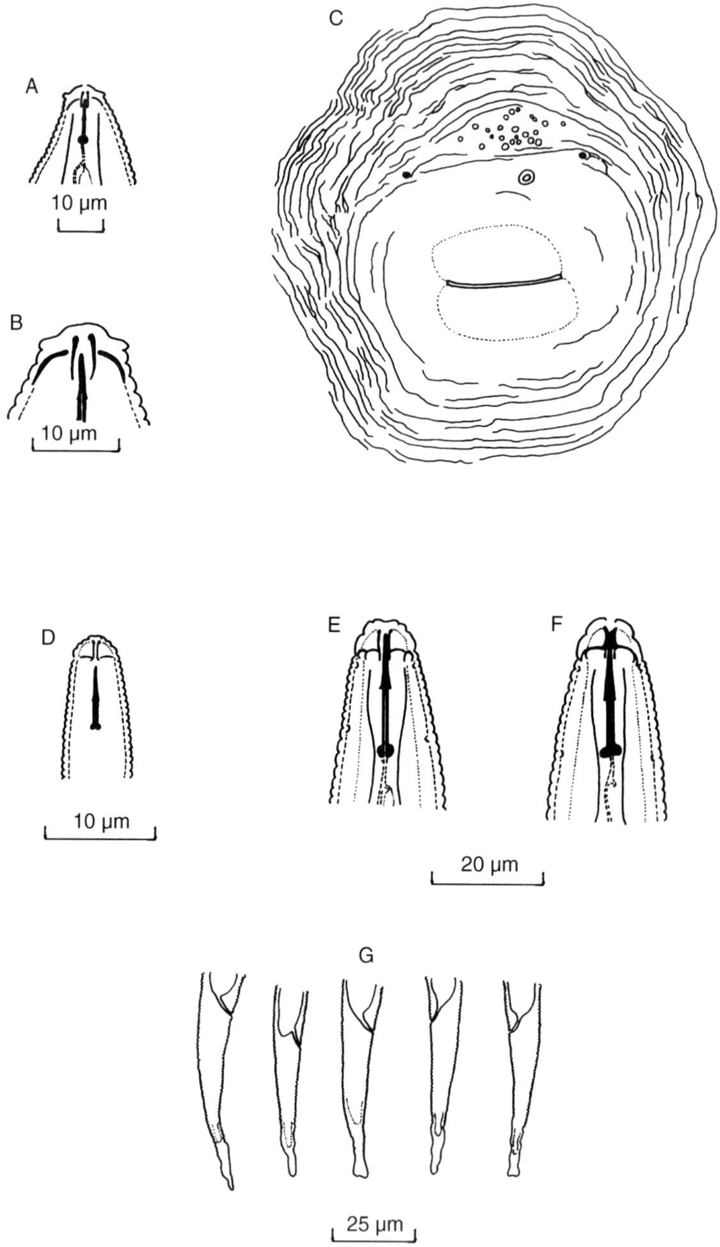

Fig. 8.7. *Meloidogyne hapla*: A, B, female head end; C, female posterior cuticular pattern; D, anterior end of second stage juvenile; E, F, males anterior end; G, five juvenile tail ends.

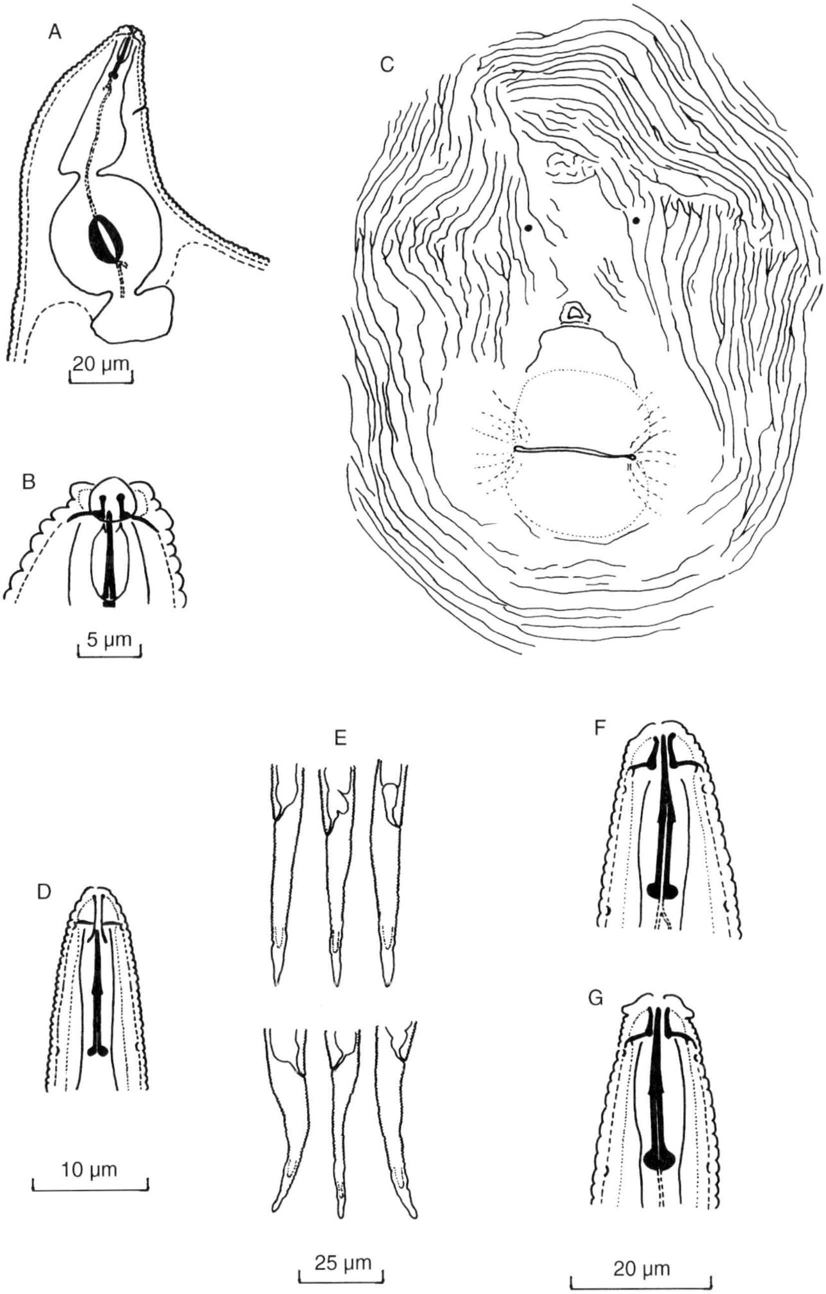

Fig. 8.8. *Meloidogyne incognita*: A, B, female anterior end; C, female posterior cuticular pattern; D, anterior end of second stage juvenile; E, six juvenile tail ends; F, G, males anterior end.

assessed, the nematode was controlled best by a groundnut–wheat rotation (Sharma *et al.*, 1980). Rice is also a poor or non-host in India but whether it is the effect of flooding or of the unsuitability of rice roots is uncertain. Numbers of the nematode were also decreased by growing *Tagetes erecta* or *T. patula* in infested soil (Prasad *et al.*, 1992). The nematode was controlled best by a rotation of *Brassica campestris*–two fallows–*Portulaca oleracea*–two fallows–fenugreek–*Zinnia elegans*–rape–okra (Alam *et al.*, 1981) and by a jute–mustard–ridge gourd–jute rotation (Saikia and Phukan, 1986).

In the USA, groundnut grown for two years decreased race 3 on susceptible cotton which followed it (Kirkpatrick and Sasser, 1984). In Italy, groundnut, maize, wheat and sorghum were poor or non-hosts of *M. incognita* (Di Muro, 1975). Pangola grass was a poor host of *M. incognita*, *M. arenaria*, *M, hapla* and *M. javanica* (Haroon and Smart, 1983). American jointvetch, hairy indigo cv. Florida 101 and cowpea cv. Iron were poor hosts of both *M. incognita* and *M. arenaria* in Alabama (Rodriguez-Kábana *et al.*, 1988b). In Florida, rye, *Sorghum bicolor* and a mixture of *S. bicolor* and Sudan grass (*S. sudanense*) were poor hosts of *M. incognita* (McSorley and Gallacher, 1992). Numbers of the nematode were less than 1 ml^{-1} soil in rotations of turnip–groundnut–turnip or snap bean (Johnson, 1985).

In Japan, *Panicum maximum* cv. Natsukaze or *Crotalaria juncea* cv. Nemakorori decreased the nematode as effectively as fallowing the soil (Araki *et al.*, 1991). Groundnut, *Crotalaria retusa*, *Stylosanthes gracilis*, *Tagetes patula* and *T. erecta* were highly resistant to race 2 (Atu, 1984). In Brazil, the nematode was significantly decreased by growing *Crotalaria spectabilis* for eight months, which decreased the number of forked roots in carrots grown after it (Huang *et al.*, 1980). *Macroptilium atropurpureum* grown for five months decreased the nematode to undetected levels (Ichinohe, 1984). Groundnut and castor were immune, velvet bean and *Crotalaria spectabilis* were resistant to *M. incognita* (Carneiro and Carneiro, 1982). Six species of *Mucuna* were poor hosts of race 3 and of *M. javanica* (Resende *et al.*, 1987). Maize and *Crotalaria juncea* were poor rather than non-hosts of race 1 and this race of *M. incognita* was controlled best by groundnut–groundnut and maize–groundnut rotations (de Moura, 1991). Intercropping mulberry with *Tagetes patula* 30 cm distant decreased infestation of mulberry roots by 75% without decreasing mulberry leaf yield (Govindaiah *et al.*,1991).

In uncultivated land in India, cultivations decreased numbers of most nematodes. Ploughing, levelling, summer fallowing and green manuring with *Crotalaria juncea* almost eliminated nematodes from the soil, including *Meloidogyne incognita*, which was kept under control by specific crop rotations (Lal *et al.*, 1983). *Meloidogyne incognita* and *M. javanica* attacking jute, *Corchorus capsularis* and *C. olitorius*, were greatly decreased by weed control, stubble clearance and growing rice and wheat for two years (Mishra *et al.*, 1987). Carrots sown late in south California were much less damaged by *M. incognita* because the lower soil temperatures decreased the nematode's activity (Roberts, 1987). Trap crops of dill or periwinkle (*Catharanthus roseus*) can greatly decrease the nematode in the soil (Dement'eva *et al.*, 1979; Patel *et al.*, 1991).

MELOIDOGYNE JAVANICA (Fig. 8.9)
In Malawi, all cotton cultivars, *Chloris gayana* cv. Katambora and cassava cv. Masangwi were immune to *M. javanica* (Saka, 1982). In Tanzania, groundnut, sesame (*Sesamum orientale*) and *Crotalaria ochroleuca* greatly decreased *M. javanica* infestation of the soil, increasing subsequent yields of tomato or tobacco (Madulu *et al.*, 1994). In India, carrot, *Capsicum* and onion were poor hosts of the nematode and controlled it in carrot–onion–okra–tomato and *Capsicum*–onion–mung bean-tomato rotations (Kanwar and Bhatti, 1992). Surprisingly, groundnut has been reported susceptible to *M. javanica* (Sakhuja and Sethi, 1986). Interplanting tomatoes with *Tagetes erecta* in Pakistan significantly decreased the nematode on tomato roots and increased tomato yield (Abid and Maqbool, 1990). *Tagetes erecta*, *T. patula* or *T. minuta* interplanted with mung bean decreased the nematode on the bean roots as effectively as *Panicum maximum* or maize, which are too competitive as intercrops for mung bean (Toida *et al.*, 1991). Of a range of legumes assessed in the field in Brazil, *Crotalaria ambrosioides* followed by *C. paulina*, *C. spectabilis*, pigeon pea (*Cajanus cajan*), *Mucuna aterrimum*, *Crotalaria juncea* and *C. retusa* decreased *M. javanica* best (Asmus and Ferraz, 1988). In India, ploughing 10 or 20 cm deep in June, followed by two months fallow reduced numbers of *M. javanica* by 78–92% (Jain and Gupta, 1990).

Resistance

Many cultivated and wild plants have been assessed, worldwide, for resistance or immunity to species of *Meloidogyne*. Resistant plants prevent the nematodes completing their life cycles in their roots. Immune plants prevent all nematode development in their tissues. Plants with useful resistance have been selected, either to replace susceptible cultivars or for use in breeding programmes to transfer resistance to cultivars with more acceptable agronomic qualities. The breeding programme may have been long and complex and for some crop plants sources of resistance have not been found. Nevertheless, many resistant crop cultivars have been produced (Appendix).

In tomato, a single gene (Mi) confers resistance to *M. arenaria*, *M. incognita* and *M. javanica* but not to *M. hapla*, whereas in *Capsicum annuum* resistance to these three species was conferred in two breeding lines by five genes (Dalmasso *et al.*, 1985). The Mi gene does not confer total resistance in tomato to any nematode isolate and, although eight isolates of *M. incognita* and five of *M. javanica* multiplied little or very little on cultivars with the gene, two other isolates of *M. javanica* multiplied appreciably on the roots of resistant tomatoes (Roberts and Thomason, 1986). Roots of the cotton cultivar Auburn 634 prevented multiplication of 12 populations of *M. incognita*, representing the four recognized host races and two cytological races (Kirkpatrick and Sasser, 1983). Resistance to *Meloidogyne* in cotton is important in lessening damage by the wilt fungus, *Fusarium oxysporum* f.sp. *vasinfectum*, because nematode infestation of the roots may render cultivars bred for resistance to the fungus alone susceptible to it. Of course, resistance to *Meloidogyne* does not guarantee freedom from crop loss due to nematode invasion of the roots. That depends on the tolerance of the cultivar.

The roots of plants resistant to a species or race of *Meloidogyne* are invaded by

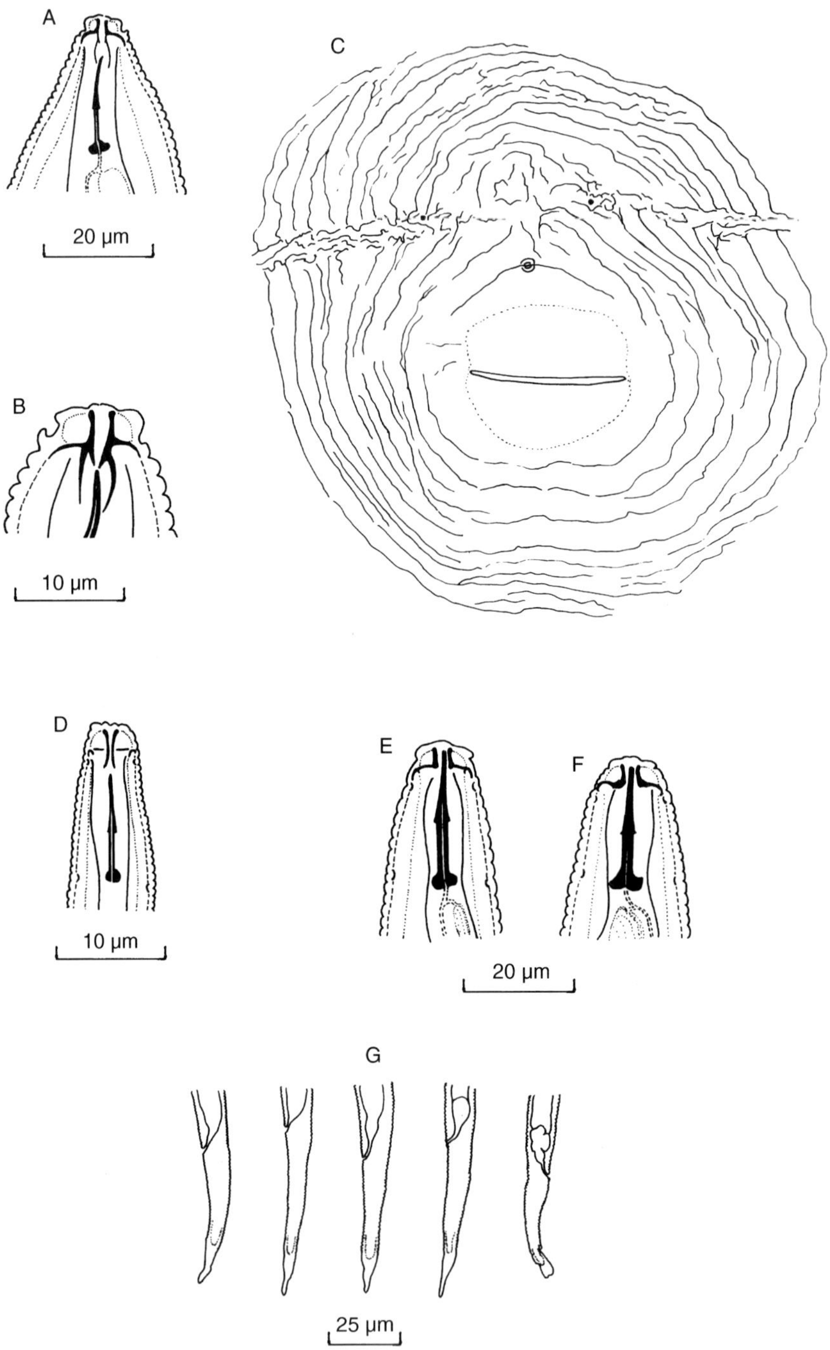

Fig. 8.9. *Meloidogyne javanica*: A, B, female head end; C, female posterior cuticular pattern; D, anterior end of second stage juvenile; E, F, males, anterior end; G, five juvenile tail ends.

second stage juveniles but the multinucleate syncytia (the 'giant' or 'transfer' cells) that form around the heads of sedentary juveniles in the roots of susceptible plants, which are essential to the developement of egg-laying females, are absent, fewer or smaller. As a result, few if any females develop to maturity, though males may develop from small syncytia. Second stage juveniles may even vacate resistant roots, soon after entering them. In roots of the resistant peach rootstock Nemaguard, the syncytia that form in response to the feeding of *M. javanica* are surrounded by a dense layer of cells, followed by degeneration of the syncytia, with the consequent failure of the females to develop (Meyer, 1978). In resistant cowpea roots invaded by *M. incognita*, sclereids formed in the root cortex and there were fewer syncytia, which were small and had few nuclei. Cells died around the juveniles in a hypersensitive reaction to invasion of the roots of one of the cowpea lines studied (Singh *et al.*, 1984).

There have been several comparative biochemical studies of roots of resistant and susceptible cultivars. Resistance to *M. graminicola* in three resistant rice cultivars was associated with an increase in the silica content of the roots as the plants aged, which did not occur in two susceptible cultivars (Swain and Prasad, 1988). Most other studies have been on tomatoes. In resistant, but not in susceptible cultivars of tomato, invasion of the roots by *M. incognita* stimulated synthesis of ascorbic acid. Reduced ascorbic acid in the roots increased their susceptibility to the nematode (Arrigoni *et al.*, 1979a). Conversely, increasing the amount of ascorbic acid in the roots of susceptible tomatoes induced resistance in them to the nematode (Arrigoni *et al.*, 1979b). Spraying tomato foliage with ascorbic acid (4000 ppm), before and after planting, significantly decreased the number of eggs of *M. incognita* produced by the roots and increased tomato fruit yield (Al-Sayed, 1990). When tomato roots were inoculated with *M. incognita*, the concentrations of phenol and orthodihydroxyphenol increased more in the roots of the resistant cv. Anahu than in the roots of the susceptible cv. Marglobe. Polyphenoloxidase increased more in the former than in the latter (Hasan and Saxena, 1979). Accelerated senescence of polyphenoloxidase isozymes was seen in susceptible cv. Pusa Ruby but not in resistant SL-120 tomato roots (Ganguly and Dasgupta, 1984). In resistant cotton roots, there is increased polyphenoloxidase and peroxidase activity involved in the necrotic reaction to invasion by *M. incognita*. Resistance also appears to be linked to the presence of gossypol, a terpenoid aldehyde, at more than 0.52% of dry root matter (Okopnyĭ *et al.*, 1983b). In sweet potato, root extracts of resistant plants following invasion by *M. incognita* and *M. javanica*, contained larger amounts of phenolics (principally chlorogenic acid, scopoletin and esculin) than did root extracts of susceptible plants (Gapasin *et al.*, 1988).

Zacheo and Bleve-Zacheo (1988) found that infestation of tomato roots by *M. incognita* led to the production of superoxide anions ($O_2\bullet$). These were scavenged by increased superoxide dismutase (SOD) activity in susceptible cultivars, whereas in resistant cultivars there was less SOD activity, so there was an increase in $O_2\bullet$ anions. They suggested this might cause cell necrosis and the hypersensitive reaction common in resistant plants. Similarly, SOD activity increased in susceptible cowpea roots inoculated with *M. incognita* but decreased in resistant cowpea roots (Ganguly and Dasgupta, 1988). In cultures of excised tomato roots, hydroxyurea at 10 or 20 ppm in the culture medium induced resistance to *M. javanica* by suppressing the syncytia (Stender *et al.*, 1986). Glyceollin accumulated (mainly in the stele) in the roots of resistant Centennial soybean two to three days after they had been inoculated with *M. incognita*. This did not occur in the roots of the related but susceptible cv. Pickett, nor in either cultivar exposed to *M. javanica*, to which both are susceptible. As glyceollin inhibited movement of *M. incognita* but not *M. javanica*, it might be a factor for resistance in Centennial (Kaplan *et al.*, 1980). Finally, terpenoid aldehydes (polyphenols), which are toxic to *M. incognita*, were produced more rapidly in the endodermis and stele of the roots of cotton cultivars resistant to the nematode (Veech, 1977, 1979). However, there was no significant negative relationship between the terpenoid aldehyde content of healthy roots and the numbers of egg masses of *M. incognita* in ten strains of cotton, whereas there was a significant negative correlation between the terpenoid aldehyde content of 17 other strains of cotton and the incidence of *Fusarium oxysporum* f.sp. *vasinfectum* (Hedin *et al.*, 1984). More recent work has shown that some resistant lines of cotton produce less terpenoid aldehydes when inoculated with *M. incognita* than susceptible lines, so terpenoid aldehydes cannot, in general, be the factor governing infestation in resistant cotton (Khoshkoo *et al.*, 1994).

Cultivar resistance to *Meloidogyne* spp. may break down when soil temperatures are high, when the roots are also attacked by certain fungi and, more commonly, when a cultivar is exposed to a species or race to which it is susceptible.

At soil temperatures in excess of 30°C, cultivars may lose their resistance to root-knot nematodes, so the species or races to which they are resistant at lower temperatures are able to multiply on the roots. Four tobacco lines highly resistant to *M. incognita* at 27°C became moderately susceptible at 35°C (Ohashi, 1977). Similarly, two tobacco cultivars resistant to *M. javanica* at 27°C lost their resistance at 35°C (Fukudome and Kamigama, 1982). In tomato cv. VFN 8, resistance to *M. incognita* decreased progressively as temperature rose from 28 to 37°C. At 34°C, 50% of invading juveniles completed their life cycles in the roots, which showed maximum galling. Peroxidase activity ceased after the nematodes had become established in the roots (Zacheo and Bleve-Zacheo, 1984). In this same cultivar exposed to *M. incognita*, polyphenoloxidase and peroxidase activity increased in infested roots at 27°C, not at 34°C. An O_2- generating system and lipid peroxidation increased at 27°C but decreased at 34°C. In inoculated roots, SOD and catalase activity, which were unchanged at 27°C, were increased at 34°C. All this suggests that the higher temperature induced susceptibility to the nematode as a result of decline in superoxides and hydrogen peroxide, due to the activity of SOD and catalase (Zacheo *et al.*, 1988). In nine other cultivars of tomato, however, little loss of resistance to *M. incognita* was observed above 30°C (Roberts and May, 1986). Although high temperatures have been reported not to affect the resistance of cotton to *M. incognita* (Jorgenson, 1979), in a later study (Carter, 1982) *M. incognita* completed its life cycle on 'Auburn 623' cotton at 35°C, at which temperature there were as many or more nematodes in the resistant roots as in the roots of susceptible Deltapine 16 cotton. Resistance to *M. incognita*, which is heat-stable at 25 and 30°C, has been transferred from *Lycopersicon peruvianum* to tomato (*L. esculentum*) hybrids, from which cultivars with heat-stable resistance may be produced (Cap *et al.*, 1991). When either *Rhizoctonia solani* or *Pythium aphanidermatum* was inoculated with or after *M. incognita* in *Capsicum annuum* cvs Jowala or Longthin Faizabadi, resistance to the nematode broke down (Hasan, 1985).

A major difficulty with the control of *Meloidogyne* spp., by the use of resistant cultivars or rotations with presumed non-host or resistant crops, is the existence of races within each species of nematode. These frequently become apparent when a resistant cultivar is grown, especially if it is grown frequently in the same soil. *Meloidogyne* spp., especially in tropical and subtropical regions, have short life cycles and so pass many generations in a growing season. Selection of virulent nematodes or host races from within a mixed field population may, therefore, be rapid. Plant breeders must necessarily work with a tiny fraction of the nematode genotype range found within a species of nematode, so breakdown of resistance is not so surprising. Eisenback *et al.* (1981) recognized four host races in *M. incognita* and two host races in *M. arenaria*, as well as cytological races in *M. arenaria*, *M. hapla* and *M. incognita* (Table 8.1). These races are widespread. All four races of *M. incognita*, for example, are found in India (Sharma and Gill, 1993). At present, eight races of *M. incognita* are recognized in south-east USA alone, so it is a matter of conjecture how many races of this and other species of root-knot nematode exist. The ability to recognize them is dependent on the number of differential hosts that are used to assess them, as already pointed out for the soybean cyst-nematode, *Heterodera glycines*.

A naturally occurring biotype ('race B') of *M. arenaria* in Senegal multiplied as much on resistant tomatoes cvs Anahu, Healani, Nemavite, Rossol, Caesar, Royal Chico, Slumac and Pannevis 42166 as on susceptible tomato cv. Roma. It was also more pathogenic to cvs Roma and Rossol than *M. javanica* (Prot, 1984). In the Pacific north-west of North America, 13 of 32 geographical isolates of *M.*

Table 8.1. Host and cytological races of three species of *Meloidogyne*.

Species	Differential host reactions		Host race
M. arenaria	Groundnut	+	1
		−	2
M. incognita	Cotton (cv. Deltapine 16)	Tobacco (cv. NC 95)	
	−	−	1
	−	+	2
	+	−	3
	+	+	4
	Cytological races		
M. arenaria	2n = 50–56		A
	2n = 34–37		B
M. hapla	n = 14–17		A
	2n = 30–31		
	3n = 43–48		B
M. incognita	2n = 32–36		B
	3n = 40–46		A

After Eisenback *et al.* (1981).

chitwoodi multiplied on lucerne, which is therefore not a reliable rotational crop to grow in potato-producing areas (Pinkerton *et al.*, 1987) and three races of this nematode are now recognized there. In India, a new biotype of *M. graminicola* (the Agartala isolate) multiplied on rice cvs SR 190, MW 10 and IR 36, which were resistant to the Cuttack and Kerala isolates (Sahu and Chawla, 1986). In a growth chamber experiment, race 2 of *M. incognita* multiplied much more on soybean cv. Centennial than did races 1, 3 or 4, which multiplied well on cv. Pickett 71. This suggested race-specific resistance to *M. incognita* in soybean (Swanson and Van Gundy, 1984). Soybean cv. Lee 74, putatively resistant to *M. incognita*, was found susceptible to all eight races of the nematode. The soybean cv. Coker 156 was resistant to races 1 and 2 and to some race 3 populations but was very susceptible to other race 3 populations and to race 4. Least multiplication of all races occurred on cv. Centennial (Windham and Barker, 1986).

A 'race B' of *M. incognita* found in the Ivory Coast, West Africa is defined as capable of multiplying on resistant tomato cv. Rossol, sweet potato cvs CDH and Chinoise and soybean cv. Forrest. It is distinguished from other populations of the nematode by having esterase phenotype p VI in vertical electrophoresis, as opposed to p I and p VII, from which it is clearly distinct. (Fargette, 1988). Populations of *M. incognita* selected for their virulence to the Mi resistance gene in tomato did not lose their virulence when cultured for 18 generations on susceptible tomatoes, i.e. they did not revert to avirulent forms when selection pressure for virulence was removed (Castagnone-Sereno *et al.*, 1993). This suggests that avirulent genomes had been eliminated in the selection process. One of two populations of *M. javanica* from Gujarat, India, reproduced on three cultivars of groundnut (D. J. Patel *et al.*, 1993). Tobacco cultivars bred for resistance to *M. javanica* (Okinawa 6 and Okinawa 12) were susceptible to this species in Gujarat (Shah *et al.*, 1985) but perhaps this is a case of resistance loss at high soil temperature, rather than due to a difference in nematode population virulence. *M. javanica* from California differed in virulence on cowpea to the same nematode from other parts of the USA (Swanson, 1985).

All these facts lead to the conclusion that resistant cultivars or rootstocks will have a shorter 'life span' than expected if exposed to conditions or nematode populations which they have not been bred to resist. Cultivars or rootstocks must be tested against local populations and conditions before their widespread release to growers and, whenever possible, resistant cultivars of annual or short-term perennial crop plants should not be grown continuously, either as monocultures or in rotations, because they will speed selection of virulent biotypes that may occur in field populations.

Biological control

In recent years, numerous experiments have shown that certain fungi (notably *Arthrobotrys irregularis* and *Paecilomyces lilacinus*) and bacteria (notably *Bacillus penetrans*) applied to infested soil can lessen root-knot nematode infestations on susceptible plants. Some of this work has been extended to the field and has resulted in the development of two commercial preparations – Royal 350 (containing *A. irregularis*) and Biocon (containing *P. lilacinus*). Some success in the control of *Meloidogyne* spp. with these products has been reported, though soil conditions, dosage rates and microbial antagonists still present obstacles to their widespread use. Their commercial viability is uncertain.

ARTHROBOTRYS IRREGULARIS

Arthrobotrys irregularis is predatory on the free-living second stage juveniles of *Meloidogyne* spp. and is active in soil at 5–30°C (optimum 25°C) but is destroyed at 35°C (Cayrol and Frankowski, 1980; Anon., 1981b). 'Royal 350' formulation of the fungus was recommended for use in soils with pH above 6.4, organic matter greater than 0.8% and salinity less than 3%, at a dosage of 140 g product m^{-2} one month before planting in soil free of previous crop debris. In heavily infested soil, R 350 should be applied in conjunction with a nematicide and a resistant cultivar (Cayrol, 1981). In plots of 25 m^2 in southern France, 140 g R 350 m^{-2} decreased the root gall index on tomatoes from 1.7 to 0.6 and increased tomato yield from 9.1 to 15.2 kg / plant (Cayrol, 1983).

PAECILOMYCES LILACINUS

Paecilomyces lilacinus was reported infesting eggs of *M. incognita* (*acrita* form) and cysts of *G. pallida* in Peru (Jatala *et al.*, 1979). The fungus appears to be a facultative parasite, because although 50% of viable eggs of *M. incognita* became infested, nearly all eggs that had been heat killed also became infected (Freire and Bridge, 1985). All isolates of *P. lilacinus*, including a Peruvian one, were able to infect *M. incognita* eggs and lessen hatching. Hatched juveniles exposed to the fungus were less able to infest tomato roots. Root gall formation was reduced by up to 82% when the roots of tomato seedlings were dipped in a suspension of the fungus prior to exposure to the nematode. Spore suspensions were more effective than mycelial suspensions as root dips (Villanueva and Davide, 1984). Dipped in a spore suspension of the fungus and inoculated with *M. incognita*, tomato root surfaces were colonized by the fungus, which also invaded epidermal and cortical cells (Cabanillas *et al.*, 1988).

Davide and Zorilla (1986) decreased *M. incognita* on okra in microplots by 67–77% (compared with 87% by isazophos) by applying *P. lilacinus* to the soil. Control was greater when the spores were applied on rice hulls or bran or on chopped water lily than when applied directly to the soil. Control of *M. incognita* was increased when *P. lilacinus* was applied with *B. penetrans* than when either was applied alone (Dubé and Smart, 1987).

At 25°C +/−2°C, the viability of *P. lilacinus* was low in soil but high on wheat grain or on diatomaceous earth and intermediate on alginate pellets. Applied to soil on one of these carriers, the fungus decreased galling of tomato by *M. incognita* by 25% in the next two years and trebled tomato yield in the second year. In the second year on plots treated with alginate pellets carrying the fungus, there were ten times as many colony-forming units of the fungus after harvest as before planting (Cabanillas *et al.*, 1989a). The best residual control of *M. incognita* was obtained with a Peruvian isolate or a mixture of isolates of the fungus (Cabanillas *et al.*, 1989b). Tomatoes were protected best from attack by the nematode when the fungus was applied to the soil ten days before and at planting, but more egg masses were infested by the fungus when *P. lilacinus* was applied to the soil at planting and 45 days after planting (Cabanillas and Barker,

1989). Large amounts of the fungus are needed to lessen the number of *M. incognita* developing on host roots. For example, Sharma and Trivedi (1989) found that 1, 5, 15 and 75 g of *P. lilacinus*-infected rice inoculum per aubergine plant decreased nematode infestation by 34, 75, 87 and 91%, respectively. Assuming there are 37,000 plants ha^{-1}, 15 and 75 g per plant are equivalent to 555 and 2775 kg ha^{-1}, i.e. impractically large amounts. The best isolate of *P. lilacinus* (Philippine strain No. 1) has been marketed in the Philippines as 'Biocon' (Davide, 1990). In Puerto Rico, applying *P. lilacinus* to field soil one week before planting *Capsicum annuum* greatly decreased its infestation by *M. incognita* and increased its yield (Vicente and Acosta, 1992). While there have been such successes in root-knot nematode control with *P. lilacinus*, there have also been failures. Significant reduction of *M. arenaria* on tomato in pots of heavily infested soil required applications equivalent to 10 or 100 g commercial product of the fungus m^{-2} (10 or 100 $\times$ 10^{11}spores) and even at a dose of 10^{13}spores m^{-2}, though all egg masses were colonized, only 50% of the eggs were parasitized (Gomes Carneiro and Cayrol, 1991). Applied on autoclaved wheat seed, the fungus failed to control *M. javanica* on tobacco. Although there were 1.2–1.3 $\times$ 10^6 spores g^{-1} soil one week after inoculating it and 1.6–2.3 $\times$ 10^4 g^{-1} soil after the second year of tobacco, there was probably insufficient fungus to effect significant nematode control (Hewlett *et al.*, 1988).

VERTICILLIUM CHLAMYDOSPORIUM

Pot experiments have suggested that *M. incognita* might be controlled by applying *V. chlamydosporium* to the soil. At low population density (250 second stage juveniles per tomato plant), 10^3 or 10^4 chlamydospores g^{-1} soil colonized nematode egg masses well, giving up to 80% control, but it had no effect on initial invasion of the roots by the juveniles, even when the roots were extensively colonized by the fungus. At first harvest, galling of tomato roots was controlled better by aldicarb at a dosage equivalent to 3.75 kg ha^{-1} than by combined use of *V. chlamydosporium* and *B. penetrans*. At the second harvest, the combined use of both organisms gave up to 92% control, whereas the aldicarb treatment gave only 55% control. *Bacillus penetrans* was able to infest females within large galls, which the fungus was not (de Leij *et al.*, 1992). If more pathogenic strains of the fungus can be found, it may be possible to reduce the effective dosage from 1000 kg of fungus-infected inoculum ha^{-1} to a more practical 25 kg ha^{-1} (Carlisle, 1987).

BACILLUS PENETRANS AND OTHER POTENTIAL BIOLOGICAL CONTROL AGENTS

This bacterium is of worldwide occurrence, found parasitizing 76 species of nematodes, nearly all of them plant parasites (Sturhan, 1985). The life cycle of an Australian isolate took 85–100 days to complete at 20°C, but only 20–30 days at 30°C. At the higher temperature, infected female *M. javanica* were destroyed before they had reached maturity (Stirling, 1981). The addition of 100 mg of finely ground dried tomato roots, that were heavily infested with *Meloidogyne* females parasitized by *B. penetrans*, to 1 kg of moist soil resulted in 99% of the *Meloidogyne* juveniles in the soil having spores attached to them after 24 h (Stirling and Wachtel, 1980). Davies *et al.* (1988) showed that *B. penetrans* spores from *Meloidogyne* populations would only attach themselves to *Meloidogyne* juveniles and in the largest numbers to the species from which they had been isolated. Adding to soil *M. incognita* juveniles with 1–15 spores attached to their cuticles resulted in a reduction of 82–93% in numbers of the second generation that developed on tomato roots. Dickson *et al.* (1992) found that *B. penetrans* increased from 0.11 spores per juvenile of *M. incognita* to about eight per juvenile in rye and *Vicia villosa* microplots two years later. In the second year, there was less *M. arenaria* in spring and autumn in plots which had been inoculated with *B. penetrans* than in uninoculated plots and the yield of groundnuts was 64% larger in inoculated than in uninoculated plots. Galling of tomato and pepper by *M. incognita* in Puerto Rico was significantly reduced in field trials when *Bacillus thuringiensis* (culture CR-371) was added to the soil (Zuckerman *et al.*, 1993). *Bacillus cereus* in Israel inhibited the invasion of tomato roots by *M. javanica* (Oka *et al.*, 1993).

Apart from vampyrellid amoebae (Homma and Kegasawa, 1984) and predatory mites (Karg, 1983), a number of 'fungal' pathogens or antagonists are potential control agents for *Meloidogyne* spp. *Dactylella oviparasitica* is an egg parasite of *Meloidogyne* spp. (Stirling, 1978) and the nematode-trapping fungus *Monacrosporidium ellipsosporum* has decreased invasion of tomato roots by *M. incognita* by 90% (Gaspard and Mankau, 1987). In two large field plots, isolate 'PD–42' of *Aspergillus niger* significantly lessened infestation of pepper and tomato by *M. incognita* (Zuckerman *et al.*, 1994). Inoculating soil, four weeks before sowing red and yellow pitaya, with the vesicular-arbuscular mycorrhiza *Glomus manihotis* decreased multiplication of *M. incognita* by 83 and 91%, respectively (Palacino and Leguizamon, 1991). A species of *Streptomyces* (culture CR-43) significantly decreased galling of pepper and tomato by *M. incognita* in field experiments. Nematicidal activity was attributed to thermostable molecules of m.wt. > 6000, produced by the streptomycete (Dicklow *et al.*, 1993).

Organic soil amendments

Root-knot nematodes have been controlled and the yields of susceptible crops increased by applying large amounts of organic materials to the soil. Such treatments seem practical only for small areas, such as nurseries and smallholdings and for land close to units producing large amounts of agricultural wastes.

Singh and Sitaramaiah (1966) found oil cakes of margosa or neem (*Azadirachta indica*), castor or groundnut at 1.8 t ha^{-1} lessened the incidence of *M. javanica* on okra and on the tomato crop that followed it and improved crop growth. Margosa at 1.9 or 2.5 t ha^{-1} and groundnut cake at 1.3 t ha^{-1} were similarly effective (Singh and Sitaramaiah, 1971a). Decomposing leaves of *Cassia occidentalis*, sawdust or urea (181 kg ha^{-1}) also lessened the incidence of root-knot on okra and tomato and improved their growth (Singh and Sitaramaiah, 1967). At 2.5 t ha^{-1}, sawdust applied with NPK fertilizer three weeks before planting lessened *M. javanica* on okra, greatly improved yields and was of residual benefit to the tomato crop which followed okra (Singh and Sitaramaiah, 1971b). Organic amendments appeared to reduce the activity of the nematode directly and also indirectly by improving plant health and changing root physiology, conferring greater resistance to infestation (Singh and Sitaramaiah, 1973). Root-knot nematodes are obligate parasites, so a healthy root system *per se* provides a better environment for their development than an unhealthy one. Recent research has been concentrated on control of *M. incognita*, and to a lesser extent *M. javanica*, with animal manures, plant by-products, oil cakes and plant, bacterial and algal extracts

MANURES AND AGRICULTURAL BY-PRODUCTS

Poultry or cattle manure, with or without urea, manipucira (a cassava flour by-product) and other by-products have been shown to decrease *M. incognita* and other species of *Meloidogyne* on susceptible plants.

Meloidogyne incognita. At 4 t ha^{-1}, poultry manure applied to infested soil in Nigeria lessened infestation of tomatoes mid-season and increased tomato yield, partly as a result of nematode control (Chindo and Khan, 1986, 1990). In Assam, 2 t poultry manure ha^{-1} reduced infestation of green gram better than 1 or 2 t mustard or neem cake ha^{-1} (Borah and Phukan, 1992). In the Philippines, chicken manure at 10 t ha^{-1} in the planting furrows controlled the nematode on tomatoes until harvest and was more effective than 10 kg carbofuran ha^{-1} (Duhaylongsod, 1988).

Amending soil with urea, cow dung or leaf mould, as with castor, mustard or neem cake, lessened infestation of tomato, increased the phenolic and amino acid content of the plants and improved their growth. In the soil, saprobiotic fungi increased and parasitic fungi decreased (Singh *et al.*, 1986). In Australia, poultry manure or sawdust at 24, 36 or 48 t ha^{-1} incorporated in infested soil with urea at 1.8 t N ha^{-1} controlled the nematode on ginger better and increased yield as much or more than the best nematicide treatment applied. The urea treatment alone did not control the nematode (Stirling, 1989). Amending soil with cocoa pod husk or cassava peelings lessened infestation of cowpea and although equally effective, rice husk carried pathogenic fungi (Egunjobi and Olaitan, 1986). Cassava peelings decreased galling of sugarcane and increased cane growth (Salawu, 1988). Infestation of *Capsicum annuum* was decreased and its growth increased when crop residues of *Tagetes erecta*, and to a lesser extent, residues of mustard (*Brassica campestris*) or sunflower were incorporated in the soil (Akhtar and Alam, 1992). At 20–80 m^3 ha^{-1}, manipueira lessened both *M. incognita* and *M. javanica* on okra (Ponte *et al.*, 1987).

Meloidogyne javanica. Large amounts of animal manures reduced galling of okra, though some were phytotoxic at 7.5% (w/w) or more in the soil (Montasser, 1991). At 10 t ha^{-1}, neem leaves greatly decreased galling of chickpea in five soil types in Haryana, India and greatly increased chickpea growth (Gupta and Kali Ram, 1981). Nematode control and growth of tomatoes were best when the leaves were allowed to decompose for six, rather than four, two or one week before planting tomato seedlings in the treated soil (Jain and Bhatti, 1988). The addition of *Tagetes patula* green manure, castor or neem cake to infested soil in Kenya reduced galling of tomato, increased tomato growth and increased egg parasitism by *Verticillium chlamydosporium* (Owino *et al.*, 1993).

OIL CAKES

Applied to infested soil at 0.4 to 4 t ha^{-1}, a number of oilseed cakes have lessened infestation of a range of tropical crop plants and increased their yields. Most experiments have been with *M. incognita*. The effects of such soil amendments include inhibited hatching (Hasan and Saxena, 1974), raised levels of phenolic compounds in roots and fewer juveniles in them (Alam *et al.*, 1980) and increase in fungal colony and *Azotobacter* spp. content of treated soil (Pandey and Singh, 1990). The addition of neem cake or sawdust plus NPK fertilizer to the soil increased the phenolic content of both the soil and the roots of tomato plants grown in it and lessened infestation of the roots by *M. javanica* (Sitaramaiah and Singh, 1978). Infestation of betel vines by *M. incognita* was decreased for at least six months by applying 1 t neem cake or 2 t sawdust ha^{-1} to the soil (Jagdale *et al.*, 1985). In another experiment the nematode was controlled for 10–12 months on betel vines by 1 t neem cake ha^{-1} (Acharya and Padhi, 1989). In both experiments, betel leaf yields were increased. Similar results were obtained on mulberry, with 1 t neem, castor, groundnut or gingilly oil cake ha^{-1} applied four times a year (Sikdar *et al.*, 1986) and on okra with 1 t castor, groundnut, karanj (*Pongamia pinnata*) or neem, 2 kg carbofuran or 1 kg carbofuran plus 0.5 t oil cake ha^{-1} (Reddy and Khan, 1991). At 2.5 t ha^{-1}, neem cake was effective on papaya (Routaray and Das, 1988) and on sugarcane (Jonathan *et al.*, 1991). Applied to soil at 1.5% (w/w), castor cake controlled *M. incognita* on transplanted coffee seedlings and improved seedling growth (Jaenh and Lambert, 1983). Similarly, soybean, cotton, groundnut or castor cake applied to soil at 1% (w/w) decreased *M. exigua* on coffee seedlings (de Moraes, 1977). At 3 t ha^{-1}, mustard cake controlled root-knot nematodes on tomato and increased tomato yield (Khan *et al.*, 1995).

The efficacy of control of root-knot nematodes with oil cakes is increased and the risk of phytotoxicity is decreased as the interval between applying them to soil and planting is increased (e.g. Hossain *et al.*, 1992). Neem or groundnut cake was more effective in unsterilized than in sterilized soil, presumably because of faster breakdown of the cake and release of more of the nematotoxic compounds (Bhattacharya and Goswami, 1987). At 400 kg ha^{-1}, neem or karanj cake, applied in furrows or as spot applications, controlled *M. incognita* on tomato, spot application being the more effective treatment (Darekar *et al.*, 1990a). Of all the oil cakes, neem (margosa) seems the most consistently effective in controlling the nematode. The yield benefits from oil cake application are due partly to nematode control and partly to other effects of their decomposition in soil. For example, neem increases *Rhizobium* and *Azotobacter* spp. as well as fungi in the soil and yield benefits may be attributable in part to these effects. The yield benefits in green gram (*Phaseolus aureus*) (Pandey *et al.*, 1991; Dwivedi and Pandey, 1992) could be due in large part to the increase in *Rhizobium* bacteria in the soil, leading to better nodulation of the roots.

PLANT, BACTERIAL AND ALGAL EXTRACTS

All *M. graminicola* were killed eight days after applying the roots and shoots of *Eclipta alba* to the soil (Prasad and Rao, 1979b). Nematotoxic principles have also been shown to occur in several other plants when *M. incognita* was exposed to extracts of their leaves, roots, flowers or seeds. Hatching and the viability of hatched juveniles were decreased by root extracts of neem or Persian lilac (*Melia azadarach*) (Siddiqui and Alam, 1989b), root and leaf extracts of *Crotalaria spectabilis* (Subramaniyan and Vadivelu, 1990), seed, leaf, root and especially flower extracts of *Tagetes lucida* (Siddiqui and Alam, 1989c) and seed coat and especially kernel extracts of neem (Mojumder and Mishra, 1991). Two nematicidal principles, eugenol and linalool, were demonstrated in the leaves of *Ocimum sanctum* and *O. basilicum* (Chatterjee *et al.*, 1982). Dipping roots of green gram (mung bean) in extracts of neem seed coat or kernel protected them from early invasion by *M. incognita* (Mojumder and Mishra, 1991). Aqueous extracts of lemon grass (*Cymbopogon flexuosus*), used as a root dip, suppressed development of the nematode in tomato and aubergine roots (Tiyagi *et al.*, 1990).

Applied to infested soil as a drench, selected strains of certain rhizobacteria, producing compounds toxic to *M. incognita*, reduced the number of galls on the roots of white clover and increased shoot and root weights (Becker *et al.*, 1988). Extracts of the seaweed *Spatoglossum schroederi* at concentrations of 0.5–1% killed *M. incognita* and *M. javanica in vitro* and in soil at 0.5–1%, *S. schroederi*, *Botryocladia occidentalis* and *Bryothamnion triquetrum* significantly decreased galling of tomato by both species and by *M. arenaria* (Paracer *et al.*, 1987).

Disinfesting soils of *Meloidogyne* spp.

Although it is virtually impossible to eradicate root-knot nematodes from soil, especially after deep-rooted crops, its infestation can be decreased dramatically by heating it to a temperature lethal to the nematodes by solarization or by steaming or by thorough fumigation. In hydroponic beds, control is easier, because the much smaller volume of the mineral rooting medium can be treated effectively with a strong disinfectant.

STEAM

Steam sterilization has long been practised in glasshouses to rid the soil of a variety of pests and diseases, including soil-borne viruses that are not controlled by soil fumigation. Thorough treatment is, however, very costly in labour and energy and only affordable for high value crops. Excellent control of root-knot nematodes in the topsoil was obtained by passing steam at 7 kg cm^{-2} (7 bar) pressure into the soil under covers for 30 min or by fumigating the soil with 244–977 kg methyl bromide ha^{-1} under gas-tight sheeting (Daulton, 1956).

SOIL FUMIGATION

A number of liquid, halogenated, aliphatic hydrocarbons and compounds or formulations that release methyl isothiocyanate gas when applied to soil have been effective in controlling root-knot nematodes.The efficacy of these and some other fumigant-type nematicides in earlier reports is first reviewed as background to an account of more recent research on fumigant nematicides that are still available for commercial use.

Carbon disulphide. Bessey (1911) found carbon disulphide used as a preplant treatment for soil more effective in controlling root-knot nematodes than formaldehyde, calcium carbide, potassium sulphocarbamate, quicklime or a number of ammonium compounds. Carbon disulphide emulsion was effective in controlling the nematodes on tomatoes in glasshouses when applied to soil free of galled roots (Anon., 1932). Out-of-doors in New Zealand, however, it was found ineffective at 646 l ha^{-1}, compared with 323 l D-D or chloropicrin ha^{-1} injected 12.5 cm deep, which were effective in controlling root-knot on tomato (Jacks, 1949).

Chloropicrin. Chloropicrin is a powerful fungicide as well as being nematicidal. It is a powerful lachrymatory compound, which makes it unpleasant, even impractical, to apply to large areas. In a fundamental study, Stark (1948) showed that chloropicrin was most effective against root-knot nematodes in porous, sandy soil in which the galls of the previous crop roots had rotted away and when the soil surface was adequately sealed immediately after application. Factors affecting nematode control in order of importance were low soil moisture, sandy soil texture and high soil temperature (29.4°C was best). Chloropicrin vapour was adsorbed increasingly with increase in the clay and organic matter content of the soil. This adsorption was decreased by increasing soil temperature and decreasing soil moisture, because dry organic matter adsorbed little of the gas. Clayton and Ellis (1948, 1949) found 374 l ha^{-1} applied seven or ten months before planting reduced galling of tomatoes from 98% in untreated to 5% in treated sandy loam soil. Similarly, 323 l ha^{-1} gave excellent control of root-knot on tomatoes in New Zealand (Jacks, 1949). Much cheaper, yet effective, spot treatment for watermelons was reported by Taylor and McBeth (1941). Injected 15 cm deep at the planting positions, followed by watering of the soil, 2 ml, 3 ml, or three lots of 1 ml chloropicrin greatly decreased infestation of watermelons planted after fumigation.

D-D mixture (1,3-dichloropropene : 1,2-dichloropropane). A new soil fumigant (Shell's D-D mixture) was first reported effective as a nematicide when Carter (1943) found that 187 l ha^{-1} injected into soil at points 30 cm apart, whether or not it was covered with asphalt-impregnated paper to delay escape of the fumigant from the soil, greatly decreased infestation of pineapples transplanted into the treated soil. McBeth (1951) showed that applying D-D for five consecutive years to soil infested with root-knot nematodes virtually eliminated infestation of susceptible crops in years four and five, without significant phytotoxicity over the five year period. Preplant application of D-D greatly decreased galling, at least temporarily, on a range of crop plants at dosages of 187 to 561 l ha^{-1}. Good control and increased tomato yields in glasshouses were obtained with 215 l ha^{-1} (Jacks, 1945) and 374 l ha^{-1} (Heeley, 1951). Out-of-doors, rates of 187–561 l ha^{-1} were effective for root-knot control on tomatoes (Clayton and Ellis, 1948, 1949; Read, 1947; Jacks, 1949; Sundaram, 1972; Cuany *et al.*, 1974). Effective control was also achieved with 281 l ha^{-1} for okra (Batista and Carneiro, 1947), 400 l ha^{-1} for potato (Mukopadhyaya, 1970), 320 l ha^{-1} for carnation (Scognamiglio, 1954), 323 l ha^{-1} for soybeans (Fukui and Yarimizu, 1953a b), 450 l ha^{-1} for jute (Hung, 1956) and 225 l ha^{-1} for cotton (Raski and Allen, 1953; Raski *et al.*, 1953). In tobacco seedbeds the aim was to ensure all seedlings were free of infestation, so larger doses were used. Naude (1949) found 674 l ha^{-1} effective. Daulton (1950) found 861 l ha^{-1} killed all juvenile root-knot nematodes in tobacco seedbed soil up to 23 cm laterally from injection points 30.5 cm apart and 23 cm deep.

As D-D does not kill nematodes within undecayed roots, control may be less effective if these are present in the soil (Jacks, 1945). In grapevine soil, D-D killed a greater percentage of the nematodes when it was applied two or three years after removal of vines, than just after uprooting, probably for the same reason (Raski and Lear, 1962). D-D was more effective as well as less noxious to apply than chloropicrin (Allen, 1946; Conroy, 1951; Fukui and Yarimizu, 1953a,b). Applying D-D in widely spaced bands of soil in which tomatoes were subsequently planted controlled the nematodes as effectively and at lower cost than treating all the topsoil (Walter and Kelsheimer, 1949). In cotton, however, 44–135 l ha^{-1} in rows 97 cm apart controlled *M. incognita* (*acrita*-type) less and increased cotton yield less than 225 l ha^{-1} applied on a broadcast basis (Raski and Allen, 1953; Raski *et al.*, 1953).

1,3-Dichloropropene. The Dow Chemical Co. introduced Telone in 1956 and more recently, Telone II, mixtures containing, respectively, 78% and 92% 1,3-dichloropropene, 8% and 7% 1,2-dichloropropane and about 14% and trace propylene dichloride. Telone did not contain impurities found in D-D able to taint potatoes (Shepherd, 1952) but the propylene dichloride in it persisted in soil and caused malformation and stunting of the ears of wheat planted after soil fumigation. This problem did not arise with Telone II, which has now replaced Telone. 1,3-dichloropropene exists as two isomers : *cis* (now Z) and *trans* (now *E*):

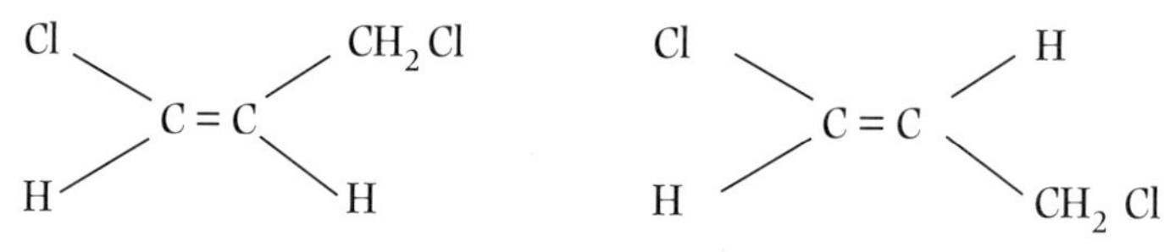

cis (Z) *trans* (*E*)

In laboratory assessments against *M. incognita*, *M. javanica* and other nematodes, the *trans* isomer was 90% as toxic as the *cis* isomer in the vapour phase and 60% as toxic in the water phase. At 25°C, only a quarter of the soil-water phase dosage of *cis*-1,3-D required at 5°C was needed to attain the LD_{95} (McKenry and Thomason, 1974). In the laboratory, only 3.7 ppm of 1,3-D gas compared with 650 ppm methyl bromide gas was required to kill all *M. incognita* (Abdalla, 1974). In the field, however, gas concentrations less than 3.7 ppm of 1,3-D maintained for a long period also gave good nematode control. Maggenti and Hart (1970) found that 1,3-D at 468 l ha^{-1} and in two applications at 749 and 187 l ha^{-1}, in soil containing 125 juvenile *M. incognita* l^{-1}, prevented any infestation of peach seedlings in the soil for up to 22 months. Deep placement of 1,3-D in soil increased nematode control (Lembright *et al.*, 1968). In a Californian vineyard, 2339 l of 1,3-D ha^{-1} virtually eradicated *M. incognita* from the soil (Kissler *et al.*, 1973). Such large amounts are needed if the soil contains infested, undecayed roots, as two to four times the dosage to kill free juveniles in soil is needed to kill the nematodes in *Ficus* roots 3.2–6.4 mm thick, whereas even ten times that dosage did not kill all root-knot nematodes in roots 25 mm thick (McKenry and Thomason, 1975). The $LD_{99.9}$ was estimated to be 180 μg *cis* 1,3-D ml^{-1} per day to kill *Meloidogyne* spp. in vine roots 3.2–13 mm thick at 20–23°C, an eight-fold increase in the dosage needed to kill infective juveniles in the soil (McKenry *et al.*, 1977). Ideal dispersion of 1,3-D occurred in uncompacted, drying, sandy loam at 15–20°C (McKenry *et al.*, 1978).

Ethylene dibromide. Lear (1947) found that root-knot nematodes were almost completely eradicated from soil in which a mixture of one part EDB to nine parts propylene dichloride had been injected at 620 l ha^{-1} 10 cm deep, in rows 25 cm apart at points staggered 25 cm apart. Solid gall tissue left in the soil after the preceding tomato crop was penetrated by the fumigant in 24 h. Root-knot nematodes and *Fusarium* wilt in cotton were almost completely controlled by preplant application of 290 and 346 l ha^{-1} of a 10% solution of EDB in Naptha B (Smith, 1948). Infestation of potato and soybean roots was decreased over 90% by treating the soil before planting with 281 l ha^{-1} of 10% EDB solution (Boock, 1950). Injury to groundnuts by *M. hapla* was prevented and galling of the roots was decreased by applying 4.9 or 9.8 ml of 40% EDB m^{-1} row before sowing in the predetermined plant rows (Cooper, 1952). EDB has been widely used to fumigate tobacco seedbeds and at 674 l ha^{-1}, Dowfume W-40 (270 l EDB ha^{-1}) was as effective as D-D (Naude, 1949). The fumigants were applied to the plough furrows, which were immediately closed and the soil rolled. Against *M. javanica* on tobacco, 225 l of 15% EDB (33.8 l EDB ha^{-1}) was more effective than several non-fumigant nematicides (Colbran and Brouwer, 1972). Eight weeks after sowing tobacco seeds, root-knot infestation was controlled equally well by EDB that had been applied in a single fumigant row (55 l of 15% EDB ha^{-1}) as in a double row (110 l ha^{-1}) or overall application (225 l ha^{-1}) and tobacco yields were equally good. Although, after harvest, root-knot control was less in row-treated than in overall-treated plots, this did not adversely affect leaf yields (Broadley, 1976b). EDB is thought to be carcinogenic and its use has been discontinued in many countries, despite its efficacy in controlling root-knot nematodes.

1,2-Dibromo-3-chloropropane. DBCP has given variable control of root-knot nematodes. Some failures may be attributed to its inability to kill nematodes inside unrotted root galls. Control of root-knot nematodes was also more effective when DBCP was injected into the topsoil than when it was applied in the seed furrows or drenched into the soil (Birat, 1965). *Meloidogyne incognita (acrita)* on cotton was controlled better by injecting the soil with 14 l DBCP ha^{-1} than 62 l EDB or 225 l D-D ha^{-1} similarly applied (Allen *et al.*, 1955). Against *M. incognita* on soybeans in Florida, 6.8 l DBCP ha^{-1} was effective (Kinloch, 1974). *Meloidogyne javanica* on tomatoes was controlled by DBCP in Sudan (Yassin, 1974) and at 40 l ha^{-1} in Haryana, India (Sundaram, 1972). The withdrawal of DBCP from general agricultural use, for toxicological reasons, is of less concern for root-knot nematode control than for the control of other nematodes by repeated applications to perennial crop plants. All other soil fumigants in current use are too phytotoxic for this purpose.

Methyl bromide. At dosages of 336–1435 kg ha^{-1} under polyethylene sheeting, MB has given good or excellent control of root-knot nematodes. The cost of treatment and the hazardous nature of this fumigant have mostly restricted its use to glasshouse crops and nursery beds. At 1076–1435 kg ha^{-1}, MB has given very good control of root-knot nematodes in tobacco seedbeds (Martin, 1950). MB was more effective than dazomet or metham-sodium in suppressing infestation of tea cuttings, over a period of 36 weeks in nursery beds (Nara *et al.*, 1973). At 244, 488 or 977 kg ha^{-1} (Johnston and Reynolds, 1954) and at 500 kg ha^{-1} (Cuany *et al.*, 1974), MB controlled root-knot very well on tomatoes. In vineyards, the need is for long-lasting nematode control in both topsoil and subsoil. Although conventional application of 336 kg ha^{-1} fully controlled *M. incognita* on tomato test plants, it did not control the nematodes deep in the soil (Kissler *et al.*, 1973). Injected about 0.8 m deep at 337–898 kg ha^{-1}, without polyethylene soil covering, MB concentrations were large enough to kill all *Meloidogyne* and other nematodes to 2.4 m deep in vineyard soil, control persisting for up to two and a half years (Raski *et al.*, 1973; Abdalla *et al.*, 1974).

Methyl isothiocyanate. A mixture of methyl isothiocyanate in D-D was a very effective soil fumigant against both *M. incognita* and *Verticillium* wilt of tomatoes (McKeen and Sayre, 1964). Peach seedlings planted in soil which had been treated with 561 l of methyl isothiocyanate (20%) : D-D (80%) ha^{-1} and immediately covered remained free from root-knot infestation for 22 months (Maggenti and Hart, 1970). Similarly, 600 l of Di-Trapex ha^{-1} was effective against *M. incognita* and *M. arenaria* on cucumber (Mancini *et al.*, 1974). On tomato, *M. incognita* was controlled and yield was increased by injecting the soil with 400 l methyl isothiocyanate (20%) in D-D ha^{-1} (D'Errico and Cancellara, 1977).

Metham-sodium (sodium *N*-methyl dithiocarbamate). Formulated as Mylone or Vapam, metham-sodium injected into soil breaks down to release methyl isothiocyanate. At 1500 l in 20,000 l water ha^{-1}, metham-sodium prevented root-knot formation on cucumber (Stettmeier, 1957). At 1.5–2 t ha^{-1}, metham-sodium controlled *M. hapla* on tomato in Kazakh SSR (Sagitov, 1971). Generally, metham-sodium is rather less effective than some other soil fumigants, being used rather as a soil fungicide.

Dazomet. Dazomet, like metham-sodium, also breaks down in soil to release methyl isothiocyanate, but unlike metham-sodium it is a solid, which is rotary cultivated into the top 10–15 cm of the soil. At 1–1.5 t ha^{-1} it controlled *M. hapla* on tomato in Kazakh SSR (Sagitov, 1971). At 750 kg ha^{-1} it killed over 94% of juvenile *M. incognita* in soil and decreased infestation of tomato plants from 81 galls per plant in untreated soil to one per plant in treated soil (Nirula and Bassi, 1965). At 200 kg ha^{-1} dazomet decreased galling of lettuce roots due to *M. hapla* by 83% (Heide, 1973). As it is costly, dazomet is only suitable for the protection of high value crops.

Depth Distribution of *Meloidogyne* spp. and the Depth of Control

Bird (1969) found that *M. incognita* could migrate to a soil depth of 120 cm with the growing roots of tomato or cucumber. Steam and methyl isothiocyanate in D-D used separately or together did not affect the nematode below 1 m deep in sandy glasshouse soil. The nematode was shown able to migrate up and down through a soil depth of 150 cm. Inoculating the top 30 cm of the soil with 4400 *M. incognita* kg^{-1} soil led to heavy infestation of the soil to a depth of 150 cm. Infestation of the top 30 cm of soil reduced yield of the first crop by 20% and of the second crop by 70%. Similar population densities 120–135 cm below the soil surface reduced tomato yield by 11% in the first crop and by 59% in the second crop (Johnson and McKeen,1973).

Both subsoiling and ethylene dibromide increased yields of tomato, though there was no extra benefit from the use of both treatments in plots heavily infested with a root-knot nematode (Hughes, 1958). He suggested that improved rooting may protect plants from drought. At 247 l ha^{-1} injected at 20 and at 45 cm deep in the soil, D-D gave deeper control of *M. incognita* than 494 l ha^{-1} injected 20 cm deep, though it did not produce greater yield from tomatoes planted in the treated soil and neither treatment protected more than one crop (Johnson, 1975). In cotton, subsoiling under the rows increased yields of cv. Coker 310 in soil infested with *M. incognita* but did not reduce infestation with *M. incognita* (Bird *et al.*, 1974). In sandy loam heavily infested with *M. incognita*, subsoiling under the rows or application of DBCP increased yields of soybeans similarly. Subsoiling reduced root-knot in unfumigated plots. The effects of combined subsoiling and fumigation were greater than those of either alone, though they were not additive (Minton and Parker, 1975). The cost : effectiveness ratio of deep soil fumigation for a number of annual crop plants would seem to be smaller than for topsoil fumigation and subsoiling.

Recent Research on Control of *Meloidogyne* spp. with Soil Fumigants

The range of soil fumigants currently in use for the control of root-knot nematodes is smaller than it used to be. DBCP has been withdrawn from use in the USA and elsewhere, EDB is a possible carcinogen and methyl bromide is possibly harmful to the ozone layer of the Earth's atmosphere. The remaining compounds are D-D and Vidden D, which are increasingly being replaced by the more refined Telone II (1,3-dichloropropene) and dazomet, metham-sodium and methyl isothiocyanate alone or mixed with D-D.

MELOIDOGYNE ARENARIA

EDB at 14–18.7 l of 90% e.c. ha^{-1} applied to the soil at planting was not phytotoxic to groundnuts, controlled the nematode as effectively as DBCP and increased yields greatly (Rodriguez-Kábana *et al.*, 1979a). The nematode was controlled just as well by 56 as by 112 l of 1,3-D ha^{-1} applied in winter or two weeks before sowing groundnuts (Kinloch and Dickson, 1991). Race 2 of the nematode on soybeans was decreased and the yields of seven cultivars were increased by fumigating the soil with 105 l EDB ha^{-1} (Weaver *et al.*, 1987). Galling of soybeans by *M. arenaria* or *M. incognita* was significantly decreased by 17.2–51.6 l of 1,3-D ha^{-1} applied to the soil at sowing or three days earlier. Juveniles of both species were fewer at harvest in plots given 34.4 or 51.6 l ha^{-1} than in unfumigated plots. The data predicted maximum yield increases from 40 l ha^{-1} before sowing and from 25 l ha^{-1} (*M. arenaria*) or 30 l ha^{-1} (*M. incognita*) at sowing (Kinloch, 1986). Applying 450 kg methyl bromide ha^{-1} to the soil plus mycorrhizae at planting controlled root-knot nematodes on Lovell peach rootstocks and increased the yield of peaches cv. Loadel grafted on it. Such treatment controlled the nematodes better than the addition of the S isolate of *Dactylella oviparasitica*, which did not improve plant growth. Heavy irrigation offset nematode damage, without decreasing nematode abundance (McKenry and Kretsch, 1987).

MELOIDOGYNE CHITWOODI

In Washington State, USA, this nematode has been controlled with D-D, 1,3-D and metham-sodium. Application of D-D reduced their numbers to 1.8 m deep in the soil (Santo *et al.*, 1983). Applied shallow or deep, D-D, Telone II or Telone C 17 (1,3-D plus 17% chloropicrin) reduced numbers of the nematode in the soil and in the tubers and increased tuber yield (Santo *et al.*, 1985a). Telone II significantly lowered P_f/P_i at harvest (Santo *et al.*, 1985b). D-D or Telone II applied in combination with ethoprophos gave excellent control of *M. chitwoodi* (Santo *et al.*, 1985c). This was confirmed by Pinkerton *et al.* (1986), who found neither nematicide alone was effective. Ethoprophos has been shown to suffer enhanced biodegradation in previously treated soil, an effect which lasted 17 months after the last application but which does not occur in sterilized soil (Mojtahedi *et al.*, 1991). In Utah, Griffin found 1,3-D soil fumigation more effective in controlling the nematode than non-fumigant nematicides. Infestation of tubers and the number of nematode generations were correlated with the number of day-degrees in the soil and this varied greatly over four years of observation (Griffin, 1985, 1989). Metham-sodium at 468 l ha^{-1} in overhead irrigation water decreased infested tubers from 82% in untreated to 1.5% in treated soil, the nematode being controlled to depths of 30, 61 and 91 cm in the soil in Washington State (Santo and Qualls, 1984).

MELOIDOGYNE HAPLA

In Virginia, USA, Soilbrom 90 (90% EDB) controlled the nematode consistently well on groundnuts and promoted large yields (Phipps and Elliott, 1984). In Shandong, China, at the optimum dosage of 67.5 kg ha^{-1}, chloropicrin injected 15 cm deep into soil which was then compacted gave up to 78% nematode control and increased groundnut yield 56% (Song, 1991). At less than 37.4 l ha^{-1}, Soilbrom 90 or Terr-O-Cide (72% EDB : 27% chloropicrin) applied at planting controlled the nematode as well on soybeans as did DBCP and without phytotoxicity (Rodriguez-Kábana *et al.*, 1979b). Infestation of potato tubers was prevented by applying 468 or 701 l metham-sodium ha^{-1} to the soil in overhead irrigation water, compared with 69% of tubers infested in untreated soil (Santo and Qualls,

1984). At 120 l ha^{-1}, metham-sodium killed all juveniles in the soil (Nesterov *et al.*, 1978). In pyrethrum nurseries in Kenya, 500 kg methyl bromide ha^{-1} controlled *M. hapla* (Ngugi *et al.*, 1993).

MELOIDOGYNE INCOGNITA

The efficacy of soil fumigants in controlling this species has been studied most on tomatoes, cotton and tobacco. In Japan, 90% of these nematodes were killed up to 10–15 cm from the injection point in soil by 1 ml Telone II (92% 1,3-D) or 2 ml D-D (55% 1,3-D) (Sano *et al.*, 1980).

Tomato plants transferred to soil treated annually with 135 kg sodium azide ha^{-1} in winter or 327 l methyl isothiocyanate–D-D mixture ha^{-1} or 490 kg methyl bromide ha^{-1} in spring remained free of the nematode to harvest over four years of observation (McCarter *et al.*, 1978). In Florida, methyl bromide applied in the first year only did not reduce root-knot galling of tomato in the following three years, so annual treatment was essential for continuous tomatoes. The nematode was controlled best by 364 kg methyl bromide (67%) : chloropicrin (33%) ha^{-1}, 210 kg EDB (54%) : chloropicrin (45%) ha^{-1} or 487 kg methyl bromide (98%) : chloropicrin (2%) ha^{-1} (Overman and Jones, 1981). The nematode was also well controlled by a single injection of metham-sodium or ethylene dibromide into the drip irrigation system, to beds already fumigated with ethylene dibromide (Overman, 1982). Also in Florida, preplant application of methyl isothiocyanate and/or C_3 chlorinated hydrocarbons suppressed *M. incognita* six months later equally as well as mixtures of methyl bromide and chloropicrin (McSorley *et al.*, 1986). In California, 187–702 l of 32.7% metham-sodium ha^{-1} in 2.5–14.2 cm of water drip irrigated into 50–75% of the central area of tomato or carrot beds decreased both *M. incognita* and *M. javanica*, as well as *Pythium ultimum* and *Fusarium* sp. Effects on the root-knot nematodes were similar after 75–144 l of 1,3-D ha^{-1} had been applied to the soil but the fungi were unaffected (Roberts *et al.*, 1988). Increasing the dosage of methyl bromide applied under plastic sheeting, from 200 to 300, 400 or 500 kg ha^{-1}, increased tomato yields and decreased infestations of *M. incognita* (Basile *et al.*, 1988).

In Mississippi, USA, each increase of 1.0 in the root-knot index on upland cotton, due to *M. incognita*, lessened lint yield 35.7 kg ha^{-1}. Although Telone II reduced the root-knot index, it did not increase lint yields and actually decreased them in three of the nine cultivars assessed (Minton and Meredith, 1987). This suggests that there may have been some residual phytotoxicity in the soil affecting the young cotton plants, offsetting the benefit of nematode control. In California, larger than conventional amounts of 1,3-D, with or without chloropicrin, suppressed *Fusarium* wilt of cotton better in the presence of *M. incognita* and increased yields more than conventional dosages (Jorgenson *et al.*, 1978).

In black tobacco seedbeds, methyl bromide has controlled the nematode better than some other soil fumigants (Teran *et al.*, 1986) and in Karnataka, India, 180 l methyl bromide (50%) : EDB (50%) or 155 l methyl bromide ha^{-1} controlled the nematode better in tobacco seedbeds than 180 l EDB ha^{-1} (Hussaini, 1989). The nematode was completely controlled in onions for transplanting by methyl bromide alone or mixed with chloropicrin (Jaworski *et al.*, 1978). The nematode was controlled better on cabbage by chloropicrin, D-D–methyl isothiocyanate mixture, methyl bromide (98%) : chloropicrin (2%) or methyl bromide (67%) : chloropicrin (32%) than by metham-sodium, PCNB or Bunema (Johnson *et al.*, 1979).

MELOIDOGYNE JAVANICA

In Brazil, 6 or 8 ml Telone C-17 per plant gave 95% control of this nematode on tomato (Covolo and Oliveira, 1982). The nematode was controlled well on tobacco in Queensland, Australia in coarse, sandy soil by 72 kg EDB ha^{-1} (Broadley, 1979c). The nematode was controlled better on tobacco by fumigating the soil with D-D or EDB than by applying non-fumigant nematicides to the soil, except when large amounts of the latter were applied (Garcia and Rich, 1983; Rich *et al.*, 1984). Similarly, 1.8 g EDB per plant was more effective in controlling *M. javanica* on tobacco than 9 kg aldicarb ha^{-1}, although both treatments increased yields of tobacco leaf very significantly (Shepherd, 1986). In California, 1,3-D at 56 l ha^{-1} decreased infestation of soybean plants from 68% in untreated soil to 2% in treated soil (Kontaxis and Hagemann, 1981).

Recent Research on Control of *Meloidogyne* spp. with Non-fumigant Nematicides

Various compounds of low vapour pressure have been found effective in the control of root-knot nematodes. Some of these, for various reasons, have been little used in commercial practice but their efficacy against root-knot nematodes warrants brief mention before a comprehensive account is given of widely used non-fumigant nematicides, particularly carbamates and organophosphates.

Nellite (*O*-phenyl-*N,N*'-dimethylphosphorodiamide). Youngson and Goring (1963) found Nellite effective in soil against hatched second stage juveniles of *M. incognita* at 1.25–2.5 ppm in irrigation water. At 2 ppm in irrigation water, Nellite was effective if it was not leached down the soil profile too quickly. It was ineffective even at 80 ppm if galled tissue occurred in the soil and it was ineffective against other nematodes or soil fungi. At 2.24 kg ha^{-1}, Nellite lessened *M. incognita* on cotton as well as did 13.5 kg DBCP ha^{-1} (Turner, 1963). At 4.5 kg in 455 l transplant water ha^{-1}, this compound gave 95% control of root-knot on transplanted tobacco in both moderately and heavily infested soil (O'Melia, 1974).

Shell SD 7727 (2,4-dichlorophenyl methanesulphonate). This was effective against root-knot nematodes at 9–22 kg ha^{-1} applied to the soil. Its mode of action was uncertain, as it was not highly toxic to the nematodes (McBeth *et al.*, 1964). SD 7727 controlled root-knot on cotton for at least ten months, which neither Nellite nor DBCP did, and cotton yields were increased in the two years that followed its application to the soil (Reynolds and O'Bannon, 1966).

Gall inhibitors. On susceptible peach rootstocks, non-phytotoxic doses of thiourea, maleic hydrazide, triiodobenzoic acid and cyclohexamide inhibited gall formation by *M. javanica* and maturation of the nematodes in the galls that did form (Kochba and Samish, 1972).

Amino acids. Some success in decreasing *M. incognita* on tomato followed drenching the soil with DL-amino acids, especially 104 mg DL-serine, 120 mg DL-threonine or 87.5 mg DL-alanine per plant. DL-Serine was also successful when applied as a foliar spray, which did not stunt the plants (Prasad *et al.*, 1979). On soybeans, foliar sprays of phenylalanine or methionine lessened galling due to *M. incognita*, without harming the *Rhizobium* nodules (Osman and Viglierchio, 1981a). Applied as a foliar spray at 1000 ppm, phenylalanine reduced galling and egg mass production by *M. incognita* on sunflower roots without phytotoxicity (Osman and Viglierchio, 1981b). DL-Aspartic acid was very effective against the same nematode when applied as a foliar spray to peas (Sharma and Tiagi, 1984). Interest in amino acids as nematicides has waned, probably because the treatments are too costly.

Lectins. Reproduction of races 1 and 3 of *M. incognita* and of *M. javanica* on Centennial and Pickett 71 soybeans was greatly decreased when the juveniles had been exposed to sialic acid or sialic acid plus *Limulus polyphemus* agglutinin, perhaps because of impaired host finding and root penetration (Davis *et al.*, 1989).

Anthelmintics. One hour exposure *in vitro* to methyridine at 0.5 ppm was lethal to hatched juveniles of *M. incognita*. Phenzidole, 5-nitro-2-tri-fluoromethyl-benzimidazole or 5-bromo-2-tri-fluoromethylbenzimidazole at 5 ppm were lethal to the nematodes after 1, 24 or 48 h, respectively (Ivanova, 1975). Nilverm at rates from 1100 down to 0.05 ppm controlled root-knot on cucumbers (Ivanova, 1976).

Action of Non-fumigant Nematicides on Root-knot Nematodes

Carbamate and organophosphate nematicides are cholinesterase inhibitors and at commercial rates of application act as temporary nematostats, i.e. temporarily paralysing the nematodes. Nelmes and Keerweewan (1970) showed that aldicarb temporarily paralysed hatched juveniles of *M. incognita in vitro*; roots dipped in aldicarb solution were protected from invasion by the second stage juveniles. Root-knot nematode development in cucumbers was prevented by treating the plants with propyl gallate, which inactivated enzymes essential to extra-intestinal digestion. As a result, gall formation was retarded, no adults developed and yield was increased (Myuge and Vasil'eva, 1971). At 20 or 50 ppm *in vitro*, hatching of *M. arenaria* was decreased or halted by aldicarb, oxamyl, fenamiphos or ethoprophos (Bergé and Cuany, 1972b). Hatching of *M. arenaria* was inhibited in egg masses containing over 10 ppm aldicarb but development within eggs was unaffected by up to 200 ppm around them (Bergé and Cuany, 1972c). Hatching of *M. hapla*, *M. incognita* and *M. javanica* was lessened by fenamiphos or thionazin at 2 ppm, whereas aldicarb had little effect on hatching at less than 8 ppm. At 2 ppm, aldicarb or thionazin prevented development of *M. javanica* in tomato roots (McLeod and Khair, 1975). Further studies of the mode of action of non-fumigant nematicides have been made on *M. incognita* and *M. javanica*, particularly with oximecarbamates and organophosphates.

MELOIDOGYNE INCOGNITA

Exposed to 0.5 μg oxamyl ml^{-1}, juveniles became hyperactive and lost their coordination. In sand, 2 μg oxamyl ml^{-1} lessened migration of juveniles and their orientation to cucumber roots was lessened by 0.5–4.0 μg ml^{-1}. At 1, 4 or 8 μg ml^{-1} in sand, oxamyl decreased the numbers of juveniles invading cucumber roots (Wright *et al.*, 1980). At 6.25 μg ml^{-1} of soil water, oxamyl was more effective in suppressing nematode development when applied soon after the juveniles had invaded roots than when applied later (Wright and Womack, 1981; Wright and Rowland, 1982). Movement of the juveniles was stopped or decreased by as little as 4 μg oxamyl ml^{-1} within 24 h (McGarvey *et al.*, 1984). In agar culture, oxamyl was transported within extending tomato roots to affect nematodes in the untreated new growth (Mugnier, 1988). Under light of 300 nm, oxamyl in a thin (solid state) film was photolysed in 2 h but, although the photoproducts were less toxic than oxamyl, they were similarly effective in suppressing juvenile invasion of cowpea roots (Dureja *et al.*, 1988).

Exposed to fenamiphos at 1.25–10 μg ml^{-1} for 1–15 days, hatching of *M. incognita* was reduced in proportion to length of exposure and chemical concentration. Exposure to as little as 0.0005–0.020 μg ml^{-1} for 24 or 48 h reduced the ability of the juveniles to penetrate cowpea roots (Meher and Sethi, 1988). Similarly, exposing the juveniles to 0.0005, 0.005 or 0.05 μg sebufos ml^{-1} for 24 or 48 h significantly reduced their ability to invade aubergine roots, even at 0.0005 μg ml^{-1} (Meher, 1990). For ethoprophos, oxamyl, fenamiphos, its sulphoxide and sulphone, aldicarb, its sulphoxide and sulphone and carbofuran, 3-hydroxy-carbofuran and 3-keto-carbofuran there was no correlation between the concentration that reduced root galling by 50% (0.2–8 μg ml^{-1}, *in vivo*) and the concentration that inhibited acetylcholinesterase activity by 50% (0.02–875 $\times 10^{-6}$M.) (Nordmeyer and Dickson, 1990). The activity of these compounds is not, therefore, simply related to acetylcholinesterase inhibition but can be modified considerably by other factors.

MELOIDOGYNE JAVANICA

Hatching of *M. javanica* was inhibited 100% by aldicarb at 12.5 or 25 ppm and 98 or 99% at 12.5 or 25 ppm, respectively, by isazophos (CGA 12223). Inhibition of hatching was reversible for aldicarb, not for isazophos (Yousif and Badra, 1979). Prevention of hatch by non-fumigant nematicides is not usually feasible in the field because the dosages of nematicide needed are impractically large. Oxamyl applied at 300 ppm to one half of a divided tomato root system, one or five days before, or five days after inoculation of the untreated half root system, significantly reduced the number of juveniles invading those roots and the size of the females. When the roots were inoculated two to five days after the cotyledons or stems had been treated with oxamyl, fewer juveniles invaded the roots of treated than untreated plants. These results therefore support the hypothesis that oxamyl is translocated basipetally and from one side of the root system to the other (Atilano and Van Gundy, 1978). An earlier split root study had shown that at 10 kg ha^{-1}, aldicarb, carbofuran, oxamyl or thionazin protected the untreated half root system from invasion by *M. incognita* in heavily infested soil for approximately 20, 12, 25 or 20 days respectively. At 45 days, however, all stages of the nematodes were found in the untreated half root systems, irrespective of which nematicide had been applied to the other half (Vovlas and Lamberti, 1976). It is now known that 1% or less of the dosage applied reaches untreated roots.

Juveniles of the nematode exposed to 2 or more μg available chlorine ml^{-1}(from sodium hypochlorite) were unable to invade or develop in roots. The nematode was controlled in a peat–vermiculite mix with 4 or more μg available chlorine ml^{-1} of nutrient solution applied four weeks before planting (Stanton and O'Donnell, 1994).

Effective Application of Non-fumigant Nematicides to Control *Meloidogyne* spp.

The ability of many non-fumigant nematicides to control root-knot nematodes has been assessed worldwide on a range of crop plants. The review that follows is restricted to those reports in which there was evidence of nematode control, not just yield increases resulting from the application of non-fumigant nematicides to infested soil. Yield responses may be due in part to effects other than nematode control. Conversely, lack of yield response does not prove that the nematodes were not affected by the treatment. The cultivar may have been tolerant of attack or the treatment may have been phytotoxic.

MELOIDOGYNE ARENARIA

Dipping the seeds of various crop plants in aqueous oxamyl (up to 24%) lessened infestations of *M. arenaria* on the plants grown from them and enhanced *Rhizobium* nodulation of dwarf beans and soybeans (Truelove *et al.*, 1977). Applied to groundnut seed at 6% (w/w) in pots, aldicarb sulphone (Patel *et al.*, 1986) or carbofuran 40F (flowable) (Prasad, 1990) decreased galling of the roots by the nematode.

The nematode was controlled better and the yields of groundnut and soybean were increased more by 1.1–3.4 kg aldicarb, oxamyl, fenamiphos or carbofuran ha^{-1} applied in bands 12.5 or 18 cm wide astride the plant rows than in the seed furrows (Rodriguez-Kábana *et al.*, 1981b). Numbers of the nematode in the soil just before groundnut harvest in Alabama were less in plots treated with 1.1 or 2.2 kg aldicarb, carbofuran, oxamyl or fenamiphos, 2.2 or 4.4 kg ethoprophos or 8.4 or 16.8 l EDB ha^{-1} (Rodriguez-Kábana and King, 1985). However, nematode control was increased when, at sowing, 1.1 or 2.2 kg aldicarb ha^{-1} was applied in two rows 20 cm apart straddling the seed furrows in addition to 17.2, 34.4 or 51.6 l of 1,3-D ha^{-1} injected 25 cm deep in two rows 20 cm apart also straddling the seed furrows (Rodriguez-Kábana *et al.*, 1985). There was some initial phytotoxicity from the 1,3-D. In India, 1.5 or 3.0 kg aldicarb or carbofuran ha^{-1} controlled the nematode on groundnuts and increased yields equally as well as 1 or 1.5 t mustard cake ha^{-1} (Grewal *et al.*, 1985). In Georgia, USA, the nematode was also controlled on groundnuts by fosthiazate applied to the soil (Minton *et al.*, 1993).

In Italy, 12 kg oxamyl ha^{-1} applied to the soil controlled the nematode on carnation (Del Bene, 1977). At 6.7 kg ha^{-1}, aldicarb or ethoprophos, applied to the soil in which tobacco was planted out, decreased galling of the roots by root-knot nematodes, including *M. arenaria*, 11 and 15 weeks later but not at harvest (Nordmeyer *et al.*, 1982). In a subsequent experiment, the same dosage of aldicarb, ethoprophos, fenamiphos, oxamyl or carbofuran lessened galling of tobacco roots by the same nematodes up to the time of harvest (Nordmeyer and Dickson, 1985).

MELOIDOGYNE CHITWOODI

In potato fields in Idaho, USA, applying aldicarb plus ethoprophos or oxamyl to the soil decreased *M. chitwoodi* in soil, roots and tubers (Hafez, 1984). Ethoprophos at 14 or 19 l a.i. ha^{-1} controlled the nematode well (Santo *et al.*, 1985d). However, in more recent work in Washington State, USA, ethoprophos, the most used non-fumigant nematicide in potatoes in north-west USA, did not reduce infestation of potato tubers, whereas cadusafos broadcast at 10.1, 13.4 or two applications of 6.7 kg ha^{-1} or applied at 6.7 kg ha^{-1} as a 20 cm band treatment controlled the nematode as effectively as 176 l of 1,3-D ha^{-1} (Santo and Wilson, 1990).

MELOIDOGYNE EXIGUA

In vitro, more than 0.1 μg aldicarb, ethoprophos or carbofuran ml^{-1} significantly inhibited hatching of *M. exigua*, while the motility and longevity of the second stage juveniles were decreased by carbofuran and especially by aldicarb. In the field, all three compounds decreased the nematode on coffee (Huang *et al.*, 1983). Drenching infested soil in bags containing freshly planted coffee seedlings with aldicarb at 200 ppm almost eliminated the nematode from seedlings transplanted 18 days later (Pereira *et al.*, 1979). Carbofuran, fensulphothion, oxamyl, ethoprophos or fenamiphos also controlled the nematode on coffee seedlings in Colombia (Baeza-Aragón and Leguizanón, 1977). At 60 or 120 mg kg^{-1} soil, oxamyl reduced gall formation on the untreated side of divided root systems of coffee seedlings. Similar amounts of aldicarb, disulphoton or phorate only affected the treated half of the root system (Pereira *et al.*, 1980).

MELOIDOGYNE GRAMINICOLA

In India, 1 kg carbofuran ha^{-1} decreased the nematode in both the nursery and in the field and doubled rice yield (Anon., 1985). In Thailand, 1.8 kg carbofuran ha^{-1} (as 3% granules) applied to plots infested with *M. graminicola* increased yields of three rice cultivars by 11.6% ('resistant' cv.), 16.2% ('moderately resistant' cv.) and 32.6% ('susceptible' cv.) (Arayarungsanit, 1987). In Vietnam, in soil infested with the nematode, the application of carbofuran or diazinon in irrigation water increased rice yield by 100–160% compared with untreated plots but also increased the yield of uninfested rice by 70–90%. The numbers of root-knot nematode galls began to increase 20 days after nematicide treatment. Continuous flooding increased rice yields similarly and the longer the period of flooding the fewer the nematode galls on the roots (Dang-ngoc Kinh *et al.*, 1982). In deep water rice in Bangladesh, despite reductions in the numbers of galls and mature females in the roots from preplant applications of 1 kg carbofuran, 3 kg aldicarb, 5 kg ethoprophos or 5 kg fenamiphos ha^{-1} to the soil, rice yields were unaffected (Rahman, 1991). Applied to rice plants as foliar spray ten days after they had been inoculated with *M. graminicola*, oxamyl at 500–1000 ppm decreased root infestation with the nematode (Prasad and Rao, 1986). Such treatment in the field would be hard to justify on human toxicological grounds.

MELOIDOGYNE HAPLA

Dipping potato seed pieces in dimethoate (800 ppm) for 24 h eliminated *M. hapla* from the tissues (Helton, 1964). Bolander and Santo (1977) found that the nematode could be eliminated from potato tuber pieces by dipping them in fenamiphos (4750 ppm) for 10 s. Dressing seeds of lucerne, carrot or tomato with oxamyl plus a polymer sticker gave useful early control of the nematode (Townshend, 1989, 1990). As seed coatings at 1–10% (w/w) (i.e. g per 100 g seed), oxamyl, carbofuran or ethoprophos protected white clover seedlings from attack by *M. hapla* (Barker *et al.*, 1993). The nematode was eliminated from the roots of rose, tomato and other plants by dipping them in thionazin (750 or 1150 ppm) (Harlan and Jenkins, 1967). Similarly effective on the roots of two-year-old roses was dipping in oxamyl (600 or 1200 ppm) (Hart and Maggenti, 1971). A 30 min dip in 0.1% ethoprophos or fenamiphos freed dormant rose plant roots of the nematode (Dale, 1973). Dipping the infested roots of dormant Kiwi fruit plants in 0.1% fenamiphos for 1 h controlled the nematode without phytotoxicity (Grandison, 1983).

Two foliar sprays six weeks apart of fenamiphos (3200 ppm) kept redcurrant roots free of *M. hapla* for 18 weeks (Santo and Bolander, 1979). A foliar spray of oxamyl (3000 or 4000 ppm) before inoculation prevented invasion of tomato roots by *M. hapla*. Applying oxamyl to the foliage after inoculation was less effective than soil applied granules or drenches (Stephan and Trudgill, 1983). The nematode was controlled better in strawberry runners by dipping them in suspensions of fenamiphos or ethoprophos than by dipping them in hot water plus prolonged cold storage (Willers and Meyer, 1979).

Meloidogyne hapla has also been controlled by non-fumigant nematicides applied to the soil at or just before planting. The nematode was controlled in this way by aldicarb or fenamiphos on strawberry in Maryland, USA (Babineau and Krusberg, 1978) and by 2 kg aldicarb ha^{-1} over groundnut rows at sowing in Virginia, USA (Phipps and Elliott, 1984). More recently, aldicarb, carbofuran, ethoprophos, fenamiphos, disulphoton or chlorpyrifos, alone or in combinations, minimized galling of groundnut roots, 1.2 kg aldicarb ha^{-1} in the seed furrows plus 2.2–3.4 kg carbofuran ha^{-1} banded in the soil increasing yields most (Phipps *et al.*, 1988). The nematode was controlled best by the combined use of a soil fumigant, Telone, Vapam or Vorlex and in-furrow application of aldicarb at sowing (Phipps and Herbert, 1990; Phipps, 1991). Similarly in carrots, the nematode was controlled best at harvest when the soil had been treated with a soil fumigant or oxamyl before sowing and oxamyl had been applied to the foliage six weeks after sowing, not where preplant treatments alone had been applied (Johnston *et al.*, 1992). Earlier work in Quebec had shown that the nematode was controlled better by 1.5 kg aldicarb or oxamyl ha^{-1} applied to the seed furrows than 6 kg of either nematicide ha^{-1} applied on a broadcast basis or as a banded treatment (Belair, 1984), no doubt because of the greater concentration around the developing tap root. Infestation of potato daughter tubers by the nematode was lessened by incorporating ethoprophos into the seedbed or mixing aldicarb with the soil after planting (Ingham *et al.*, 1991). In Poland, application of 1.5 kg aldicarb or 2 kg oxamyl ha^{-1} to the seedbed in spring minimized invasion of carrot roots by the nematode and increased yields (Berbeć and Dolna,1990). Finally, infestation of pyrethrum (*Tanacetum cinerariaefolium*) in Kenya was decreased by applying fenamiphos to nursery soil (Anyango, 1988) and aldicarb, fenamiphos or carbofuran in plots of pyrethrum decreased infestations of *M. hapla* and *Pratylenchus penetrans* and increased yields of dry flowers (Anyango *et al.*, 1991).

MELOIDOGYNE INCOGNITA

The efficacy of non-fumigant nematicides has been assesssed more on *M. incognita* than on any other root-knot nematode and has generated many publications in the last twenty years. These are reviewed below according to method of use.

Seed treatments. Dipping watermelon or dwarf bean seeds for 3 h in a 10% solution of oxamyl decreased galling of the seedlings by *M. incognita* by 90% for watermelon and 100% for dwarf bean (Rodriguez-Kábana *et al.*, 1976). Similarly, dipping seeds of various crop plants in solutions of oxamyl up to 24% a.i. lessened infestation of the plants by the nematode, without phytotoxicity and with increased *Rhizobium* nodulation of dwarf and soybeans (Truelove *et al.*, 1977). Galling of chickpea roots was decreased by coating the seeds with 1–2% aldicarb or 3% carbofuran (w/w) (Mishra and Gaur, 1984). The most effective seed treatment for *Capsicum annuum* was to dip the seeds in 2% aldicarb in water containing 0.01% gibberellic acid (Singh and Maheshwari, 1989). In other studies, the most effective treatments have been 2% dressings of carbofuran or benfuracarb to pigeon pea seed (Mishra, 1986; Dwivedi *et al.*, 1994), a 3% dressing of carbofuran to green gram seed (Borah and Phukan, 1990; Darekar *et al.*, 1990b), 1% dressings of fenamiphos to cowpea, pea and dwarf bean seeds (Parvatha Reddy, 1984), 0.5–2% dressings of triazophos to dwarf bean seed (Hong and Sethi, 1988a) and soaking dwarf bean seed for 6 h in fenamiphos (125 ppm) (Hong and Sethi, 1988b). *Meloidogyne incognita* on aubergine roots was decreased by applying 10 g fensulphothion to 500 g of the seeds (Singh and Singh, 1988). Dressing cotton seeds with 1, 2 or 4% (w/w) carbofuran, isofenphos or benfuracarb lessened galling of the roots in pots. Carbofuran was the most effective compound, though all three compounds at 4% (w/w) prevented gall formation (Chand and Jain, 1990). The nematode was lessened on lentil roots by dressing the seeds with quinalphos at 25% (w/w) (Gogoi and Phukan, 1990) and on black gram roots by dressings of ethoprophos at 3% (w/w) or carbofuran at 1% (w/w) to the seeds (Mahanta and Phukan, 1992; Kalita and

Phukan, 1993). All these seed dressings are toxic and their commercial use, especially by inadequately protected operatives is questionable. The greatest danger would be the inadvertent consumption of treated seed in times of famine.

The nematode has also been controlled by applying plant extracts to the seeds. Water soluble extracts of neem or Persian lilac applied to the seeds, especially fruit extracts, lessened galling of tomato, aubergine and okra seedlings (Siddiqui and Alam, 1987; Dash and Padhi, 1990). Applied to tomato and okra seeds, the triterpenoid nimbin from neem lessened *M. incognita* in the seedlings in inoculated soil and improved plant growth (Siddiqui and Alam, 1990c). The nematode was also controlled in pigeon pea and chickpea by dressing the seeds with the latex of *Calotropis procera*, which increased the growth of the plants and the number of pods set (Anver and Alam, 1992). The treatment was similarly effective on okra (Wani *et al.*, 1994).

Set and root dips. *Meloidogyne incognita* was eliminated from the tubers of Centennial sweet potato, without phytotoxicity, by dipping them in water at 50°C for 3–5 min (Martin, 1970). Paring banana corms to a depth of 1 cm or immersing them for 20 min in water at 55°C was also effective against the nematode (Gupta, 1975). The nematode was eliminated, with little or no damage to infested plant tissues, by dipping rhizomes of society garlic (*Tulbaghia violacea*) in water at 50°C for 2.5 min or 47.5°C for 15 min (McSorley *et al.*, 1984) and *Mentha spicata* rhizomes in water at 50°C for 10 min or 48°C for 30 min (Gotke and Mathur, 1990). Hot water has also been used to rid roots of the nematode. Lear and Lider (1959) eradicated *M. incognita* (*acrita*) and *M. javanica* from grapevine roots by immersing them at 47.8, 48.9, 51.7 or 52.8°C for 30, 10, 5 or 3 min, respectively, without plant injury. Two-year-old grapevine transplants were similarly freed of root-knot nematodes by immersing the roots in water at 53, 52 or 50°C for 5, 12.5 or 30 min, respectively, without harming the plants (Vega, 1973). *Meloidogyne incognita* was killed in tomato roots by 50, 47 or 44°C for 5, 20 or 80 min, respectively (Saigusa and Matsumoto, 1961).

The infestation of transplants has been greatly decreased by dipping their roots in aqueous solutions or suspensions of nematicides. McBeth *et al.* (1963) freed plant roots of *M. incognita* (*acrita*) by dipping them in 1,6-hexanedithiol (200 ppm) for 24 h at 20°C. Winchester (1964) found thionazin dips more effective than fensulphothion dips for ridding celery roots of the same nematode. However, both these compounds eliminated *M. incognita* from the roots of nursery plants when used as dips at up to 1000 ppm concentration (Miller and Perry, 1965). Recent effective treatments were dipping tomato roots in oxamyl at 1000 ppm for 15 min (Reddy and Singh, 1979), or 10,000 ppm for 30 min (Inderjit Singh *et al.*, 1979), dipping coffee roots in oxamyl (600 ppm) or fenamiphos (400 ppm) for 10 min (Echávez-Badel and Ayala, 1979), dipping aubergine roots in oxamyl (10,000 ppm) for 60 min or carbosulphan (1000 ppm) for 30 min (Ahuja, 1978; Venkata Rao *et al.*, 1987) and dipping ginger rhizomes in fenamiphos at 2600 ppm for 30 min or 1000 ppm for 60 min (Willers, 1991). Such dipping treatments can protect transplanted seedlings from damaging root-knot nematode attack in the critical early weeks of establishment, but they are unlikely to protect the plants for very long. Ahuja (1980) found that applying a drench of 100 ml oxamyl or aldicarb (5000 or 10,000 ppm) to the planting hole controlled the nematode more effectively on tomato than 60 min dips of four-week-old transplants in nematicide solutions and without phytotoxicity, which occurred with many of the dip treatments.

Root dips of some plant extracts have also controlled *M. incognita*. Extracts of water hyacinth (*Eichhornia crassipes*) were nematicidal and nematostatic, retarding nematode development and preventing plant damage, when they were used as bare root dips (Siddiqui and Alam, 1989c). Dipping tomato roots in extracts of neem or Persian lilac lessened their invasion and galling by the nematode (Siddiqui and Alam, 1987). 'Nimin' dip, containing neem triterpenes, was very effective as a root dip for tomatoes, protecting them from attack by the nematode (Akhtar and Mahmood, 1993). Leaf extracts of *Calotropis procera* (Akhtar *et al.*,1992) and *Galium aparinoides* (Owino and Wando, 1992) were also effective.

Foliar applied nematicides. The nematode has been controlled to some extent by spraying the foliage with nematicidal plant extracts or synthetic nematicides. Spraying the latter can be hazardous and unexpected rain may lessen control with foliar nematicides. Spraying the foliage of susceptible crop plants with 1.8 kg oxamyl in 377 l water ha^{-1} controlled *M. incognita* for 21–28 days (Radewald *et al.*, 1970). The nematode was virtually eliminated from carnations in a glasshouse by eight, monthly foliar sprays of 4.5 kg oxamyl in 378 l water ha^{-1}, whereas applying 9.8 kg granular oxamyl, carbofuran or fenamiphos ha^{-1} to the soil at planting was ineffective (Radewald *et al.*, 1974). Oxamyl applied twice in summer and twice in autumn to the leaf axils of banana in Puerto Rico decreased *M. incognita* in the roots and increased bunch weights more than the same amount applied to the soil around the base of the plant. Functional roots were increased from 50% to 78–96% by either method of applying oxamyl (Robalino *et al.*, 1983). Pea roots were fully protected from invasion by the nematode for three days by spraying the foliage with aldoxycarb at 5000 ppm. Aldoxycarb was rapidly translocated from leaves to roots and into the soil (Meher *et al.*, 1987) and such short-term control is inadequate. In the field, *M. incognita* (and *M. javanica*) was decreased and tomato yields were increased by foliar sprays of maleic hydrazide (0.001 M), salicylic acid, cinnamic acid or catechol (0.001 or 0.01 M), CCC (50 ppm) or 2,4-D (3 ppm) (Sitaramaiah and Pathak, 1981). A 0.2% aqueous mixture of oil from *Argemone mexicana* applied to the leaves of okra controlled the nematode better than when applied to the soil, and it increased plant growth (Das and Sukul, 1988). In pots, foliar sprays of ascorbic acid or riboflavin lessened galling of tomato roots by *M. incognita* (Montasser, 1990). A phloem-mobile pro-nematicide – an hydroxymethyloxamyl glucurinide (JR 522) or its methyl ester (JM 775) – applied to the foliage of transgenic tobacco, expressing *Escherischia coli* beta-glucoronidase in the root tips, controlled the nematode better than oxamyl. JR 522 is safer than oxamyl to apply and is site specific (Hsu *et al.*, 1995).

Soil applied nematicides. Many experiments on the control of *M. incognita* with non-fumigant nematicides applied to the soil have been reported. This review is restricted to adequately documented studies, many of which have used tomato or tobacco as the test plant. Some other annual crop plants and some perennials (banana, black pepper and betel vine) have also been used. There is little comparable information on control of the nematode on sugarcane.

TOMATO. *Meloidogyne incognita* on tomato is widely used as an early screen for potential new nematicides and to assess, locally, the efficacy of promising compounds. At 12.5 ppm in soil, fensulphothion was very effective against the nematode on tomato (Jarnevic and Coffee, 1965). Six weeks after planting tomatoes in infested soil treated with 4–8 kg thionazin or aldicarb ha^{-1} there was very little infestation of the roots or nematode multiplication (Reddy and Seshadri, 1971, 1974, 1977). Ethoprophos at rates up to 10 kg ha^{-1} controlled several species of *Meloidogyne*, including *M. incognita* on tomatoes and cucumbers (Grujičić, 1974b). In southern Italy, 12 kg ethoprophos ha^{-1} applied to the soil before transplanting tomatoes in April, followed by 8 kg ha^{-1} to the soil in June increased tomato yields by 60% and decreased the numbers of juveniles in the soil at harvest by 77% (D'Errico *et al.*, 1978). At 11.2 kg ha^{-1}, fenamiphos or carbofuran, applied by sprinkler irrigation, lessened multiplication of the nematode and increased tomato yields by 50 and 33%, respectively (Johnson, 1978). Fenamiphos applied at planting at 0.6 g in 250 ml water per tomato plant (equivalent to 10 kg ha^{-1}, assuming 37,000 plants ha^{-1}) kept tomato plants free of both *M. incognita* and *M. javanica* to harvest, whereas similar applications 29 or 66 days after planting were less effective (Daiber, 1990). Fenamiphos or ethoprophos at 6.7 kg ha^{-1} controlled the nematode equally as well when applied to the soil through an overhead sprinkler irrigation system as when incorporated in the top 15 cm of the soil (Johnson *et al.*, 1981). Recent results in Georgia, USA with 3.4 and 6.7 kg fenamiphos ha^{-1} applied to a field of squash in 63,500 l irrigation water ha^{-1} suggest that the nematode could be controlled with less than 6.7 kg ha^{-1} (Johnson and Young, 1994).

Oxamyl at 4.3 kg ha^{-1} applied six days and three weeks after transplanting tomatoes decreased nematode infestation of the roots by 94% and increased tomato yield by up to 80%, whereas the six day treatment alone increased yield by 42% (Cadet, 1990). Soil treatment with oxamyl before or at planting might have been even more effective. Compared with its toxicity to *M. incognita* on tomato seedlings when applied in water, oxamyl applied in ferric sulphate solution of 20 ppm or in zinc sulphate solution of 1 ppm was 11 and four times as toxic to the nematode, respectively (Korayem and El-Sisi, 1989).

Multiple applications of non-fumigant nematicides through a drip irrigation system also controlled the nematode on tomatoes and increased yields. Aldoxycarb, carbosulphan, carbofuran or oxamyl applied at 1.68 kg ha^{-1} three times at two week intervals, or at 0.37 kg ha^{-1} six times at three or six day intervals, significantly reduced galling of tomato roots at harvest in California and greatly increased tomato yields (Garabedian and Van Gundy, 1985).

In pot and field trials in the USA, avermectins B_1 and B_2 applied to soil in granular or liquid form or by low pressure drip irrigation at 0.093–0.34 kg a.i. ha^{-1} controlled *M. incognita* on tomato ten times better than 3.36 kg aldicarb ha^{-1} (Garabedian and Van Gundy, 1983). Avermectin B_{2a}-23-ketone affected juveniles of the nematode differently to oxamyl, which is an acetycholinesterase inhibitor, for it is blocked by bicuculline and less blocked by picrotoxin, which are γ-aminobutyric acid (GABA) antagonists (Wright *et al.*, 1984).

TOBACCO. Planting spot applications of 1 kg aldicarb, fensulphothion or carbofuran ha^{-1} controlled the nematode on tobacco as well as did 4 kg ha^{-1} as a row treatment (Reddy, 1975). *Meloidogyne incognita* and *M. javanica* were also well controlled on tobacco by 0.5, 1 or 2 kg oxamyl ha^{-1} applied in the transplant water (Reddy, 1976). At dosages of 0.05–0.5 kg ha^{-1}, the avermectins B_{1a}, B_{2a} and B_{2a}-23-ketone lessened galling and egg production by the nematode on tobacco by 21–86%, according to dosage and compound used, the control achieved being similar to that obtained with 6.7 kg ethoprophos or fenamiphos ha^{-1} (Sasser *et al.*, 1982). Ethoprophos or aldicarb at 6.7 kg ha^{-1} applied at planting decreased the numbers of root-knot nematodes, including *M. incognita*, on the roots 11 and 15 weeks later but, at harvest, treated plants were as heavily infested as untreated plants (Nordmeyer *et al.*, 1982). In soil with 90% *M. incognita* race 1 and 10% *M. javanica*, tobacco cultivars partially resistant to the former became infested with the latter and were less resistant to the black shank fungus (*Phytophthora parasitica* var. *nicotianae*). The nematode–fungus complex was satisfactorily controlled, however, by applying 6.7 kg fenamiphos plus 2.24 kg metalaxyl ha^{-1} in the irrigation water. This method of application was cheaper than using a sprayer and a rototiller to incorporate the chemicals into the top 15 cm of the soil (Csinos *et al.*, 1986). More recently, however, Johnson *et al.* (1992) found that a mixture of 6.7 kg fenamiphos plus 2.2 kg metalaxyl ha^{-1} was more effective against the root-knot–black shank complex when it was incorporated in the soil before planting than when it was applied by pre- or post-planting irrigation.

OTHER PLANTS. Oxamyl controlled both *M. incognita* and *M. arenaria* on carnation in Italy when 12 kg ha^{-1} was applied to the soil (Del Bene, 1977). Earlier work in France had shown that *M. arenaria* was controlled for 14 months on carnation when 4 kg aldicarb ha^{-1} was applied at planting and a further 8 kg ha^{-1} was applied to the soil in autumn (Cuany *et al.*, 1973). In microplots of pea and chickpea, galling of the roots at harvest by *M. incognita* was lessened when 3 kg carbofuran ha^{-1} had been applied to the soil at planting (Upadhyay and Dwivedi, 1987). The same treatment was highly effective against the nematode on cowpea in Nigeria (Olowe *et al.*, 1987). The nematode was well controlled on melon and lima bean by applying 9 kg fenamiphos or ethoprophos ha^{-1} to the soil before planting (Johnson and Harmon, 1974a,b). On cucumber, fenamiphos at 9 kg ha^{-1} controlled the nematode best to harvest when it was broadcast and incorporated into the top 15 cm of the soil before planting. Applied by sprinkler irrigation it controlled the nematode to mid-season. Although there was less than 1 ppm in the soil 30 days after applying fenamiphos to the soil, 1.5 ppm in the top 15 cm of the soil for 10 days after planting prevented significant injury to cucumber or squash (A.W. Johnson *et al.*, 1982). Very large doses of heterophos (600–1000 kg of 7.5% ha^{-1}) applied to the soil three to four weeks before planting controlled *M. incognita* (*acrita*) well on cotton in Tadzhikistan (Semikolenova, 1987), which is not surprising but surely uneconomic.

For many annual crop plants, injury by root-knot nematodes is much more important to the seedlings or young transplants than to the older, established plants. Provided the nematode populations are kept under control by appropriate rotations or some other means, the dosage of nematicide can be decreased to the amount needed to prevent injury to the young plants. For crops grown in widely spaced rows, and especially those also grown wide apart within the row, dosage can be greatly decreased. For example, only 30% of the broadcast dosage is needed to treat bands of soil 30 cm wide, 1 m apart and if the nematicide can be applied to soil within a 15 cm radius of each plant, grown 0.5 m apart in rows 1 m apart, only 14% of the broadcast dosage is needed. Sivakumar *et al.* (1974) obtained economic control of both *M. incognita* and *M. javanica* on tomato by spot applications of 1.4 kg aldicarb, 1 kg fensulphothion or 0.6 kg carbofuran ha^{-1}. If only very early protection is needed to widely spaced rows it may be possible to reduce dosage to 10% of the broadcast rate, without the need to restrict treatment to around each plant. Of course, such localized treatment, while stimulating early growth and increasing yield is likely to increase population density of the nematode at harvest. Control of root-knot nematode multiplication is of greater importance in perennial than in annual crop plants, if productivity is to be maintained. Achieving it may be more difficult in a perennial crop plant because the control options are more limited than with annual crop plants.

Meloidogyne incognita was controlled well in banana by applying 5 g aldicarb per mat four times over a period of 20 months (Jones, 1979). Aldicarb has also been widely and profitably applied to sugarcane, which is attacked by *M. incognita* and a wide range of other root-parasitic nematodes. Aldicarb has been most used at 1–6 kg ha^{-1} applied to the plant crop and in South Africa it is also applied to the ratoon crop at 1.5 or 2 kg ha^{-1}. Data on control of *M. incognita* on sugarcane, however, appears lacking (McLaren, 1984). In black pepper, the nematode was decreased in the first year by applying aldicarb around each cutting at planting and again three months later or by applying carbofuran to the soil at planting (Ichinohe, 1984). Applied to furrows on both sides of rows of three-year-old betel vines (*Piper betle*), aldicarb, carbofuran and especially benfuracarb, at dosages up to 5 kg ha^{-1}, decreased *M. incognita* infestation of soil and roots 120 days after treatment (Dethe and Pawar, 1987). Recently, 4 kg carbofuran ha^{-1} decreased *M. incognita* infestation of betel vines and increased leaf yield by 38% (Jonathan *et al.*, 1990).

MELOIDOGYNE JAVANICA

The effect of non-fumigant nematicides on *M. javanica* has been much less studied than on *M. incognita* and there are few convincing reports of very effective control of the former.

Seed treatments. The galling of mung bean (green gram) roots by *M. javanica* 35 days after sowing was greatly reduced when, before sowing, the seeds had been dipped for 60 s in 2 or 4% carbofuran or bendiocarb in acetone or 4% fenamiphos in acetone. Galling of black gram roots 42 days after sowing was also decreased when the seeds had been dipped for 60 s in 1, 2 or 4% carbofuran or fenamiphos. Part of the reduction in galling was due to the acetone alone (Kaushik and Bajaj, 1981). Infestation of dwarf beans 30 days after planting was also decreased when the seeds had been dipped for 5–60 min in 1.5–7.5% oxamyl in acetone (Carvalho *et al.*, 1981). As black gram can be damaged by as few as 0.5 juvenile *M. javanica* g^{-1} soil, early protection of the seedlings is essential. Used as a seed dressing at 6% a.i.(w/w), phorate 10G afforded early protection and lessened gall and egg mass formation (Gupta *et al.*, 1987).

Tuber treatments. All *M. javanica* were killed, without phytotoxicity, when infested potato tubers were dipped for 2 h in water at 46–47.5°C (Martin, 1968).

Bare root dip treatments. The nematode was eliminated from the roots of dormant grapevine transplants by dipping them in tap water at 50°C for 15 min, then in water at 16°C for 15 min to cool the plants (Barbercheck, 1986). There is generally less risk of phytotoxicity when roots are dipped in nematicidal liquids at ambient temperatures than when they are dipped in hot water. Solutions or suspensions of dimethoate, fenamiphos or plant extracts have given early control of *M. javanica* in tomato, grapevine and pineapple transplants.

Dipping their roots in dimethoate (250 ppm) for 1 h (Jain and Bhatti, 1979) or fenamiphos (Thakar and Patel, 1985) was very effective in protecting transplanted tomatoes from early infestation. Dipping uninfested tomato seedlings in 20% (w/v) neem leaf extract or 30% (w/v) *Eucalyptus tereticornis* leaf extract for 6 h controlled their infestation by *M. javanica* when planted in infested soil (Vats and Nandal, 1993a). Dipping uninfested seedlings in 40% (w/v) of either extract was more effective than similar treatment of preinfested seedlings (Vats and Nandal, 1993b). The nematode was eliminated from the roots of grapevine cuttings, without phytotoxicity, by dipping them in fenamiphos (1000 ppm) for 60 min (16°C) or 30 min (24°C) (Suatmadji, 1982). Dipping pineapple transplants in fenamiphos (600 ppm) for 10 min followed by three post-planting foliar sprays of 2.5 kg fenamiphos ha^{-1} at three, six and nine month intervals after planting controlled the nematode well in South Africa (Anon., 1979). Dipping tomato roots in crude seed extracts of neem (20,000 ppm), *Hannoa undulata* or *H. klaineana* (200 ppm) fully inhibited their invasion by juveniles of *M. javanica*. Delipified extracts were twice as effective as crude extracts (Prot and Kornprobst, 1983a).

Foliar applied treatments. Long-term control of *M. javanica* on pineapple was afforded by post-plant foliar sprays of oxamyl or fenamiphos, following preplant soil fumigation or dipping the transplants in oxamyl or fenamiphos (Milne and Villiers, 1975; Anon, 1979). The severity of infestation of tomatoes by the nematode was reduced by foliar sprays of 3 or 5 ppm 2,4-dichloropropane, 500 ppm lihocin or 1000 ppm miraculan. Used in combination with 3 or 5 kg phorate or 3 kg aldicarb ha^{-1}, control was increased (Prasadji and Sitaramaiah, 1994).

Soil treatments. Galling of tomatoes by *M. javanica* at harvest was decreased most by 8 kg fenamiphos ha^{-1} applied to the soil at planting (Brown and Turner, 1978). Infestation at harvest was also lessened by spot application of 1 kg aldicarb ha^{-1} to transplanted tomato (Jain and Bhatti, 1989). On grapevines, 4 kg aldicarb ha^{-1} decreased soil and root populations of the nematode better than interplanting them with *Asparagus racemosus*, *Crotalaria juncea*, *Tagetes erecta* or *T. patula* (Baghel and Gupta, 1987). In Zimbabwe, aldicarb at 3 kg ha^{-1} was the best non-fumigant alternative to EDB or 1,3-D for control of the nematode on tobacco (Shepherd and Way, 1986). In tomato and aubergine nursery beds in India, *M. javanica* was controlled best by fenamiphos at 6 kg ha^{-1} for tomato and 4 kg ha^{-1} for aubergine (Jain and Gupta, 1992). Applying 4 kg aldicarb ha^{-1} to the nursery soil and 1 kg aldicarb ha^{-1} to the field soil controlled the nematode better on tomatoes than applying neem leaves at 2.5 t ha^{-1} to the nursery with or without 1 kg aldicarb ha^{-1} to the field soil at transplanting. The largest yields, however, came from uninfested tomato plants transplanted in soil to which 1 kg aldicarb ha^{-1} was applied (Jain and Bhatti, 1993).

Penetration of *M. javanica* into tomato roots was greatly reduced by 1 ppm in soil of a quassinoid fraction of seeds of *Hannoa undulata*, a small, common tree in Sudan savannah. This was a mixture of three polycyclic lactones – chapparinone, klaineanone and glaucarubalone. The action of this mixture was nematostatic and at 5 ppm or more it lessened migration of the juveniles and disorientated them. Multiplication of the nematode on tomato roots was reduced by 1 ppm but not completely prevented by 10 ppm of the mixture (Prot and Kornprobst, 1983b, 1985).

Field Resistance to Non-fumigant Nematicides

There is some evidence that *M. incognita*, parasitizing grapevine in California, exposed to 10.2 kg fenamiphos or 11.4 kg carbofuran ha^{-1} applied to the soil for three consecutive years, became tolerant or resistant to the nematicides in the field (Yamashita and Viglierchio, 1987). Such a result points to the need to use nematicides infrequently in an integrated control programme whenever possible, rather than relying on continuous application of nematicides.

Promising New Non-fumigant Nematicides

Large numbers of newly synthesized compounds are screened annually against root-knot nematodes (principally *M. incognita*) by chemical manufacturers, worldwide. For many reasons, most compounds found effective against nematodes are never marketed and, at present, very few nematicides are available for commercial use. Though there is no guarantee that any of them will be marketed, a few compounds recently reported effective in the control of *M. incognita* seem worthy of mention as possible new nematicides.

In glasshouse experiments, oil geranium (*Pelargonium graveolens*) was shown to be highly resistant to several species of *Meloidogyne* (Whitehead, 1965). The nematicidal activity of the essential oil is now known to be due to its major constituents, especially geraniol, followed by citronellol and linalool (Leela *et al.*, 1992).

Eleven of 55 phenolic compounds assessed were highly toxic to *M. incognita* juveniles, high activity being associated with a quinone group (*ortho* or *para*), meta-hydroxy and coumarin groups (Mahajan *et al.*, 1992). A number of organophosphates have also shown promise as nematicides. *Meloidogyne incognita* was controlled by *O*-ethyl *S,S*-bis (1-methylpropyl) phosphorodithioate (Pelagati and Caroppo, 1988) and *O,O*-di (2,4-dimethylphenyl) *S*-ethyl phosphorothiolate (Gupta *et al.*, 1986). Of 12 new *O*-alkyl *S,S*-diaryl phosphorotrithioates, *O*-methyl *S,S*-di (4-chlorophenyl) phosphorotrithioate was the most effective, with activity similar to that of fenamiphos (Gupta *et al.*,1991). Of six nematicidal 2-chloroethyl *O*-aryl *O*-methyl phosphonates, *O*-methyl 2-chloroethyl *O*-4 thiomethyl phenyl phosphonate was the most effective against *M. incognita* on soybean (Gupta *et al.*, 1994).

Foliar sprays of DMDP (2,5-dihydroxymethy–3,4-dihydroxypyrrolidine), a basipetally translocated sugar analogue isolated from tropical legumes (*Lonchocarpus* and *Derris*) roughly halved galling of tomato roots by *Meloidogyne* spp. (Birch *et al.*, 1993).

Nacobbus aberrans (Thorne, 1935) Thorne and Allen, 1944

(Fig. 8.10)

Sher (1970) considered *N. serendipiticus* Franklin, 1959, *N. serendipiticus bolivianus* Lordello, Zamith and Boock, 1961 and *N. batatiformis* Thorne and Schuster,1956 morphologically indistinguishable from *N. aberrans*. There are, however, differences in host range between these nematodes, which are currently regarded as races of *N. aberrans*. *Nacobbus aberrans* (*sensu lato*) has been reported from many parts of the world, attacking 69 species of plant in 17 families. Important crop plants, susceptible to one or more forms of *N. aberrans*, include potato, tomato, carrot, pea, cucumber, lettuce, kale, broccoli, cauliflower, Brussels sprouts, mustard, turnip, radish and spinach. The nematode is a serious pest of potato in Peru, Argentina, Chile, Ecuador and Bolivia. It is endemic to Peru and Bolivia and is the most important nematode pest of potatoes between 2000 and 4000 m altitude in Bolivia, Argentina and south Peru. Tuber yield losses can reach 90% in heavily infested land. The nematodes can survive temperatures as low as −15°C. and have a measure of drought resistance. Reproduction is sexual and each female deposits about 250 eggs in a gelatinous matrix. The hatched second stage juveniles invading roots cause considerable necrosis. They and the adult vermiform females can, however, leave dying rootlets and infest fresh roots, which become galled. The galls are easily confused with those due to *Meloidogyne* spp., hence the common name for *N. aberrans* – 'false root-knot nematode' (Fig. 8.11). The females become saccate with increasing egg content but the males remain vermiform. The life cycle lasts 25–50 days (Mai *et al.*, 1981) according to temperature and no doubt to some extent depending on which host plant is infested. There may be two or more generations in a growing season. There appear to be two races of the nematode of differing pathogenicity that attack potatoes (Jatala and Golden, 1977).

On sugarbeet, the tolerance level for total fresh plant weight was found to be 0.8 second stage juveniles ml^{-1} soil (Inserra *et al.*, 1984a). The sugarbeet race of the nematode was described as *N. batatiformis* by Thorne and Schuster (1956). It occurs in several western states of the USA, notably in Nebraska. Although yield

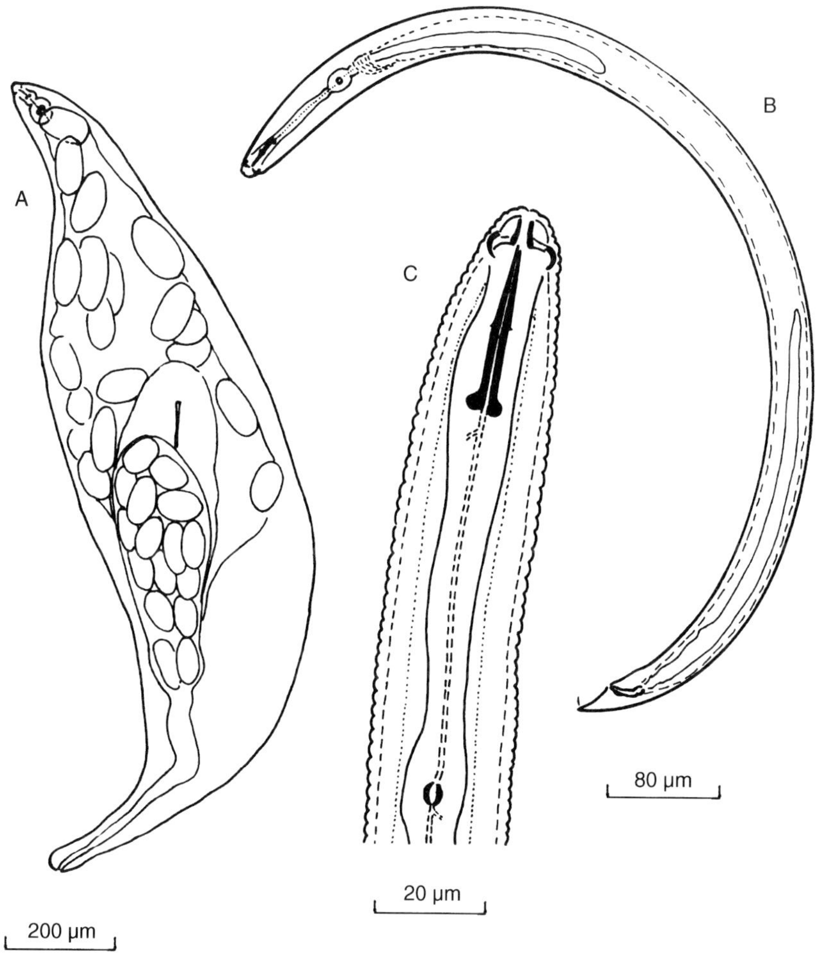

Fig. 8.10. *Nacobbus aberrans*: A, gravid female; B, male; C, male anterior end.

losses of 20% have been reported on sugarbeet, the nematode is less damaging than *H. schachtii* or *M. hapla* (Inserra *et al.*. 1984b). The sugarbeet race (*N. batatiformis*) does not attack potatoes.

Control

Although *N. aberrans* is an important pest of potatoes in S. America and to a lesser extent on sugarbeet and other crops elsewhere, research on its control has been very limited.

Crop rotations

Numbers of *N. aberrans* decrease rapidly in the absence of a host plant. However, races have wide but differing host ranges. Some *Gramineae* and many

Fig. 8.11. Sugarbeet attacked by false root-knot nematode, *Nacobbus aberrans* (Photo: D.J. Hooper).

Leguminosae are poor or non-hosts of the potato race of the nematode (Mai *et al.*, 1981). Potatoes, cereals, lucerne and onions are non-hosts of the sugarbeet race (Altman and Thomason, 1971). Bridge (1983) added okra, cotton, sunflower, sesame, melon, squash, watermelon, groundnut and soybean to the list of non-hosts of potential use in rotations to control *N. aberrans* (*sensu lato*). The efficacy of control by crop rotation is reduced if host weeds are inadequately controlled.

Resistance

Some native cultivars of *Solanum tuberosum* ssp. *andigena* are known to be resistant to *N. aberrans*, as are some wild, tuber-bearing species of *Solanum*, so control of this pest through the use of resistant cultivars looks promising (Cahuana *et al.*, 1975; PROINPA, 1991).

Nematicides

Nacobbus aberrans was controlled on sugarbeet in Utah by fumigant and non-fumigant nematicides (Inserra *et al.*, 1985). In Wyoming, galling of sugarbeet was significantly reduced by the combined use of 7.8 kg aldicarb plus 1.7 l of Rovral (iprodione) drench ha^{-1} (Beaupre *et al.*, 1990). In Peru, aldicarb or carbofuran applied to the seedbed controlled the nematode adequately in potatoes (Cornejo Quiróz, 1977). However, in other experiments in the Andes region, granular nematicides failed to decrease galling of potato roots, although yields were increased by up to 40% (Otazú *et al.*, 1985; Bascopé and López, 1987). Dipping seed potato tubers in fenamiphos or ethoprophos at 2000 ppm for 10 min freed them of the nematode, so avoiding its spread to uninfested fields (Costilla, 1989).

Conclusions and Future Prospects for Control 9

Before any control measures can be applied to an apparent nematode pest its pathogenic role must be clearly established. If it is shown that it is the cause of specific symptoms in the affected plants, which are closely correlated with poorer plant growth, one can be assured that controlling the nematode will prevent or lessen the disease. Most plant-parasitic nematodes are obligate parasites, so a close inverse relationship between the numbers of a nematode and plant yield suggests that the nematode is at least partly responsible for the loss of yield. It is not, however, conclusive proof, because nematode abundance might be closely linked to some other factor. One or more other nematodes may be involved and secondary disease organisms may be responsible for much of the yield loss. The nematode may simply be carrying the causal pathogen to the plant. Because the nematodes are easily overlooked and the symptoms they cause are often confused with nutrient deficiences and other pathogens, much nematode damage often goes undetected. Proper diagnosis is essential to the appropriate use of nematode control measures. It is also important to know the scale of the problem, i.e. how heavily infested the affected fields or plantations are, as this can affect the control strategy.

In the short-term, the aim of control is to prevent significant loss of yield and quality and in the long-term to keep nematode populations small. Reports in the literature of yield improvements on infested land following control measures that were not supported by nematode control data have not been included in this review. Similarly, less attention has been paid to pot experiments than to field experimentation, except those of special interest, because one cannot reliably extrapolate from pot to field, the two environments often differing considerably. The control measures referred to in this book can be classified as physical, biological or chemical. The effectiveness of the principal control measures (Table 9.1) and those of more limited application are now considered.

Physical Control Measures

Thermotherapy

All nematodes in soil can be killed by steaming or electrically heating it. As moist, well drained soil usually contains at least 10% water, raising its temperature to

Table 9.1. Principal measures for controlling plant-parasitic nematodes.

Nematode	*Ht. S.	*Ht. P.	Rot.	Res.	Sd.tr.	Tplt.dp	SF	NF	Fol.sp.
Anguina tritici			+	−	+				
Aphelenchoides besseyi		+		+	+	+		+	
A. fragariae		+				+		+	+
Bursaphelenchus xylophilus			−	+				+	+
Rhadinaphelenchus cocophilus			−	−					
Ditylenchus angustus			+		+			+	
D. destructor		+	+		+		+	+	
D. dipsaci	+	+	+/−	+/−	+		+/−	+	+
Belonolaimus longicaudatus	?+		−	−			+	+	
Criconemella xenoplax			−	−			+	+	
Hoplolaimus columbus			?+	−			+	+	
Longidorus elongatus			−	−			+	?+	
Longidorus spp.			−	−			+	?+	
Paratrichodorus minor			?+	−			+/−	+	
Other trichodorids			−	−			+	+	
Tylenchorhynchus spp.			−	−			+	+	
Xiphinema diversicaudatum			−	−			+		
X. americanum			+/−	−			?+	+	
X. index			?+	−			+		
Helicotylenchus dihystera			−	−			?+	+	
H. multicinctus			−	−		+		+	
Rotylenchulus reniformis	+		+	+		+	+	+	+
Tylenchulus semipenetrans		+	−	+		+	+	+	
Hirschmanniella oryzae			?+	?+		+		+	
Pratylenchus brachyurus				−		+		+	+
P. coffeae		+	+	+		+	+	+	
P. loosi			−	+			+	+	
P. penetrans		+		−	+		+	+	+
P. thornei	+		?+	−	?+		+	+	
P. vulnus		+		−		+	+	+	
P. zeae			+	−			+	+	
Radopholus similis		+	+/−	?+		+	+	+	+
R. citrophilus		+		+		+	+	+	?+
Scutellonema bradys		+	+	−		+		+	
Globodera pallida			+	+			+/−	+/−	
G. rostochiensis	+/−		+	+			+/−	+	−
Heterodera avenae			+	+	+		+	+	
H. cajani				−	+				
H. carotae	?+		?+	−			+	?+	
H. ciceri	+		+	−					
H. cruciferae			+	−			+	+	
H. glycines			+	+			?+	+/−	
H. goettingiana			+	−			+	+	
H. schachtii			+	?			+	+	
H. zeae			?+	−				+	
Meloidogyne arenaria		+	+	+	+		+	+	
M. chitwoodi			?+				+	+	

Table 9.1. *Continued.*

Nematode	*Ht. S.	*Ht. P.	Rot.	Res.	Sd.tr.	Tplt.dp	SF	NF	Fol.sp.
M. exigua			+	+			+	+	
M. graminicola			+	+			+	+	+
M. hapla			+	+	+	+	+	+	+
M. incognita	+	+	+	+	+	+	+	+	+
M. javanica	+	+	+	+	+	+	+	+	+
Nacobbus aberrans			?+		+	+	+	+/−	

+ Actually or potentially effective; ? uncertain; − inappropriate, ineffective or unavailable; *Ht., Heat therapy, S of soil, P of plant; Rot., rotation; Res., resistance; Sd.tr., seed treatment; Tplt.dp., transplant dip; SF, soil fumigation; NF, non-fumigant nematicide treatment; Fol.sp., foliar sprays.

50–55°C to kill the nematodes is expensive. In warm climates, nematodes in topsoil out-of-doors and in glasshouses are killed by solarization for 4–8 weeks, using single or, more effectively, double layered polyethylene sheeting over the soil. The increase in soil temperature achieved decreases with soil depth and even in hot climates nematodes are not affected in the subsoil, i.e. below about 30 cm. Soil solarization has been assessed against few nematodes (Table 9.1) and it is more suited to nursery beds and glasshouses than to large fields because of the costs involved and the loss of the land for up to two months. The technique does not sterilize the soil as effectively as steam, though it may kill as many nematodes in the topsoil as a soil fumigant. Part of the nematode control achieved by soil fumigation under polyethylene sheeting in warm sunny climates may be due not only to the retention of fumigant gases for longer near the soil surface but to solar heating of the soil.

Planting stock is freed of nematode infestation by heating it sufficiently to kill the nematodes it contains. Some stocks (e.g. tulip bulbs) may be seriously injured by hot water treatment and immersion time and water temperature must be carefully controlled to minimize phytotoxicity in less sensitive stocks. Nematode infestation in rice seed can be minimized but not eliminated by solar heating it without harming the grain's viability. Infestation of garlic bulbs with *Ditylenchus destructor* was decreased by high temperature storage and raising the temperature of pine wood to 60°C freed it of the pine wood nematode, *Bursaphelenchus xylophilus*. Banana corms, strawberry runners, rooted woody perennials and some yams may be essentially or completely freed of nematodes by hot water treatments, though cultivars differ in their heat tolerances.

Flooding

Flooding soil for several weeks or months has almost eradicated some nematode infestations. Nine weeks flooding in the Netherlands was effective against stem nematode, *Ditylenchus dipsaci* and there have been similar successes against the burrowing nematode, *Radopholus similis* and some root-knot nematodes, *Meloidogyne* spp. Flooding is not an option where water is scarce nor where the land slopes. Downward percolation of fine soil particles in the soil profile may impair its drainage.

Cultivations

Some nematodes may be killed mechanically, or desiccated in periods of drought, by soil cultivations. Repeated summer ploughing and discing reduces root-knot nematode infestations but speeds oxidation of soil organic matter and risks wind and water erosion of the soil. The mechanical disruption of compacted layers in the soil, allowing deeper root penetration of the soil, lessens the impact of nematodes attacking the roots during plant establishment and can enhance plant response to nematicides applied to the soil. Plants with deeper root systems are less prone to drought and can absorb nutrients from a larger volume of soil. However, they are likely to leave larger numbers of nematodes, to which they are susceptible, in the soil and to a greater depth. Nevertheless, deep cultivations or subsoiling can greatly improve crop growth in the presence or absence of harmful nematodes.

Biological Control Measures

Bare fallowing

Some nematodes decline rapidly under bare fallow, others do not. The disadvantages its use may entail are the increased risk of soil erosion and the loss of crop production over several months. Some success has been reported in controlling root-lesion nematode, *Pratylenchus penetrans*, reniform nematode, *Rotylenchulus reniformis* and burrowing nematode, *Radopholus similis* in this way. Two year bare fallows, following fumigation of land infested with citrus spreading decline nematode, *Radopholus citrophilus*, lessened or prevented spread of the nematode to adjacent healthy citrus.

Non-hosts in crop rotations

Many nematode pests of arable crops may be controlled to some degree by crop rotations that include a number of non-host crops (Table 9.1). Control is lessened if weed hosts are inadequately controlled. Controlling nematodes by crop rotation is environmentally attractive but it may be complicated by the presence of more than one target species, because a crop plant may be immune to one and susceptible to another. There may also be difficulties in marketing some non-host crops. Success is therefore dependent on knowing which harmful species are present in the nematode complex of the soil and on assessing the host status of potential rotational crops *locally*. Crop rotations can also enhance soil fertility and should be used wherever possible as a basis for nematode control in arable crops.

Cultivars and rootstocks

Cultivars and rootstocks which prevent nematode multiplication (resistant) or prevent nematode development (immune) are available for a number of important nematode pests (Table 9.1), notably root-knot nematodes (*Meloidogyne* spp.), some cyst-nematodes (*Globodera* spp. and *Heterodera* spp.) and some shoot pests (stem nematodes, *Ditylenchus dipsaci* and *D. angustus*, leaf nematode, *Aphelenchoides besseyi* and the pine wilt nematode, *Bursaphelenchus xylophilus*).

Rootstocks have been developed that are tolerant and/or resistant to citrus nematode, *Tylenchulus semipenetrans*, citrus decline nematode, *R. citrophilus* and some root-lesion nematodes, *Pratylenchus* spp. Cultivars or rootstocks resistant to root-ectoparasitic nematodes or the red ring nematode, *Rhadinaphelenchus cocophilus* have not been produced.

Resistant cultivars and rootstocks are environmentally benign and can be very effective in controlling specific nematode pests. They do not involve the grower in the extra expense entailed in nematicide applications and they control root-parasitic species to the depth of rooting. The drawbacks to the use of resistant cultivars are their often poorer yielding capacity and marketability compared with susceptible cultivars grown in uninfested or disinfested soils and to their selection of virulent pathotypes that may occur in field populations of the nematode. Resistance should be used in conjunction with other control measures to retard selection. Resistance coupled with tolerance improves yielding capacity in infested soils but may also encourage selection for virulence. Despite these caveats, nematode resistance (*sensu lato*) can play an important part in nematode control. Faster maturing cultivars may also lessen nematode population increase.

Green manures, trap cropping

The numbers of some important nematode pests in the soil have been greatly decreased by growing resistant, green manure crops or for a short period susceptible crops, both of which attract and trap the nematodes in their roots. If a susceptible crop plant is used (e.g. first early potatoes) it must be removed before the nematode life cycle is completed.

In warm, moist autumn seedbeds in northern Europe, beet cyst-nematode, *Heterodera schachtii*, has been decreased by trapping it in the roots of resistant oil radish or resistant white mustard. In south-east USA, American jointvetch and hairy indigo have greatly decreased some soil populations of soybean cyst-nematode, *Heterodera glycines*, and southern root-knot nematode, *Meloidogyne arenaria*. Heavy infestations of potato pale cyst-nematode have also been much reduced by growing susceptible potatoes (cv. Cara) for about 6 weeks in spring in England. Several green manure crops have been shown to decrease the rice root nematode, *Hirschmanniella oryzae*. The disadvantage of trap cropping is one of cost and inconvenience but the advantage is that it may decrease nematode numbers as effectively as soil fumigation with 1,3-dichloropropene.

Organic soil amendments

Large amounts of organic soil amendments (oilseed cakes, other by-products, wastes and plant extracts) added to the soil have greatly decreased numbers of spiral nematodes, *Helicotylenchus dihystera* and *H. multicinctus*, reniform nematode, *Rotylenchulus reniformis*, root-lesion nematode, *Pratylenchus brachyurus*, yam spiral nematode, *Scutellonema bradys*, brassica stunt nematode, *Tylenchorhynchus brassicae* and root-knot nematodes, *Meloidogyne* spp. Crop yields are increased in part by nematode control and in part by manurial effects. Because of the large quantities of such amendments needed for worthwhile control, their use is only feasible for small areas of land or fields or plantations close to animal rearing or crop processing units.

Antagonistic organisms

Natural biological control has been demonstrated for some plant-parasitic nematodes (Fig. 9.1). Large populations of cereal cyst-nematode, *Heterodera avenae*, are controlled, except in very droughty soils, by *Verticillium chlamydosporium* and *Nematophthora gynophila*. Similarly, beet cyst-nematode, *Heterodera schachtii*, is controlled by *V. chlamydosporium* (first generation nematodes) and *Nectria radicicola* (second generation nematodes). There has been some success in controlling *Meloidogyne* spp. in the field with *Arthrobotrys irregularis* (Royal 350), *Paecilomyces lilacinus* (Biocon), *Bacillus penetrans*, *Aspergillus niger*, the vesicular-arbuscular mycorrhiza *Glomus manihotis* and a species of *Streptomyces*. Glasshouse experiments have suggested that *V. chlamydosporium* could also be effective against *Meloidogyne* spp. The reniform nematode, *Rotylenchulus reniformis* might be controlled with *P. lilacinus*, *Streptomyces*, *Bacillus thuringiensis* or *A. niger.*

The control achieved with these organisms can be very variable, for soil conditions, especially soil moisture status and the presence of antagonists of the nematode, may be limiting factors. Another difficulty is the large quantity of inoculum currently needed to establish an effective colony of the desired control organism. Much research would be needed before widespread use of biocontrol agents could become a practical reality. They would probably work best in integrated control programmes.

Chemical Control Measures

Seed treatments

Apart from the mechanical separation of wheat seeds infested with *Anguina tritici* from healthy wheat seeds, dormant planting material may be disinfested or protected from early nematode attack by fumigation or coating it with a non-fumigant nematicide.

Methyl bromide fumigation has been successful in ridding the seeds of a number of crop plants of stem nematode, *Ditylenchus dipsaci* and white tip nematode, *Aphelenchoides besseyi*. Seed moisture is critical to successful control and lack of phytotoxicity. The gas concentration × exposure time product determines the percentage of nematodes killed in seed at optimum moisture content. The greater the seed moisture content, the greater the potential phytotoxicity. Coating seeds with nematicides can protect the seedlings and young plants from *Meloidogyne* spp., *Aphelenchoides besseyi*, *Pratylenchus penetrans*, *P. thornei* and pigeon pea cyst-nematode, *Heterodera cajani*. Dipping potato tubers in fenamiphos or ethoprophos freed them of the false root-knot nematode, *Nacobbus aberrans*. Dipping banana corms in nematicidal liquids or coating them with a clay–nematicide slurry (pralinage) protects them from early attack by burrowing nematode, *Radopholus similis* and spiral nematode, *Helicotylenchus multicinctus*.

In general, seed dressings lessen early nematode damage but most nematicides effective for this use are very toxic to vertebrates making the seed hazardous to handle and potentially lethal if eaten in times of famine.

(a)

(b)

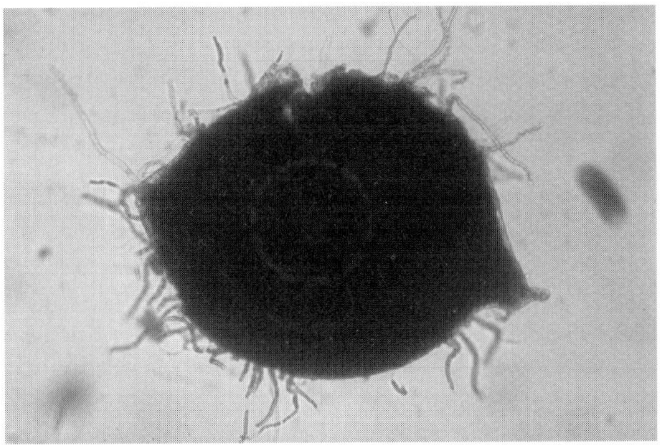

(c)

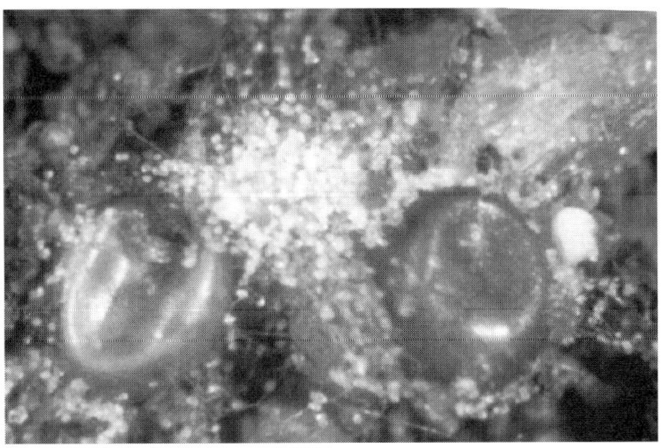

Fig. 9.1. Fungal antagonists of cyst and root-knot nematodes. (a) *Paecilomyces lilacinus* on *Meloidogyne incognita* female; (b) *Nematophthora gynophila* in *Heterodera avenae* female; (c) *Verticillium chlamydosporium* on cysts of a potato cyst-nematode (Photos: D. Crump).

Transplant root dips

Immersing the roots of transplants in liquid formulations of nematicides can rid them of nematode parasites and afford them early protection from nematode attack in the field. Such treatment can protect roots from a range of nematodes (Table 9.1). Seed extracts of *Hannoa* spp. and certain toxic leaf extracts can also free transplants of *Meloidogyne* spp. and give early protection of the roots from nematode attack after planting.

Fumigant nematicides

Currently available soil fumigants can greatly decrease numbers of plant-parasitic nematodes in the soil (Table 9.1), especially if it is covered with gas-tight sheeting. Soil fumigants work best in soils that are moist but well drained and contain little clay or organic matter. Kills in excess of 80–90% are difficult to obtain without gas-tight sheeting. Applying a second dose of fumigant to the soil after it has been ploughed to bury the top 5–10 cm can greatly increase kill in the topsoil. The recovery of nematode populations on host plants after fumigation is slow for those nematodes with long life cycles (*Belonolaimus*, *Criconemella*, *Longidorus*, *Paratrichodorus*, *Trichodorus* and *Xiphinema*) but rapid for those with shorter life cycles (*Ditylenchus dipsaci*, *Globodera*, *Heterodera* and *Meloidogyne*) and several generations per growing season. Injecting large amounts of fumigants into the soil can kill nearly all the virus vector nematodes it contains, so preventing transmission of those viruses for several years. Injecting soil fumigants at two depths can greatly increase kill at depth and large amounts of methyl bromide under gas-tight sheeting may kill nematodes to 1 m depth. Even larger amounts of 1,3-dichloropropene may kill *Radopholus citrophilus* to 2.6 m depth. Such deep fumigation may, however, contaminate drainage and ground water. Current soil fumigants do not afford long term control of nematodes on perennial crop plants and they are best employed to minimize nematode injury in the establishment of annual and perennial crop plants.

Adequate early control of root-ectoparasitic nematodes and root-knot nematodes in wide-row annual crops can be obtained by row fumigation at or before planting, using comparatively small amounts of fumigant. In this way, *Longidorus* spp. and trichodorids (except *Paratrichodorus minor*), as well as *Belonolaimus longicaudatus* and other plant-parasitic species, can be controlled sufficiently to prevent significant crop damage. Fumigating planting stations can aid the establishment of tea bushes and fruit trees, though it is labour intensive. Fumigating predetermined bands of soil in which the bushes or trees will be planted when the soil is no longer phytotoxic can be done more easily with appropriate machinery, except on hilly terrain. Very small amounts of soil fumigant applied in the seed furrows at sowing have controlled cereal cyst-nematode, *Heterodera avenae*, adequately in Australia and row fumigation with small amounts of 1,3-dichloropropene have also had some success against beet cyst-nematode, *H. schachtii*.

Non-fumigant nematicides

Many nematodes are adequately controlled by applying much smaller amounts of non-fumigant nematicides than of soil fumigants to the soil. Being non-phytotoxic and of short persistence in soil, for annual crops these compounds are best applied at planting. A single application to the topsoil or to the crop rows may be sufficient to prevent yield loss and lessen or prevent nematode increase in annual crops. A mid-season application adequately distributed in the soil may be needed to maintain control but could not be approved if unacceptable residues were left in the harvested produce. Several avermectins are effective against *Meloidogyne* spp. at rates as small as 0.05–0.5 kg a.i. ha^{-1}.

Non-fumigant nematicides are applied repeatedly to perennial crop plants at risk from damaging nematodes, such as the burrowing nematodes, *Radopholus* spp. and virus-transmitting root-ectoparasitic nematodes. Their use in perennials has increased following the demise of dibromochloropropane, the only non-phytotoxic and downwardly percolating soil fumigant. Some non-fumigant nematicides can be leached down the soil profile as much as 60–90 cm by sprinkler irrigation provided this does not contaminate drainage or ground water. The most commonly used compounds (aldicarb, oxamyl, carbofuran, fenamiphos, ethoprophos and fensulphothion) are very toxic to vertebrates and they are more safely applied to the topsoil in granular than in spray formulations.

Economics usually dictates that for annual crops these compounds are applied as row rather than broadcast topsoil treatments. For some stem nematodes they are more efficient as row or narrow band treatments over the crop rows than as wide band or broadcast treatments. Row applications of non-fumigant nematicides are effective for control of stem nematode, *Ditylenchus dipsaci* in sugarbeet, field beans, newly sown clovers and lucerne, root-ectoparasitic nematodes of sugarbeet and other row crops, cereal cyst-nematode, *Heterodera avenae*, in wheat, as vertical band treatment for sugarbeet and give early protection against root-knot nematodes, *Meloidogyne* spp. on a wide range of crop plants. Mixing these compounds in the top 10–15 cm of the soil, which is then ridged up, ensures that most of the nematicide is in the ridge soil, so affording plants grown in the ridges most protection from nematode attack in the sensitive early stages of their establishment. In widely spaced perennial crops, repeated treatments are made to the soil within the plant canopy. Repeated applications of fenamiphos around peach trees in autumn and spring have lessened 'peach tree short life' initiated by *Criconemella xenoplax*. Injecting a non-fumigant nematicide into the trunks of pine trees at the first signs of disease can control pine wilt disease, due to *Bursaphelenchus xylophilus*, though the treatment may need to be repeated to maintain control. Trunk injections can also control citrus nematode, *Tylenchulus semipenetrans* in citrus and root-lesion nematode, *Pratylenchus vulnus* in walnuts and grapevines.

The efficacy of non-fumigant nematicides may be decreased following their repeated use, due to their accelerated microbial breakdown and perhaps to increased nematode 'tolerance' of these mostly cholinesterase-inhibiting compounds.

Foliar applications of non-fumigant nematicides

Several systemic (phloem-mobile) compounds affect shoot- and root-parasitic nematodes, when they are sprayed on the foliage.

The leaf nematode *Aphelenchoides fragariae*, has been controlled by foliar sprays of thionazin or isazophos. Similarly applied, oxamyl, fenamiphos or ascorbic acid controlled the reniform nematode, *Rotylenchulus reniformis*, fenamiphos controlled root-lesion nematodes, *Pratylenchus brachyurus* and *P. penetrans* and oxamyl controlled *Radopholus similis*, *R. citrophilus* and *P. penetrans*. Except for ascorbic acid, these compounds are very toxic and must only be applied by properly protected operators. An hydroxymethyloxamyl glucurinide (JR 522) and its methyl ester (JR 775) are phloem-mobile pronematicides, which are less toxic than oxamyl, and are translocated to tobacco root tips where they are broken down to release oxamyl. Usually only a small percentage of a foliar applied nematicide reaches the roots and treatments may be spoilt by rain too soon after application to the leaves. Nevertheless, safe, effective, basipetally translocated nematicides could be very useful, especially for perennial crops. In annual plants much nematode damage is done before the plants are large enough to spray. Oxamyl controlled potato cyst-nematode, *Globodera rostochiensis*, more effectively when it was incorporated in seedbed soil at planting than when it was applied repeatedly to potato foliage. In contrast, repeated foliar applications of nematicides may be more effective than granular nematicide applied to the soil in controlling nematodes with several generations per season or in controlling nematodes on perennials.

The full benefits of nematicide usage are only obtained if appropriate application techniques are used. Efficient application techniques lessen the amounts of nematicides required, reducing cost and minimizing environmental risks.

Management of Nematode Pests

Good management of a nematode pest involves avoiding it if possible, lessening its impact on crop plants and selecting appropriate control measures, when necessary. Knowing exactly what the problem is is fundamental to good nematode management.

At a national level, important exotic nematode pests should be excluded by effective quarantine procedures, e.g. insisting that pine wood from countries affected by pine wilt nematode is heat treated to kill the nematodes, that unrooted cuttings from rooted plants are grown on in a sterile medium in quarantine and that dormant stocks are carefully checked for nematode infestation or contamination before they are released for field use. Stopping the spread of an exotic pest once it has entered a country is much more difficult if not impossible. Spread of the pine wilt and red ring nematodes can be checked by rigorous control of their beetle vectors and destruction *in situ* of nematode infested trees. Whenever possible, uninfested or lightly infested fields should be chosen for the cultivation of crop plants at risk from nematode attack. Care should be taken to ensure that the planting stock is uninfested and that nematode infestation is not spread to uninfested land on farm machinery or in irrigation water.

The impact of a nematode pest may be lessened by ensuring the plants receive adequate soil nutrients and water and, if possible, sowing or planting when conditions are less favourable to the nematode. The soil should be freed of infested roots and trash before the next crop is planted and weeds that could compete with the crop or are nematode hosts should be controlled. Yield losses in bananas affected by nematode-induced root rots can be decreased by propping up the fruiting stems at bunching. Delay in harvesting a crop may increase nematode infestation of the soil. Damage to peach trees by winter injury and bacterial canker initiated by *Criconemella xenoplax* is exacerbated by pruning the trees in December rather than in March. These are just a few of the measures that can be taken to lessen the damage done to crop plants by nematodes.

When control of the nematode is unavoidable, the measures employed must be appropriate. A combination of measures (integrated control) is preferable to reliance on any single control measure, repeated use of which may select nematodes able to survive it. The worst option is to grow a crop plant continuously on the same land without controlling damaging nematode pests with which it is afflicted. Which measures are appropriate will depend on the nematode, the crop plant(s) at risk, the environment and the economics of the control options available to the grower. Although it would seem preferable to use measures which are specific to the nematode pest to be controlled, the grower is more likely to opt for measures which control a range of crop pests. In view of the high cost of developing new pesticides, which must meet ever more stringent environmental standards, it seems most unlikely that manufacturers will target the nematode market alone, let alone produce nematicides specific to certain nematodes. Despite its cost, there are obvious attractions in controlling a wide range of pests and diseases simultaneously with a broad spectrum soil fumigant, where its use is practical. Where only a single nematode pest needs to be controlled, a resistant cultivar or a non-fumigant nematicide applied to the crop rows may be more appropriate.

Future Prospects for Control of Plant-parasitic Nematodes

Future developments in plant nematode control cannot be predicted but it is possible to suggest desirable objectives which might be met in future. Some plant-parasitic nematodes can be adequately controlled with current methods appropriately applied. The control of others is inadequate or needs improvement, especially some cyst-nematodes, *Globodera pallida*, *Heterodera cajani*, *H. ciceri* and *H. glycines*, citrus burrowing nematode, *Radopholus citrophilus* and red ring nematode, *Rhadinaphelenchus cocophilus*, as well as a number of nematodes that attack perennials, including those that transmit plant virus diseases. Nematode control in general would be increased by developments in problem recognition, crop husbandry, soil solarization, crop rotations, crop plant resistance, nematicide use and integrated control.

Problem recognition

Much nematode damage to crop plants is not recognized because of a lack of expertise on farms and plantations. More trained nematologists are needed *in the field* to assess crop losses due to nematodes and reveal hitherto unrecognized problems.

Crop husbandry

How a crop plant is grown can affect its vulnerabilty to a nematode pest and its efficiency as a host, so changes in crop husbandry could ease some nematode problems. Shallower rooting might be encouraged in citrus by keeping the topsoil moist by sprinkler irrigation through the season. This would make the control of the citrus nematode, *Tylenchulus semipenetrans*, and the burrowing nematode, *Radopholus citrophilus*, easier to achieve with nematicides. It might, of course, render the trees more prone to toppling by wind. The effect of changes in crop husbandry on the impact of nematode pests needs to be assessed.

Soil solarization

This environmentally benign control technique could be more widely applied, especially in hot climates. The development of machinery to apply and recover the polyethylene sheeting could make the technique more attractive and cost effective. Small, cheap machines would allow the solarization of smallholdings and nurseries. Another possibility would be the use of rapidly degradable sheeting.

Crop rotation

Soils may be infested with more than one nematode pest. The effect of crop rotations on the nematode pest complex of the soil needs to be assessed *locally* before specific rotations are introduced. The viability of the selected rotations would also need to be assessed in relation to other pests and diseases, soil fertility and the marketability of the crops produced. Crop rotations could, however, greatly ease nematode problems, especially where more sophisticated control measures are inappropriate. The introduction of new crops offers the possibility of improved nematode control through crop rotation.

Crop plant resistance

Although there have been notable successes in the control of some nematode pests with resistant cultivars and rootstocks, there are two major problems that remain. These are the often poorer yield and quality of resistant cultivars compared with susceptible cultivars in uninfested soil, making them generally unattractive to growers, and the often narrow base of their resistance making them prone to selection of virulent nematode biotypes or sibling species present in field populations. More broadly based resistance would be useful, though crop plants *immune* to nematode species would be even more valuable. Resistant cultivars and rootstocks need to be tolerant of attack, though this might increase selection for virulence. Marketability of resistant or immune cultivars would be greatly improved if the necessary genes could be introduced without altering the desired phenotypes. More widespread field testing of resistant cultivars and rootstocks before their release to growers would better indicate their usefulness. Virus-resistant rootstocks could control nematode-transmitted virus diseases of woody perennials better than repeated nematicide usage to control the nematode vectors.

Nematicide use

The arsenal of soil fumigants still available for use is small – methyl bromide for specialist contractor use, 1,3-dichloropropene mixtures and methyl isothiocyanate liberators, such as metham-sodium and dazomet. The possible withdrawal of methyl bromide on environmental grounds would be serious but, apart from its use as a seed disinfestant, it could be replaced by soil fumigation with 1,3-dichloropropene, with or without dazomet, under polyethylene sheeting in glasshouses and nurseries. In hot climates and in glasshouses in summer, soil fumigation under clear polyethylene sheeting could result in improved nematode control partly through soil solarization of the topsoil. Deep soil fumigation would also be possible with 1,3-D. Nematodes may be controlled at depth in soil with non-fumigant nematicides leached downwards, but they have to be applied repeatedly and will not control nematodes in soil much below their depth of incorporation without irrigation. Furthermore, non-fumigant nematicides are nematostatic, so they do not decrease the numbers of inactive or dormant nematodes in the soil, which soil fumigants do. For these reasons it is important that 1,3-D and the methyl isothiocyanate liberators remain available for farm use.

The discovery and development of less toxic, effective, phloem-mobile nematicides would be very helpful in the control of nematode pests of perennial crops, as would new, less toxic non-fumigant nematicides for annual crops. Further work is needed on the control of nematodes of woody perennials by trunk injection. Finally, the use of improved methods of applying nematicides and better training and protection of operatives should achieve better nematode control with less nematicide and with less risk to the operatives and the environment.

Nematode management in subsistence farming

In a recent review, Bridge (1996) stresses the value and need for simple and cheap control measures for nematode pests in subsistence farming. In particular, preventing nematode spread, use of nematode-free planting material, crop rotations, soil cultivations, use of organic soil amendments and maintaining or increasing crop diversity by multiple cropping and use of multiple cultivars to increase resistance or tolerance can go a long way towards nematode control, without recourse to the use of nematicides. If low-income subsistence agriculture is to move towards a higher-income, market-orientated agriculture, however, in which produce must meet specific and exacting standards, nematode control may be more difficult to achieve and require the use of all effective measures.

Integrated control of nematodes

Much can be done to improve the control of nematode pests by using a combination of currently available control measures properly and appropriately applied. Further research on integrated control programmes, incorporating new methods of control as they come forward, should keep the grower in control of the nematode and not the reverse.

Bibliography

Abawi, G.S. and Mai, W.F. (1971) Root diseases of fruit trees in New York III. Control of *Pratylenchus penetrans* in roots of apple, peach, pear and cherry by foliar sprays with a nematicide–insecticide. *Plant Disease Reporter* 55, 617–620.

Abawi, G.S. and Mai, W.F. (1975) Effect of foliar applications of oxamyl on movement of *Pratylenchus penetrans* in and outside host roots. *Plant Disease Reporter* 59, 795–799.

Abawi, G.S. and Mai, W.F. (1978) Root diseases of fruit trees in New York State. X. Toxicity of oxamyl to *Pratylenchus penetrans*. *Plant Disease Reporter* 62, 637–641.

Abawi, G.S. and Mai, W.F. (1980) Effects of initial population densities of *Heterodera schachtii* on yield of cabbage and table beets in New York State. *Phytopathology* 70, 481–485.

Abdalla, N.A. (1974) Distribution, persistence and nematicidal activity of monobromomethane and a pesticide containing 1,3-dichloropropene in soil. *Dissertation Abstracts International* 35B, 296.

Abdalla, N.A., Raski, D.J., Lear, B. and Schmitt, R.V. (1974) Movement, persistence and nematicidal activity of a pesticide containing 1,3-dichloropropene in soils treated for nematode control in replant vineyards. *Plant Disease Reporter* 58, 562–566.

Abid, M. and Maqbool, M.A. (1990) Effects of intercropping of *Tagetes erecta* on root-knot disease and growth of tomato. *International Nematology Network Newsletter* 7, 41–42.

Abrego, L. (1974) Ensayos de selectividad de nematicidas en il combate del *Pratylenchus coffeae* en almacigueras de cafe. *Nematropica* 4, 17.

Abrego, L. and Cardona, L.M. (1976) Ensayos de selectividad de nematicidas en il combate del *Pratylenchus coffeae* en almacigueras de cafe. *Revista Cafetalera* No. 157, 17–20.

Abrego, L. and Holdeman, Q.L. (1961) Nematodos del cafe en El Salvador. *Instituto Salvadoreno invest. cafe, Santa Tecla, El Salvador, Bol. inf. suplemento* 8, 1–16.

Abu-Gharbieh, W.I., Saleh, H. and Abu-Blan, H. (1991a) Use of black plastic for soil solarization and post plant mulching. *FAO Plant Production and Protection Paper* No. 109, 229–242.

Abu-Gharbieh, W.I., Saleh, W.I.H. and Al-Banna, L. (1991b) Application of solar-heated water for soil solarization. *FAO Plant Protection Paper* No. 109, 69–77.

Abylova, N.A. and Vasilevskii, V.N. (1983) [The potato stem nematode at the collective farm 'Sokh' in the Fergana valley and its control.] In: *Steblevye nematody sel skokhozyaistrennykh kul'tur i mery bor'by s nimi. (Materialy Simpozuana, Voronezh, 27–29 Sentyabrya, 1983 g.)*, 59–61.

Acharya, A. and Padhi, N.N. (1989) Effect of neem oilcake and sawdust against root-knot nematode, *Meloidogyne incognita* on betel vine (*Piper betle* L.). *Indian Journal of Nematology* 18, 105–106.

Acosta, N. and Ayala, A. (1976) Hot water and chemical dips for nematode control in tubers of *Dioscorea rotundata* Poir. *Journal of Agriculture of the University of Puerto Rico* 60, 395–402.

Adamová, B. and Rotnekl, J. (1991) Vliv moreni osiva vojesky na napadeni hadatkem zhoubnym *Ditylenchus dipsaci* (Kühn) Filipjev. *Sbornik UVTIZ, Ochrana Roślin* 27, 167–176.

Adams, R.E. and Hickman, C.E. (1970) Influence of nematicidal treatments and fungicidal sprays on yield of strawberries. *Plant Disease Reporter* 54, 923–926.

Adams, R.E. and True, R.P. (1962) The response of Norway spruce to soil fumigation for the control of nematodes. *Phytopathology* 52, 1216.

Adesiyan, S.O. (1976) Host range studies of the yam nematode, *Scutellonema bradys*. *Nematropica* 6, 60–63.

Adesiyan, S.O. (1977) Studies on the effect of gamma radiation (from Cobalt 60 source) on storage life of white yam (*Dioscorea rotundata* var. Efon) infected with *Scutellonema bradys*. *Annals of Applied Biology* 86, 213–218.

Adesiyan, S.O. (1981) *In vitro* toxic effect of two organophosphates on the yam nematode *Scutellonema bradys* (Steiner and Le Hew). *African Journal of Agricultural Sciences* 8, 1–10.

Adesiyan, S.O. and Adeniji, M.O. (1976) Studies on some aspects of yam nematode (*Scutellonema bradys*). *Ghana Journal of Agricultural Sciences* 9, 131–136.

Adesiyan, S.O. and Badra, T. (1982) Granular nematicides for control of the yam nematode, *Scutellonema bradys*, and relevant residues in raw tubers. *Journal of Nematology* 14, 213–216.

Ahuja, S. (1978) The effect of root-dip treatments on infestation of brinjal by *Meloidogyne incognita*. *Nematologia Mediterranea* 6, 133–134.

Ahuja, S. (1980) Comparison of nematicides applied as root-dips or soil drenches for the control of *Meloidogyne incognita* on tomato. *Nematologia Mediterranea* 8, 199–200.

Akhtar, M. and Alam, M.M. (1990) Effect of bare-root dip treatment with extracts of castor on root-knot development and growth of tomato. *Nematologia Mediterranea* 18, 53–54.

Akhtar, M. and Alam, M.M. (1992) Effect of crop residues amendments to soil for the control of plant-parasitic nematodes. *Bioresource Technology* 41, 81–83.

Akhtar, M. and Mahmood, I. (1993) Control of plant-parasitic nematodes with 'Nimin' and some plant oils by bare-root dip treatment. *Nematologia Mediterranea* 21, 89–92.

Akhtar, M., Wain, A.H. and Alam, M.M. (1992) Control of root-knot nematode with bare-root dip in leaf extracts of Persian lilac and *Calotropis*. *Current Nematology* 3, 41–44.

Alam, M.M. (1986) Possible use of weeds as soil amendment for the management of root-knot and stunt nematodes attacking eggplant. *Agricultural Wastes* 16, 97–102.

Alam, M.M., Ahmad, M. and Khan, A.M. (1980) Effect of organic amendments on the growth and chemical composition of tomato, egg plant and chilli and their susceptibility to attack by *Meloidogyne incognita*. *Plant and Soil* 57, 231–236.

Alam, M.M., Saxena, S.K. and Khan, A.M. (1981) Soil populations of plant parasitic nematodes under various cropping sequences. *Nematologia Mediterranea* 9, 49–56.

Aleksandrova, I.A. (1985) [Efficacy of irradiation of rice seed infested with *Aphelenchoides besseyi* Christie, 1942 with ^{60} Co γ rays.] *Byulleten 'Vsesoyuznogo Instituta Gel'mintologii im. K.I. Skryabina* No. 41, 5–8.

Allen, M.W. (1946) Control of root-knot nematode with D-D mixture and chloropicrin. *California Agricultural Experiment Station, Circular* No. 365, 62–65.

Allen, M.W. (1957) A review of the nematode genus *Trichodorus* with descriptions of ten new species. *Nematologica* 2, 32–62.

Allen, M.W. and Raski, D.J. (1950) Chemical control of nematodes. Soil type important limiting factor in control of certain plant parasitic nematodes with volatile soil fumigants. *California Agriculture* 4, 5.

Allen, M.W., Burton, V. and George, A. (1955) Control of nermatodes on cotton. Investigations indicate preplanting fumigation of cotton land effective treatment of root knot nematodes. *California Agriculture* 9, 10, 14.

Allen, W.R. and Marks, C.F. (1977) Chemical control and population studies of *Pratylenchus penetrans* on fruit tree understocks. *Plant Disease Reporter* 61, 84–87.

Almeida, V.F. de and Campos, V.P. (1991a) Reproductividade de *Meloidogyne exigua* em plantas antagonistas e em culturas de interesse economico. *Nematologia Brasileira* 15, 24–29.

Almeida, V.F. de and Campos, V.P. (1991b) Alternancia de culturas e sobre vivencia de *Meloidogyne exigua* em areas de cafezal infestado e erradicado. *Nematologia Brasileira* 15, 30–42.

Alphey, T.J.W., Cooper, J.I. and Harrison, B.D. (1975) Systemic nematicides for the control of trichodorid nematodes and of potato spraing disease caused by tobacco rattle virus. *Plant Pathology* 24, 117–121.

Alphey, T.J.W., Bromilow, R.H. and Abdalla, A.M. (1985) Phloem translocation of foliar applied oxamyl and its role in plant protection. *Nematologica* 31, 468–477.

Alphey, T.J.W., Woodford, J.A.T. and Gordon, S.C. (1986) Field and laboratory studies on the control of nematode and aphid virus vectors in potatoes by pesticides applied as side bands. *Crop Protection* 5, 114–121.

Alphey, T.J.W., Phillips, M.S. and Trudgill, D.L. (1988) Integrated control of potato cyst nematodes using small amounts of nematicide and potatoes with partial resistance. *Annals of Applied Biology* 113, 545–552.

Al-Sayed, A.A. (1990) The inhibitory effect of ascorbic acid on *Meloidogyne incognita* infesting tomato. *Annals of Agricultural Science, Moshotor* 28, 1737–1740.

Al-Sayed, A.A. (1991) *Rotylenchulus reniformis* and soybean response to ascorbic acid. *Annals of Agricultural Science, Moshotor* 29, 1227–1232.

Altman, J. and Thomason, I.J. (1971) Nematodes and their control. In: Johnson, R.T., Alexander, J.T., Rush, G.E., and Hawkes, J.R. (eds) *Advances in Sugar Beet Production: Principles and Practices.* Iowa State University Press, Ames, Iowa, USA, pp. 335–370.

Ambrogioni, L. and Marinari-Palmisano, A. (1976) Effetto di avvicendamenti colturali su *Heterodera carotae* (Nematoda: Heteroderidae) e sulla produzione di carota in terreno infestato. *Redia* 59, 355

Amsing, J.J. (1988) Results obtained with Nemacur in the control of the root-lesion nematode, *Pratylenchus vulnus*, on roses as a green-house crop. *Pflanzenschutz-Nachrichten Bayer* 41, 239–258.

Amsing, J.J. (1990) Bestrijding *Pratylenchus vulnus* in steenwo wortelaaltjes makkelijk de baas. *Vakblad voor de Bloemisterij* 45, 38–39.

Anand, S.C. (1986) Sources of resistance to the soybean cyst nematode. In: Lamberti, F. and Taylor, C.E. (eds) *Cyst Namatodes.* Plenum Press, New York, pp. 269–276.

Anand, S.C. and Shumway, C.R. (1985) Response of secondary selection on soybean cyst nematode reproduction on resistant soybean lines. *Crop Protection* 4, 231–234.

Anand, S.C., Sharma, S.B., Rao-Arelli, A.P. and Wrather, J.A. (1994) Variation in parasitic potential of *Heterodera glycines* populations. *Crop Science* 34, 1452–1454.

Anand, S.C., Koenning, S.R. and Sharma, S.B. (1995) Performance of blends of soybean cyst nematode resistant and susceptible cultivars *Crop Science* 35, 524–528.

Andersen, S. (1960) Havreal problemer. *Sonderdr. Ans. Tolumands Bladet* 11, 45.

Andersson, S. (1975) Aktuellt om potatiscystnematoden-var viktigasts karantanskadegorare. *Vaxtskyddsnotiser* 39, 70–76.

Andersson, S. (1987) Den gula potatiscystnematodens (*Globodera rostochiensis*) populationsminskning vid odling av icke-vardvaxter och nematodresistent potatis-nagra preliminara resultat. *Vaxtskyddsnotiser* 51, 5–6, 145–150, 154.

Anon. (1932) Eradication of nematodes in green-house soil by the use of chemicals. *Massachusetts Agricultural Experimental Station, Bulletin* No. 280, 192.

Anon. (1963) Stem eelworm on cereals and other farm crops. *Ministry of Agriculture Fisheries and Food Advisory Leaflet* 178, Her Majesty's Stationery Office, London, UK, 6 pp.

Anon. (1974) *India Coffee Board, 33rd Annual Report 1972–1973*. Bangalore, India, 269 pp.

Anon. (1978) *Annual Report on Research and Technical Work*, 1977. N. Ireland Department of Agriculture, Belfast, 263 pp.

Anon. (1979) Effect of systemic nematicides on pineapples. *Citrus and Subtropical Fruit Research Institute, Nelspruit, South Africa, Information Bulletin* No. 82, 3–4.

Anon. (1981a) Pegletta – ein Olrettich als Feind-und Fandpflanze. Ein neuer Weg zur Nematoden bekampfung. *Zuckerrube* 30, 120–121.

Anon. (1981b) Une methode de lutte contre les nematodes à galles des racines appartenant au genre *Meloidogyne*. *Defense des Vegetaux* No. 207, 85–88.

Anon. (1984) Nematodes on strawberry. *Ministry of Agriculture, Fisheries and Food ADAS leaflet* 414, Her Majesty's Stationery Office, London, UK.

Anon. (1985) *Annual Report, ICAR Research Complex for North Eastern Hills Region, India, 1984*. 97–99.

Anon. (1986) Quick and slow wilt diseases of black pepper. *Report, India, Central Plantation Crops Research Institute, Research Highlights 1986, Kerala, India*,2.

Anon. (1990) [Study on the control of root-rot nematodes in ramie using Rugby.] *China's Fiber Crops* No. 3, 21–25.

Anon. (1994) Phytosanitary procedure No. 52. Treatment method for strawberry plants. Methode phytosanitaire *Aphelenchoides besseyi*. Methode de traitement des plants de fraisier. *Bulletin OEPP/ EPPO* 24, 322–323.

Anon. (1996) *Potato Variety Handbook, Including NIAB Recommended Lists of Potatoes 1997*. National Institute of Agriculural Botany, Cambridge, UK, 60pp.

Anver, S. and Alam, M.M. (1992) Effect of latex seed dressing on interacting root-knot and reniform nematodes. *Afro-Asian Journal of Nematology* 2, 17–20.

Anyango, J.J. (1988) The effect of root-knot (*Meloidogyne hapla*) and lesion nematodes (*Pratylenchus penetrans*) of pyrethrum seedlings in Kenya. *Acta Horticulturae* No. 218, 355–358.

Anyango, J.J., Ngugi, C.W. and Ikahu, J.M. (1991) Effect of granular nematicides on the control of root-knot and lesion nematodes in pyrethrum *Chrysanthemum cinerariaefolium* Vis. *Pyrethrum Post* 18, 100–103.

Apt, W.J. (1979) Control of the reniform nematode on pineapple in Hawaii with phenamiphos. *Proceedings of the American Phytopathological Society*, Abstract No. 612.

Apt, W.J. and Gould, C.J. (1961) Control of the root-lesion nematode, *Pratylenchus penetrans* on narcissus. *Plant Disease Reporter* 45, 290–295.

Araki, M., Inoue, T. and Sudo, M. (1991) [The effects of marketed forage crops on the population of root-knot nematode, *Meloidogyne incognita*, under field conditions.] *Proceedings of the Association for Plant Protection of Kyushu* 37, 130–132.

Arayarungsanit, L. (1987) Yield ability of rice varieties in fields infested with root-knot nematodes. *International Rice Research Newsletter* 12, 14.

Arneson, P.A. and Mai, W.F. (1976) Root diseases of fruit trees in New York State VII. Costs and returns of preplant soil fumigation in a replanted apple orchard. *Plant Disease Reporter* 60, 1054–1057.

Arntzen, F.K., Visser, J.H.M. and Hoogendoorn, J. (1994) The effect of the potato cyst nematode *Globodera pallida* on *in vitro* root growth of potato genotypes, differing in tolerance. *Annals of Applied Biology* 124, 59–64.

Arrigoni, O., Zacheo, G., Arrigoni-Liso, R., Bleve-Zacheo, T and Lamberti, F. (1979a) In: Lamberti, F. and Taylor, C.E. (eds) *Root-knot Nematodes* (Meloidogyne *Species*) *Systematics, Biology and Control.* London, Academic Press, pp. 469–470.

Arrigoni, O., Zacheo, G., Arrigoni-Liso, R., Bleve-Zacheo, T. and Lamberti, F. (1979b) Relationship between ascorbic acid and resistance in tomato plants to *Meloidogyne incognita. Phytopathology* 69, 579–581.

Aruntyunov, A.V. (1985) [The use of solar energy in the control of gall nematodes from the genus *Meloidogyne.*] *Izvestiya Akademii Nauk Turkmenskoi SSR, Biologicheskikh Nauk* No. 4, 52–56.

Asmus, R.M.F. and Ferraz, S. (1988) Antagonismo de algumas especies vegetais, principalmente leguminosas, a *Meloidogyne javanica. Fitopatologia Brasileira* 13, 20–24.

Atanasov, P. and Collingwood, C.A. (1962) Effects of metham sodium and other chemicals on potato root eelworm. *Plant Pathology* 2, 153–157.

Atilano, R.A. and Van Gundy, S.D. (1978) Systemic activity of oxamyl to *Meloidogyne javanica* at the root surface of tomato plants. In: Harley, J.L. and Russell, E.S. (eds) *The Soil-Root Interface.* Academic Press, London, pp. 339–349.

Atkins, J.G. and Fielding, M.J. (1956) A preliminary report on the response of rice to soil fumigation for the control of stylet nematodes, *Tylenchorhynchus martini. Plant Disease Reporter* 40, 488–489.

Atu, U.G. (1984) Effect of cover plants in fallow lands on root knot nematode population. *Beitrage zur Tropischen Landwirtschaft und Veterinarmedizin* 22, 275–280.

Atu, U.G. and Duru, N. (1990) Control of stem borer (*Sesamia calamistis* Hmps.) and lesion nematode (*Pratylenchus brachyurus* (Godfrey) Filipjev and Stekhoven) of late season maize with carbofuran. *Beitrage zur Tropischen Landwirtschaft und Veterinarmedizin* 28, 69–73.

Aubert, V. (1986) Contribution à l'étude de la biologie d'un parasite de la carotte: le nematode à kyste, *Heterodera carotae* (Jones). Université de Neuchatel, Institut de Zoologie, Neuchâtel 1986. Baillod SA, Boudry, Switzerland, 224 pp.

Aung, T. (1990) Effect of carbofuran on *Hirschmanniella* spp. infestation in irrigated rice and partitioning of rice yield increase. *International Rice Research Newsletter* 15, 36–37.

Aung, T. and Prot, J.C. (1990) Effect of crop rotation on *Pratylenchus zeae* and on yield of rice cultivar UPL Ri-5. *Revue de Nématologie* 13, 445–447.

Ayala, A. (1971) Rotation with pangola grass and sugarcane beneficial for pineapple production in Puerto Rico. *Nematropica* 1, 6–7.

Ayala, A. and Acosta, N. (1971) Observations on yam (*Dioscorea alata)* nematodes. *Nematropica* 1, 39–40.

Ayala, A., Román, J. and Gonzalez Tejera, E. (1967) Pangola grass as a rotation crop for pineapple nematode control. *Journal of Agriculture of the University of Puerto Rico* 51, 94–96.

Babatola, J.O. (1988) Nematode control in the rehabilitation of plantain *Musa* spp. *Pakistan Journal of Nematology* 6, 25–30.

Babich, A.G., Linnick, L.I. and Gus'kova, L.A. (1991) [Intermediate crops against the beet nematode.] *Sakharnaya Svekla: Proizvodstvo i Pereabotka* No 1, 23–25.

Babineau, D.E. and Krusberg, L.R. (1978) Efficacy of nematicides for control of *Meloidogyne hapla* on strawberry. *Proceedings of the American Phytopathological Society* 4, 213–214.

Bachthaler, G. and Hien, L. (1976) [Results of long-term cereal rotations with varying proportions of maize.] *Bayerisches Landwirtschaftliches Jahrbuch* 53, 387–403.

Badra, T. and Caveness, F.E. (1979) Chemotherapy of *Dioscorea alata* for disinfestation of *Scutellonema bradys*. *Nematropica* 9, 135–137.

Badra, T. and Caveness, F.E. (1981) Effect of gibberellic acid alone and in combination with aldicarb on plantains infested with *Meloidogyne javanica* and *Helicotylenchus multicinctus*. In: *Proceedings Third Planning Conference on root-knot nematodes, Meloidogyne* spp., *Regions IV 7 V, 16–20 Nov. 1981*, Ibadan, Nigeria (*International Meloidogyne Project*), pp. 189–201.

Badra, T., Saleh, M.A. and Oteifa, B.A. (1979) Nematicidal activity and composition of some organic fertilizers and amendments. *Revue de Nematologie* 2, 29–36.

Baeza-Aragón, C.A. and Leguizanón, C.J.E. (1977) Evaluación de nematicides para el control de *Meloidogyne exigua* Goeldi en plantas de *Coffea arabica* var. Caturra. *Cenicafé* 28, 108–116.

Baghel, P.P.S. and Gupta, D.C. (1987) Effect of intercropping on root-knot nematode (*Meloidogyne javanica*) infesting grapevine (var. Perlette). *Indian Journal of Nematology* 16, 283–284.

Baheti, B.L. and Yadav, S.M. (1993) Seed soaking of black-gram (*Vigna mungo* L.) in pesticides for control of root-knot nematode, *Meloidogyne incognita*. *Indian Journal of Mycology and Plant Pathology* 23, 173–174.

Baines, R.C. (1950) Nematodes on citrus. *California Agriculture* 4, 7.

Baines, R.C. and Clarke, O.F. (1952) Some effects of the citrus-root nematode on the growth of orange and lemon trees. *Phytopathology* 42, 1.

Baines, R.C., Foote, F.J., Stolzy, L.H., Small, R.H. and Garber, M.J. (1959) Factors influencing control of the citrus nematode in the field with D-D. *Hilgardia* 29, 359–381.

Baines, R.C., Martin, J.P., De Wolfe, T.A., Boswell, S.B. and Garber, M.J. (1962a) Effect of high doses of D-D on soil organisms and the growth and yield of lemon trees. *Phytopathology* 52, 723.

Baines, R.C., Klotz, L.J., de Wolfe, T.A. and Garber, M.J. (1962b) Comparison of the fungicidal and nematocidal properties of a number of compounds in field soils. *Phytopathology* 52, 723.

Bajaj, H.K. and Gupta, D.C. (1994) Existence of host races in *Heterodera zeae* Koshy *et al.* *Fundamental and Applied Nematology* 17, 389–390.

Bajaj, H.K., Gupta, D.C. and Dahiya, R.S. (1987) Development of *Heterodera zeae* Koshy on wheat and maize. *Nematologica* 32, 209–215.

Bakker, J. (1988) Protein variation in cyst nematodes. Thesis, Landbouwuniversiteit te Wageningen, Netherlands, 156 pp.

Baksik, A. (1989) Zwalczanie niszczyka zladiwego w uprawie cebuli. *Zeszyty Problemswe Postepow Nauk Rolniczych* 358, 63–74.

Baldwin, J.G., O'Bannon, J.H. and Huettel, R.N. (1978) Scanning electron microscopy of *Radopholus similis*. *Nematropica* 8, 2–3.

Barbercheck, M. (1986) Control of *Meloidogyne javanica* in dormant grapevine nursery stock. *Phytophylactica* 18, 39–40.

Barker, G.M., Addison, P.J. and Watson, R.N. (1993) Seed treatment for protection of seedling white clover from nematodes. *Proceedings 46th New Zealand Plant Protection Conference, Christchurch and Wellington, August, 1992 and 1993, pp.* 228–232.

Bascopé, J.B. and López, B. (1987) Ensayo preliminar para el control químicó de *Nacobbus* sp. en la zona de chaqueli. In: *Control Integrado de Plagas, Cochabamba, Bolivia,* 7–10.

Bashkardin, Y.I. (1971) [Control of *Ditylenchus destructor* infection on seed potatoes by the use of cuttings.] *Trudy Nauchno-Issledovatel'skogo Instituta Kartofel'nogo Khozyaistva* No. 9, 141–143.

Basile, M., Lamberti, F., Melillo, V.A. and Basile, A.C. (1986) Influenza del metodo di somministrazione e della qualita della cpertura sull'efficacia del bromuro di metile nei confronti di nematodi Longidoridae e sulle concentrazioni di bromuro inorganico nel terreno. *Revista della Ortoflorofrutticoltura Italiana* 70, 193–203.

Basile, M., Lamberti, F. and Carella, A. (1988) La fumigazione con bromuro di metile sotto plastich de tipo differente nella lotta contro *Meloidogyne incognita* su pomodoro. *Informatore Fitopatologico* 38, 69–71.

Batista, A.C. and Carneiro, H. (1947) D-D como nematicida, para a hortcultura. *Bol. Sec. Agric., Pernambuco* 14, 147.

Batterby, S., LePatourel, G.N.J. and Wright, D. (1980) Effects of aldicarb persistence on field populations of *Heterodera schachtii*. *Annals of Applied Biology* 95, 105–113.

Beaupre, C.M.S., Gray, F.A., Vincelli, P.C. and Miller, S.D. (1990) Effects of aldicarb *Rhizoctonia* disease and weeds on the severity of the false root-knot nematode of sugar beet. *Fungicide and Nematicide Tests, American Phytopathological Society* 45, 161.

Becker, J.O., Zavaleta-Mejia, E., Colbert, S.F., Schroth, M.N., Weinhold, A.R., Hancock, J.G. and Van Gundy, S.D. (1988) Effects of rhizobacteria on root-knot nematodes and gall formation. *Phytopathology* 78, 1466–1469.

Becker, W.F. (1988) Avaliacao de residuos de Temik e Furadan aplicados em 3 dosagens em epocas de plantio de alho, 1986. *Horticultura Brasileira* 6, 35.

Belair, G. (1984) Nonfumigant nematicides for control of northern root knot nematode in muck-grown carrots. *Canadian Journal of Plant Science* 64, 175–179.

Belair, G. (1992) Effects of cropping sequences on population densities of *Meloidogyne hapla* and carrot yield in organic soil. *Journal of Nematology* 24, 450–456.

Belcher, J.V. and Hussey, R.S. (1977) Influence of *Tagetes patula* and *Arachis hypogaea* on *Meloidogyne incognita*. *Plant Disease Reporter* 61, 525–528.

Belloni, V. (1954) Comparsa in Italia di una anguillulosi del frumento e prove di lotta. *Notiziario sulle Malattie delle Piante*. No. 27 Milan, 3–6.

Berbeć, E. and Dolna, J. (1990) Skutecznosé nematocydów granulowanych w zwalczanin quzaka pólnocnego w zaziewach machwi. *Zeszyty Problemowe Postpów Nauk Rólniczch* No. 391, 25–32.

Bergé, J.B. and Cuany, A. (1972a) Détermination des conditions optimales d'emploi d'un nématicide endothérapique. Essai d'application fractionnée d'aldicarbe sur rosiers parasité par *Pratylenchus vulnus* Allen and Jensen, 1951. *Revue de Zoologie Agricole et de Pathologie Végétale (1972) Part 3*, INRA, Antibes, France, 85–94.

Bergé, J.B. and Cuany, A. (1972b) Action de quelques produits nématicides sur les oeufs de *Meloidogyne arenaria*. *Mededelingen van de Faculteit Landbouwwettenschappen, Rijksuniversiteit, Gent* 37, 808–815.

Bergé, J.B. and Cuany, A. (1972c) Activite de l'aldicarbe sur les oeufs de *Meloidogyne arenaria*. In: *Proces-verbal de la Seance de l'Academie d'Agriculture de France 15 Mars, 1972*, 371–376.

Bergeson, G.B. (1971) Response of muskmelon to fumigation for control of *Meloidogyne incognita* following one year of a non host crop. *Plant Disease Reporter* 55, 55–56.

Bessey, E.A. (1911) Root-knot and its control. *USDA Bureau of Plant Industry*, Bulletin No. 217.

Beugnon, M. and Vilardebó, A. (1974) Étude comparée des nématicides DBCP, prophos et phenamiphos à Madagascar. *Fruits* 29, 785–789.

Bhagwati, B. and Bora, L.C. (1993) Managing ufra disease in deepwater rice (DWR) in Assam India. *International Rice Research Notes* 18, 30.

Bhattacharya, D. and Goswami, B.K. (1987) A study of the comparative efficacy of neem and groundnut oil cakes against root-knot nematodes, *Meloidogyne incognita* as influenced by microorganisms on sterilized and unsterilized soil. *Indian Journal of Nematology* 17, 81–83.

Biehn, W.L., Hawkins, A. and Miller, P.M. (1971) Effect of 1,3-dichloropropene and a combination of 1,3-dichloropropene and chloropicrin on infection of potatoes by *Verticillium albo-atrum*. *Plant Disease Reporter* 55, 968–971.

Bindra, O.S., Chhabra, H.K., Chadha, K.L. and Mehrotra, N.K. (1967) A study on the correlation of citrus nematode population with decline of citrus. *Journal of Research Punjab Agricultural University* 4, 543–546.

Bingefors, S. (1961) Stem nematode in lucerne in Sweden. *Kungl. Lantbrukshogskolans Annaler* 27, 385–398.

Bingefors, S. (1967) Studier over vardvaxtkretsen for rodklover-och lusernraserna av stjalk-nematod, *Ditylenchus dipsaci*. *Lantsbruks-hogskolans meddelanden Serie A No. 79*, Uppsala 1967, 63 pp.

Bingefors, S. (1969) The use of nematode-resistant varieties of grasses and legumes. *Herbage Abstracts*, 39, 107–111.

Bingefors, S. (1985) Improved resistance to stem nematodes in Swedish cultivars of lucerne and red clover and its value. *Zeitschrift für Pflanzenzuchtung* 95, 164–172.

Birat, R.B.S. (1965) To evaluate methods of applying 1,2-dibromo-3-chloropropane for the control of nematode root galls. *Indian Phytopathology* 18, 82–84.

Birat, R.B.S. (1970) Influence of cropping pattern on the population of *Meloidogyne javanica* (Treub, 1885) Chitwood, 1949. *Indian Phytopathology* 22, 391–393.

Birch, A.N.E., Robertson, W.M., Geoghegan, I.E., McGavin, W.J., Alphey, T.J.W., Phillips, M.S., Fellows, L.E., Watson, A.A., Simmonds, M.S.J. and Porter, E.A. (1993) DMDP – a plant-derived sugar anologue with systemic activity against plant-parasitic nematodes. *Nematologica* 39, 521–535.

Birchfield, W. (1954) The hot water treatment of infested nursery stock. *Proceedings of the Florida State Horticultural Society* 67, 94–96.

Birchfield, W. (1957) Observations on the longevity without food of the burrowing nematode. *Phytopathology* 47, 161–162.

Birchfield, W. (1962) Host-parasite relations of *Rotylenchulus reniformis* on *Gossypium hirsutum*. *Phytopathology* 52, 862–865.

Birchfield, W. (1968) Evaluation of nematocides for control of reniform nematodes on cotton. *Plant Disease Reporter* 52, 786–789.

Birchfield, W. and Martin, W.J. (1968) Evaluation of nematocides for controlling reniform nematodes on sweet potatoes. *Plant Disease Reporter* 52, 127–131.

Bird, G.W. (1969) Depth of migration of *Meloidogyne incognita* (Nematoda) associated with greenhouse tomato and cucumber roots. *Canadian Journal of Plant Science* 49, 90–92.

Bird, G.W. and Vitosh, M.L. (1978) Effects of chemical control of *Pratylenchus penetrans* on potato varieties grown at three levels of nitrogen fertilization. *Journal of Nematology* 10, 281–282.

Bird, G.W., Brooks, O.L., Perry, C.E., Futral, J.G., Caverday, T.D. and Boswell, F.C. (1974) Influence of subsoiling and soil fumigation on the cotton stunt disease complex *Hoplolaimus columbus* and *Meloidogyne incognita*. *Plant Disease Reporter* 58, 541–544.

Blackmon, C.W. and Musen, H.L. (1974) Control of the Columbia (lance) nematode, *Hoplolaimus columbus* on soybeans. *Plant Disease Reporter* 58, 641–645.

Blair, G.P. (1969) The problem of control of red ring disease. In: Peachey, J.E. (ed.) *Nematodes of Tropical Crops*. Technical Communication Commonwealth Bureau of Helminthology, No. 40, pp. 99–108.

Blake, C.D. (1961) Root rot of bananas caused by *Radopholus similis* (Cobb) and its control in New South Wales. *Nematologica* 6, 295–310.

Blake, C.D. (1962a) The etiology of tulip-root disease in susceptible and in resistant varieties of oats infested by the stem nematode, *Ditylenchus dipsaci* (Kuhn) Filipjev I. Invasion of the host and reproduction by the nematode. *Annals of Applied Biology* 50, 703–712.

Blake, C.D. (1962b) The etiology of tulip-root disease in susceptible and in resistant varieties of oats infested by the stem nematode, *Ditylenchus dipsaci* (Kuhn) Filipjev II. Histopathology of tulip-root and development of the nematode. *Annals of Applied Biology* 50, 713–722.

Blake, C.D. (1966) The histological changes in banana roots caused by *Radopholus similis* and *Helicotylenchus multicinctus*. *Nematologica* 12, 129–137.

Boag, B. (1984) Effect of aldicarb and dichloropropene on the plant parasitic nematode *Longidorus elongatus* and the yield of a grass/clover sward. *Annals of Applied Biology (Tests of Agrochemicals and Cultivars, 5)* 104 (Suppl.), 20–21.

Bobrysheva, M.N. (1972) [Production of healthy planting material of strawberry in the Kraznodar territory.] *Sbornik Nauchnykh Rabot, Nauchno-Issledovatel'skogo Zonal'nyogo Instituta Sadovodstva Nechernozemnoi Polosy (Vyrashchivanic bezvirusnog posadochnogo material plodovykh i yagodnykh kul'tur)* Moscow, USSR, 5, 155–159.

Bockel, P. (1977) Bestrijding van de T-ziekte door verzwaring van de grond. *Rapport, Institut voor Bodumvruchtbaarheid* No. 7–77, 24 pp.

Boer, R.F. de and Kollmorgen, J.F. (1988) Effects of cultivation and stubble retention on soil- and stubble-borne pathogens of wheat in Victoria – an overview. *Plant Protection Quarterly* 3, 3–4.

Boer, R.F. de, Kollmorgen, J.F., Macauley, B.J. and Franz, P.R. (1991) Effects of cultivation on *Rhizoctonia* root rot, cereal cyst nematode, common root rot and yield of wheat in the Victorian Mallee. *Australian Journal of Experimental Agriculture* 31, 367–372.

Bolander, W.J. and Santo, G.S. (1977) Elimination of *Meloidogyne hapla* in potato seed pieces with a phenamiphos dip treatment. *Plant Disease Reporter* 61, 743–745.

Boldyrev, M.I. and Borzykh, G.I. (1979) [Development of agrotechnical methods of controlling nematodes – carriers of viruses.] *Sbornik Nauchnych Trudov Vsesoyuznogo Nauchno-Issledovatel'skogo Instituta Sadovodstva* No. 29, 84–88.

Boldyrev, M.I. and Borzykh, G.I. (1983) [Plants in the control of nematodes.] *Zashchita Rastenii*, Moscow No. 9, 30.

Boock, O.J. (1950) O'Dowfume W-10 no combate nos nematoides que parsitam as plantas de soja. *Revista Agricultura*, Sao Paulo 25, 297–304.

Boontjes, J., Mantel, P. and Brinkman, H. (1978) Werking van aldicarb (Temik 10G) tegen wortellesie-aaltjes in lelieplantgoed. *Bloembollencultuur* 88, 994–995.

Bora, L.C. and Medhi, B.N. (1992) Resistance of deep water rice (DWR) varieties to ufra disease in Assam. *International Rice Research Newsletter* 17, 12.

Borah, A. and Phukan, P.N. (1990) Efficacy of certain chemicals as seed treatment for the control of *Meloidogyne incognita* on green gram. *Indian Journal of Nematology* 20, 219–220.

Borah, A. and Phukan, P.N. (1992) Effect of different organic amendments for the management of *Meloidogyne incognita* on green gram. *Current Nematology* 3, 49–50.

Bossis, M. (1986) Observations on the population dynamics and control of *Heterodera carotae* in western France. In: Lamberti, F. and Taylor, C.E. (eds) *Cyst Nematodes*. NATO ASI Series A : Life Sciences Vol. 121. Plenum Press, New York, 349–353.

Bossis, M. (1988) Control of *Heterodera carotae* Jones (1980): effectiveness of available methods and prospects. In: *Progress on pest management in field vegetables* Rotterdam, Netherlands : D.A. Balkema, pp. 187–193.

Bossis, M. and Mugniery, D. (1989a) *Heterodera carotae* Jones, 1950. 3. Nuisibilite en conditions de culture de l'ouest de la France. *Revue de Nematologie* 12, 177–180.

Bossis, M. and Mugniery, D. (1989b) *Heterodera carotae* Jones 1950. 5. Effet phytostimulant du DD. *Revue de Nematologie* 12, 432–434.

Bossis, M., Cavelier, A. and Mugniery, D. (1989) *Heterodera carotae* Jones 1950. 4. Interets et limites de la lutte chimique. *Revue de Nematologie* 12, 343–350.

Boswell, T.E. and Grichar, W.J. (1983) Nematicide trial for control of lesion nematodes, 1980 *Fungicide and Nematicide Tests, American Phytopathological Society* 38, 3.

Botseas, D.D. and Rowe, R.C. (1994) Development of potato early dying disease in response to infestion by two pathotypes of *Verticillium dahliae* and co-infection by *Pratylenchus penetrans*. *Phytopathology* 84, 275–282.

Boubals, D. and Dalmasso, A. (1968) Résultats d'essais de désinfection de sols à vigne du sud de la France. *Progrès Agriculture Viticulture* 168, 29–37.

Boyd, F.T. and Perry, V.G. (1970) The effect of sting nematodes on establishment, yield and growth of forage grasses on Florida sandy soils. *Proceedings of the Soil Crop Science Society of Florida* 29, 288–300.

Boyd, F.T., Schroder, V.N. and Perry, V.G. (1972) Interaction of nematodes and soil temperature on growth of three tropical grasses. *Agronomy Journal* 64, 497–500.

Braasch, H. (1994) Intra-und interspezifische Kreuzungsexperimente mit einem deutschen und einem sibirischen Isolat von *Bursaphelenchus mucronatus* (Nematoda : Aphelenchoididae) und verwandten *Bursaphelenchus*-Arten. *Nachrichtenblatt des Deutschen Pflanzenschutzdienstes* 46, 276–281.

Braithwaite, C.W.D. (1974) Effect of crop sequence and fallow on populations of *Rotylenchulus reniformis* in fumigated and untreated soil. *Plant Disease Reporter* 58, 259–261.

Bridge, J. (1983) Nematodes. In: *Pest Control in Tropical Tomatoes.* Centre for Overseas Pest Research, Overseas Development Administration, London, pp. 69–84.

Bridge, J. (1996) Nematode management in sustainable and subsistence agriculture. *Annual Reviews Phytopathology* 34, 201–225.

Broadley, R.A. (1976a) Nematode control in bananas. *Australian Plant Pathology Society Newsletter* 5, (1, Suppl.) Abstract 101.

Broadley, R.A. (1976b) Low rates of EDB for root-knot control in north Queensland tobacco. *Australian Plant Pathology Society Newsletter* 5, (1, Suppl.) Abstract 102.

Broadley, R.A. (1979a) Non-volatile nematicides for control of burrowing nematode in banana plantations in north Queensland. *Australian Journal of Experimental Agriculture and Animal Husbandry* 19, 626–630.

Broadley, R.A. (1979b) Nematicide treatment of banana planting material. *Australian Journal of Experimental Agriculture and Animal Husbandry* 19, 631–633.

Broadley, R.A. (1979c) Control of root-knot of tobacco with ethylene dibromide in north Queensland. *Australian Journal of Experimental Agriculture and Animal Husbandry* 19, 616–619.

Brodie, B.B. (1983) Control of *Globodera rostochiensis* in relation to method of applying nematicides. *Journal of Nematology* 15, 491–495.

Brodie, B.B. and Cooper, W.E. (1964) Relation of plant parasitic nematodes to post-emergence damping-off of cotton. *Phytopathology* 54, 1023–1027.

Brodie, B.B., Good, J.M. and Jaworski, C.A. (1970) Population dynamics of plant nematodes in cultivated land : effect of summer cover crops in newly cleared land. *Journal of Nematology* 2, 217–222.

Bromilow, R.H. and Lord, K.A. (1979) Distribution of nematicides in soil and its influence on control of cyst-nematodes (*Globodera* and *Heterodera* spp.). *Annals of Applied Biology* 92, 93–104.

Brown, E.B. (1958) Pea root eelworm in the eastern counties of England. *Nematologica* 3, 257.

Brown, E.B. (1969) Assessment of the damage caused to potatoes by potato cyst eelworm, *Heterodera rostochiensis* Woll. *Annals of Applied Biology* 63, 493–502.

Brown, E.B. and Sykes, G.B. (1973) Control of tobacco rattle virus (spraing) in potatoes. *Annals of Applied Biology* 75, 462–464.

Brown, E.B. and Sykes, G.B. (1983) Assessment of the losses caused to potatoes by the potato cyst nematodes, *Globodera rostochiensis* and *G. pallida*. *Annals of Applied Biology* 103, 271–276.

Brown, R.H. (1973) Chemical control of the cereal cyst nematode (*Heterodera avenae*) – a comparison of methods and rates of two systemic nematicides. *Australian Journal of Experimental Agriculture and Animal Husbandry* 13, 587–592.

Brown, R.H. (1981) Low volume in-row application of fumigant nematicides for control of *Heterodera avenae* in wheat. *Phytopathology* 70, 565–572.

Brown, R.H. (1982) Cultural practices and their effects on *Heterodera avenae* and grain yields of wheat in Victoria, Australia. *Bulletin EPPO/OEPP* 12, 477–484.

Brown, R.H. (1983a) Control of cereal cyst nematode of wheat in Australia, 1981. *Fungicide and Nematicide Tests, American Phytopathological Society* 38, 20.

Brown, R.H. (1983b) Control of cereal cyst nematode in wheat, 1981. *Fungicide and Nematicide Tests, American Phytopathological Society* 38, 20.

Brown, R.H. (1983c) Evaluation of some pesticides in the control of cereal cyst nematode in wheat, 1981. *Fungicide and Nematicide Tests, American Phytopathological Society* 38, 20–21.

Brown, R.H. (1984) Cereal cyst nematode and its chemical control in Australia. *Plant Disease* 68, 922.

Brown, R.H. and Pye, D.L. (1981) The effect of nematicide application and time of sowing on cereal cyst nematode, *Heterodera avenae*, and the subsequent yield of wheat. *Australian Plant Pathology* 10, 17–18.

Brown, R.H. and Turner, P.L. (1978) Chemical control of root-knot nematode, *Meloidogyne javanica*, in processing tomatoes. *Australian Plant Pathology* 7, 23–24.

Brown, R.H., Pye, D.L. and Stratford, G.T. (1982) A comparison of low volume in-row applications of nematicides at seeding, for control of the cereal cyst nematode (*Heterodera avenae*) in wheat. *Nematologia Mediterranea* 10, 9–19.

Brzeski, M.W. (1973a) Dobór preparatów do zwalcczania węgorka niszczyka (*Ditylenchus dipsaci* Kühn) na czosnku. In: *Materialy Ogólnoposkeig Zjazdu Warzywniczego Skierniewice, June 14–15, 1973*, 83–84.

Brzeski, M. W. (1973b) Zwalczanie węgorka niszczyka *Ditylenchus dipsaci* (Kühn) na cebuli. In: *Materialy Ogólnoposkiego Zjazdu Warzywniczego, Skierniewice, June 14–15, 1973*, 78–82.

Brzeski, M.W. (1986) Zwalçzanie niszczyka zladiwego na cibuli nasiennej. *Ochrana Roślin* 30, 17–18.

Brzeski, M.W., Mercik, S. and Bakowski, J. (1973) [Preliminary experiments with DD for control of *Heterodera schachtii* Schm. on late cabbage varieties.] In: *Materialy Ogólnopolskiego Zjazdu Warzywniczego, Skierniewice, June 14–15, 1973*, 65–67.

Bunt, J.A. (1972) De systemische werking van D-1410, S-methyl 1-((dimethylcarbamoyl)-N-(methylcarbamoyl) oxy) thioformimidate, tegen *Heterodera rostochiensis* bij aardappel en *Pratylenchus penetrans* bij *Rosa canina*. *Mededelingen van de Faculteit Landbouwwetenschappen, Rijksuniversiteit, Gent* 37, 798–807.

Burgt, A.M. van der, Schouten, H.J. and Mulder, A. (1993) Resistance of potato varieties to field populations of the potato cyst nematode *Globodera pallida* in the northeastern part of the Netherlands. *Netherlands Journal of Plant Pathology* 99, 189–195.

Burgt, A.M. van der, Schouten, H.J. and Bakker, J. (1995) Virulentie van aardappelcystaaltjes en teelt van resistente rassen in Nooroost-Nederland van 1973 tot 1991 *Gewasbescherming* 26, 3–9.

Burnett, W.M., Burnett, R.M. and Inglis, I.M. (1974) The development of Nemacur for control of nematodes in Australian banana plantations. *Pflanzenschutz-Nachrichten Bayer* 27, 233–266.

Busey, P., Giblin-Davis, R.M., Riger, C.W. and Zaenker, E.I. (1991) Susceptibility of diploid St. Augustine grass to *Belonolaimus longicaudatus*. *Journal of Nematology* 23, 604–610.

Butler, E.J. (1919) The rice worm (*Tylenchus angustus*) and its control. *Memoirs of the Department of Agriculture in India. Botanical Series* 10, 1–37.

Cabanillas, E. and Barker, K.R. (1989) Impact of *Paecilomyces lilacinus* inoculum level and application time on control of *Meloidogyne incognita* on tomato. *Journal of Nematology* 21, 115–120.

Cabanillas, E., Barker, K.R. and Daykin, M.E. (1988) Histology of the interactions of *Paecilomyces lilacinus* with *Meloidogyne incognita* on tomato. *Journal of Nematology* 20, 362–365.

Cabanillas, E., Barker, K.R. and Nelson, L.A. (1989a) Survival of *Paecilomyces lilacinus* in selected carriers and related effects on *Meloidogyne incognita* on tomato. *Journal of Nematology* 21, 121–130.

Cabanillas, E., Barker, K.R. and Nelson, L.A. (1989b) Growth of isolates of *Paecilomyces lilacinus* and their efficacy in biocontrol of *Meloidogyne incognita* on tomato. *Journal of Nematology* 21, 164–172.

Cadet, P. (1990) Influence de quelques nematicides sur les rendements et la multiplication des nematodes parasites et la tomate a la Martinique. *Bulletin Agronomique des Antilles et de la Guyane* No. 10, 42–48.

Cadet, P. and Quénéhervé, P. (1994) Fluctuations naturelles de *Scutellonema bradys* (Nematoda: Hoplolaimidae) au cours de la croissance et du stockage de l'igname (*Dioscorea alata*) à la Martinique. *Nematologica* 40, 587–600.

Cahuana, A., Cornejo, O.W. and Huanco, S.V. (1975) Sources of resistance to the false root-knot nematode (*Nacobbus* sp.) among 50 potato clones. *Fitopatologia* 10, 71.

Canto-Saenz, M. and de Scurrah, M.M. (1977) Races of the potato cyst nematode in the Andean region and a new system of classification. *Nematologica* 23, 340–349.

Cap, G.B., Roberts, P.A., Thomason, I.J. and Murashige, T. (1991) Embryo culture of *Lycopersicon esculentum* × *L. peruvianum* hybrid genotypes possessing heat-stable resistance to *Meloidogyne incognita*. *Journal of the American Society for Horticultural Science* 116, 1082–1088.

Carlisle, D. (1987) Farmers fight their foes with fungi. *New Scientist* No. 1558, 35.

Carlsson, B. and Videgård, G (1971) Sanering av potatiscystnematoden med Ljälp av fångstgröda. *Potatis*, Year 1971, 30–32.

Carneiro, R.G. and Carneiro, R.M.D.G. (1982) Seleção preliminar de plantas para rotação de culturas em areas infestados por *M. incognita* nos anos de 1979 e 1980. In: *Trabalhos Apresantados a VI Reuniao Brasileira de Nematologia, 8–12 Fevereiro de 1982, Fortaleza. Publiçãcao* No. 6, 141–148.

Carroll, J. and Mc Mahon, E. (1939) Experiments on trap cropping with potatoes as a control measure against potato eelworm (*Heterodera schachtii*). *Journal of Helminthology* 17, 101–112.

Carter, W.W. (1943) A promising new soil amendment and disinfestant. *Science* 97, 383.

Carter, W.W. (1982) Influence of soil temperature on *Meloidogyne incognita* in resistant and susceptible cotton, *Gossypium hirsutum*. *Journal of Nematology* 14, 343–346.

Carter, W.W. and Nieto, S. Jr. (1975) Population development of *Meloidogyne incognita* as influenced by crop rotation and fallow. *Plant Disease Reporter* 59, 402–403.

Cartia, G. and Greco, N. (1987) Effetti della 'solarizzazione del suelo' su una coltura di peperone in serra. *Colture Protette* 16, 61–65.

Carvalho, J.M.F.C., Ferraz, S., Cardosa, A. and Dhingra, O.D. (1981)Tratamento de sementes de feijão (*Phaseolus vulgaris* L.) com oxamil dissolvido em acetona ou etanol para controle de fitonematóides. *Revista Ceres* 28, 580–587.

Casamayor, R., Seidel, D. and Decker, H. (1967) Tratamiento con aqua caliente contra nematodos parásitos en plátano. *Beitrage Tropischen and Subtropischen Landwirtschaftliche Tropenveterinaar Medezin* 5, 41–48.

Castagnone-Sereno, P. (1988) Désinfestion des semences d'igname par thermo ou chimiothérapie: efficacité nématicide et conséquences agronomiques. *Turrialba* 38, 337–340.

Castagnone-Sereno, P., Bongiovanni, M. and Dalmasso, A. (1993) Stable virulence against the tomato resistance Mi gene in the parthenogenetic root-knot nematode *Meloidogyne incognita*. *Phytopathology* 83, 803–805.

Castillo, M.B., Bajet, N.B. and Harwood, R.R. (1976) Nematodes in cropping patterns. I. Population of *Rotylenchulus reniformis* on successively monocultured crops. *Philippine Agriculturist* 59, 288–294.

Caswell, E.P., Sarah, J.L. and Apt, W.J. (1990) Nematode parasites of pineapple. In: Luc, M., Sikora, R.A. and Bridge, J. (eds) *Plant Parasitic Nematodes in Subtropical and Tropical Agriculture.* CAB International, Wallingford, UK, pp. 519–537.

Caubel, G. (1968) Données nouvelles sur la lutte contre le nématode des bulbes en culture d'oignon et d'échalote. *Pépiniéristes-Horticulteurs-Maraîchers* No. 140, Paris, 1–4.

Caubel, G. (1973a) Problèmés posés par le nématode des tiges (*Ditylenchus dipsaci* (Kühn)) en culture de maïs. *Académie d'Agriculture de France. Extrait du Procès-verbal de la Séance du 9 Mai 1973*, 627–640.

Caubel, G. (1973b) Influence de la date du semis du maïs sur la verse précoce causée par le nématode des tiges (*Ditylenchus dipsaci*). *Sciences Agronomiques Rennes* (1973), 101–107.

Caubel, G. (1983) Effets de l'application de pesticide sur des nématodes phytoparasites. *Sciences Agronomiques Rennes* (1983), 36–44.

Caubel, G. and Muchembled, C. (1991) Lutte contre *Heterodera schachtii* au moyen de nématicides ou de plantes-pièges de la famille des crucifers. *Congrès d'Hiver – Institut International de Recherches Betteravières* No. 54, 211–220.

Caubel, G., Bossis, M., Genier, G. and Guy, P. (1977) Mise au point d'un test de selection de luzernes resistants à *D. dipsaci. Sciences Agronomiques Rennes* (1977), 25–32.

Caubel, G., Hemery, J. and Le Bohec, F. (1978) Observations sur l'effet de *Tagetes patula* L. et *T. erecta* L. sur *Pratylenchus penetrans* Cobb. *Extrait de la Revue La Défense des Végétaux* No. 194, 1–10.

Caubel, G., Curvale, J.P., Hemery, F. and Le Bohec, J. (1981) Observations écologiques et experimentations des méthodes de lutte contre *Pratylenchus penetrans*, nématode nuisible à l'artichaut en Bretagne. In *3º Congresso Internazionale di Studi sul Carciofo, 1979, Bari, Italy*, 985–1002.

Caubel, G., Ducom, P. and Marre, R. (1985) La fumigation au bromure de methyle dirigée contre le nématode des tiges, *Ditylenchus dipsaci*, contenu dans des lots de semences et de bulbes. *Bulletin OEPP/EPPO* 15, 17–22.

Caveness, F.E. and Badra, T. (1980) Control of *Helicotylenchus multicinctus* and *Meloidogyne javanica* in established plantain and nematode survival as influenced by rainfall. *Nematropica* 10, 10–14.

Cayrol, J.C. (1981) Possibilités de lutte biologique contre les *Meloidogyne* à l'aide du champignon nématophage 'R 350' (*Arthrobotrys irregularis*). *Défense des Végétaux* 35, 279–282.

Cayrol, J.C. (1983) Lutte biologique contre les *Meloidogyne* au moyen d'*Arthrobotrys irregularis. Revue de Nématologie* 6, 265–273.

Cayrol, J.C. and Frankowski, J.P. (1980) Connaissances nouvelles sur le champignon nématophage *Arthrobotrys irregularis* (Royal 350). *Pépiniéristes Horticulteurs Maraîchers – Revue Horticole* No. 203, 33–38.

Cenis, J. and Argemi, J. (1988) Modelización del efecto de la disinfección solar en la supervivencia de *Meloidogyne javanica* (Treub) Chit.. In: Llobet, L.G. (ed) *Communicaciones des III Congreso Nacional de Fitopatologia.* Pierto de la Cruz (Tenerife, Islas Canarias) 29 Oct.–2 Nov., 1984, pp. 152–160.

Cetas, R.C. and Harrison, W.B. (1963) Evaluation of soil fumigants for the control of early maturity wilt of potatoes on Long Island. *Phytopathology* 53, 347–348.

Chand, T. and Jain, R.K. (1990) Chemical control of root-knot nematode *(Meloidogyne incognita*) through seed dressing in cotton (*Gossypium hirsutum*). *Indian Journal of Nematology* 20, 224–227.

Chapman, P.J. (1983) The effectiveness of 1,3-dichloropropene for controlling virus vector nematodes with reference to the MAFF certification scheme for strawberry nursery stock. *Plant Pathology* 32, 273–279.

Chatterjee, A., Sukul, N.C., Laskar, S. and Ghoshmajumdar, S. (1982) Nematicidal principles from two species of Lamiaceae. *Journal of Nematology* 14, 118–120.

Chen, T.A. and Rich, A.E. (1962) The role of *Pratylenchus penetrans* in the development of strawberry black root-rot. *Plant Disease Reporter* 46, 839–843.

Chengzu, M., Weiquan, H. and Zhaoliang, L. (1987) [Biology and control of reniform nematode (*Rotylenchulus reniformis*) on some vegetable crops.] *Acta Agriculturae Shanghai* 3, 49–58.

Chhabra, H.K. and Bindra, O.S. (1972) Screening of nematicides against citrus nematode, *Tylenchulus semipenetrans* Cobb. *Indian Journal of Entomology* 33, 472–473.

Chhabra, H.K. and Dhaliwal, (1979) Population fluctuations and chemical control of *Hirschmanniella oryzae*. *Indian Journal of Nematology* 8, 163–165.

Chhabra, H.K. and Mahajan, R. (1976) *Pratylenchus coffeae*, the root-lesion nematode in groundnut and its control by granular nematicides. *Nematologia Mediterranea* 4, 241–242.

Chhabra, H.K., Sajjan, S.S. and Singh, J. (1974) Occurrence and control of the stunt nematode, *Tylenchorhynchus brassicae* Siddiqi, 1961 infesting rice nursery in Punjab. *Current Science* 43, 632.

Chhabra, H.K., Bindra, O.S. and Inderjit Singh (1978) Further studies on the control of the citrus nematode (*Tylenchulus semipenetrans* Cobb) infesting *Citrus maximus* Merrill. *Haryana Journal of Horticultural Science* 7, 136–138.

Chindo, P.S. and Khan, F.A. (1986) Effect of soil organic amendment with poultry manure on the damage caused by the root-knot nematode *Meloidogyne incognita* on tomato. *International Nematology Network Newsletter* 3, 30–33.

Chindo, P.S. and Khan, F.A. (1990) Control of root-knot nematodes, *Meloidogyne* spp. on tomato with poultry manure. *Tropical Pest Management* 36, 332–335.

Chitwood, B.G. and Krusberg, L.R. (1977) Pectolytic enzymes in three populations of *Ditylenchus dipsaci*. *Journal of Nematology* 9, 187–191.

Cho, S.S., Han, M.J. and Yang, J.S. (1987) [Chemical control of rice white-tip nematode (*Aphelenchoides besseyi*) by seed disinfectant and in the paddy field.] *Korean Journal of Plant Protection* 26, 107–111.

Choi, I.H., Yoo, J.K., Na, S.Y. and Han, S.C. (1989) [Chemical control effect of rice white-tip nematode, *Aphelenchoides besseyi* C. by some different treatment methods.] *Research Reports of the Rural Development Administration, Crop Protection Korea Republic* 31, 24–29.

Christie, J.R. (1959) *Plant Nematodes their Bionomics and Control.* Agricultural Experiment Station, University of Florida, Gainesville, Florida, 256 pp.

Christie, J.R. and Perry, V.G. (1951) A root disease of plants caused by a nematode of the genus *Trichodorus*. *Science* 113, 491–493.

Chukantseva, N.K. (1976) [Some results of nematicide trials on the control of *Ditylenchus dipsaci* on potato.] In: *VIII Vsesoyuznoe soveshchanie no nematodym boleznyam sel'skokozyaistrennykh kul'tur. Tezisy dokladov i soobshchenii.* Kishinev, USSR; Izdatel'stvo' Shtiinsta 89.

Chukantseva, N.K. (1983) [Some aspects of the potato stem nematode in the Central Chernozem zone of the RSFSR.] In : *Steblevye nematody sel'skokhozyaistvennykh kul'tur i mery bor'by s nemi. (Materialy simpoziuma, Voronezh, 27–29 Sentyabrya 1983g.)* Voronezh, USSR Vserossiiskii NII Zaschity Rastenii (1983), 11–27.

Chumakova, E.I., Guschin, B.E., Doroshenko, N.D. and Popova, M.B. (1985) [Etaphos – an effective preparation against *Aphelenchoides besseyi* on rice.] *Byulleten' Vsesyuznogo Instituta Gel'mintologii im K.I. Skryabina* No.41, 80–84.

Cíndea, E. (1979) Contributii la combaterea nématodului *Ditylenchus dipsaci* Kühn la usturri. *Analele Institului de Cerietari pentru Legumicultura si Floricultura* 5, 267–273.

Cíndea, E. (1981) Recherches sur les moyens de lutte contre le nématode *Ditylenchus dipsaci* Kühn chez l'ail. *Bulletin de l'Académie des Sciences Agricoles et Forestières* No. 9, 95–100.

Cíndea, E. (1982) Eficacitatea unor pesticid in combaterea paianjenului (*Rhizoglyphus echinopus*) bulbilor de usturoi. *Productia Vegetata, Horticultura* 31, 17–18.

Clamot, G. (1985) Amélioration de la résistance de l'avoine au nématode à kyste des céréales (*Heterodera avenae* Woll.) et au nématode des tiges (*Ditylenchus dipsaci* (Kühn) Fil.) en Belgique. *Comptes Rendus des Séances de l'Academie d' Agriculture de France* 71, 751–760.

Clausen, R., Graf, A., Joseph, E., Keller, E., Leichti, H. and Savary, A. (1964) Bericht der Arbeitsgemeinschaft zur Bekämpfung des Rüben-kopfalchens (*Ditylenchus dipsaci*) über die im Jahre 1963 durchgeführten Untersuchungen und Versuche. *Mitteilungen fur die Schweizerische Landwirtschaft* No. 9, 134–142

Clayton, C.N. and Ellis, D.E. (1948) Soil treatments reduce vegetable root-knot. *Research and Farming* 6, 58.

Clayton, C.N. and Ellis, D.E. (1949) Soil treatments with D-D, chloropicrin and Uramon for the control of the root-knot nematode. *Phytopathology* 39, 583-589.

Coates-Beckford, P.L. (1977) Comparison of various treatments for the control of *Pratylenchus coffeae* in yams. *Nematropica* 7, 20–26.

Cohn, E. and Auscher, R. (1971) Seasonal occurrence of *Longidorus vineacola* on celery in Israel and its control. *Israel Journal of Agricultural Research* 21, 23–25.

Cohn, E. and Minz, G. (1961) Citrus nematode on American persimmon in Israel. *Plant Disease Reporter* 45, 505.

Cohn, E., Minz, G. and Monselise, S.P. (1965) The distribution, ecology and pathogenicity of the citrus nematode in Israel. *Israel Journal of Agricultural Research* 15, 187–200.

Cohn, E. Feder, W.A. and Mordechai, M. (1968a) The growth response of citrus to nematocide treatments. *Israel Journal of Agricultural Research* 18, 19–24.

Cohn, E., Krikun, J. and Yisraeli, U. (1968b) Further observations on the effect of *Longidorus vineacola* on onion. *Plant Disease Reporter* 52, 525–527.

Colbran, R.C. (1979) Problems in tree replacement. III The effects of cultural practices and soil fumigation on root-lesion nematodes and on the growth of apple trees. *Australian Journal of Agricultural Research* 30, 113–123.

Colbran, R.C. and Brouwer, H.M. (1972) Nematocide trials with tobacco in the 1968–69 and 1969–70 seasons. *Queensland Journal of Agriculture and Animal Sciences* 29, 265–268.

Colbran, R.C. and Saunders, G.W. (1961) Nematode root-rot of bananas. *Queensland Agricultural Journal* 87, 22–24.

Cole, C.S. and Howard, H.W. (1962) Further results from a field experiment on the effect of growing resistant potatoes on a potato root eelworm (*Heterodera rostochiensis*) population. *Nematologica* 7, 57–61.

Collins, R.J. and Feldman, A.W. (1965) Penetration of nematocides for control of *Radopholus similis* and for destruction of citrus roots in the deep sands of central Florida. *Phytopathology* 55, 1103–1107.

Compos, P., Sivapalan, P. and Gnanpragasam, N.C. (1990) Nematode parasites of coffee, cocoa and tea. In: Luc, M., Sikora, R.A. and Bridge, J. (eds) *Plant Parasitic Nematodes in Subtropical and Tropical Agriculture.* CAB International, Wallingford, UK, pp. 387–430.

Conroy, R.J. (1951) The use of soil fumigants for root knot control. *Agricultural Gazette of New South Wales* 62, 411–414.

Cook, R. and Evans, D.R. (1988) Observations on resistance in white clover (*Trifolium repens* L.) to the stem nematode (*Ditylenchus dipsaci* (Kühn) Filipjev). *Journal of Agricultural Science*, 110, 145–154.

Cook, R. and York, P.A. (1985) Nematode problems of grassland. In: Brookman, J.S. (ed.) *Weeds, Pests and Diseases of Grasses and Herbage Legumes.* British Grassland Society Occasional Symposium No. 18. British Crop Protection Council Monograph 29, Croydon, UK, 180–187.

Cook, R., Evans, D.R., Williams, T.A. and Mizen, K.A. (1992a) The effect of stem nematode on establishment and early yields of white clover. *Annals of Applied Biology* 120, 83–94.

Cook, R., Mizen K.A., Plowright, R.A. and York, P.A. (1992b) Observations on the incidence of plant parasitic nematodes in grassland in England and Wales. *Grass and Forage Science* 47, 274–279.

Cooke, D.A. (1984a) The effect of soil-applied granular pesticides and irrigation on the yield of sugar beet in fields infested with *Trichodorus viruliferus. Annals of Applied Biology* 105, 253–261.

Cooke, D.A. (1984b) The relationship between numbers of *Heterodera schachtii* and sugar beet yields on a mineral soil, 1978–1981. *Annals of Applied Biology* 104, 121–129.

Cooke, D.A. (1985) The effect of resistant cultivars of catch crops on the hatching of *Heterodera schachtii. Annals of Applied Biology* 106, 111–120.

Cooke, D.A. (1987) Beet cyst nematode (*Heterodera schachtii* Schmidt) and its control on sugar beet. *Agricultural Zoology Reviews* 2, 135–183.

Cooke, D.A. (1989) Damage to sugarbeet crops by ectoparasitic nematodes, and its control by soil-applied pesticides. *Crop Protection* 8, 63–70.

Cooke, D.A. (1991) The effect of beet cyst nematode, *Heterodera schachtii*, on the yield of sugar-beet in organic soils. *Annals of Applied Biology* 118, 153–160.

Cooke, D.A. and Draycott, A.P. (1969) The effect of soil fumigation and nitrogen on Docking disorder of sugar beet. *Proceedings 5th. British Insecticides and Fungicide Conference, Brighton, November 17–20, 1969*, Vol. 1, 260–264.

Cooke, D.A. and Thomason, I.J. (1979) The relationship between population density of *Heterodera schachtii*, soil temperature and sugar beet yields. *Journal of Nematology* 11, 124–128.

Cooke, D.A., Dunning, R.A. and Winder, G.H. (1970) *Annual Report for 1969, Rothamsted Experimental Station*, Part 1, pp. 319–321.

Cooke, D.A., Dunning, R.A. and Winder, G.H. (1973) Control of sugar-beet Docking disorder: trials comparing spring application of fumigants and aldicarb. *Annals of Applied Biology* 73, 460.

Cooke, D.A., Thomason, I.J., McKinney, H.E., Bendexen, W.E. and Hagemann, R.W. (1979) Chemical control of *Heterodera schachtii* on sugar beet in California. *Journal of Nematology* 11, 205–206.

Cooper, B.A. (1953) Eelworm problems in North Fenland with special reference to crop rotation. *Horticultural Education Association. Annual Report*, 106–115.

Cooper, J.I. and Harrison, B.D. (1973) The role of weed hosts and the distribution and activity of vector nematodes in the ecology of tobacco rattle virus. *Annals of Applied Biology* 73, 53–66.

Cooper, J.I. and Thomas, P.R. (1971) Chemical treatment of soil to prevent transmission of tobacco rattle virus to potatoes by *Trichodorus* spp. *Annals of Applied Biology* 69, 23–34.

Cooper, W.E. (1952) Control of peanut root knot by soil fumigation and by crop rotation. *Phytopathology* 42, 282–283.

Cooper, W.E. and Sasser, J.N. (1962) Cynem – an effective control for sting nematodes on peanuts. *Plant Disease Reporter* 46, 640–642.

Cornejo Quiróz, W. (1977) Control quimico de *Nacobbus aberrans* y *Globodera* spp. *Nematropica* 7, 6.

Costilla, M.A. (1973) El nematode *Pratylenchus zeae* Graham en caña de azucar. *Revista Industrial y Agricola de Tucumán* 50, 39–43.

Costilla, M.A. (1989) Comportamiento e importancia de tres poblaciones del falso nematodo del nudo *Nacobbus aberrans* (Thorne, 1935) Thorne y Allen 1944 como plaga del tomate y pimento en 3 localidades del Noveste Argentino. *Programas y Resúmenes XXI Reunión Anual ONTA*, Tucumán, Argentina, 6–10 November, 18.

Courtney, W.D. (1962) Stem nematode of red clover in the Pacific Northwest. *Bulletin of the Washington Agricultural Experimental Station*, 640.

Covolo, G. and Oliveira, S.J.B. de (1982) Comportamento de alguns nematicidas no controle do nematóide *Meloidogyne javanica* (Treub, 1885) Chitwood, 1949 em tomateiro (*Lycopersicon esculentum* Mill.). In: *Trabalhos Apresentados a V Reunião Brasileira de Nematologia, 9–13 fevereiro de 1981, Londrina, PR, Brasil, Publicação No. 5, 1981*, Piracicaba, SP, Brazil, pp. 145–151.

Cox, P.G. (1980) A theoretical model for ufra control. *International Rice Research Newsletter* 5, 16–17.

Cralley, E.M. (1952) Control of white tip of rice. *Arkansas Farm Research* 1, 6.

Cralley, E.M. (1954) Controlling white tip of rice. *Arkansas Farm Research* 3, 8.

Crump, D.H. (1991) Biological control of the beet cyst nematode. *British Sugar Beet Review* 59, 54–55.

Crump, D.H. and Irving, F. (1992) Selection of isolates and methods of culturing *Verticillium chlamydosporium* and its efficacy as a biological control agent of beet and potato cyst nematodes. *Nematologica* 38, 367–374.

Crump, D.H. and Kerry, B.R. (1987) Studies on the population dynamics and fungal parasitism of *Heterodera schachtii* in soil from a sugar-beet monoculture. *Crop Protection* 6, 49–55.

Csinos, A.S., Johnson, A. W. and Golden, A.M. (1986) Metalaxyl and fenamiphos applied through irrigation water to control black shank/root knot complex on tobacco. *Plant Disease* 70, 210–213.

Csinos, A.S., Johnson, A.W. and Golden, A.M. (1994) Management of the tobacco black shank–root-knot complex with combinations of soil fumigants and metalaxyl. *Plant Disease* 78, 565–568.

Cuadra, R., Aguilera, C. and Pérez, J.A. (1990) Control de *Meloidogyne incognita* con los nematicidas DD y heterophos. *Ciencias de la Agricultura* No. 40, 36–42.

Cuany, A., Berge, J.B. and Scotto La Massese, C. (1973) Observations sur l'emploi des nématicides endotherapiques contre le genre *Meloidogyne* en horticulture florale. *Bulletin OEPP/EPPO* 3, 75–87.

Cuany, A., Lavergne, J. C., Clerjeau, M. and Nourrisseau, J.G. (1974) Action du bromure de methyle et du dazomet sur un complexe parasitaire associé à la tomate. *Pépiniéristes Horticulteurs Maraîchers* 47, 37–39, 41–42.

Cuc, Nguyen-ti and Giang, Le-Than (1982) Relative humidity and nematode number and survival in rice seeds. *International Rice Research Newsletter* 7, 14.

Daiber, K.C. (1990) Effect of fenamiphos on population dynamics of *Meloidogyne incognita* and *M. javanica* and on yield of tomatoes. *Phytophylactica* 22, 307–310.

Dale, M.F.B. (1989) Resistance to and tolerance of potato cyst nematode as objectives within a breeding programme. In: *Parental Line Breeding and Selection in Potato Breeding. Proceedings of the Joint Conference of the EAPR Breeding Section and the EUCARPIA Potato Section, Wageningen, Netherlands 11–16 December, 1988*, 66–70.

Dale, M.F.B. and Philipps, M.S. (1985) Variation for the degree of susceptibility to the potato cyst-nematode (*Globodera pallida*) within *Solanum tuberosum* ssp. *tuberosum*. *Potato Research* 28, 55–64.

Dale, P.S. (1973) Elimination of root-knot nematodes from roses by chemical bare-root dips. *New Zealand Journal of Experimental Agriculture* 1, 121–122.

Dale, W.T. and Hendy, A.F. (1967) Control of arabis mosaic virus in strawberry with nematicides. *Plant Pathology* 16, 187–191.

Dalmasso, A., Cardin, M.C., Pachard, E. and Daunay, M.C. (1985) Pouvoir pathogène des nématodes *Meloidogyne* et génétique de la résistance chez quelques solanacées maraîchères. *Comptes Rendus des Séances de l'Académie d'Agriculture de France* 71, 751–760.

Damadzadeh, M. and Hague, N.G.M. (1979) Control of stem nematode (*Ditylenchus dipsaci*) in narcissus and tulip by organophosphate and organocarbamate pesticides. *Plant Pathology* 28, 86–90.

Dang-ngoc Kinh, Huong, N.M. and Ut, N.V. (1982) Root-knot disease of rice in the Mekong Delta, Vietnam. *International Rice Research Newsletter* 7, 15.

Darekar, K.S., Mhase, N.L. and Shelke, S.S. (1990a) Effects of placement of non-edible oilseed cakes on the control of root-knot nematodes on tomato. *International Nematology Network Newsletter* 7, 5–7.

Darekar, K.S., Mhase, N.L. and Shelke, S.S. (1990b) Effect of green-gram seed treatment with certain nematicides on root-knot nematodes and crop yields. *International Nematology Network Newsletter* 7, 4–5.

Darling, H.M. (1959) Control of the potato rot nematode in Wisconsin. *Plant Disease Reporter* 43, 239–242.

Darling, H.M., Adams, J. and Norgren, R.L. (1983) Field eradication of the potato rot nematode, *Ditylenchus destructor*: a 29-year history. *Plant Disease* 67, 422–423.

Das, S. and Sukul, N.C. (1988) Nematicidal effect of the oil from the seeds of *Argemone mexicana*. *Environment and Ecology* 6, 194–197.

Dash, B.K. and Padhi, N.N. (1990) Response of root knot nematode infecting tomato to seed treatment by some neem products. *Orissa Journal of Agricultural Research* 2, 148–150.

Daulton, R.A.C. (1950) Soil pests. A. Root knot nematode. *Annual Report of the Trelawney Tobacco Research Station for 1949, Tobacco Research Board Publication No. 12*, Southern Rhodesia, 45–52.

Daulton, R.A.C. (1956) Nematode and weed control in tobacco seedbeds. *Rhodesian Tobacco* No. 13, 4–5.

Daulton, R.A.C. (1963) Controlling *M. javanica* in Southern Rhodesia. *Rhodesia Agricultural Journal* 60, 150–152.

Davaine, C.J. (1857) *Recherches sur l'Anguillule du Blé Niellé Considerée au Point de Vue de l'Histoire Naturelle et de l'Agriculture*. Paris, chez J.B. Baillière. Libraire de l'Academie Imperiale de Medécine, 80 pp.

Davide, R.G. (1990) Biological control of nematodes using *Paecilomyces lilacinus* in the Philippines. In: Hahn, S.K. and Caveness, F.E. (eds) *Integrated Pest Management for Tropical Root and Tuber Crops: Proceedings of the Global Status of and Prospects for Integrated Pest Management of Root and Tuber Crops in the Tropics. 25–30 October, 1987, Ibadan, Nigeria*, pp. 156–163.

Davide, R.G. and Zorilla, R.A. (1986) Evaluation of a fungus *Paecilomyces lilacinus* for the biological control of root-knot nematode, *Meloidogyne incognita* on okra compared with a nematicide isazophos. *International Nematology Network Newsletter* 3, 32–33.

Davies, K.G., Kerry, B.R. and Flynn, C.A. (1988) Observations on the pathogenicity of *Pasteuria penetrans*, a parasite of root-knot nematodes. *Annals of Applied Biology* 112, 491–501.

Davies, T.G. (1960) A spring oat variety resistant to the stem eelworm. *Nature* No. 4727, 813.

Davis, E.L., Kaplan, D.T., Dickson, D.W. and Mitchell, D.J. (1989) Reproduction of lectin-treated *Meloidogyne* spp. in two related soybean cultivars. *Revue de Nématologie* 12, 257–263.

Decker, H. and Dowe, A. (1991) Some investigations with *Heterodera schachtii* Schmidt on different varieties of rape. *Bulletin SROP* 14, 223–228.

Decker, H., Casamayor Garcia, R. and Seidel, D. (1970) Untersuchungen über Wirtspflanzen und Bekämpfungsmöglichkeiten von *Radopholus similis* in Cuba. *Zeszyty Problemowe Postepów Nauk Rolniczych* No. 92, 447–464.

Del Bene,G. (1977) Prove di lotta chimica contro nematodi galligeni su garafano. *Redia* 60, 193–210.

Dement'eva, S.P., Nesterov, P.I. and Cherevatova, Z. Y. (1979) [Agrotechnical control of *Meloidogyne* in glasshouse vegetable growing.] In: *Gallovye nematody sel' skokozyaistrennykh kul'tur i mery bor'by S nimi*, Dushanbe, USSR, 'Donish', 67–68.

Deng, D.L. (et al.) (1993) [Study on the biological characters of citrus nematode, *Tylenchulus semipenetrans*.] *Southwest China Journal of Agricultural Sciences* No. 6 (Suppl.), 79–82.

Deproft, M. (1989) Dégradation accelerée d'insecticides du sol suite a leur applications repetées. *Mededelingen Rijksuniversiteit, Gent* 54, 141–150.

Dern, R. (1968) Erfahrungen bei der Nematodenbekämpftung mit Terracur P. *Comptes Rendus du Huitième Symposium International de Nématologie*, 1965 Antibes, France, 123–124.

D'Errico, F.P. and Cancellara, I. (1977) Orientamenti di lotta chimica contro *Meloidogyne incognita* su pomodoro. *Informatore Fitopatologico* 27, 17–21.

D'Errico, F.P., Di Maio, F. and Lamberti, F. (1978) Prove di lotta guidata contro i nematodi galligeni del pomodoro nel Salernitano. Il contibuto. *Nematologia Mediterranea* 6, 223–238.

Dethe, M.D. and Pawar, V.M. (1987) Control of root-knot nematode on betelvine by granular carbamate pesticides. *International Nematology Network Newsletter* 4, 6–7.

Devakumar, C., Goswami, B.K. and Mukerjee, S.K. (1985) Nematicidal principles from neem (*Azadirachta indica* A. Juss.) part 1. Screening of neem kernel fractions against *Meloidogyne incognita*. *Indian Journal of Nematology* 15, 121–124.

D'Herde, J., Brande, J. van den and Gillard, A. (1960) Control of *Pratylenchus penetrans* (Cobb), causal agent of root-rot in lilies. *Nematologica Suppl II*, 64–67.

Diab, K.A. and El-Eraki, S. (1968) Plant-parasitic nematodes associated with olive decline in the United Arab Republic. *Plant Disease Reporter* 52, 150–154.

Dick, J. and Harris, R.H.G. (1971) Recent investigations on nematodes in sugarcane fields. *South African Sugar Technologists Association. Annual Congress (45th) Mount Edgecombe, Natal, 7–11 June, 1971. Proceedings and Index 1952–1970*, 249–254.

Dicklow, M.B., Acosta, N., and Zuckerman, B.M. (1993) A novel *Streptomyces* species for controlling plant-parasitic nematodes. *Journal of Chemical Ecology* 19, 159–173.

Dickson,D.W., Oostengorp, M. and Mitchell, D.J. (1992) Development of *Pasteuria penetrans* on *Meloidogyne arenaria* race 1 in the field. In: Gommers, F.J. and Maas, P.W. (eds) *Nematology from Molecule to Ecosystem; Proceedings of the 2nd. International Nematology Congress, 11–17 August 1990, Veldhoven, the Netherlands*, pp. 213–218.

Di Edwardo, A.A. (1963) *Report of the Florida Agricultural Experimental Station* 1963, 109.

Di Maio, F. (1979) Il nematode *Tylenchulus semipenetrans* Cobb su radici di loto. *Informatore Fitopatologico* 29, 13–14.

Di Muro, A. (1975) Root knot nematode (*Meloidogyne incognita*) control by two- and three-year crop rotations. *Annali dell' Istituto Sperimentale per il Tabacco.* 2, 89–91.

Di Vito, M. and Greco, N. (1986) The pea cyst nematode. In: Lamberti, F. and Taylor, C.E. (eds) *Cyst Nematodes* Plenum Press, New York, pp. 321–332.

Di Vito, M. and Lamberti, F. (1976a) Prove di lotta chimica contro *Heterodera göttingiana* su pisello in Puglia. *Nematologia Mediterranea* 4, 121–131.

Di Vito, M. and Lamberti, F. (1976b) Efficacia di nematocidi sistemici nella lotta contro *Heterodera göttingiana* su pisello. *Nematologia Mediterranea* 4, 63–69.

Di Vito, M., Lamberti, F. and Inserra, R. (1973) Prove preliminari di lotta chimica contro *Heterodera göttingiana* Liebscher, grave parasita del pisello nell'Italia meridionale. *Nematologia Mediterranea* 1, 125–138.

Di Vito, M., Greco, N., Singh, K.B. and Saxena, M.C. (1988) Response of chickpea germplasm lines to *Heterodera ciceri* attack. *Nematologia Mediterranea* 16, 17–18.

Di Vito, M., Greco, N. and Saxena, M.C. (1991) Effectiveness of soil solarization for control of *Heterodera ciceri* and *Pratylenchus thornei* on chickpea in Syria. *Nematologia Mediterranea* 19, 109–111.

Doncaster, C.C. (1952) In: Peters, B.G. Nematology Department. *Report of the Rothamsted Experimental Station for 1951*, Harpenden , UK, pp. 93–99.

Dong, J.M. (1988) [Research advances on soybean cyst nematode in Shanxi Province.] *Shanxi Agricultural Science* No. 4, 31–34.

Dowe, A. and Decker, H. (1981) Zur Populationsentwicklung des Kartoffelzystenalchen und zum Ertagsverlauf in einem langjahrigen kartoffelmonokulturversuch bei organischer und mineralischer Düngung. *Archiiv für Phytopathologie und Pflanzenschutz* 17, 105–113.

Doyle, A.D., McLeod, R.W., Wong, P.T.W., Hetherington, S.E. and Southwell, R.J. (1987) Evidence for the involvement of the root lesion nematode *Pratylenchus thornei* in wheat yield decline in northern New South Wales. *Australian Journal of Experimental Agriculture* 27, 563–570.

Du Charme, E.P. (1968) Burrowing nematode decline in citrus, a review. In: Smart, G.C. and Perry, V.G. (eds) *Tropical Nematology*. University of Florida Press, Gainesville, Florida, pp. 20–37.

Du Charme, E.P. and Price, W.C. (1966) Dynamics of multiplication of *Radopholus similis*. *Nematologica* 12, 113–121.

Dubé, B. and Smart, G.C. Jr. (1987) Biological control of *Meloidogyne incognita* by *Paecilomyces lilacinus* and *Pasteuria penetrans*. *Journal of Nematology* 19, 222–227.

Ducusin, A.R. and Davide, R.G. (1972) *Meloidogyne incognita*: its effect on tomato yield and some methods of control. *Philippine Agriculturist* 55, 261–281.

Duggan, J.J. (1959) On the number of generations of beet eelworm, *Heterodera schachtii* Schmidt, produced in a year. *Nematologica* 4, 241–244.

Duggan, J.J. (1969) Leaf and bud eelworms of strawberries. *Farm Research News* 10, 134-136.

Duggan, J.J. (1971) Control of stem nematode (*Ditylenchus dipsaci*) in onions. *Proceedings of the 6th British Insecticides and Fungicides Conference, Brighton, UK*, pp. 150–153.

Duggan, J.J. (1978) Integrated control of the potato cyst eelworm (*Globodera rostochiensis*). In: Duggan, J.J. (ed.) *Proceedings of a Seminar on Biological Control, Royal Irish Academy, Irish National Committee for Biology, Republic of Ireland, 17–18 February, 1977*, pp. 206–211.

Duhaylongsod, R.D. (1988) Populations of root-knot and reniform nematodes in soil amended with fresh and composted organic materials. *International Nematology Network Newsletter* 5, 24–26.

Duncan, L.W. (1989) Effect of fenamiphos placement on *Tylenchulus semipenetrans* and yield in a Florida citrus orchard. *Journal of Nematology* 21, 703–706.

Duncan, L.W. and Cohn, E. (1990) Nematode parasites of citrus. In: Luc, M, Sikora, R.A.

and Bridge, J. (eds) *Plant Parasitic Nematodes in Subtropical and Tropical Agriculture.* CAB International, Wallingford, UK, pp. 321–346.

Duncan, L.W., Inserra, R.N., O'Bannon, J.H. and El-Morshedy, M.M. (1994) Reproduction of a Florida population of *Tylenchulus semipenetrans* on resistant citrus rootstocks. *Plant Disease* 78, 1067–1071.

Dunning, R.A. and Winder, G.H. (1969) Trials on the control of Docking disorder in sugar beet; seed and row treatments with nematicides. *Proceedings of the 5th British Insecticides and Fungicides Conference, Brighton, UK* 1, 265–273.

Dunning, R.A. and Winder, G.H. (1974) Effects of aldicarb and some other nematicides on growth of sugarbeet in *Heterodera schachtii*-infested soil. *Plant Pathology* 23, 1–8.

Dureja, P., Prasad, D., Walia, S. and Mukerjee, S.K. (1988) Photodegradation of methyl N,N′,-dimethyl-N-((methyl-carbamoyl)-oxy)-1-thiooxamimidate and nematicidal activity of its photoproducts. *Toxicological and Environmental Chemistry* 17, 215–222.

Dwinell, L.D. (1990) Heat treating and drying southern pine lumber infested with pinewood nematodes. *Forest Products Journal* 40, 53–56.

Dwivedi, B.K. and Pandey, G. (1992) Effect of organic amendments on root-knot nematode with its development of egg-sac, egg, juvenile, male and female in green gram (*Vigna radiata) Current Nematology* 3, 209–210.

Dwivedi, R.K., Upadhyay,K.D. and Dwivedi, K. (1994) Influence of seed treatment on seedling emergence of pigeon pea (*Cajanus cajan* L.) and larval penetration of *Meloidogyne incognita* in presence of fungus, *Fusarium oxysporum* var. *udum. Indian Journal of Nematology* 22, 146–148.

Echávez-Badel, R. and Ayala, A. (1979) Combate químico de los nematodos en viveros de café. *Indian Journal of Nematology* 9, 8.

Edmunds, J.E. (1969) Plant nematode problems of the Windward Islands. In: Peachey, J.E. (ed) *Nematodes of Tropical Crops. Technical Communication Commonwealth Bureau of Helminthology* No. 40, 142–148.

Edwards, D.I. and Noel, G.R. (1981) Update: Soybean cyst nematode. *Illinois Research* 23, 4.

Edwards, D.I. and Wehunt, E.J. (1971) Host range of *Radopholus similis* from banana areas of Central America with indications of additional races. *Plant Disease Reporter* 55, 415–418.

Efremenko, V.P. and Burshtein, K.S. (1975) [Penetration routes of *Ditylenchus destructor* Thorne, 1945, into potatoes and agrotechnical control of this infection.] *Byulleten' Vsesoyuznogo Instituta Gel' mintologii im. K.I. Skryabina* No. 15, 47–51.

Efremenko, V.P. and Klimakova, E.I. (1971) [Principles and development of an integrated method of control of *Heterodera rostochiensis*.] *Byulleten' Vsesoyuznogo Instituta Gel' mintologii im K.I. Skryabina* No. 6, 13–21.

Egunjobi, O.A. (1977) On the possible utilisation of discarded cocoa (*Theobroma cacao* L.) pod husks as fertilizer and nematicide. In: *Proceedings of the 5th International Cocoa Research Conference, Ibadan, Nigeria 1–9 September, 1975*, pp. 541–547.

Egunjobi, O.A. and Olaitan, J.O. (1986) Response of *Meloidogyne incognita*-infested cowpea to some agro-waste soil amendments. *Nematropica* 16, 33–43.

Ehrenpfordt, V., Klügling, R. and Sachse, B. (1985) Populationsdynamik des Getreidezystenalchens (*Heterodera avenae*). Untersuchungen in einem statischen Versuch mit alternierenden Getreidenbau. *Archiv für Acker-und Pflanzenbau und Bodenkunde* 29, 113–118.

Eisenback, J.D., Hirschmann, H., Sasser, J.N. and Triantaphyllou, A.C. (1981) *A guide to the four common species of root-knot nematodes* (Meloidogyne *spp.) with a pictorial key.* International Meloidogyne Project, North Carolina State University, Raleigh, USA.

Elgin, J.H. Jr., Evans, D.W. and Faulkner, L.R. (1977) Response of resistant and susceptible alfalfa cultivars to regional isolates of stem nematode. *Crop Science* 17, 957–959.

Elgindi, D.H., Van Gundy, S.D. and Small, R.H. (1978) Dispersion and persistence of aldicarb in soil water of nematode-infested citrus soils. *Revue de Nématologie* 1, 207–215.

Ellenby, C. (1945a) The influence of crucifers and mustard oil on the emergence of larvae of the potato-root eelworm, *Heterodera rostochiensis* Wollenweber. *Annals of Applied Biology* 32, 67–70.

Ellenby, C. (1945b) Control of the potato-root eelworm *Heterodera rostochiensis* Wollenweber, by allyl isothiocyanate. *Nature*, 155 (3940), 544.

Ellenby, C. (1951) Mustard oils and control of the potato-root eelworm, *Heterodera rostochiensis* Wollenweber. Further field and laboratory experiments. *Annals of Applied Biology* 38, 859–875.

Ellenby, C. (1954) Tuber forming species and varieties of the genus *Solanum* tested for resistance to the potato root eelworm *Heterodera rostochiensis* Wollenweber. *Euphytica* 3, 195–202.

Elston, D.A., Philipps, M.S. and Trudgill, D.A. (1991) The relationship between initial population density of potato cyst nematode *Globodera pallida* and the yield of partially resistant potatoes. *Revue de Nématologie* 14, 221–229.

EPPO (1992) Distribution of *Globodera rostochiensis* and *G. pallida*. *Reporting Service 523/16*, EPPO Secretariat, Paris.

Eremin, G.V. and Bobrysheva, M.N. (1972) [Strawberry propagation in the Krasnodar Territory.] In: *Kul'tura zemlyaniki v SSR. Doklady simpoziuma, (28 ijunya–1 ijulya 1971) Moscow, USSR; 'Kolos' (1972)*, pp. 401–402.

Eriksson, K.B. (1965) Crossing experiments with races of *Ditylenchus dipsaci* on callus tissue cultures. *Nematologica* 11, 244–248.

Evans, K. (1983) Hatching of potato cyst-nematodes in root diffusates collected from twenty five potato cultivars. *Crop Protection* 2, 97–103.

Evans, K. (1987) The interaction of potato cyst-nematodes and *Verticillium dahliae* on early and maincrop potato cultivars. *Annals of Applied Biology* 110, 329–339.

Evans, K. and Brodie, B.B. (1980) The origin and distribution of the golden nematode and its potential in the USA. *American Potato Journal* 57, 79–89.

Evans, K. and Haydock, P.P.J. (1990) A review of tolerance by potato plants of cyst nematode attack, with consideration of what factors may confer tolerance and methods of assaying and improving it in crops. *Annals of Applied Biology* 117, 703–740.

Evans, K. and Stone, A.R. (1977) A review of the distribution and biology of the potato cyst- nematodes *Globodera rostochiensis* and *G. pallida*. *PANS* 23, 178–189.

Evans, K., Franco, J. and de Scurrah, M.M. (1975) Distribution of species of potato cyst-nematode in South America. *Nematologica* 21, 365–369.

Fargette, M. (1988) Présence de race B du nématode phytoparasite *Meloidogyne incognita* en Côte d'Ivoire et leur caractérisation par l'elecrophorèse des esterases. *Comptes Rendus de l'Academie des Sciences, III (Science de la Vie)* 306, 437–440.

Farley, J.D. and Riedel, R.M. (1971) Effect of *Pratylenchus penetrans* on *Verticillium* wilt of tomato. *Plant Disease Reporter* 55, 673–675.

Fassuliotis, G. (1971) Tolerance of *Hoplolaimus columbus* to high osmotic pressures, desiccation and high soil temperatures. *Journal of Nematology* 3, 309–310.

Fassuliotis, G. (1976) *Hoplolaimus columbus. Commonwealth Institute of Helminthology Descriptions of Plant-Parasitic Nematodes Set 6, N. 81*, Commonwealth Institute of Helminthology, St Albans, UK.

Fassuliotis, G., Rau, G.J. and Smith, F.H. (1968) *Hoplolaimus columbus*, a nematode parasite associated with cotton and soybeans in South Carolina. *Plant Disease Reporter* 52, 571–572.

Feldmesser, J., Rebois, R.V. and Taylor, A.L. (1959) Progress report on growth response of burrowing nematode infected citrus following chemical treatments under greenhouse conditions. *Plant Disease Reporter* 43, 261–263.

Fenwick, D.W. (1951) On the varying nematicidal effects of different samples of D-D against the potato root eelworm *Heterodera rostochiensis*. *Journal of Helminthology* 25, 166.

Fenwick, D.W. (1962) Penetration of coconut tissues by *Rhadinaphelenchus cocophilus* (Cobb, 1919) Goodey, 1960. *Annual Report of the Research Station, Trinidad and Tobago Coconut Research Ltd., 1961*, pp. A1–15.

Fenwick, D.W. (1969) Red ring disease of the coconut palm. In: Peachey, J.W. (ed) *Nematodes of Tropical Crops. Technical Communication, Commonwealth Bureau of Helminthology* No. 40, 89–98.

Ferris, H. and Greco, N. (1992) Management strategies for *Heterodera goettingiana* in a vegetable cropping system in Italy. *Fundamental and Applied Nematology* 15, 25–33.

Fichtner, E., Koppen, D. and Korschens, M. (1984) Bezeihung zwischen Zuckerrübenanteil in der Fruchtfolge, Ertrag und Populationsdynamik van *Heterodera schachtii* Schm. auf dem Löss-Schwartzende-Standort Bad Lauchstädt. *Archiv für Acker-und Pflanzenbau und Bodenkunde* 28, 617–623.

Fielding, M.J. (1951) Observations on the length of dormancy in certain plant infecting nematodes. *Proceedings of the Helminthological Society of Washington* 18, 110–112.

Figueroa, A. (1980) Efecto de los nematicidas aldicarb, aldoxycarb y DBCP en el control de nematodos y la producción del bananas, *Nematropica* 10, 15–20.

Firoza, K. and Maqbool, M.A. (1995) Numerical threshold for infection of the spiral nematode, *Helicotylenchus dihystera* (Cobb, 1893) Sher, 1961 on brinjal, tomato and wheat. *Pakistan Journal of Nematology* 13, 93–97.

Flegg, J.J.M. (1968) Life cycle studies of some *Xiphinema* and *Longidorus* species in south-eastern England. *Nematologica* 14, 197–210.

Fluiter, H.J. de (1947) Het aaltjesprobleem in de koffiecultuur. *Tijdschrift PlantenZiekten* 53, 101–109.

Ford, H.W. (1952) The effect of spreading decline on the root distribution of citrus. *Proceedings of the Florida State Horticultural Society* 65, 47–50.

Ford, H.W. (1953) Effect of spreading decline disease on the distribution of feeder roots of orange and grapefruit trees on rough lemon rotstocks. *Journal of the American Society for Horticultural Science* 61, 68–72.

Fortuner, R. and Merny, G. (1979) Root parasitic nematodes of rice. *Revue de Nématologie* 2, 79–102.

Francl, L.J. and Dropkin, V.H. (1986) *Heterodera glycines* population dynamics and relation of initial population and soybean yield. *Plant Disease* 70, 791–795.

Franco, J. and Evans, K. (1978) Multiplication of some South American and European populations of potato cyst nematodes on potatoes possessing the resistance genes H_1, H_2, H_3. *Plant Pathology* 27, 1–6.

Frandsen, K.J. (1966) Observations on the attack by populations of *Ditylenchus dipsaci* on strains of red clover. *Suom. maatal Seur. Julk.* 107, 18–29.

Fraser, W.D. and Lindley, C.D. (1962) Field experiments with a new organo-phosphorus nematicide against potato root eelworm. *Proceedings of the 1st British Insecticide and Fungicide Conference 1961*, 1, 27–36.

Freire, F.C.O. and Bridge, J. (1985) Parasitism of eggs, females and juveniles of *Meloidogyne incognita* by *Paecilomyces lilacinus* and *Verticillium chlamydosporium*. *Fitopatologia Brasileira* 10, 577–596.

French, J.V. and Timmer, L.W. (1979) Control of rust mite and reduction of citrus nematode populations on Texas oranges with Temik. *Journal of the Rio Grande Valley Horticultural Society* 33, 63–70.

French, N. and Wilson, W.R. (1976) Influence of crop rotation, weed control and nematicides on spraing in potatoes. *Plant Pathology* 25, 167–172.

Fujimura, T., Ichita, T. and Kimura, T. (1989) [Occurrence of potato-rot nematode, *Ditylenchus destructor* Thorne, in garlic and control 1. Evaluation of treatments applied before planting and after harvest for control.] *Japanese Journal of Nematology* 18, 22–29.

Fukuda, K., Hogetsu, T. and Suzuki, K. (1992) Cavitation and cytological changes in xylem of pine seedlings inoculated with virulent and avirulent isolates of *Bursaphelenchus xylophilus* and *B. mucronatus*. *Journal of the Japanese Forestry Society* 74, 289–299.

Fukudome, N. and Kamigama, K. (1982) [Resistance of tobacco to *Meloidogyne javanica* II. Effect of soil temperature on the manifestation of resistance.] *Japanese Journal of Nematology* 11, 13–18.

Fukui, J. and Yarimizu, H. (1953a) [Control of the nematode *Heterodera marioni* (Cornu) Goodey on soybean. I. Effect of D-D (dichloropropane-dichloropropene) as worm killer.] *Journal of the Kanto-Tosan Agricultural Experimental Station* 4, 19–22.

Fukui, J. and Yarimizu, H. (1953b) [Control of the nematode *Heterodera marioni* (Cornu) Goodey on soybean. II. Residual effectiveness of D-D (dichloropropane-dichloropropene) application.] *Journal of the Kanto-Tosan Agricultural Experimental Station* 4, 23–26.

Fukuzawa, A., Furusaki, A., Ikura, M. and Masamune, T. (1985) Glycinoeclepin A, a natural hatching stimulus for the soybean cyst nematode. *Journal of the Chemical Society, Chemical Communications*, 222.

Ganguly, S. and Dasgupta, D.R. (1984) Sequential development of polyphenol oxidase (E.C. 1.14.18.1) in resistant and susceptible tomatoes inoculated with the root knot nematode, *Meloidogyne incognita*. *Nematologia Mediterranea* 12, 15–22.

Ganguly, Y.A.K. and Dasgupta, D.R. (1988) Superoxide dismutase (E.C. 1.15.1.1.) activity in resistant and susceptible cowpea cultivars inoculated with root-knot nematode, *Meloidogyne incognita*. *Indian Journal of Nematology* 18, 322–325.

Gapasin, R.M., Valdez, R.B. and Mendoza, E.M.I. (1988) Phenolic involvement in sweet potato resistance to *Meloidogyne incognita* and *M. javanica*. *Annals of Tropical Research* 101, 63–73.

Garabedian, S. and Van Gundy, S.D. (1983) Use of avermectins for the control of *Meloidogyne incognita* on tomatoes. *Journal of Nematology* 15, 503–510.

Garabedian, S. and Van Gundy, S.D. (1985) Effects of non-fumigant nematicides applied through low-pressure drip irrigation on control of *Meloidogyne incognita* on tomatoes. *Plant Disease* 69, 138–140.

Garcia, M.R. and Rich, R. (1983) Efficacy of selected fumigant and non fumigant nematicides to control *Meloidogyne javanica* in Florida tobacco. *Nematologica* 13, 125–134.

Garibaldi, A. (1987) Impiego dei materiali plastici nella solarizzazione del terreno. *Colture Protette* 16, 25–28.

Gaspard, J.T. and Mankau, R. (1987) Density dependence and host specificity of the nematode-trapping fungus *Monacrosporidium ellipsosporum*. *Revue de Nématologie* 10, 241–246.

Gaur, H.S. and Dhingra, A. (1991) Management of *Meloidogyne incognita* and *Rotylenchulus reniformis* in nursery beds by soil solarization and organic soil amendment. *Revue de Nématologie* 14, 189–195.

Gaur, H.S. and Khan, E. (1994) Long-term anhydrobiotic survival and lipid utilization of *Pratylenchus zeae* in soil. *Annals of Plant Protection Services (Indian Agricultural Research Institute, New Delhi)* 2, 27–32.

Gazaway, W. (1994) Nematicides for reniform nematode control in cotton, 1993. *Fungicide and Nematicide Tests. American Phytopathological Society* 49, 179.

Gazaway, W. and Edmiston, K. (1992a) Effect of Temik applied in-the-furrow on reniform nematodes in cotton. *Fungicide and Nematicide Tests. American Phytopathological Society* 47, 159.

Gazaway, W. and Edmiston, K. (1992b) An evaluation of various Temik and Telone rates for controlling reniform nematodes in cotton. *Fungicide and Nematicide Tests. American Phytopathological Society* 47, 159.

Gazaway, W. and Rodriguez-Kábana, R. (1991) Nematicide effect on cotton yields and reniform nematodes. *Fungicide and Nematicide Tests. American Phytopathological Society* 46, 190.

Georg, D., Dubé, A.J. and Wayte, S.E. (1983) Nematicide control of cereal cyst nematode damage to wheat, 1981. *Fungicide and Nematicide Tests. American Phytopathological Society* 38, 21.

Georg, D.G., Fisher, J.M., Dubé, A.J. and Wayte, S.E. (1989) The relation between time of sowing of cereals and cereal cyst nematode (*Heterodera avenae* Woll.) *Australian Journal of Experimental Agriculture* 29, 647–653.

Gergon, E.B. and Prot, J.C. (1993) Effect of benomyl and carbofuran on *Aphelenchoides besseyi* on rice. *Fundamental and Applied Nematology* 16, 563–566.

Germani, G., Reversat, G. and Luc, M. (1983) Effect of *Sesbania rostrata* on *Hirschmanniella oryzae* in flooded rice. *Journal of Nematology* 15, 269–271.

Giebel, J. (1967) Azydek sodu jako srodek nicieniobójczy. *Pr. nauk. Inst. Ochr. Rośl.* 9, 117–126.

Giebel, J., Kornobis, S., Czaplicki, E. and Lewanska, A. (1982) Azydek sodu jako nematocyd. In: *Materialy XXII i XXIII Sesji Naukowej Instytutu Ochrony Roślin, Poznan, 1982*, 77–88.

Gienke, W., Ort, O. and Stark, H. (1989) Total synthesis of the azoxy antibiotic jietacin A. *Liebigs Annalen der Chemie* No. 7, 671–676.

Gill, J.S. and Swarup, G. (1977) Effect of interaction between *Heterodera avenae* Woll. 1924, *Fusarium moniliforme* and *Helminthosporium gramineum* on barley plants and nematode reproduction. *Indian Journal of Nematology* 7, 42–45.

Gilman, D.F., Jones, J.E., Williams, C. and Birchfield, W. (1978) Cotton-soybean rotation for control of reniform nematodes. *Louisiana Agriculture* 21, 10–11.

Gintzberger, G. and Prosperi, J.M. (1987) D'autres luzernes annuelles. *Bulletin. Fnams Semences* No. 101, 23–26.

Glaba, B. and Kuś, J. (1990) Wplyw udzialu zbóz w strukturge zasiewów na zasiediemie gieby prezez matwika. *Parniethik Pulawski* No. 94, 145–159.

Gladkaya, R.,M. (1973) [Elimination of foci of potato cyst nematode by a combination of control methods.] In: *Materialy Vsesoyuznogo simpoziuma po bor'be s kartofel'noi nematodoi, Tartu, 3–5 iyulya 1973, Tartu, Estonia*, 20–22.

Gladkaya, R.M. and Ponin, I.Ya. (1976) [New nematicides against the potato nematodes.] *Zaschita Rastenii*, 1, 138–142.

Gnanapragasam, N.C. (1982) Growth performance of nursery plants in soils infested with the root-lesion nematode, *Pratylenchus loosi* and treated with dazomet 98% G. and methyl bromide. *Tea Quarterly* 51, 4–6.

Gnanapragasam, N.C. (1987) Integrated strategies to manage nematodes in tea. *Proceedings of the International Conference on Tea Quality and Human Health, China (Hangzhou), November, 1987*, 117–183.

Gogoi. B.B. and Phukan, P.N. (1990) Efficacy of certain chemicals as seed treatment against *Meloidogyne incognita* on lentil. *Indian Journal of Nematology* 20, 53–56.

Golden, A.M. and Shafer, T. (1959) Host–parasite relationships of various plants and the sugar-beet nematode (*Heterodera schachtii*). *Plant Disease Reporter* 43, 1258–1262.

Golden, A.M., Epps, J.M., Riggs, R.D., Duclos, R.D., Fox, L.A. and Bernard, R.L. (1970)

Terminology and identity of infraspecific forms of the soybean cyst nematode (*Heterodera glycines*). *Plant Disease Reporter* 54, 544.

Gomes Carneiro, R.M.D. and Cayrol, J.C. (1991) Relationship between inoculum density of the nematophagous *Paecilomyces lilacinus* and control of *Meloidogyne arenaria* on tomato. *Revue de Nématologie* 14, 629–634.

Good, J.M., Jr., Robertson, W.K. and Thompson, L.G., Jr.(1954) Effect of crop rotation on the populations of meadow nematode, *Pratylenchus leiocephalus*, in Norfolk fine sand. *Plant Disease Reporter* 38, 178–180.

Goodey, J.B. (1949) The control of *Anguillulina dipsaci* on the seeds of teazle and red clover by fumigation with methyl bromide. *Journal of Helminthology* 23, 171–174.

Goodey, J.B. and Brown, E.B. (1956) Stem eelworm attacking carrots. *Journal of Helminthology* 29, 187–192.

Goodey, J.B. and Hooper, D.J. (1958) Observations on the effects of *Ditylenchus dipsaci* and *Anguina tritici* on certain wheat and barley varieties. *Nematologica* 3, 24–29.

Goodey, J.B. and Hooper, D.J. (1962) Observations on the attack by *Ditylenchus dipsaci* on varieties of oats. *Nematologica* 8, 33–38.

Goodey, J.B., Franklin, M.T. and Hooper, D.J. (1965) *The Nematode Parasites of Plants Catologued under their Hosts.* Commonwealth Agricultural Bureaux, Farnham Royal, England, 214 pp.

Goodey, T. (1933) *Plant Parasitic Nematodes.* Methuen and Co., London, pp. 25–26.

Goodey, T. (1935) Observations on a nematode disease of yams. *Journal of Helminthology* 13, 173–190.

Goodey, T. (1937) Observations on the susceptibility of certain varieties of oats to 'tulip-root' caused by the stem eelworm, *Anguillulina dipsaci*. *Journal of Helminthology* 15, 203–214.

Goodey, T. (1945) *Anguillulina dipsaci* on onion seed and its control by fumigation with methyl bromide. *Journal of Helminthology* 21, 45–59.

Gosse, E. (1975) Untersuchungen zur Schadwirkung von *Heterodera avenae* Woll. In: *Vortragstugung (1) zu Aktuellen Problemen der Phytonematologie am 29.5.1975 in Rostock. Manuskriptdruck der Vorträge,* Biologische Gesellschaft der DDR, Section Phytopathologie und Universität Rostock, DDR, 33–43.

Gostick, K.G. (1963) Control of seed-borne lucerne stem eelworm with phorate. *Annals of Applied Biology* 51, 503–507.

Gotke, N. and Mathur, V.K. (1990) Thermal therapy for the eradication of *Mentha spicata* roots infested with *Meloidogyne incognita*. *FAO Plant Protection Bulletin 8*, 213–215.

Gotke, N. and Mathur, V.K. (1993) Treatment schedule for denematization of seeds of *Setaria italica* and *Panicum miliaceum* infested with *Aphelenchoides besseyi*. *Nematologica* 39, 274–276.

Gotke, N. and Swarup, G. (1985a) Development of *Heterodera avenae* in susceptible and resis tant barley cultivars in relation to larval inoculum and penetration. *Indian Journal of Nematology* 14, 128–134.

Gotke, N. and Swarup, G. (1985b) Relationship between initial inoculum levels of *Heterodera avenae* and penetration of juveniles in the roots of susceptible and resistant barley cultivars. *Indian Journal of Nematology* 14, 95–96.

Gotke, N. and Swarup, G. (1989) Effect of neem (*Azadirachta indica*) seed kernel powder treatment on penetration and development of *Anguina tritici* on wheat. *Indian Journal of Nematology* 18, 149.

Gottlieb, Y., Cohn, E. and Spiegel-Roy, P. (1986) Biotypes of the citrus nematode (*Tylenchulus semipenetrans* Cobb) in Israel. *Phytoparasitica* 14, 193–198.

Govindaiah, S.B.D., Tomy Philip and Datta, R.K. (1991) Effect of marigold (*Tagetes patula*) intercropping against *Meloidogyne incognita* infesting mulberry. *Indian Journal of Nematology* 21, 96–99.

Graf, A. and Meyer, H. (1973) Bedeutung des Rübenkopf-alchens (*Ditylenchus dipsaci*) in der Schweiz und seine Bekämpfungsmöglichkeiten. *IIRB* 6, 117–126.

Graf, A., Keller, J.E., Leichti, H. and Savary, A. (1964) Essais de lutte contre *Ditylenchus dipsaci* Kühn nématode agent de la pourriture du collet des betteraves. *Agriculture Romande 1964 III Serie A*, 4–7.

Graham, C.W. (1980) The effects of rainfall and soil type on the population dynamics of cereal cyst nematode (*Heterodera avenae*) on spring barley (*Hordeum vulgare*) and spring oats (*Avena sativa*). *Annals of Applied Biology* 94, 243–253.

Grandison, G.S. (1983) Root-knot nematode control on kiwifruit (*Actinidia chinensis*) by chemical bare-root dip. *Plant Disease* 67, 899–900.

Graney, L.S. and Miller, L.I. (1980) Differentiation of five isolates of *Heterodera schachtii* as races of the sugar beet cyst nematode. *Journal of Nematology* 12, 233.

Gray, F.A. and Soh, D.H. (1989) A nematicide treatment to control *Ditylenchus dipsaci* on seedling alfalfa. *Journal of Nematology* 21, 184–188.

Greco, N. and Brandonisio, A. (1980) Relationship between *Heterodera carotae* and carrot yield. *Nematologica* 26, 497.

Greco, N. and Lamberti, F. (1977a) Ricerca di dosi ottimali di alcuni nematocidi per la lotta contro *Heterodera carotae*. *Nematologia Mediterranea* 5, 25.

Greco, N. and Lamberti, F. (1977b) Ulteriori saggi di lotta chimica contro *Heterodera carotae* in Puglia. *Informatore Fitopatologia* 27, 15.

Greco, N., Lamberti, F. and Brandonisio, A. (1976) La lotta chimica contro *Heterodera carotae* in Puglia. *Nematologia Mediterranea* 4, 133–137.

Greco, N., Brandonisio, A. and de Marinis, G. (1982a) Tolerance limit of the sugarbeet to *Heterodera schachtii*. *Journal of Nematology* 14, 199–202.

Greco, N., Brandonisio, A. and de Marinis, G. (1982b) Investigation on the biology of *Heterodera schachtii* in Italy. *Nematologia Mediterranea* 10, 201–214.

Greco, N., Brandonisio, A. and Elia, F. (1985) Control of *Ditylenchus dipsaci*, *Heterodera carotae* and *Meloidogyne javanica* by solarization. *Nematologia Mediterranea* 13, 191.

Greco, N., Di Vito, M. and Lamberti, F. (1986) Studies on the biology of *Heterodera goettingiana* in Southern Italy. *Nematologia Mediterranea* 14, 23.

Greco, N., Di Vito, M., Saxena, M.C. and Reddy, M.V. (1988a) Investigation on the root lesion nematode *Pratylenchus thornei*, in Syria. *Nematologia Mediterranea* 16, 101–105.

Greco, N., Di Vito, M., Saxena, M.C. and Reddy, M.V. (1988b) Effect of *Heterodera ciceri* on yield of chickpea and lentil and development of this nematode on chickpea in Syria. *Nematologica* 34, 98–114.

Greco, N., D'Addabbo, T., Brandonisio, A. and Zweep, A. (1990) Combined effect of soil solarization and 1,3-dichloropropene for control of *Heterodera carotae*. *Nematologica* 18, 261–264.

Greco, N., Ferris, H. and Brandonisio, A. (1991a) Effect of *Heterodera goettingiana* population densities on the yield of pea, broad bean and vetch. *Revue de Nématologie* 14, 619–624.

Greco, N., Di Vito, M. and Saxena, M.C. (1991b) Soil solarization for control of *Pratylenchus thornei* on chickpea in Syria. *FAO Plant Production and Protection Paper* No. 109, 182–188.

Greco, N., Brandonisio, A. and Elia, F. (1992) Efficacy of SIP 5561 and soil solarization for management of *Meloidogyne incognita* and *M. javanica* on tomato. *Nematologia Mediterranea* 20, 13–16.

Green, C.D., Williamson, K., Dennis, E.B. and McBurney, T. (1981) The effect of oxamyl on the growth of peas attacked by pea cyst-nematode. *Annals of Applied Biology* 97, 303–309.

Grewal, P.S., Chhabra, H.K. and Kaul, V.K. (1985) Control of *Meloidogyne arenaria* on groundnut with nematicides and oil-cakes. *Journal of Oilseeds Research* 2, 22–28.

Griffin, G.D. (1977a) Pathological differences in *Heterodera schachtii* populations on sugar beet. *Proceedings of the American Phytopathological Society* 4, 206.

Griffin, G.D. (1977b) Effects of soil moisture on control of *Heterodera schachtii* with aldicarb. *Journal of Nematology* 9, 211–215.

Griffin, G.D. (1981a) The relationship of plant age, soil temperature and population density of *Heterodera schachtii* on the growth of sugar beet. *Journal of Nematology* 13, 184–190.

Griffin, G.D. (1985) Host–parasite relationship of *Meloidogyne chitwoodi* on potato. *Journal of Nematology* 17, 395–399.

Griffin, G.D. (1987a) Effects of environmental factors and cultural practices on parasitism of alfalfa by *Ditylenchus dipsaci*. *Journal of Nematology* 19, 267–276.

Griffin, G.D. (1987b) Efficacy of using split and postplant applications of aldicarb for control of *Heterodera schachtii* in sugar beet. *Annals of Applied Nematology* 1, 119–122.

Griffin, G.D. (1989) Comparison of fumigant and non-fumigant nematicides for control of *Meloidogyne chitwoodi* on potatoes. *Journal of Nematology* 21, 640–644.

Griffin, G.D. (1990) Pathological relationship of *Ditylenchus dipsaci* and *Fusarium oxysporum* f.sp. *medicaginis* on alfalfa. *Journal of Nematology* 22, 333–336.

Griffin, G.D., Hunt, O.J. and Murray, J.J. (1975) Pathogenicity of *Ditylenchus dipsaci* to sainfoin (*Onobrychis viciaefolia* Scop.). *Journal of Nematology* 7, 93–94.

Griffith, R. and Koshy, P.K. (1990) Nematode parasites of coconut and other palms. In: Luc, M., Sikora, R. and Bridge, J. (eds) *Plant Parasitic Nematodes in Subtropical and Tropical Agriculture.* CAB International, Wallingford, UK, pp. 363–386.

Grinstein, A., Orion, D., Greenberger, A. and Katan, J. (1979) Solar heating of the soil for control of *Verticillium dahliae–Pratylenchus thornei* complex disease in potatoes. *Nematropica* 9, 95–110.

Grujičič, G. (1974a) Dejstro preparata Temik 10G (aldicarbe) na inakulacionu sposobnost invasionih larvi repine nematode (*Heterodera schachtii* Schmidt). *Zaštita Bilja* 25, 163–167.

Grujičič, G. (1974b) [Root-knot nematodes (*Meloidogyne* spp.) on garden plants and influence of non-phytotoxic nematicides.] *Zaštita Bilja* 25, 186.

Grundbacher, F.J. (1962) Testing alfalfa seedlings for resistance to the stem nematode *Ditylenchus dipsaci* (Kühn) Filipjev. *Proceedings of the Helminthological Society of Washington* 29, 152–158.

Grymaszewska, G. and Golinowski, W. (1987) Changes in the structure of whcat (*Triticum aestivum* L.) roots in varieties susceptible and resistant to infestation by *Heterodera avenae* Woll. *Acta Societatis Botanicorum Poloniae* 56, 381–389.

Guérout, R. (1970) Étude de trois nouveaux nématicides en banancraie. *Fruits D'Outre Mer* 25, 767–779.

Guérout, R. (1972) Relation entre les populations de *Radopholus similis* Cobb et la croissance du bananier. *Fruits* 27, 331–337.

Guérout, R. (1974) Quatre nématicides récents et leurs possibilités d'utilisation en bananeraie. *Fruits* 29, 339–347.

Guérout, R. and Pinon, A. (1973) Utilisation du prophos dans la lutte contre *Radopholus similis* Cobb (Nematoda, Tylenchidae) en bananeraie. *Fruits* 28, 751–756.

Guérout, R., Lassoudière, A. and Vilardebó, A. (1976) Efficacité des nématicides sur deux types de sols à caracteristiques particulières en Côte d'Ivoire. *Fruits* 31, 427–436.

Gul, S., Gul, A. and Saeed, M. (1988) Efficacy of *Paecilomyces lilacinus* against golden nematode (*Globodera rostochiensis*) in Pakistan. *International Nematology Network Newsletter* 5, 20–22.

Gupta, D.C. and Kali Ram (1981) Studies on the control of *Meloidogyne javanica* infecting chickpea in different types of soil. *Indian Journal of Nematology* 11, 77–80.

Gupta, D.C., Verma, K.K. and Paruthi, I.J. (1987) Studies on root-knot nematode, *Meloidogyne javanica*, on black gram (*Vigna mungo*). *Indian Journal of Nematology* 17, 177–179.

Gupta, J.C. (1975) Evaluation of various treatments and varietal resistance for the control of banana nematodes. *Haryana Journal of Horticultural Sciences* 4, 152–156.

Gupta, M.C. (1987) Population dynamics of *Helicotylenchus* spp. in soil amended with carbonaceous and nitrogenous materials. *Indian Journal of Mycology and Plant Pathology* 17, 189–190.

Gupta, R.L., Roy, N.K., Khazanchi, R. and Prasad, D. (1986) Efficacy of diaryl S-ethyl phosphorothiolates against root-knot nematodes. *Indian Journal of Agricultural Sciences* 56, 597–605.

Gupta, R.L., Roy, N.K. and Prasad, D. (1991) Nematotoxicity of O-alkyl S,S-diaryl phosphorotrithioates against root-knot nematode (*Meloidogyne incognita*) and relationships between chemical structure and activity. *Pesticide Research Journal* 3, 129–134.

Gupta, R.L., Prasad, D. and Roy, N.K. (1994) Nematotoxicity of O,O-diaryl O-methyl phosphonothioates against the root-knot nematode, *Meloidogyne incognita*. *Annals of Plant Protection Sciences* 2, 39–42.

Gurner, P.S., Dubé, A.J. and Fisher, J.M. (1980) Chemical control of cereal cyst-nematode (*Heterodera avenae*) on wheat by a new low-volume applicator. *Nematologica* 26, 448–454.

Gurr, G.M. (1987) Testing potato varieties for resistance to and tolerance of the white potato cyst nematode (PCN), *Globodera pallida*. *Journal of the National Institute of Agricultural Botany, UK* 17, 365–369.

Gurr, G.M. (1992) Control of potato cyst nematode (*Globodera pallida*) by host plant resistance and nematicide. *Annals of Applied Biology* 121, 167–173.

Gus'kova, L.A. (1967) [A control method] *Zashchita Rasteniĭ ot Vrediteleiĭ Bolezneĭ* No. 1, 30.

Gus'kova, L.A. and Makovskaya, S.A. (1982) [Results of studies of *Globodera* populations in the north-western Chernozem zone.] In: Gus'kova, L.A. and Chumakova, A.E. (eds) *Sbornik Nauchnykh Trudov Vsesoyuznogo Nauchno-Issledovatel'skogo Instituta Zashchity Rastenii (Biologicheskie osnovy bor'by s nematodami)*. Leningrad, USSR, VIZR, pp. 58–68.

Gus'kova, L.A., Sveshnikova, N.M. and Gladkaya, R.M. (1969) [Combined control method.] *Zashchita Rasteniĭ ot Vrediteleiĭ Bolezneĭ* No. 8, 7.

Gus'kova, L.A., Chakaeva, A.S. and Vasil'Ev, S.V. (1982) [Prediction of losses caused by *Heterodera schachtii* and its control.] In: Gus'kova, L.A. and Chumakova, A.E. (eds) *Sbornik Nomchnykh Trudov Vsesoyuznogo Nauchno-Issledovatel'skogo Instituta Zashchity Rastenii (Biologicheskie osnovy bor'by s nematodami)*. Leningrad, USSR, VIZR, pp. 7–18.

Hackney, R.W. and Dickerson, O.J. (1975) Marigold, castor bean and chrysanthemum as controls of *Meloidogyne incognita* and *Pratylenchus alleni*. *Journal of Nematology* 7, 84–90.

Hafez, S.L. (1984) Testing systemic nematicides as alternatives to fumigation for potato nematode control, 1982. *Fungicide and Nematicide Tests. American Phytopathological Society* 39, 95–96.

Haglund, W.A. (1983) Efficacy of selected nonvolatile nematicides on control of *Ditylenchus destructor* in iris. *Journal of Nematology* 15, 92–96.

Hague, N.G.M. (1959) Control of plant parasitic nematodes 1. Susceptibility of potato root eelworms and seed potato tubers to the vapour of sulphur dioxide. *Nematologica* 4, 110–114.

Hague, N.G.M. (1968) Fumigation of shallot seed with methyl bromide to control stem eelworm, *Ditylenchus dipsaci. Plant Pathology* 17, 127–128.

Hague, N.G.M. and Pain, B.F. (1970) Some observations on the effect of 'Temik' on the potato cyst eelworm, *Heterodera rostochiensis* Woll. *Plant Pathology* 19, 69–71.

Hague, N.G.M. and Sood, U. (1963) Soil sterilization with methyl bromide to control soil nematodes. *Plant Pathology* 12, 88–90.

Hague, N.G.M. (1975) Nematicides – past and present. *Proceedings of the 8th. British Insecticide and Fungicide Conference, 17–20 November, 1975, Brighton*, 3, 837–851.

Hague, N.G.M., Lubatti, O.F. and Page, A.B.P. (1964) Methyl bromide fumigation of nematodes in soil under gas-proof sheets. *Horticultural Research* 3, 84–101.

Hague, N.G.M., Damadzadeh, M., Garabedian, S.K. and Radwan, K.H. (1983) Methods of assessing the efficacy of non-volatile nematicides. *Pesticide Science* 14, 587–595.

Halbrendt, J.M. and Brown, D.J.F. (1994) Inter- and intraspecific variation in wild-type and single female-derived populations of *Xiphinema americanum* group of nematodes. *Journal of Nematology* 26, 212–221.

Halbrendt, J.M. and Jing, G. (1994) Nematode suppressive rotation crops for orchard renovation. *Acta Horticulturae* No. 363, 49–56.

Hall, T.J., Walduck, G.D. and Walker, R.W. (1993) *Digitaria milanjiana* (Rendle) Stapf. (finger grass) cv. Jarra. *Tropical Grasslands* 27, 55–57.

Hamid, G.A., Van Gundy, S.D. and Lovatt, C.J. (1989) Phenologies of the citrus nematode and citrus roots treated with oxamyl. In: *Proceedings of the 6th International Citrus Congress, Middle East, Tel Aviv, Israel, 6–11 March, 1988*, Vol.2. Weikersheim, GFR, Margraf Scientific Publishers, pp. 993–1004.

Hanks, G.R. (1995) Prevention of hot-water treatment damage in narcissus bulbs by prewarming. *Journal of Horticultural Science* 70, 343–355.

Hannon, C.I. (1963) Longevity of *Radopholus similis* under field conditions. *Plant Disease Reporter* 47, 812–816.

Hannon, C.I. (1964) Control of the citrus nematode, *Tylenchulus semipenetrans*, in microplot experiments. *Plant Disease Reporter* 48, 471–475.

Harlan, D.P. and Jenkins, L. (1967) Elimination of root-knot nematodes from plants by chemical bare-root dips or soil drenches. *Plant Disease Reporter* 51, 103–107.

Haroon, S. and Smart, G.C. Jr. (1983) Effects of Pangola digitgrass on *Meloidogyne arenaria, M. javanica* and *M. hapla. Journal of Nematology* 15, 649–650.

Harranger, J. (1971) Les nématodes des cultures maraîchères. In: *Les Nématodes des Cultures*. Paris, Ed. ACTA/FNGPC, 351–375.

Harries, V., Adolphi, H., Kiehs, K. and Neumann, U. (1980) Insektizide und nematizide Wirkung Eines neuen carbamate fur den einsatz in Mais und anderen lanwirtschaftlichen Kulturen. *Mededelingen van de Faculteit Landbouwwettenschappen Rijksuniversiteit Gent*, 45, 739–748.

Harris, A.R. (1986) Comparison of some nematicides on *Vitis vinifera* cv. Sultana in Victoria, Australia. *American Journal of Enology and Viticulture* 37, 224–227.

Harris, P.G.W. and Evans, K. (1988) Field investigations of the responses to nematicide treatment of three winter oilseed rape cultivars infested by *Heterodera cruciferae. Crop Protection* 7, 137–142.

Harrison, B.D., Peachey, J.E. and Winslow, R.D. (1963) The use of nematicides to control the spread of arabis mosaic virus by *Xiphinema diversicaudatum* (Micol.) *Annals of Applied Biology* 52, 243–255.

Harrison, M.B. (1956) Moisture and other environmental effects in relation to the control of nematodes by fumigation with special reference to the golden nematode. *Dissertation Abstracts* 16, 193.

Harrison, M.B. (1971) Evidence of the mode of action of a systemic nematicide. *Journal of Nematology* 3, 311–312.

Harrison, M.B. (1974) Control of golden nematode (*Heterodera rostochiensis*) with systemic nematicides. *Nematropica* 4, 1–2.

Hart, W.H. and Maggenti, A.R. (1971) Control of root-knot nematode, *Meloidogyne hapla*, on 2-year-old field-grown roses, *Rosa multiflora japonica*. *Plant Disease Reporter* 55, 89–92.

Hart, W.H. and Maggenti, A.R. (1974) The use of systemic nematicides on ornamental bulb crops. *Journal of Helminthology* 6, 142.

Hartwig, E.E., Buehring, N. and Young, L.D. (1982) Effect of the soybean cyst nematode upon soybean production under different management systems. *Research Report, Mississippi Agricultural and Forestry Experiment Station* 7, 4 pp.

Hartwig, J. and Sikora, R.A. (1991) Mode-of-action of the carbamate nematicides cloethocarb, aldicarb and carbofuran on *Heterodera schachtii* 1. Contact activity. *Revue de Nématologie* 14, 525–530.

Hasan, A. (1985) Breaking resistance in chilli to root knot nematode by fungal pathogens. *Nematologica* 31, 210–217.

Hasan, N. and Saxena, S.K. (1974) Effect of extracts of soil amended with oilcakes on hatching of *Meloidogyne incognita* (Kofoid and White) Chitwood. *Proceedings of the 61st Indian Science Congress, Nagpur, 1974 Part III*, Calcutta.

Hasan, N. and Saxena, S.K. (1979) Effect of inoculating tomatoes with root-knot nematode, *Meloidogyne incognita* on the phenolic content and polyphenoloxidase activity. In: Mahadevan, A. (ed) *Physiology of Host–Pathogen Interactions. (Current Trends in Life Sciences, Vol. 6)* New Delhi, India, Today and Tomorrow Publishers, pp. 199–203.

Hashmi, G., Hashmi, S., Krusberg, L.R. and Huettel, R.N. (1994) Resistance in *Zea mays* to *Heterodera zeae*. *Journal of Nematology* 25, (4 Supp), 820–823.

Hasing-Lama, M., Franco, M.K. and Escobar-Goyes, J. (1976) Effects of aldicarb, UC 21865, and DBCP on *Radopholus similis* (Cobb) infestations and on yields in a banana plantation in Machala, Ecuador. *Acta Horticulturae* No. 57, 213–221.

Hasselbach, O.E. (1973) Some first impressions on the curative use of phenamiphos in bananas in the coastal area of Surinam. *Surinaamse Landbouw* 21, 3–11.

Hawn, E.J. (1963) Transmission of bacterial wilt of alfalfa by *Ditylenchus dipsaci* (Kühn). *Nematologica* 9, 65–68.

Hawn, E.J. (1971) Mode of transmission of *Corynebacterium insidiosum* by *Ditylenchus dipsaci*. *Journal of Nematology* 3, 420–421.

Hawn, E.J. and Hanna, M.R. (1967) Influence of stem nematode infestation on bacterial wilt reaction and forage yield of alfalfa varieties. *Canadian Journal of Plant Science* 47, 203–208.

Heald, C.M. (1978) Effect of the reniform nematode on vegetable yields. *Plant Disease Reporter* 62, 902–904.

Heald, C.M. and Robinson, A.F. (1987) Effects of soil solarization on *Rotylenchulus reniformis* in the lower Rio Grande Valley of Texas. *Journal of Nematology* 19, 93–103.

Hedin, P.A., Shepherd, R.L. and Kappelman, A.J. Jr. (1984) Evaluation of cotton polyphenols as factors of resistance to root-knot nematode and *Fusarium* wilt. *Journal of Agricultural and Food Chemistry* 32, 633–636.

Heeley, W. (1951) D-D a soil fumigant against *Heterodera marioni* in the United Kingdom. *Proceedings, International Congress of Crop Protection, 1949, London*, 193–202.

Heide, A. (1973) Bekämpfung von Wurzelgallenälchen und wandernden Wurzelnematoden mit verminderten Aufwandmengen von Nematin und Basamid. In: *Vorträge der 12 Tagung über Problem der Phytonematologie, Gross Lüsewitz, 1 Juni, 1973*, pp. 85–100.

Heide, A. (1976) Bekämpfung von Blattälchen in Erdbeervermehrungsbestanden. *Gartenbau* 23, 5–6.

Heijbroek, W. (1975) [The significance of Temik 10G for the sugar beet crop.] *Bedrijfsontwikkeling* 6, 255–260.

Heijbroek, W. (1982) The influence of resistant cruciferous green manure crops on beet cyst nematodes. *Mededelingen van het Institut voor Rationele Suikerproductie, Bergen-op-Zoom*, Nr. 8, 20 pp.

Helton, A.W. (1964) Chemotherapeutic control of a root-knot nematode (*Meloidogyne hapla*) with dimethoate under greenhouse conditions. *Plant Disease Reporter* 48, 881–885.

Hendro, M.E., Prot, J.C. and Madamba, C.P. (1992) Population dynamics of *Hirschmanniella mucronata* and *H. oryzae* on *Sesbania rostrata, Aeschynomene afrasperma* and rice cv. IR 58. *Fundamental and Applied Nematology* 15, 167–172.

Herne, D.H.C., Simpson, C.M. and Townshend, J.L. (1973) Aldicarb: a systemic soil pesticide useful for establishing apple plantings. *Proceedings of the Entomological Society of Ontario* 103, 27–31.

Hesling, J.J. (1959) The emergence of larvae of *Heterodera rostochiensis* Woll. from single cysts. *Nematologica* 4, 126–131.

Hesling, J.J. (1966a) Narcissus eelworm – yesterday and today. *The Daffodil and Tulip Year Book 1966*, 1–9.

Hesling, J.J. (1966b) Biological races of stem eelworm. *Report of the Glasshouse Crops Research Institute, 1965 (1966)*, Littlehampton, UK, pp. 132–141.

Hesling, J.J. (1967) Vertical migration of the narcissus stem eelworm in soil. *Plant Pathology* 16, 1–5.

Hesling, J.J. (1970) Some aspects of the biology of narcissus eelworm *Ditylenchus dipsaci*. *Proceedings of the 9th International Nematology Symposium (Warsaw, 1967) Zeszyty Problemowwe Postępów Nauk Rolniczych* No. 92, 183–184.

Hesling, J.J. (1971) Narcissus eelworm, *Ditylenchus dipsaci* : some aspects of its biology and control by thionazin. *Acta Horticulturae* 23, 249–254.

Hesling, J.J. (1972) Host plants of tulip stem eelworm (*Ditylenchus dipsaci*). *Plant Pathology* 21, 109–111.

Hewitt, W.B., Raski, D.J. and Goheen, A.C. (1958) Nematode vector of soil-borne fanleaf virus of grapevines. *Phytopathology* 48, 586–595.

Hewlett, T.E., Dickson, D.W., Mitchell, D.J. and Kannwischer-Mitchell, M.E. (1988) Evaluation of *Paecilomyces lilacinus* as a biocontrol agent of *Meloidogyne javanica* on tobacco. *Journal of Nematology* 20, 578–584.

Hijinck, M.J. (1963) A relation between stem infection by *Phoma solanicola* and *Ditylenchus dipsaci* on potato. *Netherlands Journal of Plant Pathology* 69, 318–321.

Hijinck, M.J. and Suatmadji, R.W. (1967) Influence of different *Compositae* on population density of *Pratylenchus penetrans* and some other root-infesting nematodes. *Netherlands Journal of Plant Pathology* 73, 71–82.

Hirling, W. (1974) Warmwasserbehandlung zur Gewinnung blattalchenfreien Erdbeerpflanzgutes in Baden-Württemberg – Beobachtunggenzur Technik der Untersuchung warmwasserbehandelter Pflanzen auf Blattälchen (*Aphelenchoides* spp.). *Nachrichtenblatt des Deutschen Pfanzenschutzdienstes* 26, 118–121.

Hoestra, H. (1976) Effect of benomyl on the potato cyst nematode, *Heterodera rostochiensis*. *Netherlands Journal of Plant Pathology* 82, 17–23.

Hoestra, H. and Oostenbrink, M. (1962) Nematodes in relation to plant growth IV. *Pratylenchus penetrans* (Cobb) on orchard trees. *Netherlands Journal of Agricultural Science* 10, 286–296.

Hoff, J.K. and Mai, W.F. (1962) Pathogenicity of the stubby root nematode to onion. *Plant Disease Reporter* 46, 24–25.

Homeyer, B. (1973) Results with fenamiphos in potatoes and vegetable crops. *Proceedings of the 7th British Insecticide and Fungicide Conference, Brighton, UK, November 19–22, 1973* 2, pp. 415–423.

Homeyer, B., Marczok, P. and Rall, J. (1987) Results of long-term field trials with Nemacur and Curaterr and resistant cultivars in potato crops. *Pflanzenschutz-Nachrichten Bayer* 40, 306–315.

Homma, Y. and Kegasawa, K. (1984) Predation on larvae of plant-parasitic nematodes by soil vampyrellid amoebae. *Japanese Journal of Nematology* 14, 1–7.

Hong, L.T.T. and Sethi, C.L. (1988a) Effect of nematicidal seed treatment on *Meloidogyne incognita* in French bean (*Phaseolus vulgaris*). *Indian Journal of Nematology* 18, 219–222.

Hong, L.T.T. and Sethi, C.L. (1988b) Effect of nematicidal seed soaking on penetration and development of *Meloidogyne incognita* in French beans. *Indian Journal of Nematology* 18, 212–215.

Hoogendoorn, J.M., Dellaert, L.M.W., Gelder, W.J.M. van, Eeuwijk, F.A. van, Jonker, H.H., Keizer, L.C.P. and Vinke, J.H. (1992). Glycoalkaloids and resistance to potato cyst nematodes (*Globodera* spp.). In: *Proceedings of the Joint Conference of the EAPR Breeding and Varietal Assessment Section of the EUCARPIA Potato Section, Landerneau, France, 12–17 January, 1992*, pp. 208–210.

Hooper, D.J. (1961) A redescription of *Longidorus elongatus* (de Man, 1876) Thorne and Swanger, 1936 (Nematoda, Dorylaimidae) and descriptions of five new species of *Longidorus* from Great Britain. *Nematologica* 6, 237–257.

Hooper, D.J. (1972) *Ditylenchus dipsaci. CIH Descriptions of Plant Parasitic Nematodes.* Set 2, no. 14. 4 pp. Commonwealth Agricultural Bureaux, Slough, UK.

Hooper, D.J. (1984) Observations on stem nematode *Ditylenchus dipsaci*, attacking field beans, *Vicia faba*. In: *Rothamsted Experimental Station Report for 1983*, pp. 239–260.

Hossain, M.S., Zahid, M.I. and Mian, I.H. (1992) Effect of decomposition period on the efficacy of two oil cakes for control of root-knot nematode (*Meloidogyne incognita*). *Japanese Journal of Nematology* 22, 1–10.

Hough, A. and Thomason, I.J. (1975) Effects of aldicarb on the behaviour of *Heterodera schachtii* and *Meloidogyne javanica*. *Journal of Nematology* 7, 221–229.

Howard, H.W. (1969) Breeding potatoes resistant to cyst-nematode. *Proceedings of the 5th British Insecticide and Fungicide Conference 17–20th. November, 1969, Brighton, UK*, 1, pp. 159–163.

Hsu, F.C., Sun, K., Kleier, D.A. and Fielding, M.J. (1995) Phloem mobility of xenobiotics VI. A phloem-mobile pro-nematicide based on oxamyl exhibiting root-specific activation in transgenic tobacco. *Pesticide Science* 44, 9–19.

Huang, C.S., Charchar, J.M. and Tenente, R.C.V. (1980) Controle de nematoide das galhas em cenoura atraves de rotacao. *Fitopatologia Brasileira* 5, 329–336.

Huang, S.P., Resende, I.C., De Souza, P.E. and Campos, V.P. (1983) Effect of aldicarb, ethoprop and carbofuran on control of coffee root-knot nematode, *Meloidogyne exigua*. *Journal of Nematology* 15, 510–514.

Huettel, R.N., Kaplan, D.T. and Dickson, D.W. (1986) Characterization of a new burrowing nematode population, *Radopholus citrophilus*, from Hawaii. *Journal of Nematology* 18, 50–54.

Hughes, M.B. (1958) Subsoiling of tomatoes and its relation to nematocide treatment (tests covered the years 1955–1956). *Proceedings of the Association of Southern Agricultural Workers, 55th Annual Convention (1958)*, pp. 152–153.

Huijsman, C.A. (1961) The influence of resistant potato varieties on the soil population of *Heterodera rostochiensis* Woll. *Nematologica* 6, 177–180.

Hung, C.H. (1956) [Experiments on kenaf and jute root-knot nematode control.] *Agricultural Research, Taiwan* 6, 25–28.

Hung, C.L.P. and Jenkins, W.R. (1969) *Criconemoides curvatum* and the peach tree decline problem. *Journal of Nematology* 1, 12.

Hussaini, S.S. (1989) Comparative efficacy of Durofume and other fumigants against root-knot nematode and weeds in FCV tobacco nurseries. *Indian Journal of Nematology* 19, 14–17.

Hussey, R.S. (1977) Effects of subsoiling and nematicides on *Hoplolaimus columbus* populations and cotton yield. *Journal of Nematology* 9, 83–86.

Hussey, R.S. and Boerma, H.R. (1992) Tolerance in soybean. In: Riggs, R.D. and Wrather, J.A. (eds) *Biology and Management of the Soybean Cyst Nematode.* St Paul, Minnesota, USA.

Hutton, D.G., Wahab, A.H. and Murray, H. (1982) Yield response of yellow yam (*Dioscorea cayenensis*) after disinfesting planting material of *Pratylenchus coffeae. Turrialba* 32, 493–496.

Ibrahim, S.K. and Perry, R.N. (1994) Infectivity and population dynamics of the rice stem nematode, *Ditylenchus angustus. Nematologica* 40, 412–422.

Ichinohe, M. (1984) Integrated control of the root-knot nematode, *Meloidogyne incognita*, on black pepper plantations in the Amazonian region. *Agriculture, Ecosystems and Environment* 12, 271–283.

Imran, M. and Saxena, S.K. (1993) Effect of plant materials on the population of tomato root-knot nematode (*Meloidogyne incognita*). *Indian Journal of Mycology and Plant Pathology* 23, 180–182.

Inagaki, H. (1979) Race status of five Japanese populations of *Heterodera glycines. Japanese Journal of Nematology* 9, 1.

Inderjit Singh, Sakhuja, P.K., Sharma, N.K. and Mahajani, R. (1979) Evaluation of Vydate and Lannate as dip treatment against root-knot nematode, *Meloidogyne incognita* on tomato. *Haryana Journal of Horticultural Sciences* 8, 160–161.

Ingham, R.E., Morris, M. and Newcomb, G.B. (1991) Effects of incorporation method of ethoprop and addition of aldicarb on potato tuber infection by *Meloidogyne hapla. Journal of Nematology* 23, 686–692.

Inomoto, M.M. and Monteiro, A.R. (1991) Tratamento quimicode mudas de bananeira Nanicao visando a erradicação de nematoides fitoparasitos. *Nematologia Brasileira* 15, 85–93.

Inserra, R.N., Vovlas, N. and O'Bannon, J.H. (1980) A classification of *Tylenchulus semipenetrans* biotypes. *Journal of Nematology* 12, 283–287.

Inserra, R.N., Di Vito, M. and Ferris, H. (1984a) Influence of *Nacobbus aberrans* densities on growth of sugarbeet and *Kochia* in pots. *Journal of Nematology* 16, 393–395.

Inserra, R.N., Griffin, G.D., Vovlas, N., Anderson, J.L. and Kerr, E.D. (1984b) Relationship between *Heterodera schachtii*, *Meloidogyne hapla* and *Nacobbus aberrans* on sugar beet. *Journal of Nematology* 16, 135–140.

Inserra, R.N., Griffin, G.D. and Anderson, J. (1985) The false root-knot nematode *Nacobbus aberrans. Utah Agricultural Experiment Station, Logan, Utah Research Bulletin* 510, 14 pp.

Inserra, R.N., Vovlas, N., O'Bannon, J.H. and Esser, R.P. (1988) *Tylenchulus graminis* n.sp. and *T. palustris* n.sp. (Tylenchulidae) from native flora of Florida, with notes on *T. semipenetrans* and *T. furcus. Journal of Nematology* 12, 283–287.

Ivanova, B.P. (1971) [The role of selection in disinfection of potato from *Ditylenchus destructor.*] *Byulletin' Vsesoyuznogo Instituta Gel'mintologii im. K.I. Skryabina* No.6, 27–31.

Ivanova, I.V. (1975) [The effect of some anthelmintics on *Meloidogyne incognita* larvae *in vitro.*] *Byulletin' Vsesoyuznogo Instituta Gel'mintologii im. K.I. Skryabina* No. 15, 72–75.

Ivanova, I.V. (1976) [The effect of Nilverm on gall nematode larvae *in vivo* and *in vitro.*] In:

VIII Vsesoyuznoe soveshchanie no nematodnym boleznyam sel' skokhozyaistvennykh kul'tur. Tezisy dokladov i soobshchenii. Kishinev, USSR; Izdatel'stvo 'Shiintsa' Moscow, pp. 102–103.

Jacks, H. (1945) Soil disinfection. IV. Chemical treatment of glasshouse soil. *New Zealand Journal of Science and Technology, Agricultural Research Section* 27, 250–255.

Jacks, H. (1949) Soil disinfection IX. Control of eelworm in outdoor soil. *New Zealand Journal of Science and Technology, Agricultural Research Section* 30, 123.

Jacks, H. (1951) DD a soil fumigant against *H. marioni* in the United Kingdom. *Proceedings of the International Congress of Crop Protection, London,* 1949, 193.

Jacobsen, B.J., MacDonald, D.H. and Bissonette, H.L. (1979) Interaction between *Meloidogyne hapla* and *Verticillium albo-atrum* in the verticillium disease of potato. *Phytopathology* 69, 288–292.

Jaenh, A. and Lambert, M.S. (1983) Uso de la torta de liguerillo como nematicida en almacigos de cafe. *Revista Cafetalera* No. 232, 28–29.

Jaffee, B.A. and Muldoon, A.E. (1989) Suppression of cyst nematode by natural infestation of a nematophagous fungus. *Journal of Nematology* 21, 505–510.

Jagdale, G.B., Pawar, A.B. and Darekar, K.S. (1985) Effect of organic amendments on root-knot nematodes infecting betel vine. *International Nematology Network Newsletter* 2, 7–10.

Jain, B.K. and Bhatti, D.S. (1979) Bare root treatment with systemics for controlling root-knot nematode in tomato transplants. *Indian Journal of Nematology* 8, 19–24.

Jain, R.K. and Bhatti, D.S. (1988) Effect of degradation of neem leaves on incidence of root-knot nematode in tomato. *International Nematology Network Newsletter* 5, 7–9.

Jain, R.K. and Bhatti, D.S. (1989) Integrated control of root-knot nematode *Meloidogyne javanica* infesting tomato. *Indian Journal of Nematology* 18, 30–34.

Jain, R.K. and Bhatti, D.S. (1993) Evaluation of the effective integrated methods for the control of root-knot nematode (*Meloidogyne javanica*) in tomato. *Indian Journal of Nematology* 21, 107–112.

Jain, R.K. and Gupta, D.C. (1990) Integrated management of root-knot nematode (*Meloidogyne javanica*) infecting tomato and okra. *Indian Journal of Nematology* 20, 14–18.

Jain, R.K. and Gupta, D.C. (1992) Evaluation of chemicals as nursery bed treatment for control of root-knot nematode (*Meloidogyne javanica*) infesting tomato and brinjal. *Haryana Agricultural University Journal of Research* 22, 86–90.

Jakobsen, J. and Rasmussen, A.N. (1976) Anvendelse af aldicarb og thioxamyl til bekaempelse af *Pratylenchus penetrans* og *P. vulnus* i rosenkulturer i vaeksthuse. *Tidsskrift voor Planteavl* 80, 510–514.

James, G.L. and Hague, N.G.M. (1967) Methyl bromide fumigation of soil infested by the fungus causing brown root rot complex of tomatoes and by the potato cyst-nematode. *Phytopathology* 16, 148–152.

Janýška, A. (1958) Hád'átko zhoubné (*Ditylenchus dipsaci* Kühn, 1858) na cesneku. *Sbornik Ceskoslovenské Akademie Zemědělských Véd. Rostlinná Výroba Ročník 4 (31) Císlo 4,* 395–410.

Jarnevic, N.B. and Coffee, E.G. (1965) Control of root-knot nematodes with organo-phosphate insecticides. *Plant Disease Reporter* 49, 603–604.

Jatala, P. and Golden, A.M. (1977) Taxonomic status of *Nacobbus* species attacking potatoes in South America. *Nematropica* 7, 9–10.

Jatala, P., Kaltenbach, R. and Bocangel, M. (1979) Biological control of *Meloidogyne incognita acrita* and *Globodera pallida* on potatoes. *Journal of Nematology* 11, 303.

Javed, M., Ahmad, R., Inam-ul-Haq, M. and Mukhtar, T. (1994) Effect of soil solarization on the population of root-knot nematode, *Meloidogyne incognita* and growth of tomato plants. *Pakistan Journal of Phytopathology* 6, 115–119.

Jaworski, C.A., McCarter, S.M., Johnson, A.W. and Williamson, R.E. (1978) Response of onions grown from transplants to soil fumigation. *Journal of the American Society for Horticultural Science* 103, 385–388.

Jellema, P. and Brinkman, H. (1986) De invloed van het tijdstip van aardappelopslagbestrijding op de lokking en vermenigvuldigning van aardappelcystenaaltjes. *Verslagen en Medelingen, Plantenziektenkundige Dienst, Wageningen*, No. 165, 71–74.

Jensen, H.J. and Horner, C.E. (1957) Peppermint decline caused by *Longidorus sylphus* can be controlled by soil fumigation. *Phytopathology* 47, 18.

Johnson, A.B., Kim, K.S., Riggs, R.D. and Scott, H.D. (1993) Location of *Heterodera glycines*-induced syncytia in soybean as affected by soil water regimes. *Journal of Nematology* 25, 422–426.

Johnson, A.W. (1978) Effect of nematicides applied through overhead irrigation on control of root-knot nematodes on tomato transplants. *Plant Disease Reporter* 62, 48–51.

Johnson, A.W. (1985) Specific crop rotation effects combined with cultural practices and nematicides. In: Sasser, J.N. and Carter, C.C. (eds) *An Advanced Treatise on* Meloidogyne *Vol. 1: Biology and Control*, pp. 283–301.

Johnson, A.W. and Campbell, G.M. (1977) Influence of cropping systems and a nematicide on *Meloidogyne* species in tomato transplant production. *Journal of Nematology* 9, 273–274.

Johnson, A.W. and Chalfont, R.B. (1972) Control of nematodes and corn earworm on sweet corn. *Journal of Nematology* 4, 227–228.

Johnson, A.W. and Harmon, S.A. (1974a) Cantaloup yield and grade increased by chemical control of *Meloidogyne incognita*. *Plant Disease Reporter* 58, 746–749.

Johnson, A.W. and Harmon, S.A. (1974b) Lima bean yield increased by chemical control of *Meloidogyne incognita*. *Plant Disease Reporter* 58, 749–753.

Johnson, A.W. and Littrell, R.H. (1970) Pathogenicity of *Pythium aphanidermatum* to chrysanthemum in combined inoculations with *Belonolaimus longicaudatus* or *Meloidogyne incognita*. *Journal of Nematology* 2, 255–259.

Johnson, A.W. and Wright, W.C. (1988) Evaluation of soil nematicides for nematode control growth and yield increase of field corn, 1987. *Fungicide and Nematicide Tests American Phytopathological Society* 43, 166.

Johnson, A.W. and Young, J.R. (1994) Efficacy of fenamiphos formulations applied through irrigation for control of *Meloidogyne incognita* on squash. *Journal of Nematology* 26, 697–700.

Johnson, A.W., Dowler, C.C. and Hauser, E.W. (1975) Crop rotation and herbicide effects on population densities of plant-parasitic nematodes. *Journal of Nematology* 7, 158–168.

Johnson, A.W., McCarter, S.M., Jaworski, C.A. and Williamson, R.E. (1979) Chemical control of nematodes and soil-borne plant-pathogenic fungi on cabbage transplants. *Journal of Nematology* 11, 138–144.

Johnson, A.W., Young, J.R. and Mullinix, B.G. (1981) Applying nematicides through an overhead sprinkler irrigation system for control of nematodes. *Journal of Nematology* 13, 154–159.

Johnson, A.W., Rohde, W.A. and Wright, W.C. (1982) Soil distribution of fenamiphos applied by overhead sprinkler irrigation to control *Meloidogyne incognita* on vegetables. *Plant Disease* 66, 489–491.

Johnson, A.W., Csinos, A.S., Golden, A.M. and Glaze, N.C. (1992) Chemigation for control of black-shank–root knot complex and weeds in tobacco. *Journal of Nematology* 24, 648–655.

Johnson, C.E., Lancaster, D.M., Young, W.A. and Overstreet, C. (1987) Postplant use of

Nemacur for control of ring nematodes in peaches, 1982–1986. *Fungicide and Nematicide Tests. American Phytopathological Society* 42, 161.

Johnson, C.E., Lancaster, D.M., Young, W.A. and Overstreet, C. (1988) Evaluation of Nemacur for control of ring nematodes in peaches, 1982–1987. *Fungicide and Nematicide Tests. American Phytopathological Society* 43, 168.

Johnson, J.T. and Dickson, D.W. (1973) Evaluation of methods and rates of application of three nematicide-insecticides for control of the sting nematode on corn. *Proceedings of the Soil and Crop Science Society of Florida* 32, 171–173.

Johnson, P.W. (1975) Effect of rate and depth of application of nematicide on nematode vertical distribution and tomato production in a sandy loam greenhouse soil. *Canadian Journal of Plant Science* 55, 1017–1021.

Johnson, P.W. and McClanahan, R.J. (1974) Nematode control and other effects of soil treatment of greenhouse rosebeds with aldicarb and oxamyl. *Plant Disease Reporter* 58, 730–732.

Johnson, P.W. and McKeen, C.D. (1973) Vertical movement and distribution of *Meloidogyne incognita* (Nematodea) under tomato in a sandy loam greenhouse soil. *Canadian Journal of Plant Science* 53, 837–841.

Johnson, P.W., Dicks, V.A., Prec, D.J. and Marks, C.F. (1982) Systemic nematicides for control of *Pratylenchus penetrans* during apple orchard establishment in Ontario. *Plant Disease* 66, 299–301.

Johnston, F.E.Jr. and Reynolds, R.B. (1954) Field experiments on control of root knot and weeds with soil fumigants. *Proceedings of the Association of Southern Agricultural Workers* 51, 153.

Johnston, S.A., Probasco, P.R. and Phillips, J.R. (1992) Evaluation of fumigants and nematicides for control of root-knot nematodes on carrot. *Fungicide and Nematicide Tests. American Phytopathological Society* 47, 158–159.

Jonathan, E.I. and Pandiarajan, P. (1991) Effect of organic amendments in controlling rice root nematodes. *International Rice Research Newsletter* 16, 25.

Jonathan, E.I. and Velayutham, B. (1984) Nursery application of carbofuran for control of rice root nematode. *International Rice Research Newsletter* 9, 27.

Jonathan, E.I. and Velayutham, B. (1987) Evaluation of yield loss due to rice root nematode, *Hirschmanniella oryzae*. *International Nematology Network Newsletter* 4, 8–9.

Jonathan, E.I., Nagalakshmi, S. and Padmanabhan, D. (1990) Estimation of yield losses in betelvine due to *Meloidogyne incognita*. *International Nematology Network Newsletter* 7, 26.

Jonathan, E.I., Krishnamoorthy, S.V., Mandharan, M.L. and Muthukrishnan, K. (1991) Effect of organic amendments on the control of sugar cane nematodes. *Bharatiya Sugar* 16, 37, 39–40.

Jones, F.G.W. (1950) Observations on the beet eelworm and other cyst forming species of *Heterodera*. *Annals of Applied Biology* 37, 407–440.

Jones, F.G.W. (1956) Soil populations of beet eelworm (*Heterodera schachtii* Schm.) in relation to cropping II. Microplot and field plot results. *Annals of Applied Biology* 44, 25–56.

Jones, F.G.W. (1970a) The control of the potato cyst-nematode. *Journal of the Royal Society of Arts* March 1970, 179–196.

Jones, F.G.W. (1970b) Integrated control of the potato cyst nematode. *Proceedings of the 5th British Insecticide and Fungicide Conference, 1969 Brighton, UK*, 3, pp. 646–656.

Jones, F.G.W. (1978) Plant nematodes: a neglected group of pests. *University of Leeds Review* 21, 89–116.

Jones, F.G.W., Meaton, V.H., Parrott, D.M., Shepherd, A.M. and King, J.M. (1965) Population studies on pea cyst-nematode, *Heterodera goettingiana* Liebs. *Annals of Applied Biology* 55, 13–23.

Jones, J.M. (1973) Trials on the chemical control of the potato cyst eelworm in the Yorks. and Lancs. Region. *Proceedings of the 7th. British Insecticide and Fungicide Conference, 1973, Brighton, UK*, pp. 425–429.

Jones, J.M. and Uncles, J.J. (1976) Economics of nematicide use for the control of potato cyst nematode in ware crops. *Granular Pesticides. British Crop Protection Council Monograph* No. 18, 35–36.

Jones, J.M., Griffiths, D.J. and Holden, J.H.W. (1955) Varietal resistance in oats to attacks by the stem and bulb eelworm. *Plant Pathology* 4, 35–43.

Jones, R.K. (1979) Control of *Helicotylenchus multicinctus* and *Meloidogyne incognita* infecting bananas by two granular systemic nematicides. *Plant Disease Reporter* 63, 744–747.

Jorgenson, E.C. (1969) Control of the sugarbeet nematode, *Heterodera schachtii*, on sugarbeets with organophosphate and carbamate nematicides. *Plant Disease Reporter* 53, 625–628.

Jorgenson, E.C. (1979) Comparative penetration, development and parasitism of *Meloidogyne incognita* in tolerant, resistant and susceptible cotton at different temperatures. *Journal of Nematology* 11, 304.

Jorgenson, E.C., Hyer, A.H., Garber, R.H. and Smith, S.N. (1978) Influence of soil fumigation on the fusarium–root-knot nematode disease complex of cotton in California. *Journal of Nematology* 10, 228–231.

Joubert, J., and Dalmasso, A. (1974) État des travaux sur le virus du rattle. Méthodes de lutte. In: *Commun. Journ. Etud. Probl. Phytosanit. Actuel. Cult Pomme de Terre. 5 Mars, 1974, Paris, France*, pp. 26–31.

Joubert, J., Dalmasso, A. and Munck-Cardin, M.C. (1971) Étude et contrôle d'un complex pathologique établi sur pomme de terre dans la Sarthe. *Comptes Rendus (Hebdomadaires) des Séances de l'Academie d'Agriculture de France* 57, 1433–1444.

Jung, H.F. and Iwaya, K. (1971) Terracur P, an insecticide and nematocide especially suitable for use in rice crops. *Pflanzenschutz-Nachrichten Bayer* 24, 486–502.

Kaai, C. (1968) Control of stem nematode attack in onions with O,O-diethyl O-2 pyrazinyl' phosphorothioate ('Zinophos') and O-phenyl N,N′ dimethylphosphorodiamide ('Nellite'). *Nematologica* 13, 605–616.

Kaai, C. and Pronk, C.P. (1965) Invloed van de hoeveelheid water, waarmee O,O-diethyl O-2 pyrazinylfosforothioaat wordt toegediend op zijn werking tegen stengelaaltjes in plantuien. *Mededelingen van de Landbouwhogeschool en de Opzrekingsstations van de Staat te Gent* 30, 1475–1487.

Kalita, D.N. and Phukan, P.N. (1993) Efficacy of four chemicals as seed treatment for the control of *Meloidogyne incognita* on black gram. *Current Nematology* 4, 149–151.

Kanwar, R.S. and Bhatti, D.S. (1992) Controlling root-knot nematode, *Meloidogyne javanica* through crop rotation. *Current Nematology* 3, 65–70.

Kaplan, D.T. (1982) Eradication of *Pratylenchus coffeae* in 'Carrizo' citrange seedlings. *Proceedings of the Florida State Horticultural Society* 95, 70–72.

Kaplan, D.T. and O'Bannon, J.H. (1985) Occurrence of biotypes in *Radopholus citrophilus*. *Journal of Nematology* 17, 158–162.

Kaplan, D.T., Keen, N.T. and Thomason, I.J. (1980) Association of glyceollin with the incompatible response of soybean roots to *Meloidogyne incognita*. *Physiological Plant Pathology* 16, 309–318.

Karg, W. (1983) Verbreitung und Bedeutung von Raubmilben des Cohors Gamasina als Antagonisten von Nematoden. *Pedobiologia* 25, 419–432.

Kasuya, S., Sakura, T. and Kishi, Y. (1990) [Selective breeding of resistant pines against pinewood nematode: resistance of the seedlings and the graftings from resistant mother trees.] *Bulletin of the Tokyo University Forests No. 83*, 19–30.

Kaul, R.K. and Sethi, C.L. (1988) Effect of time of application of certain systemic nematicides on final cyst production and fecundity of *Heterodera zeae*. *Indian Journal of Nematology* 18, 134–135.

Kauri-Pääsuke, M. (1973a) Biologisk bekampning av jortrötthet i plantskdejord. *Lantbrukshögskolans Meddelanden Serie A* No. 190, 11 pp.

Kauri-Pääsuke, M. (1973b) Täckgröder som värdväxter för rotsårnnematode *Pratylenchus penetrans*. *Lantsbrukshogskolans Meddelanden Serie A* No. 189,17 pp.

Kaushal, K.K. and Seshadri, A.R. (1986) Evaluation of chemicals as seed-dressers against *Heterodera avenae* on wheat. *Indian Journal of Nematology* 16, 140.

Kaushik, H.D. and Bajaj, H.K. (1981) Control of root-knot nematode *Meloidogyne javanica* infesting mung bean and gram by seed treatment. *Haryana Agricultural University Journal of Research* 11, 106–108.

Keefer, R.F., Hickman, C.E. and Adams, R.E. (1978) The response of strawberrry yields to soil fumigation and nitrogen fertilization. *Horticultural Science* 13, 51–52.

Keetch, D.P. and Webster, G.S. (1977) Experimental control of pineapple nematodes. *Information Bulletin, Citrus and Subtropical Fruit Research Institute* No. 55, 14–16.

Keetch, D.P., Reynolds, R.E. and Mitchell, I.A. (1976) An evaluation of pre- and post-plant nematicides for the control of plant parasitic nematodes in bananas. *Citrus and Subtropical Fruit Journal* No. 506, 5–7.

Kermarrec, A., Degras, L. and Anais, A. (1988) Une grave maladie parasitaire de l'igname Grosse-Caille (*Dioscorea cayenensis rotundata)* due au nématode des lesions racinaires (*Pratylenchus coffeae*). *Bulletin Agronomique Antilles-Guyane* No. 7, 36–38.

Kerr, A. (1963) Fumigation of nursery soil. *Tea Quarterly* 34, 150–151.

Kerr, A. and Vythilingam, M.K. (1967) The occurrence and control of the root-lesion eelworm (*Pratylenchus loosi*) in nurseries. *Tea Quarterly* 38, 22–28.

Kerry, B.R. (1978) Natural control of the cereal cyst-nematode by parasitic fungi. *Agricultural Research Council, UK, Research Review* 4, 17–21.

Kerry, B.R. and Crump, D.H. (1980) Two fungi parasitic on females of cyst-nematodes (*Heterodera* spp.). *Transactions of the British Mycological Society* 74, 119–125.

Kerry, B.R., Crump, D.H. and Mullen, L.A. (1982a) Studies of the cereal cyst-nematode, *Heterodera avenae*, under continuous cereals, 1974–1978. I. Plant growth and nematode multiplication. *Annals of Applied Biology* 100, 477–487.

Kerry, B.R., Crump, D.H. and Mullen, L.A. (1982b) Studies of the cereal cyst-nematode, *Heterodera avenae*, under continuous cereals, 1975–1978 II. Fungal parasitism of nematode females and eggs. *Annals of Applied Biology* 100, 489–499.

Khan, M., Ayub, M., Jahangir and Mohibullah (1995) Efficacy of different organic amendments in the control of root knot nematodes on tomato in Malakand Agency. *Pakistan Journal of Nematology* 13, 25–28.

Khan, T.A., Azam, M.F. and Husain, S.I. (1985) Effect of fungal filtrates of *Aspergillus niger* and *Rhizoctonia solani* on penetration and development of root-knot nematodes and the plant growth of tomato var. Marglobe. *Indian Journal of Nematology* 14, 106–109.

Kheir, A.M., Farnhat, A.A. and Abadir, S.K. (1989) Studies of the corn cyst nematode (CCN) *Heterodera zeae* in Egypt. IV. Variation in development and reproduction of four different populations in some corn cultivars. *Pakistan Journal of Nematology* 7, 69–73.

Khoshkoo, N., Hedin, P.A. and McCarty, J.C. Jr. (1994) Terpenoid aldehydes in root-knot nematode susceptible and resistant cotton plants. *Journal of Agricultural and Food Chemistry* 42, 204–208.

Khurramov, Sh. Kh. (1990) [The use of solarization in control of gall-forming nematodes.] *Zaschita Rastenii (Moskva)* No. 4, 12–13.

Kiel, W. (1969) Populationsdynamik des Kartoffelnematoden (*Heterodera rostochiensis* Woll.) bei mehrmaligem Anbau nematoden-resistenter Kartoffeln zur Festellung der

Biotypenentwicklung. *Nachrichtenblatt für den Deutschen Pflanzenschutzdienstes* 23, 189–192.

Kim, D.G. (1986) Heat-treating wood chips: a possible solution to pinewood nematode contamination. *Tappi Journal* 69, 97–98.

Kim, D.G., Riggs, R.D. and Kim, K.S. (1992) Ultrastructure of *Heterodera glycines* parasitized by Arkansas fungus 18. *Phytopathology* 82, 429–433.

Kim, J.I. and Han, S.C. (1988) [Effect of solarization for control of root-knot nematode (*Meloidogyne* spp.) in vinyl house.] *Korean Journal of Applied Entomology* 27, 1–15.

Kim, Y.H. (1986) Structural modifications associated with resistance and susceptibility of selected plants infested with *Heterodera glycines* Ichinohe. *Dissertation Abstracts International B. (Sciences and Engineering)* 46, 3293B.

Kimpinski, J. and Sanderson, J.B. (1989) Effect of aldicarb, aldoxycarb and oxamyl on potato tuber yields and root lesion nematodes. *Canadian Journal of Plant Science* 69, 611–615.

Kimura, Y., Mori, M., Suzuki, A. and Kobayashi, A. (1981) Isolation and identification of two nematicidal substances from roots of *Erigeron philadelphicus* L. and nematicidal activities of their related compounds. *Agricultural and Biological Chemistry* 45, 2915–2917.

Kinloch, R.A. (1974) Responses of soybean cultivars to nematicidal treatments of soil infested with *Meloidogyne incognita*. *Journal of Nematology* 6, 7–11.

Kinloch, R.A. (1986) Soybean yield as related to rates of 1,3-dichloropropene applied at planting for management of root-knot disease. *Journal of Nematology* 18, 464–467.

Kinloch, R.A. and Dickson, D.W. (1991) Comparison of winter and spring soil fumigation with 1,3-D for the management of *Meloidogyne arenaria* on peanut. *Journal of Nematology* 23, 682–685.

Kirkpatrick, T.L. and Sasser, J.N. (1983) Parasitic variability of *Meloidogyne incognita* populations on susceptible and resistant cotton. *Journal of Nematology* 15, 302–307.

Kirkpatrick, T.L. and Sasser, J.N. (1984) Crop rotation and races of *Meloidogyne incognita* in cotton root-knot management. *Journal of Nematology* 16, 323–328.

Kiryanova, E.S. and Krall, E.L. (1971) *Plant Parasitic Nematodes and their Control, Vol II.* Nauka Publisher, Leningrad Section, Leningrad (English translation, Amerind Publishing Co. Pvt. Ltd., New Delhi).

Kissler, J.J., Lider, J.V., Raabe, R.D., Raski, D.J., Schmitt, R.V. and Hurlimann, J.H. (1973) Soil fumigation for control of nematodes and oak root fungus in vineyard replants. *Plant Disease Reporter* 57, 115–119.

Kiyohara, T. (1984) Pine wilt resistance induced by prior inoculation with avirulent isolate of *Bursaphelenchus xylophilus*. In: Dropkin, V. (ed.) *Proceedings of the US–Japanese Seminar: The Resistance Mechanisms of Pines against Pine Wilt Disease, 7–11 May 1984 East-West Center, Honolulu, Hawaii*, pp. 178–186.

Kiyohara, T. and Tokushige, Y. (1971) [Inoculation experiments of a nematode, *Bursaphelenchus* sp., onto pine trees.] *Journal of the Japanese Forestry Society* 53, 210–218.

Klinkenberg, C.H. (1955) Nematode diseases of strawberries in the Netherlands. *Plant Disease Reporter* 39, 603.

Kobayashi, Y. (1973) [Control of *Meloidogyne* sp. by flooding under high temperature conditions.] In: *17th Annual Meeting of the Japanese Society of Applied Entomology and Zoology, 3–5 April, 1973, Nagano, Japan*, 147.

Kochba, J. and Samish, R.M. (1972) Effect of growth inhibitors on root-knot nematodes in peach roots. *Journal of the American Society for Horticultural Science* 97, 178–180.

Koenning, S.R., Schmitt, D.P. and Barker, K.R. (1993) Effects of cropping systems on population density of *Heterodera glycines* and soybean yield. *Plant Disease* 77, 780–786.

Konova, M.E. and Vinnichuck, R.I. (1959) The disinfection of rice seed from *Aphelenchoides oryzae* Yokoo. *Trudi Gel'mintologicheski Laborotoii Akademiya Nauk SSSR* 9, 130–132.

Kontaxis, D.G. and Hagemann, R.W. (1981) Management of root-knot nematode in Rillito soybeans. *California Agriculture* 35, 18.

Kontaxis, D.G. and Thomason, I.J. (1978) Chemical control of *Heterodera schachtii* and sugar beet production in Imperial Valley, California. *Plant Disease Reporter* 62, 79–82.

Korayem, A.M. and El-Sisi, A.G. (1989) Iron and zinc as activator elements to oxamyl toxicity against the root-knot nematode, *Meloidogyne incognita. Pakistan Journal of Nematology* 7, 27–31.

Kornobis, S. (1989) Wplyw azydku sodu na nicienie glebowe w warunkach polowych. *Zeszyty Problemowe Postępów Nauk Rolniczych*, No. 358, 75–84.

Kort, J. (1972) Nematode diseases of cereals of temperate climates. In: Webster, J.M. (ed.) *Economic Nematology*, Academic Press, London, 97–126.

Kort, J., Ross, H., Rumpenhorst, H.J. and Stone, A.R. (1977) An international scheme for identifying and classifying pathotypes of potato cyst-nematodes *Globodera rostochiensis* and *G. pallida*. *Nematologica* 23, 333–339.

Korunić, Z. (1975) Suzbijanje kupusne nematode – *Heterodera cruciferae* F. nemaatocidima u rasadniku kupusa. *Agronomski Glasnik* 37, 35–40.

Koshy, P.K. and Jay, T. (1991) Host preference of the burrowing nematode, *Radopholus similis* populations from India. *Indian Journal of Nematology* 21, 39–51.

Koshy, P.K. and Sosamma, V.K. (1987) Pathogenicity of *Radopholus similis* on coconut (*Cocos nucifera* L.) seedlings under greenhouse and field conditions. *Indian Journal of Nematology* 17, 108–118.

Koshy, P.K. and Sosamma, V.K. (1988) Studies on *Radopholus similis* on coconut. *Annual Report for 1987. Central Plantation Crops Research Institute, Kasaragod. Kerala, India*, 79 pp.

Koshy, P.K. and Swarup, G. (1971) Investigations on the life history of the pigeon-pea cyst nematode, *Heterodera cajani*. *Indian Journal of Nematology* 1, 44–51.

Koshy, P.K., Sundararaju, P., Sosamma, V.K. and Ravikumar, K. (1986) Efficacy of four systemic nematicides against *Radopholus similis* in coconut nursery. *Indian Journal of Nematology* 15,148–151.

Koura, F.H. and El-Sharkawi, F. (1978) Nematicidal efficacy of aldicarb (Temik 10G) in the control of *Hoplolaimus columbus* Sher, 1963 infection on Egyptian cotton. *Proceedings of the 4th Congress of Pest Control, 30 Sept–3 Oct., part II*, Cairo, 926–930.

Krishnaprasad, K.S. and Krishnappa, K. (1981) Effect of bare-root dip treatment with pesticides on *Rotylenchulus reniformis* affecting brinjal. *Indian Journal of Nematology* 11, 154–158.

Kühn, J. (1881) Bericht über die Ergebnisse der im Auftrage des Vereins für Rübenzücker-Industrie des Deutschen Reiches ausgefuhrten Versuche zur Ernittelung der Ursache der Rübenmüdigkeit des Bodens und zur Erforschung der Natur der nematoden. *Rer. physiol. Lab. Univ. Hallevd.* 3, 1–153.

Kumar, A.C. (1982) Evaluation of Nemacur 5G. against the coffee root-lesion nematode, *Pratylenchus coffeae*. *Journal of Coffee Research* 12, 1–7.

Kumar, S. and Nanjan, K. (1984) Control of spiral nematodes in Patchouli, 1981. *Fungicide and Nematicide Tests. American Phytopathological Society* 39, 93.

Kuroda, K. and Ito, S. (1992) [Migration speed of pine wood nematodes and activities of other microbes during the development of pine-wilt disease in *Pinus thunbergii*.] *Journal of the Japanese Forestry Society* 74, 383–389.

Kuroda, K., Yamada, T. and Ito, S. (1991) *Bursaphelenchus xylophilus* induced pine wilt: factors associated with resistance. *European Journal of Forest Pathology* 21, 430–438.

Küthe, K. (1978) Auswirkungen von Saatreihen-behandelung mit Isektiziden/ Nematiziden bei Rüben und Mais. *Gesunde Pflanzen* 30, 35–41.

La Mondia, J.A. and Brodie, B.B. (1984) Control of *Globodera rostochiensis* by solar heat. *Plant Disease* 68, 474–476.

Lal, A., Sanwal, K.C. and Mathur, V.K. (1983) Changes in the nematode population of undisturbed land with the introduction of land development practices and cropping sequences. *Indian Journal of Nematology* 13, 133–140.

Lamberti, F., Vovlas, N. and Tirro, A. (1976) Infettivita e patogenicita di tre popolazioni italiane di *Tylenchulus semipenetrans* su agrumi ed altri ospiti. *Nematologia Mediterranea* 4, 85–91.

Lamberti, F., Ekanayake, H.M.R.K., Sasanelli, N. and Zacheo, F. (1993) Control of root-knot nematodes in Sri Lanka. *Nematologia Mediterranea* 21, 227–234.

Lamers, J.G. and Roosjen, J. (1993) De waardplantgeschiktheid van groenbemestings-gewassen voor het Noordelijk wortl-knobbelaaltje. *Verlag-Proefstation voor de Akkerbouw en de Groenteteelt in de Vollegrond* No. 163, 40 pp.

Lane, A. (1984) *Bulb pests.* Reference Book 51, Ministry of Agriculture, Fisheries and Food, HMSO, London.

Lane, A. and Saynor, M. (1983) Evaluation of chemical dip treatments for the control of stem nematodes in tulips. *Annals of Applied Biology* 102 (Suppl), 32–33.

Lawn, D.A. and Noel, G.R. (1986) Field interrelationships among *Heterodera glycines, Pratylenchus scribneri* and three other nematode species associated with soybean. *Journal of Nematology* 18, 98.

Lear, B. (1947) Ethylene dibromide – a promising new soil fumigant. *Phytopathology* 37, 14.

Lear, B. and Johnson, D.E. (1962) Treatments for eradication of *Ditylenchus dipsaci* in cloves of garlic. *Plant Disease Reporter* 46, 635–639.

Lear, B. and Lider, L.A. (1959) Eradication of root-knot nematodes from grapevine rootings by hot water. *Plant Disease Reporter* 43, 314–317.

Lear, B., Mai, W.F., Feldmesser, J. and Spruyt, F.J. (1952) Soil fumigation experiments on Long Island, New York to control golden nematode of potatoes. *Phytopathology* 42, 193–196.

Lear, B., Sciaroni, R.H., Atlee, C.B. Jr. and Hart, W.H. (1961) Yield response of Brussels sprouts to soil fumigation for control of sugar-beet nematode, *Heterodera schachtii. Plant Disease Reporter* 45, 739–741.

Lear, B., Johnson, D.E., Miyagawa, S.T. and Sciaroni, R.H. (1965) Response of Brussels sprouts to applications of nematicides in transplant water for control of sugarbeet and cabbage root nematode. *Plant Disease Reporter* 49, 900–902.

Lear, B., Johnson, D.E., Miyagawa, S.T. and Sciaroni, R.H. (1966a) Yield response of Brussels sprouts associated with control of sugar beet nematode and cabbage root nematode in combination with the club root organism, *Plasmodiophora brassicae. Plant Disease Reporter* 50, 133–135.

Lear, B., Miyagawa, S.T., Johnson, D.E. and Atlee, C.B. Jr. (1966b) The sugar beet nematode associated with reduced yields of cauliflower and other vegetable crops. *Plant Disease Reporter* 50, 611–612.

Lebbink, G., Proper, B. and Nipshagen, A. (1989) Accelerated degradation of 1,3-dichloropropene. *Acta Horticulturae* No. 255, 361–371.

Leclercq, D. and Caubel, G. (1991) Résistance variétale de la luzerne au nématode des tiges *Ditylenchus dipsaci* (Kühn) Filipjev; test d'evaluation et application en sélection. *Agronomie* 11, 603–612.

Lee, Y.B., Park, J.S. and Han, S.C. (1972) [Studies on the chemical control of white-tip nematode, *Aphelenchoides besseyi* Christie, before transplanting.] *Korean Journal of Plant Protection* 11, 37–40.

Lee, Y.B., Han, S.C. and Park, J.S. (1976) [On the control effects of some chemicals to rice white-tip nematode (*Aphelenchoides besseyi*) after transplanting of rice.] *Korean Journal of Plant Protection* 15, 193–197.

Leela, N.K., Khan, R.M., Reddy, P.P. and Nidiry, E.S.J. (1992) Nematicidal activity of essential oil of *Pelargonium graveolens* against the root-knot nematode *Meloidogyne incognita Nematologia Mediterranea* 20, 57–58.

Leij, F.A.A.M. de, Davies, K.G. and Kerry, B.R. (1992) The use of *Verticillium chlamydosporium* Goddard and *Pasteuria penetrans* (Thorne) Sayre and Starr alone and in combination to control *Meloidogyne incognita* on tomato plants. *Fundamental and Applied Nematology* 15, 235–242.

Lembright, H.W., Healy, M.J. and Norris, M.G. (1968) Improved soil fumigant utilization through deep placement. *Biokemia* no. 17, 33–35.

Lewis, G.C., Clements, R.O. and Whitehead, A.G. (1985) Effect on white clover establishment of carbofuran or aldicarb. *Annals of Applied Biology* 106, *Tests of Agrochemicals and Cultivars* 6, 122–123.

Lewis, G.D. and Mai, W.F. (1958) Chemical control of *Ditylenchus dipsaci* (Kühn) Filipjev in organic soils of southern New York. *Plant Disease Reporter* 42, 1360–1363.

Lewis, G.D. and Mai, W.F. (1963) Temperature and moisture in relation to the toxicity of 1,3-dichloropropene to *Ditylenchus dipsaci* (Kühn) Filipjev in organic soil. *Plant Disease Reporter* 47, 1097–1101.

Lewis, S. (1979) Control of stem nematode (*Ditylenchus dipsaci*) in strawberry by granular nematicides. *Plant Pathology* 28, 191–196.

Lewis, S. and Stone, L.E.W. (1964) The use of Trapex for control of potato root eelworm out-of-doors. *Plant Pathology* 13, 110–111.

Lewis, S.A. and Fassuliotis, G. (1982) Lance nematodes, *Hoplolaimus* spp. in the southern region of the United States. *Southern Cooperative Service Bulletin 276, University of Arkansas, Fayetteville*, 127–138.

Lewis, S.A., Smith, F.H. and Powell, W.M. (1976) Host-parasite relationships of *Hoplolaimus columbus* on cotton and soybean, *Journal of Nematology* 8, 141–145.

Li, Z.Z., Wang, Z.Y. and Peng, L.H. (1994) Experiments on the toxicity of several insecticides on *Bursaphelenchus xylophilus* Nickle. *Forest Pest and Disease No. 1*, 23–26.

Libman, G., Leach, J.G. and Adams, R.E. (1964) Role of certain plant-parasitic nematodes in infection of tomatoes by *Pseudomonas solanacearum*. *Phytopathology* 54, 151–153.

Liebscher, G. (1878) Die Rübenmudigkeit des Ackers herforgerufen durch *Heterodera schachtii*. *Zeitschrift Ver. Rübenzucker-Industrie Zollver.* 28, 893–895.

Liebscher, G. (1890) Eine Nematoden als Ursache der Erbsennündigklit des Bodens. *Deutsch Landwirtschaftliche Presse*, 17, 477.

Linford, M.B. and Yap, F. (1940) Some host plants of the reniform nematode in Hawaii. *Proceedings of the Helminthological Society of Washington* 7, 35–42.

Liu, H.Q. (*et al.*) (1988) [Present status of the occurrence, injury and research of soybean cyst nematodes in Heilongjiang Province.] *Soybean Science* 6, 141–150.

Liu, W., Liu, Y., Duan, Y. and Wu, A. (1993) [Selection effect of resistant soybeans on the race 3 of soybean cyst nematode *Heterodera glycines*.] *Acta Phytophylactica Sinica* 20, 135–137.

Locascio, S.J., Smart, G.C. Jr. and Marvel, M.E. (1967) Control of bud nematode on strawberry. *Proceedings of the Florida State Horticultural Society* 79, 170–175.

Loos, C.A. (1961) Eradication of the burrowing nematode, *Radopholus similis*, from bananas. *Plant Disease Reporter* 45, 457–461.

Loos, C.A. and Loos, S.B. (1960) Preparing nematode-free banana 'seed'. *Phytopathology* 50, 383–386.

Lordello, L.G.E. and Brancalion, A.M. (1986) Avaliaçao de nematicida Furadan 350 FMC no controle do nematóide *Rotylenchulus reniformis* atacando algodoeiro. *Nematologia Brasileira* 10, 103–107.

Lordello, R.R.A., Sawazaki, E., Lordello, A.I.L. and Aloisi Sobrinho, J. (1983) Controle de *Pratylenchus* spp. em milho, com nematicidas sistémicos com torta de mamona. In: *Trabalhos Apresentados a VII Reunião Brasileira de Nematologia, Brasilia, DF, 21–25 de Fevereiro de 1983. Publicação* No. 7, 241–250.

Lownsbery, B.F. (1959) Studies of the nematode, *Criconemoides xenoplax*, on peach. *Plant Disease Reporter* 43, 913–917.

Lownsbery, B.F. and Sher, S.A. (1957) Soil fumigation for the control of the root lesion disease of walnuts. *Phytopathology* 47, 527.

Lownsbery, B.F. and Sher, S.A. (1958) Root lesion nematode on walnut. *California Agriculture* 12, 7, 12.

Lownsbery, B.F., Hart, H.W. and Martin, G.C. (1969) Procedures for controlling root-lesion nematodes. *Diamond Walnut News* 51, 5–6.

Lownsbery, B.F., English, H., Moody, E.H. and Schick, F.J. (1973) *Criconemoides xenoplax* experimentally associated with a disease of peach trees. *Phytopathology* 63, 994–997.

Lownsbery, B.F., Martin, B.C., Forde, H.I. and Moody, E.H. (1974a) Comparative tolerance of walnut species, walnut hybrids and wingnut to the root-lesion nematode, *Pratylenchus vulnus*. *Plant Disease Reporter* 58, 630–633.

Lownsbery, B.F. Moody, E.H. and Brown, A.J. (1974b) Plant-parasitic nematodes in California prune orchards. *Plant Disease Reporter* 58, 633–636.

Luc, M. and Vilardebó, A. (1961) Les nématodes associés aux bananiers cultivés dans l'ouest Africain. *Fruits* 16, 205–219.

Lucas, L.T. (1979) Control of *Belonolaimus longicaudatus* on bermuda grass golf greens. *Phytopathology* 69, IA6 (abstract).

Lucas, L.T. (1982) Population dynamics of *Belonolaimus longicaudatus* and *Criconemella ornata* and growth response of bermuda grass and overseeded grasses on golf greens following treatment with nematicides. *Journal of Nematology* 14, 358–363.

Luedders, V.D. and Dropkin, V.H. (1983) Effect of secondary selection on cyst nematode reproduction on soybeans. *Crop Science* 23, 263–264.

Lundin, P. (1969) Breeding of lucerne for resistance to stem nematode and *Verticillium* wilt. *Sveriges Utsädesförenings Tijdskrift* 79, Suppl. 133–137.

Lung, G. (1987) Wirtspflanzen-Parasit-Interaktion bei zystenbildende Nematoden. Neue Einblicke mit Ubersicht über die verschiedenen Rezistenzmechanismen. *Mededelingen van de Faculteit Landbouwwetenschappen, Rijksuniversiteit, Gent* 52, 593–606.

Lung, G. and Moltmann, E. (1984) Neue Aspekte uber Resistenzmechanismen von Getreidearten gegüber dem Getreiezystenälchen (*Heterodera avenae*). *Mitteilungen, Biologische Bundesanstalt für Land- und Forstwirtschaft, Berlin-Dahlem* No. 223, 205–206.

Maas, P.W.T. (1975) Soil fumigation and crop rotation to control spraing disease in potatoes. *Netherlands Journal of Plant Pathology* 81, 138–143.

Madulu, J.D., Trudgill, D.L. and Philipps, M.S. (1994) Rotational management of *Meloidogyne javanica* and effects on *Pasteuria penetrans* and tomato and tobacco yields. *Nematologica* 40, 438–455.

Maggenti, A.R. and Hart, W.H. (1970) Soil treatments for root-knot nematode control in fruit tree nursery growing grounds. *Plant Disease Reporter* 54, 1014–1016.

Maggenti, A.R. and Hart, W.H. (1975) Carbamate and phosphate nematicidal granules, drenches and dips for the control of *Ditylenchus destructor* on bulbous iris, variety Wedgewood. *Plant Disease Reporter* 59, 233–235.

Maggenti, A.R., Hart, W.H. and Lenz, J.V. (1967) Soil treatments for the control of root-lesion nematode (*Pratylenchus penetrans*) on Easter lily. *Plant Disease Reporter* 51, 549–552.

Mahajan, R., Kaur, D.J. and Bajaj, K.L. (1992) Nematicidal activity of phenolic compounds against *Meloidogyne incognita*. *Nematologia Mediterranea* 20, 217–219.

Mahanta, B. and Phukan, P.N. (1992) Efficacy of certain chemicals as seed treatments for the control of *Meloidogyne incognita* on black gram. *Current Nematology* 3, 27–28.

Mahmood, I. and Saxena, S.K. (1986) Relative susceptibility of different cultivars of tomato to *Rotylenchulus reniformis* in relation to changes in phenolics. *Revue de Nématologie* 9, 89–91.

Mahmood, I. and Saxena, S.K. (1987) Effect of water extract of oil-cakes and de-oiled cakes on the mortality of *Rotylenchulus reniformis* Linford and Oliveira, 1940. *Indian Journal of Plant Pathology* 5, 184–187.

Mai, W.F. (1972) Lesion nematodes and tree fruit production in New York State. *Proceedings of the Ohio State Horticultural Society* 125, 91–94.

Mai, W.F. (1977) Worldwide distribution of potato cyst-nematodes *Heterodera rostochiensis* and *H. pallida* and their importance in crop production. *Journal of Nematology* 9, 30–34.

Mai, W.F. and Lear, B. (1952) Yearly D-D treatment and continuous potato production in relation to the golden nematode population of the soil. *Phytopathology* 42, 481.

Mai, W.F. and Parker, K.G. (1956) Evidence that the nematode *Pratylenchus penetrans* causes losses in New York State cherry orchards. *Phytopathology* 46, 19.

Mai, W.F. and Parker, K.G. (1967) Root diseases of fruit trees in New York State. I. Populations of *Pratylenchus penetrans* and growth of cherry in response to soil treatment with nematicides. *Plant Disease Reporter* 51, 398–401.

Mai, W.F. and Parker, K.G. (1970) Controlling nematodes increases growth and yield of apples and cherries. *Proceedings of the New York State Horticultural Society* 115, 207–209.

Mai, W.F. and Parker K.G. (1972) Root diseases of fruit trees in New York State. IV. Influence of preplant treatment with a nematicide, charcoal from burned brush, and complete mineral nutrition on growth and yields of sour cherry trees and numbers of *Pratylenchus penetrans*. *Plant Disease Reporter* 56, 141–145.

Mai, W.F., Bloom, J.R. and Chen, T.A. (eds.) (1977) Biology and ecology of the plant-parasitic nematode *Pratylenchus penetrans*. *Bulletin of the Pennsylvania State University Agricultural Experiment Station* No. 815, 66 pp.

Mai, W.F., Brodie, B.B., Harrison, M.B. and Jatala, P. (1981) Nematode diseases of potatoes. In: Hooker, W.J. (ed.) *Compendium of Potato Diseases*. American Phytopathological Society, St Paul, Minnesota, pp. 93–101.

Malaguti, G. (1953) 'Pudricion de Cogollo' de la palmera do aceite Africana (*Elaeis guineensis* Jacq.) en Venezuela. *Agronomia Tropical* 3, 13–21.

Malek, R.B. (1969) Population fluctuations and observations of the life cycle of *Xiphinema americanum* associated with cotton wood (*Populus deltoides*) in South Dakota. *Proceedings of the Helminthological Society of Washington* 36, 270–274.

Malek, R.B. and Jenkins, W.R. (1964) Aspects of the host-parasite relationships of nematodes and hairy vetch. *Bulletin of the New Jersey Agricultural Experiment Station* No. 813, 31 pp.

Mamiya, Y. (1972) Pinewood nematode, *Bursaphelenchus lignicolus* Mamiya and Kiyohara, as a causal agent of pine wilting disease. *Review of Plant Protection Research* 5, 46–60.

Mamiya, Y. (1984) The pinewood nematode. In: Nickle, W.R. (ed.) *Plant and Insect Nematodes*. Marcel Dekker Inc., New York, pp. 589–626.

Mancini, G., Villani, R., Rameri, G. and Russo, L. (1974) Prova di lotta contro i nematodi galligeni del cetriolo. *Annali della Facoltà di Scienze Agraric Università di Napoli, Portici, IV*, 8, 8 pp.

Mani, A., Prakash, K.S. and Zidgali, T.A. (1993) Comparative effects of soil solarization and nematicides of three nematode species infesting potato. *Current Nematology* 4, 65–70.

Manoharan, A., Vijayaraghavan, S. and Samboornaraman, S. (1978) Effect of treatment of seed tubers of potato to make the potato cyst nematode non-viable – new method of disinfestation. In: *Approaches Towards Increasing the Potato Production in Developing Countries, 20–23 Nov. 1978, Simla, India*, Abstract 57.

Marks, C.F. and Davidson, T.R. (1973) Effects of preplant and postplant nematicides on populations of nematodes in the soil and on the growth of fruit trees in the Niagara Peninsula. *Canadian Plant Disease Survey* 53, 170–174.

Marks, C.F. and Elliot, J.M. (1976a) Chemical treatment of soil for control of the root-lesion nematode *Pratylenchus penetrans* on flue-cured tobacco. *Tobacco International* 178, 142–144.

Marks, C.F. and Elliot, J.M. (1976b) Control of *Pratylenchus penetrans* in flue-cured tobacco with foliar applications of oxamyl. *Plant Disease Reporter* 60, 172–174.

Marks, C.F., Elliot, J.M. and Tu, C.M. (1972) Effect of deep fumigation on *Pratylenchus penetrans*, flue-cured tobacco, and soil nitrate content. *Canadian Journal of Plant Science* 52, 425–430.

Marks, G.C., Wonoto-Suatmadji, R. and Smith, I.W. (1987) Effects of nematode control on shoot, root and mycorrhizal development of *Pinus radiata* seedlings growing in a nursery soil infested with *Pratylenchus penetrans*. *Australian Forestry Research* 17, 1–10.

Martin, G.C. (1950) Eelworm and weed control combined in seedbeds. Efficiency of methyl bromide. *Rhodesian Tobacco Journal* 2, 35, 37, 39.

Martin, G.C. (1968) Control of *Meloidogyne javanica* in potato tubers and *M. hapla* in the roots of young rose bushes by means of heated water. *Nematologica* 14, 441–446.

Martin, W.J. (1970) Elimination of root-knot and reniform nematodes and scurf infections from rootlets of sweet potato plants by hot water treatment. *Plant Disease Reporter* 54, 1056–1058.

Martin, W.J. and Birchfield, W. (1977) Effects of organophosphate and carbamate nematicides on nematodes, insects and sweet potato yields. *Proceedings of the American Phytopathological Society* 3, 244.

Martins, J.F. da S., Gonçalo, J.F.P. and Lordello, L.G.E. (1976) [Chemical treatment of rice seeds to eradicate *Aphelenchoides besseyi*.] *O Solo* 68, 58–61.

Masamune, T., Anetai, M., Taksugi, M. and Katsui, N. (1982) Isolation of a natural hatching stimulus, glycinoeclepin A, for the soybean cyst nematode. *Nature* 297, 496.

Mathur, B.N., Handa, D.K., Swarup, G., Sethi, C.L., Sharma, G.L. and Yadav, B.D. (1986) On the loss estimation and chemical control of 'Molya' disease of wheat caused by *Heterodera avenae* in India. *Indian Journal of Nematology* 16, 152–159.

Mathur, V.K. and Prasad, S.K. (1973) Survival and host range of the rice root nematode, *Hirschmanniella oryzae*. *Indian Journal of Nematology* 3, 88–93.

Matsuura, K. (1984) [Preventive and therapeutic effects of certain systemic nematicides upon the pine wilt disease caused by pine-wood nematodes, *Bursaphelenchus xylophilus* (I) Chemotherapeutic effect of six systemic nematicides.] *Journal of the Japanese Forestry Society* 66, 1–9.

Matsuura, K., Fujishita, A. and Kishi, Y. (1975) [The effect of trunk injection of chemicals on the pine wilting disease caused by *Bursaphelenchus lignicolus*.] *Transactions of the 86th Meeting, Japanese Forestry Society*, pp. 309–310.

Matsuura, K., Kawasaki, T., Kobayashi, T., Zinno, Y., Mamiya, Y., Tamura, H. and Sasaki, K. (1976) [Trunk injection of chemicals for the control of pine wilting disease caused by *Bursaphelenchus lignicolus*. I. Effect of injected chemicals on the disease development of nematode inoculated pine tree.] *Transactions of the 87th Meeting, Japanese Forestry Society, pp.* 265–266.

Matsuzoe, N., Okubo, H. and Fujieda, K. (1992) Resistance of tomato plants grafted on *Solanum* rootstocks to bacterial wilt and root-knot nematode. *Journal of the Japanese Society for Horticultural Science* 61, 865–872.

May, H.C. and Osborne, R. (1973) Experiments with dazomet for the control of potato cyst eelworm in potato crops. *Proceedings of the 7th British Insecticides and Fungicides Conference, November 19–22, 1973, Brighton, UK*, 2, pp. 409–413.

McBeth, C.W. (1951) Observations in the repeated application of D-D. *Plant Disease Reporter* 35, 243.

McBeth, C.W. (1960) New techniques for application of soil fumigants in the US. In: *First Inter-African Plant Nematology Conference, CCTA / CSA Publication* No. 86, 19–20.

McBeth, C.W., Larson, J.E. and Ichikawa, S.T. (1963) 1,6-hexanedithiol acetate as a dip nematocide for bare rooted nursery stock. *Plant Disease Reporter* 47, 432–436.

McBeth, C.W., White, L.V. and Ichikawa, S.T. (1964) A new residual nematicide: 2,4-dichlorophenyl methanesulphonate. *Plant Disease Reporter* 48, 634–635.

McCann, J., Luedders, V.D. and Dropkin, V.H. (1982) Selection and reproduction of soybean cyst nematodes on resistant soybeans. *Crop Science* 22, 78–80.

McCarter, S.M., Jaworski, C.A. and Johnson, A.W. (1978) Effect of continuous plant culture and soil fumigation on soil borne plant pathogens and on growth of tomato transplants. *Phytopathology* 68, 1475–1481.

McCulloch, J. (1978) Strawberry crimp. *Queensland Agricultural Journal* 104, 345–347.

McElroy, F.D. (1973) Control of *Pratylenchus penetrans* in raspberry root cuttings. *Plant Disease Reporter* 57, 492–495.

McElroy, F.D. (1975) Nematode control in established red raspberry plantings. In: Lamberti, F., Taylor, C.E. and Seinhorst, J.W. (eds) *Nematode Vectors of Plant Viruses.* Plenum Press, London, pp. 445–446.

McElroy, F.D. (1977) Effect of two nematode species on establishment, growth, and yield of raspberry. *Plant Disease Reporter* 61, 277–279.

McElroy, F.D. (1985) Post-plant control of nematodes in bare root nurseries. In: *Western Forest Nursery Council–Intermountain Nurseryman's Association Combined Meeting, 14–16 August, 1984, Coeur d'Aleve, Idaho. General Technical Report, Intermountain Research Station, USDA Forest Service (1985)* No. Int-185, 99–101.

McGarvey, B.D., Potter, J.W. and Chiba, M. (1984) Nematostatic activity of oxamyl and N,N-dimethyl-1-cyanoformamide (DMCF) on *Meloidogyne incognita* juveniles. *Journal of Nematology* 16, 328–332.

Mckeen, C.D. and Sayre, R.M. (1964) Control of *Verticillium dahliae* and *Meloidogyne incognita* in greenhouse tomatoes by soil fumigation. *Canadian Journal of Plant Science* 44, 466–470.

McKenry, M.V. and Kretsch, J. (1987) Peach tree and nematode responses to various soil treatments under two irrigation regimes. *Nematologica* 33, 343–354.

McKenry, M.V. and Thomason, I.J. (1974) 1,3-dichloropropene and 1,2-dibromoethane compounds: I Movement and fate as affected by various conditions in several soils. II Organism dosage-response studies in the laboratory with several nematode species. *Hilgardia* 42, 393–438.

McKenry, M.V. and Thomason, I.J. (1975) The dosage-response of *Meloidogyne* infected *Ficus* roots to *cis*-1,3-dichloropropene. In: *International Plant Protection Congress (8th), August 21–26, 1975. Reports and Information, Section III, part II*, Moscow, USSR, pp. 507–514.

McKenry, M.V., Thomason, I.V. and Naylor, P. (1977) Dosage-response of root-knot nematode-infected grape roots to *cis*-1,3-dichloropropene. *Phytopathology* 67, 709–711.

McKenry, M.V., Thomason, I.J., Johnson, D.E., Neja, R. and Swanson, F. (1978) The movement and toxicity of preplant soil fumigants for nematode control. Preplant fumigations with 1,3-D nematicides. *California Agriculture* 32, 12–13.

McLaren, M.K. (1984) Control of nematodes and insects results in higher monetary return in sugar cane. *Revue Agricole et Sucrière de l'Ile Maurice* 63, 233–237.

McLean, K.S. and Lawrence, G.W. (1994) Localized influence of *Heterodera glycines* on sudden death syndrome of soybean. *Journal of Nematology* 25, 674–678.

McLeod, K., Golightly, W.H. and Price, C.D. (1954) 'Tulip root' of oats in Durham. *Agriculture, London* 61, 229–231

McLeod, R.W. and Khair, G.T. (1975) Effects of oximecarbamates, organophosphates and benzimidazole nematicides on life cycle stages of root-knot nematodes, *Meloidogyne* spp. *Annals of Applied Biology* 79, 329–341.

McSorley, R. and Gallacher, R.N. (1992) Managing plant-parasitic nematodes in crop sequences. *Proceedings of the Soil and Crop Science Society of Florida* 51, 42–47.

McSorley, R. and Parrado, J.L. (1983) Control of nematodes on white-fleshed sweet potato with Vydate, 1982. *Fungicide and Nematicide Tests. American Phytopathological Society* 38, 1.

McSorley, R. and Philipps, M.S. (1993) Modelling population dynamics and yield losses and their use in nematode management. In: Evans, K., Trudgill, D.L. and Webster, J.M. (eds) *Plant Parasitic Nematodes in Temperate Agriculture.* CAB International, Wallingford, UK, 61–85.

McSorley, R., McMillan, R.I. Jr. and Parrado, J.L. (1984) *Meloidogyne incognita* on society garlic and its control. *Plant Disease* 68, 166–167.

McSorley, R., McMillan, R.I. Jr. and Parrado, J.L. (1986) Soil fumigants for tomato production on Rockdale soils. *Proceedings of the Florida State Horticultural Society* 98, 232–237.

McSorley, R., Dickson, D.W., Brito, J.A.de, Hewlett, T.E. and Frederick, J.J. (1994) Effect of tropical rotation crops on *Meloidogyne arenaria* population densities and vegetable yields in microplots. *Journal of Nematology* 26, 175–181.

Meagher, J.W., Brown, R.H. and Rovira, A.D. (1978) The effects of cereal cyst nematode (*Heterodera avenae*) and *Rhizoctonia solani* on the growth and yield of wheat. *Australian Journal of Agricultural Research* 29, 1127–1137.

Meher, H,C. (1990) Effect of sebufos on invasion of *Meloidogyne incognita* in brinjal. *Indian Journal of Nematology* 20, 25–28.

Meher, H.C. and Sethi, C.L. (1988) Effect of phenamiphos on hatching and invasion of *Meloidogyne incognita* in cowpea. *Indian Journal of Nematology* 18, 186–191.

Meher, H.C., Sethi, C.L. and Agrihotri, N.P. (1987) Persistence and basipetal movement of aldoxycarb in *Pisum sativum. Indian Journal of Nematology* 17, 33–37.

Mehta, U.K., Natesan, N. and Sundararaj, P. (1994) Screening of sugarcane cultivars to *Pratylenchus zeae* for commercial release. *Afro-Asian Journal of Nematology* 4, 109–111.

Meijnecke, C.A.R.(1959) Soil exhaustion in tree nurseries. *Span* 1, 2–5.

Melin, P. and Vilardebó, A. (1973a) Efficacité de quelques nématicides en bananeraie dans les sols volcaniques de la région du Mungo (Cameroun). *Fruits* 28, 3–17, 80–82.

Melin, P. and Vilardebó, A. (1973b) Nématicide et désinfection à l'eau chaude dans la lutte contre *Radopholus similis* en bananeraie. *Fruits* 28, 843–849, 909–912.

Melin, P. and Vilardebó, A. (1976) Chemical control of *Radopholus similis* on plantains. *Nematropica* 6, 49–50.

Mel'nik, M. and Knaub, V. (1977) [Presoaking of garlic bulbs inoculated with *Ditylenchus dipsaci.*] In: *Treta Natsionalna Konferentsiya po Parasitologiya, 12–14 October, 1977, Albena, Bulgaria*, pp. 102–103.

Mel'nik, M.V. and Knaub, V.Y. (1976) [Experimental control of *Ditylenchus dipsaci* on seed garlic in Moldavia.] In: *VIII Vsesoyuznoe soveshchanie no nematodnym boleznyam sel'skokozyaistrennykh kul'tur. Tezisy dokladov i soobshchenii.* Kishinev, USSR; *Izdatel'stvo 'Shtiintsa', pp.* 81–82.

Metcalfe, G. (1940) *Bacterium rhaponticum* (Millard) Dowson, a cause of crown-rot disease of rhubarb. *Annals of Applied Biology* 27, 502–508.

Metlitskii, O.Z. and Pokhodenko, A.P. (1986) [Combined treatment of berries against nematodes.] *Byulleten Vsesyuznogo Instituta Gel'mintologii im K.I. Skryabina* No. 41, 80–84.

Meyer, A.J. (1978) Why is 'Nemaguard' peach rootstock resistant to root-knot nematodes (*Meloidogyne* spp.) and 'Kakamas' susceptible ? *Deciduous Fruit Grower* 28, 66–72.

Meyer, J.R., Zehr, E.I., Meagher, R.L. Jr. and Salvo, S.K. (1992) Survival and growth of peach trees and pest populations in orchard plots managed with experimental ground covers. *Agriculture, Ecosystems and Environment* 41, 353–363.

Millar, R.M. (1981) Tulip nematode control by hot water treatment. In: *Annual Review 1980, Kirton Experimental Horticulture Station, Boston, UK*, pp. 12–16.

Miller, H.N. and Perry, V.G. (1965) Elimination of nematodes from nursery plants by chemical bare-root dips. *Plant Disease Reporter* 49, 51–53.

Miller, L.I. (1966) Variations in development of two morphologically different isolates of *Heterodera glycines* obtained from the same field. *Phytopathology* 55, 1068.

Miller, L.I. (1986) Economic importance of cyst nematodes in North America. In: Lamberti, F. and Taylor, C.E. (eds) *Cyst Nematodes. NATO ASI Series, Series A: Life Sciences Vol. 121*, Plenum Press, New York, USA, pp. 373–385.

Miller, P.M. (1977) Interaction of plastic, hay and grass mulches, and metam-sodium on control of *Pratylenchus penetrans* in tomatoes. *Journal of Nematology* 9, 350–351.

Miller, P.M. and Hawkins, A. (1969) Long term effects of preplant fumigation of potato fields. *American Potato Journal* 46, 387–397.

Miller, P.M. and Kring, J.B. (1970) Reduction of nematode and insect damage to potatoes by band application of systemic insecticides and soil fumigation. *Journal of Economic Entomology* 63, 186–189.

Miller, P.M. and McIntyre, J.L. (1976) Activated charcoal inhibits the entry of *Pratylenchus penetrans* and *Heterodera tabacum* into tomato and tobacco roots. *Journal of Nematology* 8, 297–298.

Miller, P.M. and Taylor, G.S. (1958) Superior control of the tobacco stunt nematodes with a nematicide mixture. *Phytopathology* 48, 264.

Milne, D.L. (1974) Citrus seedbeds: methyl bromide and mycorrhizae. *Citrus and Subtropical Fruit Journal* No. 488, 9–11.

Milne, D.L. and Villiers, E.A. de (1975) Foliar systemic nematicides for pineapple nematode control. *Journal of Nematology* 7, 327.

Milne, D.L. and Villiers, E.A. de (1978) Control of citrus nematode, *Tylenchulus semipenetrans*, using fenamiphos. *Citrus and Subtropical Fruit Journal* No. 534, 8–9.

Milne, D.L. and Willers, P. (1979a) Yield and nutritional responses due to phenamiphos treatment of citrus infested with citrus nematode. *Citrus and Subtropical Fruit Journal* No. 547, 22–24.

Milne, D.L. and Willers, P. (1979b) Yield and nutritional responses due to phenamiphos treatment of citrus infested with citrus nematode. *Indian Journal of Mycology and Plant Pathology* 8, 6.

Minton, E.B. and Meredith, W.R. Jr. (1987) Root knot nematode effect on nine cotton cultivars in Mississippi. *Crop Science* 27, 1001–1004.

Minton, N.A. and Gillenwater, H.B. (1973) Methyl bromide fumigation of *Pratylenchus brachyurus* in peanut shells. *Journal of Nematology* 5, 147–149.

Minton, N.A. and Parker, M.B. (1975) Interaction of four soybean cultivars with subsoiling and a nematicide. *Journal of Nematology* 7, 60–64.

Minton, N.A., Brenneman, T.B., Bondari, K. and Harrison, G.W. (1993) Activity of fosthiazate against *Meloidogyne arenaria, Frankliniella* spp. and *Sclerotium rolfsii* in peanut. *Peanut Science* 20, 66–70.

Minz, G., Ziv, D. and Strich-Harari, D. (1960) Decline of banana plantations caused by spiral nematodes in the Jordan Valley and its control by DBCP. *Ktavim* 10, 147–157.

Mishra, C., Singh, B. and Laha, S.K. (1987) Integrated approach for root-knot nematode management in jute. *Indian Journal of Nematology* 17, 285–287.

Mishra, S.D. (1986) Control of root-knot nematode, *Meloidogyne incognita*, infesting pigeon pea by nematicidal seed treatment. In: Capoor, V.N., Misra, S.L. and Malhotra, S.K. (eds) *Proceedings of the National Symposium on New Dimensions in Parasitology, 18–19 January, 1986, University of Allahabad*, pp. 34–36.

Mishra, S.D. and Gaur, H.S. (1984) Control of plant parasitic nematodes on chick-pea by seed treatment with nematicides. *Indian Journal of Entomology* 46, 382–384.

Mitchell, B.L., Blair, B.W. and Martin, G.C. (1971) Potatoes in Rhodesia. Part II. Pests of potatoes. *Rhodesia Agricultural Journal Technical Bulletin* No. 11, 25–31.

Moel, C.P. de (1986) Cyst-aaltjes en de teelt van bloemkool. *Groentenen en Fruit* 41, 76–77.

Mohandas, C. and Ramana, K.V. (1987) Slow wilt disease of black pepper and its control. *Indian Cocoa, Arecanut and Spices Journal* 11, 10–11.

Mohandas, C., Rao, Y.S. and Sahu, S.C. (1981) Cultural control of rice root nematodes (*Hirschmanniella* spp.) with *Sphenoclea zeylanica*. *Proceedings of the Indian Academy of Sciences (Animal Sciences)* 93, 373–376.

Mojtahedi, H., Santo, G.S. and Pinkerton, J.N. (1991) Efficacy of ethoprop on *Meloidogyne hapla* and *M. chitwoodi* and enhanced biodegradation in soil. *Journal of Nematology* 23, 372–379.

Mojtahedi, H., Santo, G.S. and Ingham, R.E. (1993a) Suppression of *Meloidogyne chitwoodi* with Sudan grass cultivars as green manure. *Journal of Nematology* 25, 303–311.

Mojtahedi, H., Santo, G.S., Wilson, J.H. and Hang, A.N. (1993b) Managing *Meloidogyne chitwoodi* on potato with rapeseed as green manure. *Plant Disease* 77, 42–46.

Mojtahedi, H., Santo, G.S., Brown, C.R., Ferris, H. and Williamson, V. (1994) A new host race of *Meloidogyne chitwoodi* from California. *Plant Disease* 78, 1010.

Mojumder, V. and Mishra, S.D. (1991) Effect of aqueous extracts of neem seeds on hatchability of eggs and penetrability of hatched juveniles of *Meloidogyne incognita* into roots of mung bean. *Current Nematology* 2, 117–120.

Mondal, A.H., Rahman, M.L. and Miah, S.A. (1990) Incorporation of infested stubble and nematicides in soil and incidence of ufra disease of rice. *Bangladesh Journal of Plant Pathology* 6, 25–27.

Monke, B.J., Zeck, W.M. and Noegel, K.A. (1992) Citrus trunk application of fenamiphos to control *Tylenchulus semipenetrans*. *Proceedings of the Brighton Crop Protection Conference, Pests and Diseases November 23–26, 1992*, 3, pp. 1175–1180.

Montasser, S.A. (1990) Efficacy of certain vitamins in controlling the root-knot nematode, *Meloidogyne incognita* on tomato. *Pakistan Journal of Nematology* 8, 101–105.

Montasser, S.A. (1991) The efficacy of some organic manures in the control of root-knot nematode on okra. *Pakistan Journal of Nematology* 9, 139–143.

Moraes, M.V. de (1977) Test preliminar para a determinação do poder nematicida das tortas. In: *Trabalhos Apresentados a II Réunião de nematologia, 14–16 Setembro, 1976, Sociedade Brasileira de Nematologia et da Escola Superior de Agricultura 'Luiz de Queroz' USP, Publicação* No. 2, pp. 193–196.

Morgan, G.T. (1964) Effects of spring preplanting fumigation in strawberry soil in New Brunswick. *Canadian Journal of Plant Science* 44, 170–174.

Morgan, H.G. (1964) *Xiphinema diversicaudatum* and arabis mosaic virus in south-west England. *Nematologica* 10, 70.

Moriarty, F. (1963a) A population experiment with *Heterodera goettingiana* Lieb. and a relatively inefficient host, broad bean. *Nematologica* 9, 152–156.

Moriarty, F. (1963b) The decline of a pea root eelworm (*Heterodera goettingiana* Liebscher) population in the absence of host plants. *Journal of Agricultural Science* 61, 135–138.

Morin, J.P., Lucchini, F., Araujo, J.C.A. de, Ferreira, J.M.S. and Fraga, L.S. (1986)

Rhyncophorus palmarum control using traps made from oil palm cubes. *Oleagineux* 41, 57–62.

Moss, S.R., Crump, D. and Whitehead, A.G. (1975) Control of potato cyst-nematodes, *Heterodera rostochiensis* and *H. pallida*, in sandy, peaty and silt loam soils by oximecarbamate and organophosphate nematicides. *Annals of Applied Biology* 81, 359–365.

Moss, S.R., Crump, D. and Whitehead, A.G. (1976) Control of potato cyst-nematodes, *Globodera rostochiensis* and *G. pallida*, in different soils by small amounts of oxamyl or aldicarb. *Annals of Applied Biology* 84, 355–359.

Motsinger, R.E., Moody, E.H. and Gay, C.M. (1977) Reaction of certain French marigold (*Tagetes patula*) cultivars to three *Meloidogyne* spp. *Journal of Nematology* 9, 278.

Mountain, W.B. and Boyce, H.R. (1957) Parasitic nematodes in relation to the peach replant problem in Ontario. *Phytopathology* 47, 313.

Mountain, W.B. and Boyce, H.R. (1958) The peach replant problem in Ontario. IV. The relation of *Pratylenchus penetrans* to the growth of young peach trees. *Canadian Journal of Botany* 36, 135–151.

Moura, R.M. de (1991) Dois anos de rotação de cultura em campos de cana-de-açúcar para controle da Meloidoginose 1. Efeito dos tratamentos na população do nematóide. *Nematologia Brasileira* 15, 1–7.

Mueller, J.D. and Sullivan, M.J. (1988) Response of cotton to infection by *Hoplolaimus columbus*. *Annals of Applied Nematology (Suppl. to the Journal of Nematology)* 2, 86–89.

Mueller, J.D., Martin, S.B. and Davies, R.M. (1992) Use of Telone II and Temik 15G for Columbia lance nematode control on cotton. *Fungicide and Nematicide Tests. American Phytopathological Society* 47, 162.

Mugnier, J. (1988) Transport of the nematicide oxamyl in roots transformed with *Agrobacterium rhizogenes*. *Annals of Applied Nematology* 2, 29–33.

Mugniery, D. and Balandras, C. (1984) Examen des possibilités d'éradication du nématode à kyste, *G. pallida* Stone. *Agronomie* 4, 773–778.

Mukhopadhyaya, M.C. (1970) Efficacy of D-D, EDB, DBCP and V-C 13 in controlling plant parasitic nematodes. *Journal of Research. Punjab Agricultural University* 7 (4, Suppl.), 659–662.

Müller, J. (1987) Grubrache mit resistenten Olrettich zur Bekampfung des Rubennematoden (*Heterodera schachtii*). *Nachrichtenblatt des Deutschen Pflanzenschutz dienstes* 39, 150–151.

Müller, J. (1991) Catch cropping for population control of *Heterodera schachtii*. *Proceedings of the 54th Winter Congress, International Institute for Sugar Beet Research*, pp. 179–196.

Muller, P.J. and Aartrijk, J. van (1989) Flooding reduces the soil population of the stem nematode *Ditylenchus dipsaci* (Kühn) in sandy soils. *Acta Horticulturae* No. 255, 261–264.

Mulrooney, R.P. (1990) Evaluation of Temik to enhance yield and control of soybean cyst nematode on a resistant and susceptible cultivar. *Fungicide and Nematicide Tests. American Phytopathological Society* 45, 159.

Murdoch, G. and Murdoch, S.M. (1963) Soil sterilants in the control of potato root eelworm on tomatoes. *Experimental Horticulture* 8, 27–37.

Murphy, W.S., Brodie, B.B. and Good, J.M. (1974) Population dynamics of plant nematodes in cultivated soil: effects of combinations of cropping systems and nematicides. *Journal of Nematology* 6, 103–107.

Muthukrishnan, T.S., Rajendran, G., Ramamurthy, V.V. and Chandrasekaran, J. (1979) Pathogenicity and control of *Hirschmanniella oryzae*. *Indian Journal of Nematology* 7, 8–16.

Myuge, S.G. and Vasil'eva, N.P. (1971) [Effect of propyl gallate treatment of cucumbers on

Meloidogyne infection.] *Trudy Gel'mintologicheskoi Laboratorii (Voprosy Biologii, Fiziologii i Biokhimii Gel'mintov Zhivotnykh i Rastenii)* 21, 190–191.

Nachmias, A. and Krikun, J. (1985) *Verticillium* wilt of potato in Israel. *Amercian Potato Journal* 62, 201–205.

Nagahama, M. (1987) [The inductive resistance of surviving pines in large damaged areas to pine wilt disease.] *Bulletin of Fukuoka-ken Forest Experiment Station, Japan* No. 34, i + 23 pp.

Naganathan, T.G., Arumugam, R., Kulasekaram, M. and Vadivelu, S. (1988) Effect of antagonistic crops as intercrops on the control of banana nematodes. *South Indian Horticulture* 36, 268–269.

Nagase, A., Kuwahara, Y., Tomenaga, Y. and Sugawara, R. (1982) Nematicidal activity of alkylamine against the pinewood nematode, *Bursaphelenchus lignicolus*. *Agricultural and Biological Chemistry* 46, 167–172.

Nagy, J. (1987) Increased efficiency of seed-dressing agents by ultrasound exposure of rice nematodes. *International Agrophysics* 3, 291–300.

Nakane, K. and Kimura, Y. (1992) Assessment of pine forest damage by blight based on Landsat TM data and correlation with environmental factors. *Ecological Research* 7, 9–18.

Nakanishi, Y. (1979) [Host range and the control of *Ditylenchus destructor* in bulbous iris.] *Proceedings of the Kansai Plant Protection Society* No. 21, 12–20.

Nakasono, K. (1973) [Relationship between cultivation of *Tagetes* plants and changes in population of the reniform nematode, *Rotylenchulus reniformis*.] *Japanese Journal of Nematology* 3, 38–41.

Nara, J., Wargadipura, R., Sukandis, K. and Ruswanda (1973) [The efficacy of several nematicides in the soil media of tea cutting nurseries.] *Menara Perkebunan Batavia* 41, 297–301.

Naude, T.J. (1949) Research on the control of insect pests. *Farming in South Africa* 24, 574–581.

Neal, C.D. (1954) The reniform nematode and its relationship to the incidence of *Fusarium* wilt of cotton at Baton Rouge, Louisiana. *Phytopathology* 44, 447–450.

Neate, S.M. (1988) Effect of tillage on disease of cereals caused by *Gaumannomyces graminis, Rhizoctonia solani* and *Heterodera avenae*; a review. *Plant Protection Quarterly* 3, 5–7.

Needham, T. (1744) A letter concerning certain chalky tubulous concretions called Malm: with some microscopical observations on the farina of the red lily, and of worms discovered in smutty corn. *Philosophical Transactions of the Royal Society* 42, 173–174, 634–641.

Nelmes, A.J. (1970) Behavioural responses of *Heterodera rostochiensis* larvae to aldicarb and its sulfoxide and sulfone. *Journal of Nematology* 2, 223–227.

Nelmes, A.J. and Keerweewan, S. (1970) The mechanism of action of aldicarb in controlling root-knot nematodes *Meloidogyne incognita* on tomatoes. *International Congress of Plant Protection (7th) , September 21–25, 1970, Paris*, pp. 182–183.

Nesbitt, R.B. (1956) New host plants of plant parasitic nematodes in California. *Plant Disease Reporter* 40, 276.

Nesmith, W.C. and Dowler, W.M. (1974) Some effects of soil fumigation related to peach tree short life. *Proceedings of the American Phytopathological Society* 1, 164.

Nesterov, P.I., Dement'eva, S.P. and Lisetskaya, L.F. (1978) [Changes in the host–parasite relationships between tomato and *Meloidogyne* under the influence of trace elements.] In: *I Vsesyuzonyi s"ezd parasitotsenologov (Poltava Sentyabr' 1978). Tezisy dokladov. Chast'2 Kiev, USSR, "Naukova Duneka"*, pp. 31–32.

Netscher, C. (1974) L'arachide et le controle biologique des nématodes *Meloidogyne* spp.

dans les cultures maraîchères du Sénégal. *Comptes Rendus des Séances de l'Académie d'Agriculture de France* 60, 1332–1339.

Ngugi, C.W., Ikahu, J.M. and Koech, N.K. (1993) Effectiveness of methyl bromide in the control of nematodes, fungi and weeds in pyrethrum propagation nurseries. *Pyrethrum Post* 19, 2–7.

Ngundo, B.W. (1979) Crop rotation to control *Meloidogyne hapla* on pyrethrum. *IX International Congress Plant Protection, August 1979, Washington, DC, USA*, 132.

Nijboer, H. and Parleviet, J.E. (1990) Pathotype-specificity in potato cyst nematodes, a reconsideration. *Euphytica* 49, 39–47.

Nirula, K.K. and Bassi, K.K. (1965) Observations on the control of the root-knot nematode, *Meloidogyne incognita* (Kofoid and White, 1919) Chitwood 1949, with Mylone and Nellite. *Indian Journal of Entomology* 27, 368–369.

Nishizawa, T. (1953a) On the prevention of the rice nematode disease by Folidol. *Botyu-Kagaku* 18, 1–6.

Nishizawa, T. (1953b) [Studies on the varietal resistance of rice plant to the rice nematode disease 'senchû-shingare byô'.] *Bulletin Kyushu Agricultural Experiment Station*, 1, 339–349.

Noe, J.P. (1993) Damage functions and population changes of *Hoplolaimus columbus* on cotton and soybean. *Journal of Nematology* 25, 440–445.

Noel, G.R. (1986) The soybean cyst nematode. In: Lamberti, F. and Taylor, C.E. (eds) *Cyst Nematodes. NATO ASI Series, Series A: Life Sciences Vol. 121*, Plenum Press, New York, 257–268.

Noling, J. and Bird, G.W. (1979) Joint action of nematicides and nitrogen fertilization on control of *Pratylenchus penetrans* and growth of *Solanum tuberosum*. *Journal of Nematology* 11, 293–317.

Nollen, H.M. and Mulder, A.A. (1970) A practical method for economic control of potato cyst nematode (*Heterodera rostochiensis*). *Proceedings of the 5th British Insecticide and Fungicide Conference, Brighton*, 3, pp. 671–674.

Nordmeyer, D. and Dickson, D.W. (1985) Management of *Meloidogyne javanica*, *M. arenaria* and *M. incognita* on flue-cured tobacco with organophosphate, carbamate and avermectin nematicides. *Plant Disease* 69, 67–69.

Nordmeyer, D. and Dickson, D.W. (1990) Biological activity and acetylcholinesterase inhibition by non-fumigant nematicides and their degradation products on *Meloidogyne incognita*. *Revue de Nématologie* 13, 229–232.

Nordmeyer, D., Rich, J.R. and Dickson, D.W. (1982) Effect of ethoprop, carbofuran and aldicarb on flue-cured tobacco infested with three species of *Meloidogyne*. *Nematropica* 12, 199–204.

Novaretti, W.R.T. (1992) Nematóides em cana-de-açucar e seu controle. *Informe Agropecuario (Belo Horizonte)* 16, 37–42.

Novaretti, W.R.T. and Nelli, E.J. (1985) Use of nematicide and filtercake for control of nematodes attacking sugar cane in Sao Paulo State. *Nematologia Brasileira* 9, 175–184.

Novaretti, W.R.T., Carderan, J.O., Carpanezi, A. and Rodrigues, J.C.S. (1988) Comportamento de três variedades de cana-de-açucar em relaçao ao nematóide das lesões das raizes *Pratylenchus zeae* Graham, 1951. *Nematologia Brasileira* 12, 110–120.

Nusbaum, C.J. (1960) Soil fumigation for nematode control in flue-cured tobacco. *Down to Earth*, Summer 1960, 1–4.

Nyczepir, A.P. (1989) Peach tree short life: a nematode associated disease. *Nematology Circular, Division of Plant Industry, Florida Department of Agriculture and Consumer Services* No. 163, 4 pp.

O'Bannon, J.H. and Reynolds, H.W. (1967) The effects of chemical treatment on

Tylenchulus semipenetrans and citrus tree response during 8 years. *Nematologica* 13, 131–136.

O'Bannon, J.H. and Selhime, A.G. (1981) Evaluation of multiple foliar applications of oxamyl for control of citrus pests. *Proceedings of the Florida State Horticultural Society* 93, 50–52.

O'Bannon, J.H. and Tarjan, A.C. (1973) Preplant fumigation for citrus nematode control in Florida. *Journal of Nematology* 5, 88–95.

O'Bannon, J.H. and Tarjan, A.C. (1979) Management of *Radopholus similis* infesting citrus with DBCP or phenamiphos. *Plant Disease Reporter* 63, 456–460.

O'Bannon, J.H. and Taylor, A.L. (1967) Control of nematodes on citrus seedlings by chemical bare-root dip. *Plant Disease Reporter* 51, 995–998.

O'Bannon, J.H. and Tomerlin, A.T. (1977) Control of the burrowing nematode, *Radopholus similis*, with DBCP and oxamyl. *Plant Disease Reporter* 61, 450–454.

O'Bannon, J.H., Leather, C.R. and Reynolds, H.W. (1967) Interactions of *Tylenchulus semipenetrans* and *Fusarium* species on rough lemon (*Citrus limon*). *Phytopathology* 57, 414–417.

O'Bannon, J.H., Tarjan, A.C. and Bistline, F.W. (1974) Control of *Pratylenchus brachyurus* on citrus and tree response to chemical treatment. *Proceedings of the Soil and Crop Science Society of Florida* 33, 65–67.

O'Brien, P.C. and Fisher, J.M. (1978) Studies on the mechanism of resistance of wheat to *Heterodera avenae*. *Nematologica* 24, 463–471.

Ogawa, S. (1988) [Biological and physical control of the pine wilt disease.] *Bulletin of the Fukuoka-ken Forest Experimental Station* No. 35, 56–73.

Ohashi, Y. (1977) [Reactions of *Nicotiana* species to the root-knot nematode *Meloidogyne incognita*.] *Japanese Journal of Breeding* 27, 193–200.

Oka, Y., Chet, I. and Spiegel, Y. (1993) Control of the root-knot nematode, *Meloidogyne javanica* by *Bacillus cereus*. *Biocontrol Science and Technology* 3, 115–126.

Okopnyĭ, N.S., Chebotar, L.N. and Ladygina, L.M. (1983a) [On the physiological and biochemical mechanisms of resistance of potato to the stem nematode.] In: *Steblevye nematody sel'skokhozyaistrennikh kul'tur i mery bor'by s nimi (Materiali simpoziuma, Voronezh, 27–29 Senyabrya 1983 g.)* Voronezh, USSR, pp. 98–104.

Okopnyĭ, N.S., Malyanov, O.M. and Norbaev, Z. (1983b) [Mechanisms of resistance to root knot nematodes in cotton.] *Uzbeksii Biologicheskii Zhurnal* No. 6,45–47.

Olien, W.C., Graham, C.J., Hardin, M.E. and Bridges, W.C. Jr. (1995) Peach rootstock differences in ring nematode tolerance related to effects on tree dry weight, carbohydrate and prunasin contents. *Physiologia Plantarum* 94, 117–123.

Oliff, K.E. (1966) The control of plant-parasitic nematodes by water dispersed nematicides IV. The effect of dipping narcissus bulbs infested with *Ditylenchus dipsaci* (Kühn) Filipjev in thionazin. *Horticultural Research*, Edinburgh 6, 85–90.

Oliff, K.E., Hague, N.G.M. and Kondrollochis, M. (1969) The control of stem nematode in tulips by organophosphorus compounds. *Plant Pathology* 18, 69.

Olmedo, A., Zabala, M. and Voky, C. (1980) Dosis frecuencia y residualidad de nematicidas granulados en citricos. *Fitopatologia Colombiana* 9, 48–54.

Olowe, T., Dina, S.O., Oladiran, A.O. and Olunuga, B.A. (1987) The control of weed, pest and disease complexes in cowpea (*Vigna unguiculata* (L.) Walp.) by the application of pesticides singly and in combination. *Crop Protection* 6, 222–225.

Olsen, L.G. and Nelson, J.W. (1988) Nematode control in blueberries in Michigan, 1986/87. *Fungicide and Nematicide Tests. American Phytopathological Society* 43, 167.

Olsson, E. (1985) Die Kartoffelnematoden *Globodera rostochiensis* (Woll.) Behrens und *G. pallida* (Stone) Behrens und ihre pathotype in Schweden. *Potato Research* 28, 497–506.

Olthof, T.H.A. (1987) Effects of fumigants and systemic pesticides on *Pratylenchus penetrans* and potato yield. *Journal of Nematology* 19, 424–430.

Olthof, T.H.A. (1989a) Effect of fumigant and nonfumigant nematicides on *Pratylenchus penetrans* and yield of potato. *Journal of Nematology* 21, 645–649.

Olthof, T.H.A. (1989b) Control of *Pratylenchus penetrans* on potato with metam-sodium applied in irrigation water. *Journal of Nematology* 21, 693–696.

Olthof, T.H.A., Potter, J.W. and Peterson, E.A. (1974) Relationship between population densities of *Heterodera schachtii* and losses in vegetable crops in Ontario. *Phytopathology* 64, 549–554.

O'Melia, F.C. (1974) Nellite nematicide as a transplant water nematicide for early season control of root-knot nematodes. *Down to Earth* 29, 18–20.

Oostenbrink, M. (1951) Het erwtencystenaaltje, *Heterodera goettingiana* Liebscher, in Nederland. *Tijdschrift Plantenziekten* 57, 52–64.

Oostenbrink, M. (1955) Een inoculatieproef met het erwtencystenaaltje, *Heterodera goettingiana* Liebscher. *Tijdschrift Plantenziekten* 60, 76.

Oostenbrink, M. (1958) An inoculation trial with *Pratylenchus penetrans* in potatoes. *Nematologica* 3, 30–33.

Oostenbrink, M. (1961) Nematodes in relation to plant growth III. *Pratylenchus penetrans* (Cobb) in tree crops, potatoes and red clover. *Netherlands Journal of Agricultural Science* 9, 189–209.

Oostenbrink, M. (1966) Major characteristics of the relation between nematodes and plants. *Mededeelingen van de Landbouwhogeschool. Wageningen*, No. 66–4, 46 pp.

Oostenbrink, M., Kuiper, K. and S'Jacob, J.J. (1957) *Tagetes* als feindpflanzen von *Pratylenchus* arten. *Nematologica* 2, Suppl., 424–433.

Oostendorp, M. and Sikora, R.A. (1989) Seed treatment with antagonistic rhizobacteria for the suppression of *Heterodera schachtii* early root infection of sugar beet. *Revue de Nématologie* 12, 77–83.

Orion, D. and Shlevin, E. (1989) Nematicide seed dressing for cyst and lesion nematode control in wheat. *Journal of Nematology* 21, 629–631.

Osborne, P. (1973) The effect of aldicarb on the hatching of *Heterodera rostochiensis* larvae. *Nematologica* 19, 7–14.

Osman, A.A. and Viglierchio, D.R. (1981a) *Meloidogyne incognita* development on soybean treated with selected amino acids by alternate methods. *Revue de Nématologie* 4, 172–174.

Osman, A.A. and Viglierchio, D.R. (1981b) Foliar spray effects of selected amino acids on sunflower infected with *Meloidogyne incognita*. *Journal of Nematology* 13, 417–419.

Osseni, B. (1985) Les cultures associées à la culture de l'ananas: action sur le nématode *Pratylenchus brachyurus*, la croissance et le rendement de l'ananas Cayenne Lisse. *Fruits* 40, 709–718.

Otazú, V., Hoopes, R., Caero, G. and Huayta, I. (1985) El Rosario de la papa causado por *Nacobbus aberrans* (Thorne 1935) Thorne y Allen 1944. Su efecto en el rendimento y algunos spectos que incidu en su propagición y prevalencia en Bolivia. *Fitopatologia* 20, 65–70.

Oteifa, B.A. and Osman, A.A. (1974) Host-parasite relations of *Rotylenchulus reniformis* on *Lycopersicon esculentum* In: *Simposio Internacional (XII) de Nematologia, Sociedad Europea de Nematologos, 1–7 Sept. 1974, Granada, Spain*, pp. 78–79.

Oteifa, B.A. and Taha, A. (1964) Significance of plant-parasitic nematodes in maize deterioration problem II. Incidence of root rot caused by the root-lesion nematode, *Pratylenchus zeae* Graham. *Technical Bulletin No. 74, Bahtim Experiment Station*, Egypt Agricultural Organisation, 26 pp.

Oteifa, B.A., Gibrail, M.A. and Sedky, E.M. (1970) Effect of certain carbamate pesticides on the soil population density of *Rotylenchulus reniformis* of cotton, *Gossypium barbadense. Agricultural Research Review, Cairo* 48, 129–131.

Oteifa, B.A., Salem, A.A., Botros, A.B. and Sedky, E.M. (1976) Comparative effectiveness of fumigant and systemic nematicides in controlling the cotton reniform nematodes, *Rotylenchulus reniformis. Agricultural Research Review, Cairo* 54, 159–162.

Ouden, H. den (1960a) A note on parthenogenesis and sex determination in *Heterodera rostochiensis* Woll. *Nematologica* 5, 215–216.

Ouden, H. den (1960b) Periodicity in spontaneous hatching of *Heterodera rostochiensis* in the soil. *Nematologica* (Suppl II, 1960),101–105.

Ouden, H. den (1967) The influence of volunteer potato plants in oats on the population density of *Heterodera rostochiensis. Nematologica* 13, 325–335.

Ouden, H. den (1968) Bestrijding van het aardappelcystenaaltje met de systemische nematiciden Temik (2-methyl-2-(methylthio) propionaldehyde O-(methyl-carbamoyl) oxim) en Lannate (methyl O-(methylarbamoyl thioacetohydrox-amaat). *Mededelingen Rijksfaculteit Landouwwetenschappen, Gent* 33, 719–728.

Ouden, H. den (1970) De afname van de bevolkingsdichtheid van het aardappelcystenaaltje bij afwezjigheid van aardappelen. *Jaarverlag 1969 Instituut voor Plantenziektenkundig Onderzoek, Wageningen*, 133.

Ouden, H. den (1971) De ontwikkeling van *Heterodera rostochiensis* larven in aardappelwortel bij behandeling met systemische nematiciden in het bijzonder aldicarb. *Mededelingen Rijksfaculteit Landbouwwetenschappen, Gent* 36, 889–903.

Ouden, H. den (1974) Afname van de Bevolkingsdichtheid van het aardappelcystenaaltje bij teelt van resistente aardappelen en bij braak. *Jaarverslag 1973 Instituut voor Plantenziektenkundig Onderzoek, Wageningen*, 101.

Ouden, H. den and Beuzenberg, M.P. (1972) Vergelijking van de werking van nematiciden tegen *Pratylenchus vulnus*. In: *Jaarverslag 1971. Proefstation voor de Bloemistrij in Nederland te Aalsmeer*, pp. 115–117.

Ouden, H. den and Seinhorst, J.W. (1964) De invloed van enkele systemische nematiciden op de vermeerdering van *Heterodera rostochiensis* op aardappel en van *Tylenchorhynchus dubius* op stoppelknollen. *Mededelingen van de Opzoekingsstations van de Staat te Gent* 29, 810–817.

Overman, A.J. (1964) Efficacy of certain nematicides in strawberry production. *Proceedings of the Florida State Horticultural Society* 76, 114–119.

Overman, A.J. (1982) Soil fumigation via drip irrigation under full-bed mulch culture for row crops. *Proceedings of the Soil and Crop Science Society of Florida* 41, 153–155.

Overman, A.J. and Howard, C.M. (1984) Influence of Nemacur and Vydate on root-knot, sting and stunt nematodes and strawberry yield, 1982. *Fungicide and Nematicide Tests. American Phytopathological Society* 39, 98.

Overman, A.J. and Jones, J.P. (1981) Efficacy of methyl bromide-chloropicrin mixtures for control of nematodes and *Verticillium* wilt of tomato. *Proceedings of the Florida State Horticultural Society* 93, 248–250.

Overman, A.J. and Jones, J.P. (1987) Soil solarization, reaction and fumigation effects on double-cropped tomato under full-bed mulch. *Proceedings of the Florida State Horticultural Society* 99, 315–318.

Overman, A.J., Howard, C.M. and Albregts, E.E. (1987) Soil solarization for strawberries. *Proceedings of the Florida State Horticultural Society* 100, 236–239.

Owino, P.O. and Wando, S.W. (1992) Medicinal plants of Kenya: effects on *Meloidogyne incognita* and the growth of okra. *Afro-Asian Journal of Nematology* 2, 64–66.

Owino, P.O., Wando, S.W. and Sikora, R.A. (1993) Biological control of *Meloidogyne javanica* in Kenya: effect of plant residues, benomyl and decomposition products of mustard (*Brassica campestris*). *Nematologica* 39, 127–134.

Øydvin, J. (1978) Seleksjon av ein andigena resistens-brytande rase av gul potetycyst-nematode *Globodera rostochiensis* (Woll.): eit 12-arig feltforsok med Kontinverleg dyrking av resistent potetkultivar. *Forskning og Forsok i Landbruket* 29, 437–445.

Padhi, N.N. and Misra, R.P. (1987) Control of *Rotylenchulus reniformis* on French bean (*Phaseolus vulgaris* L.). *Indian Journal of Nematology* 17, 130–131.

Page, A.B.P., Hague, N.G.M., Jakobsens, V. and Goldsmith, R.E. (1959) Fumigation of lucerne seed with methyl bromide for the control of the stem eelworm *Ditylenchus* (*Anguillulina*) *dipsaci*. *Journal of Science of Food and Agriculture* 10, 461–467.

Pal, A.K. (1970) Survey on nematode infection of paddy and successful treatment with hexadrin. *Proceedings of the Indian Science Congress Association* 57, 496.

Palacino, J.H. and Leguizamon, C.J. (1991) Interaccion entre *Glomus manihotis* y *Meloidogyne incognita* en pitaya amarillo y roya, bajo condiciones de vivero. *Fitopatologia Colombiana* 15, 9–17.

Palanichamy, L. (1973) Nematode problems of coffee in India. *Indian Coffee* 37, 99–100.

Palanisamy, S. and Balasubramanian, M. (1983) Assessment of avoidable yield loss in cotton (*Gossypium barbadense* L.) by fumigation with metham sodium. *Nematologia Mediterranea* 11, 201–202.

Pandey, G. and Singh, K.P. (1990) Effect of organic amendments on soil microflora and nematode fauna with special reference to *Meloidogyne incognita* in chick pea. *New Agriculturist* 1, 65–70.

Pandey, G., Dwivedi, B.K. and Singh, R.B. (1991) Effect of organic amendments on the root-knot disease of green gram and its rhizosphere microflora. *New Agriculturist* 2, 203–206.

Paracer, S., Tarjan, A.C. and Hodgson, L.M. (1987) Effective use of marine algal products in the management of plant parasitic nematodes. *Journal of Nematology* 19, 194–200.

Pariselle, A. and Rinaudo, G. (1988) Étude des interactions entre *Sesbania rostrata*, *Hirschmanniella oryzae* et les rendements du riz. *Revue de Nématologie* 11, 83–87.

Parker, K.G. and Mai, W.F. (1974) Root diseases of fruit trees in New York State VI. Damage caused by *Pratylenchus penetrans* to apple trees in the orchard growing on different rootstocks. *Plant Disease Reporter* 58, 1007–1011.

Parker, M.B., Minton, N.A., Brooks, O.L. and Perry, C.E. (1975) Soybean yields and lance nematode populations as affected by subsoiling, fertility and nematicide treatments. *Agronomy Journal* 67, 663–666.

Parrott, D.M. (1972) Mating of *Heterodera rostochiensis* pathotypes. *Annals of Applied Biology* 71, 271–273.

Paruthi, I.J. and Bhatti, D.S. (1992) Efficacy of different practices for the control of ear-cockle in wheat caused by *Anguina tritici*. *Current Nematology* 3, 1–4.

Parvatha Reddy, P. (1984) Efficacy of seed treatment with three nematicides for the control of *Meloidogyne incognita* infecting cowpea, French bean and peas. *Indian Journal of Nematology* 14, 39–40.

Patel, D.J., Patel, B.A. and Patel, H.V. (1993) Pathotypes of *Meloidogyne javanica* in India. *Nematologia Mediterranea* 21, 207–208.

Patel, H.R., Vaishnav, M.V. and Dhruj, I.V. (1986) Efficacy of aldicarb sulphone and carbofuran flowable seed treatment on plant growth and against *Meloidogyne arenaria* on groundnut. *Pesticides* 20, 29–31.

Patel, H.R., Patel, D.J., Patel, C.C. and Thakar, N.A. (1991) Management of root-knot nematodes by periwinkle. *Nematologia Mediterranea* 19, 65–66.

Patel, H.R., Makwana, M.G. and Patel, B.N. (1993) Effect of *Catharanthus roseus* on reniform nematode. In: Chari, M.S. and Ramaprasad, G. (eds) *Botanical Pesticides in Integrated Pest Management: Proceedings of National Symposium, January 21–22, 1990, Central Tobacco Research Institute, Rajahmundry, India*, pp. 427–430.

Pattabhiraman, T.V. (1949) The coffee eelworm – *Anguillulina pratensis* (de Man, 1881). *Bulletin of the Indian Coffee Board Research Department* 13, 112–115.

Peachey, J.E. (1962) Investigations on nematicidal activity and crop responses to chemical soil sterilants. *Proceedings of the British Insecticide and Fungicide Conference, November 6–9, 1961, Brighton, UK* 2, pp. 349–351.

Peachey, J.E. and Brown, E.B. (1965) Ridding a glasshouse soil of dagger nematode before planting with roses. *Experimental Horticulture* No. 13, 45–48.

Peachey, J.E., Gander, M.C., Atkinson, K.M., Collingwood, C.A., Davies, T.G., Golightly, W.G., Graham, C.W., Guile, C.T., Stone, L.E.W. and Osborne, P. (1963a) Chemical control of potato-root eelworm in glasshouses: experiments in 1961. *Experimental Horticulture* No. 9, 9–15.

Peachey, J.E., Rao, G,N, and Chapman, M.R. (1963b) Field tests of experimental and commercial soil sterilants against the potato-root eelworm, *Heterodera rostochiensis*. *Annals of Applied Biology* 52, 19–31.

Peachey, J.E., Collingwood, C.A. and Rao, G.N. (1965) Control of potato root eelworm in a glasshouse soil with a formulation of methyl isothiocyanate. *Plant Pathology* 12, 84–87.

Pelagati, O. and Caroppo, S. (1988) [Evaluation of the nematicidal effect of a new active ingredient against the root-knot nematode (*Meloidogyne incognita*) on tomato.] *Redia* 71, 247–253.

Pelsmaeker, M. de (1987) Effect of Vydate 10G on trichodorids, on tobacco rattle virus disease and on potato yields. *Revue de l'Agriculture* 40, 251–262.

Pereira, L.V., Ferraz, S. and De Oliveira, L.M. (1979) Eficiência de alguns nematicidas carbamato como tratamento curativo de mudas de cafeeiero infestadas por *Meloidogyne exigua* Goeldi, 1887. *Fitopatologia Brasileira* 4, 473–476.

Pereira, L.V., Ferraz, S. and De Oliveira, L.M. (1980) Translocaçao de nematicidas e insecticidas nas raízes do cafeeiro (*Coffea arabica* L.) e seus efeitos sobre *Meloidogyne exigua* Goeldi, 1887. *Revista Ceres* 27, 249–255.

Perez, J., Garcia, O., Fernandez, E., Duarte, E. and Jiminez, E. (1987) Evaluación de isazophos 10% G. y de carbofuran 5% G. contra *Radopholus similis* en plátano (*Musa* sp.). *Agrotéchnia de Cuba* 19, 103–108.

Perry, R.N., Clarke, A.J. and Beane, J. (1980) Hatching of *Heterodera goettingiana in vitro*. *Nematologica* 26, 493–495.

Perry, V.G. (1953) Return of nematodes following fumigation of Florida soils. *Proceedings of the Florida State Horticultural Society* 66, 112–114.

Perry, V.G. (1974) The nematode problems of turfgrasses and their control. In: Braithwaite, C.W.D., Phelps, R.H. and Bennett, F.D. (eds) *Proceedings of a Symposium on the Protection of Horticultural Crops in the Caribbean, University of the West Indies, St. Augustine, Trinidad, 8–11 April 1974*, pp. 131–134.

Peters, B.G. (1955) The combined use of nematicidal soil fumigants and solubilised chemicals. *Journal of Helminthology* 29, 81–86.

Pfister, J.A. and Mittnacht, A. (1992) Bekämpfung des Rübenkopfälchens (*Ditylenchus dipsaci*) in Zuckerrüben in Wassschutzgebieten. *Gesunde Pflanzen* 44, 146–150.

Philipps, M.S. (1984) The effect of initial population density on the reproduction of *Globodera pallida* on partially resistant potato clones derived from *Solanum vernei*. *Nematologica* 30, 57–65.

Philipps, M.S. and Trudgill, D.L. (1993) A possible different introduction of *Globodera pallida* into Britain. *Crop Protection in Northern Britain 1993: Proceedings of a Conference, 23–25 March, University of Dundee*, pp. 335–339.

Philis, I. (1988) Control of the citrus nematode (*Tylenchulus semipenetrans* Cobb) in Valencia orange groves in Cyprus. *Nematologia Mediterranea* 16, 159–161.

Philis, I. (1993) Use of nematicides through sprinkler irrigation to control the citrus nematodes in established lemon trees. *Cyprus Miscellaneous Reports – Agricultural Research Institute, Ministry of Agriculture and Natural Resources*, Nicosia No. 59, 1.

Philis, J. (1994) Cultural practices for the control of *Xiphinema index* in Cyprus. *Nematologia Mediterranea* 22, 125–126.

Phipps, P.M. (1987a) Efficacy of nematicides for control of soybean cyst nematode, 1986. *Fungicide and Nematicide Tests. American Phytopathological Society* 42, 168.

Phipps, P.M. (1987b) Performance of soybean cyst nematode susceptible and resistant cultivars with and without applications of Temik 15G in a naturally infested field, 1986. *Biological and Cultural Tests for Control of Plant Diseases. American Phytopathological Society* 2, 37.

Phipps, P.M. (1991) Response of peanut to nematicides for control of northern root-knot and ring nematodes. *Fungicide and Nematicide Tests. American Phytopathological Society* 46, 191.

Phipps, P.M. and Elliott, A.P. (1983) Response of two soybean cultivars to chemical control of soybean cyst nematode, 1982. *Fungicide and Nematicide Tests. American Phytopathological Society* 38, 11.

Phipps, P.M. and Elliott, A.P. (1984) Control of northern root-knot nematode on peanut, 1983. *Fungicide and Nematicide Tests. American Phytopathological Society* 39, 94–95.

Phipps, P.M. and Herbert, D.A. Jr. (1990) Response of peanuts to nematicides for control of northern root-knot and ring nematodes. *Fungicide and Nematicide Tests. American Phytopathological Society* 45, 157.

Phipps, P.M., Bailey, J.E. and Imbriani, J.L. (1988) Control of northern root-knot nematode on peanut, 1987. *Fungicide and Nematicide Tests. American Phytopathological Society* 43, 169.

Pinkerton, J.N., Santo, G.S., Ponti, R.P. and Wilson, J.H. (1986) Control of *Meloidogyne chitwoodi* in commercially grown Russet Burbank potatoes. *Plant Disease* 70, 860–863.

Pinkerton, J.N., Mojtahedi, H. and Santo, G.S. (1987) Reproductive efficiency of Pacific Northwest populations of *Meloidogyne chitwoodi* on alfalfa. *Plant Disease* 7, 345–348.

Pinochet, J. (1979) Comparison of four isolates of *Radopholus similis* from Central America on Valery bananas. *Nematropica* 9, 40–43.

Pinochet, J. (1988) Nematode problems in *Musa* spp: pathotypes of *R. similis* and breeding for resistance. In: *Nematodes and the Borer Weevil in Bananas. Present Status of Research and Outlook; International Network for the Improvement of Banana and Plantains*, Montpellier, France, pp. 66–70.

Pinochet, J. (1992) Breeding bananas for resistance against lesion forming nematodes. In: Gommers, F.J. and Maas, P.W. (eds) *Nematology from Molecule to Ecosystem. Proceedings Second International Nematology Congress, 11–17 August, 1990, Veldhoven, the Netherlands.* Dundee, UK, pp. 157–169.

Pitcher, R.S. and Flegg, J.J.M. (1965) Observations on root feeding by the nematode *Trichodorus viruliferus* Hooper. *Nature* 207, 317.

Pitcher, R.S. and McNamara, D.G. (1973) The control of *Xiphinema diversicaudatum*, the vector of arabis mosaic virus in hops. *Annals of Applied Biology* 75, 468–469.

Ploetz, R.C. (1994) Panama disease: return of the first banana menace. *International Journal of Pest Management* 40, 326–336.

Plowright, R.A., Matias, D.M., Aung,T. and Mew, T.W. (1988) The effect of *Pratylenchus zeae* on the growth and yield of rice. *Philippine Journal of Crop Science* 13 (suppl. 1), 27.

Polychronopoulos, A.G. and Lownsbery, B.F. (1968) Effect of *Heterodera schachtii* on sugar beet seedlings under monoxenic conditions. *Nematologica* 14, 526–534.

Polychronopoulos, A.G., Houston, B.R. and Lownsbery, B.F. (1969) Penetration and development of *Rhizoctonia solani* in sugar beet seedlings infested with *Heterodera schachtii*. *Phytopathology* 59, 482–485.

Ponin, I. Y., Gladkaya, R.M. and Timofeev, N.N. (1978) [The golden potato nematode in Belorussia.] *Zaschita Rastenii* No. 10, 50–51.

Ponte, J.J. da, Franco, A. and Pontes, A.E.L. (1987) Estudo sobre a utiliziaçao da manipueira, como nematicida, em condições de campo. *Nematologia Brasileira* 11, 42–47.

Popova, M.B., Zelenskii, G.L. and Subbotin, S.A. (1994) An assessment of resistance in cultivars of *Oryza sativa* L. to *Aphelenchoides besseyi* Christie, 1942. *Russian Journal of Nematology* 2, 41–44.

Poucher, C., Ford, H.W., Suit, R.F. and Du Charme, E.P. (1967) Burrowing nematode in citrus. *Florida Department of Agriculture, Division of Plant Industry Bulletin* No. 7, 63 pp.

Powell, D.F. (1974) Fumigation of field beans against *Ditylenchus dipsaci*. *Plant Pathology* 23, 110–113.

Powell, D.F. (1975) The effect of methyl bromide fumigation on the germination of onion seed and vigour of the seedlings. *Plant Pathology* 24, 237–240.

Prasad, D. (1985) Seed soaking: a prophylactic treatment of pigeon pea against *Heterodera cajani*. *International Nematology Network Newsletter* 2, 17–18.

Prasad, D. (1990) Control of root-knot nematode, *Meloidogyne arenaria* on groundnut by chemical seed pelleting. *Current Nematology* 1, 31–34.

Prasad, D., Nagla, D.K., Kumar, S. and Saini, M.L. (1992) Marigold plants for management of nematode poopulations in fields. *Current Nematology* 3, 15–18.

Prasad, J. and Varaprasad,K.S. (1992) Elimination of white-tip nematode, *Aphelenchoides besseyi*, from rice seed. *Fundamental and Applied Nematology* 15, 305–308.

Prasad, J.S and Rao, Y.S. (1979a) Effect of soil application of carbofuran in split doses on the control of the rice root nematode. *Current Science* 48, 547.

Prasad, J.S. and Rao, Y.S. (1979b) Nematicidal properties of the weed *Eclipta alba* Hass (Compositae). *Rivista di Parasitologia* 40, 87–90.

Prasad, J.S., Panwar,, M.S. and Rao, Y.S. (1986a) Effect of seed-soaking with chemicals on the parasitic nematodes of rice. *Indian Journal of Nematology* 16, 119–121.

Prasad, J.S., Panwar, M.S. and Rao, Y.S. (1986b) Chemical control of the root nematode (*Hirschmanniella mucronata*) in rice. *Beitrage zur Tropischen Landwirtschaft und Veterinarmedizin* 24, 65–69.

Prasad, K.S. and Rao, Y.S. (1986) Effect of foliar application of systemic pesticides on the development of *Meloidogyne graminicola* on rice. *Indian Journal of Nematology* 14, 125–127.

Prasad, K.S.K., Setty, K.G.H. and Govindu, H.C. (1979) The effect of amino acids on root knot nematodes (*Meloidogyne incognita*) and tomato. *Indian Journal of Plant Protection* 5, 153–156.

Prasadji, J.K. and Sitaramaiah, K. (1994) Effect of plant growth regulators alone and in combination with nematicides on root-knot severity and yield in tomato. *Indian Journal of Nematology* 22, 77–81.

PROINPA (Programa de Investigacion de la Papa) (1991) Projecto de Investigacion de la Papa. *Informe Anual 1990–1991. Compendio*, Cochabamba, Bolivia, pp. 47–56.

Prot, J.C. (1984) A naturally occurring resistance breaking biotype of *Meloidogyne arenaria* on tomato. Reproduction and pathogenicity on tomato cultivars Roma and Rossol. *Revue de Nématologie* 7, 23–28.

Prot, J.C. and Kornprobst, J.M. (1983a) Effects of *Azadirachta indica, Hannoa undulata* and *Hannoa klaineana* seed extracts on the ability of *Meloidogyne javanica* juveniles to penetrate tomato roots. *Revue de Nématologie* 6, 330–332.

Prot, J.C. and Kornprobst, J.M. (1983b) Étude preliminaire de l'action des quassinöides extraits de *Hannoa undulata* sur les juvéniles du nématode *Meloidogyne javanica. Comptes Rendus des Séances de l'Academie des Sciences, Série III* 296, 555–557.

Prot, J.C. and Kornprobst, J.M. (1985) Effects of quassinoids extracted from *Hannoa undulata* seed on the penetration and reproduction of *Meloidogyne javanica* on tomato. *Revue de Nématologie* 8, 383–389.

Prot, J.C. and Matias, D.M. (1995) Effects of water regime on the distribution of *Meloidogyne graminicola* and other root-parasitic nematodes in a rice field toposequence and pathogenicity of *M. graminicola* on rice cultivar UPL R 15. *Nematologica* 41, 219–228.

Prot, J.C., Soriano, I.R., Matias, D.M. and Savary, S. (1992) Use of green manure crops in control of *Hirschmanniella mucronata* and *H. oryzae* in irrigated rice. *Journal of Nematology* 24, 127–132.

Purnell, R.E. and Hague, N.G.M. (1965) The control of plant-parasitic nematodes by water-dispersed nematicides. III. Field experiments on the control of *Ditylenchus dipsaci* (Kühn) Filipjev in narcissus bulbs. *Horticultural Research*, 5, 65–75.

Qamar, F., Saeed, M., Khan, A. and Seema, N. (1989) Control of anguinosis of wheat using plant extracts. *Pakistan Journal of Scientific and Industrial Research* 32, 260–261.

Qiu, J., Westerdahl, B.B., Buchner, R.P. and Anderson, C. (1994a) Refinement of hot water treatment for management of *Aphelenchoides fragariae* in strawberry. *Journal of Nematology* 25 (4 Suppl.), 795–799.

Qiu, J., Westerdahl, B.B., Giraud, D. and Anderson, C.A. (1994b) Evaluation of hot water treatments for management of *Ditylenchus dipsaci* and fungi in daffodil bulbs. *Journal of Nematology* 25, 686–694.

Radewald, J.D., Osgood, J.W., Mayberry, K.S., Paulus, A.O., Otto, H.W. and Shibuya. F. (1969) *Longidorus africanus* Merny nematode found pathogen of Imperial lettuce, southern California. *Plant Disease Reporter* 53, 381–384.

Radewald, J.D., Shibuya, F., Nelson, J. and Birens, J. (1970) Nematode control with 1410, an experimental nematicide–insecticide. *Plant Disease Reporter* 54, 187–190.

Radewald, J.D., Hall, B.J., Shibuya, F. and Nelson, J. (1971a) Results of preplant fumigation trial for the control of sugarbeet nematode on cabbage. *Plant Disease Reporter* 55, 841–845.

Radewald, J.D., Hall, B.J., Shibuya, F. and Nelson, J. (1971b) Preplant fumigation for sugar beet nematode on cabbage. *California Agriculture* 25, 12–14.

Radewald, J.D., Besemer, S.T. and Shibuya, F. (1974) Control of the root-knot nematode, *Meloidogyne incognita*, on carnations with multiple foliar sprays of oxamyl and preplant applications of granular nematicides. *Journal of Nematology* 6, 148–149.

Rahim. M.A.A. (1988) Nematodes associated with paddy seeds in Malaysia. *Proceedings of the Symposium, Movement of Pests and Control Strategies, 16–17 December, 1987, Kuala Lumpur, Malaysia*, pp. 187–199.

Rahman, M.L. (1987) Source of ufra-resistant deepwater rice. *International Rice Research Newsletter* 12, 8.

Rahman, M.L. (1990) Effect of different cropping sequences on root-knot nematode, *Meloidogyne graminicola*, and yield of deepwater rice. *Nematologia Mediterranea* 18, 213–217.

Rahman, M.L. (1991) Evaluation of nematicides to control root-knot nematode (*Meloidogyne graminicola*) in deep water rice. *Current Nematology* 2, 93–98.

Rahman, M.L. and Taylor, B. (1983) Nematode pests associated with deepwater rice in Bangladesh. *International Rice Research Newsletter* 8, 20–21.

Rahman, M.L., Sharma, N.R. and Miah, S.A. (1981) Incidence and chemical control of ufra in bor fields. *International Rice Research Newsletter* 6, 12.

Rajagopalan, P., Chinnarajan, A.N. and Vadivelu, S. (1977) Control of banana nematodes by rhizome treatment. *South Indian Horticulture* 25, 133–137.

Ramakrishnan, S., Varadharajan, G. and Sutharsan, P.D. (1984) TKM 9 is resistant to rice-root nematode. *International Rice Research Newsletter* 9, 20.

Ramana, K.V. (1986) Slow wilt disease of black pepper and the role of plant parasitic nematodes in its etiology. *Proceedings of the 2nd. Group Discussion on Nematological Problems of Plantation Crops, April 24–25, 1985, Central Coffee Research Station, Balehonnur, Karnataka, India*, pp. 17–21.

Ramsdell, D.C., Bird, G.W., Gillett, J.M. and Rose, L.M. (1983) Superimposed shallow and deep soil fumigation to control *Xiphinema americanum* and peach rosette mosaic virus (PRMV) reinfection in a Concord vineyard. *Plant Disease Reporter* 67, 625–627.

Rao, A.B.V.K. , Padhi, N.N. and Acharya, A. (1987) Effect of different nematicides, oilcakes and urea in the control of *Rotylenchulus reniformis* on okra. *Indian Journal of Nematology* 17, 171–173.

Raschke, K. and Pieroh, E.A. (1962) Ergebnisse von Freilandversuchen zur Bekämpfung des Kartoffelnematoden (*Heterodera rostochiensis*) mit Methylisothicyanate. *Zeitschrift für Pflanzenkrankheiten* 69, 667–680.

Raski, D.J. and Allen, M.W. (1953) The control of root-knot nematode on cotton. *Plant Disease Reporter* 37, 193–196.

Raski, D.J. and Goheen, A.C. (1988) Comparison of 1,3-dichloropropene and methyl bromide for control of *Xiphinema index* and grapevine fanleaf degeneration complex. *American Journal of Enology and Viticulture* 39, 334–336.

Raski, D.J. and Lear, B. (1962) Influence of rotation and fumigation on root-knot nematode populations on grape replants. *Nematologica* 8, 143–151.

Raski, D.J. and Schmitt, R.V. (1972) Progress in the control of nematodes by soil fumigation in nematode-fanleaf infected vineyards. *Plant Disease Reporter* 56, 1031–1035.

Raski, D.J., Allen, M.W. and McCallum, R.D. (1951) Sugar-beet nematode. Chemical control trials test two methods of applying soil fumigants. *California Agriculture* 5, 6, 15.

Raski, D.J., Allen, M.W. and Burton, V.E. (1953) Nematode on cotton. Root-knot nematode control by fumigation profitable in Kern County. *California Agriculture* 7, 8, 13.

Raski, D.J., Sher, S.A. and Jensen, F.N. (1956) New host records of the citrus nematode in California. *Plant Disease Reporter* 40, 1047–1048.

Raski, D.J., Schmitt, R.V., Luvisi, D.A. and Kissler, J.J. (1973) 1,3-D and methyl bromide for control of root-knot and other nematodes in vineyard replants. *Plant Disease Reporter* 57, 619–623.

Rasmussen, A.N. and Lindhardt, K. (1980) Stem nematodes in tulips. *Acta Horticulturae* No. 109, 443–446.

Rawsthorne, D. and Hague, N.G.M. (1986) The effect of *Heterodera avenae* on the root system of susceptible and resistant oat seedlings. *Annals of Applied Biology* 108, 89–98.

Raymundo, S.A., Alcazar, J. and Salas, R. (1986a) Effectiveness of soil solarization and dazomet in controlling root knot nematode and weeds in potato fields. *Philippine Agriculturist* 69, 89–97.

Raymundo, S.A., Alcazar, J. and Salas, R. (1986b) Effects of soil solarization, dazomet and bromomethane on root-knot nematode, *Meloidogyne incognita* (Kofoid and White) Chitwood and yield of potato under seedbed conditions. *CMU Journal of Agriculture Food and Nutrition* 8, 75–87.

Read, W.H. (1947) An experiment on the control of root-knot eelworm (*Heterodera marioni*) with D-D. *Report of the Experimental Research Station, Cheshunt, UK* 32, 62.

Rebandel, Z. and Szyzygiel, A. (1989) Wplyw chemieznego odkazania gleby na populacje korzeniaka szkodliwego oraz na wzrost i Plonowanie maliñy. *Zeszyty Problemowe Postępów Nauk Rolniczych* 358, 85–95.

Reddy, D.D.R. (1975) Comparisons of method of application and rate of granular nematicides for the control of *Meloidogyne* spp. on tobacco. *Plant Disease Reporter* 59, 83–85.

Reddy, D.D.R. (1976) Control of root-knot nematode on tobacco with oxamyl through transplant water. *Plant Disease Reporter* 60, 430–431.

Reddy, D.D.R. and Seshadri, A.R. (1971) Studies on some systemic nematicides. I. Evaluation for systemic and contact action against the root-knot nematode, *Meloidogyne incognita. Indian Journal of Nematology* 1, 199–208.

Reddy, D.D.R. and Seshadri, A.R. (1974) Studies on some systemic nematicides. II. Further studies on the action of thionazin and aldicarb on *Meloidogyne incognita* and *Rotylenchulus reniformis. Indian Journal of Nematology* 2, 182–190.

Reddy, D.D.R. and Seshadri, A.R. (1977) Elimination of root-knot nematode infestation from tomato seedlings by chemical bare-root dips or soil application. *Indian Journal of Nematology* 5, 170–175.

Reddy, P.P. and Khan, R.M. (1991) Integrated management of root-knot nematodes infecting okra. *Current Nematology* 2, 115–116.

Reddy, P.P. and Singh, D.B. (1979) Control of root-knot nematode, *Meloidogyne incognita* on tomato by chemical bare-root dips. *Indian Journal of Nematology* 9, 40–43.

Reighard, G.L., Watson, W.A., Coston, D.C. and Ridley, J.D. (1990) Survival of own-rooted and budded 'Redhaven' trees on a peach tree short life site. *Horticultural Science* 25, 359.

Resende, I.C., Ferraz, S. and Condé, A.R. (1987) Efeito de seis variedades de *Mucuna* (*Stizolobium* spp.) sobre *Meloidogyne incognita* raça 3 e *M. javanica. Fitopatologia Brasileira* 12, 310–313.

Reynolds, H.W. and O'Bannon, J.H. (1963) Decline of grapefruit trees in relation to citrus nematode populations and tree recovery after chemical treatment. *Phytopathology* 53, 1011–1015.

Reynolds, H.W. and O'Bannon, J.H. (1966) The efficacy and residual nature of experimental chemicals for controlling *Meloidogyne incognita acrita* on Deltapine cotton in Arizona. *Plant Disease Reporter* 50, 512–515.

Rhoades, H.L. (1964) Effects of fallowing and flooding on root-knot in peat soil. *Plant Disease Reporter* 48, 303–306.

Rhoades, H.L. (1969) Nematicide efficacy in controlling sting and stubby-root nematodes attacking onions in Central Florida. *Plant Disease Reporter* 53, 728–730.

Rhoades, H.L. (1971) Chemical control of the sting nematode, *Belonolaimus longicaudatus*, on direct-seeded cabbage. *Plant Disease Reporter* 55, 412–414.

Rhoades, H.L. (1974) Comparison of two methods of applying granular nematicides for control of sting nematodes on snap beans, sweet corn and field corn. *Proceedings of the Soil and Crop Science Society of Florida* 33, 77–79.

Rhoades, H.L. (1976) Effect of *Indigofera hirsuta* on *Belonolaimus longicaudatus, Meloidogyne incognita* and *M. javanica* and subsequent crop yields. *Plant Disease Reporter* 60, 384–386.

Rhoades, H.L. (1977) Comparison of granular and transplant water applications of non-fumigant nematicides for controlling the sting nematode, *Belonolaimus longicaudatus*, on cabbage. *Proceedings of the Soil and Crop Science Society of Florida* 36, 205–206.

Rhoades, H.L. (1979) Effect of non fumigant nematicides on the sting nematode, *Belonolaimus longicaudatus*, and yield of sweet corn. *Proceedings of the Soil and Crop Science Society of Florida* 37, 218–220.

Rhoades, H.L. (1980) Effect of nematicides and application methods on sting nematode control, root nodulation and yield of soybeans. *Proceedings of the Soil and Crop Science Society of Florida* 39, 90–92.

Rhoades, H.L. (1981a) Influence of nematicides and methods of their application on growth and yield of cabbage in sting nematode infested soil. *Proceedings of the Florida State Horticultural Society* 93, 246–247.

Rhoades, H.L. (1981b) Effect of nematicides on yield of pepper in *Belonolaimus longicauda-*

tus infested soil. *Nematropica* 11, 121–127.

Rhoades, H.L. (1985) Effect of fumigant and non fumigant nematicides on plant nematode populations and yield of sweet corn produced on a fine sandy soil in Florida. *Nematropica* 15, 127–134.

Rhoades, H.L. and Forbes, R.B. (1986) Effects of fallow, cover crops, organic mulches, and fenamiphos on nematode populations, soil nutrients and subsequent crop growth. *Nematropica* 16, 141–151.

Rhode, R.A. and Jenkins, W.R. (1957a) Effect of temperature on the life cycle of stubby-root nematodes. *Phytopathology* 47, 29.

Rhode, R.A. and Jenkins, W.R. (1957b) Host range of a species of *Trichodorus* and its host–parasite relationships on tomato. *Phytopathology* 47, 295–298.

Ricavarren-Herrera, T.A., Viera-Huiman, M.T. and Liceras-Zarate, L. (1976) Chemical control of *Radopholus similis* (Cobb) and *Cosmopolites sordidus* Germar with aldicarb and aldrin in a banana plantation, Variety Isla, in Tingo Maria, Peru. *Acta Horticulturae* No. 57, 223–231.

Rich, J.R., Hodge, C. and Johnson, J.T. (1984) Population development and pathogenicity of *Meloidogyne javanica* on flue-cured tobacco as influenced by ethoprop and DD. *Journal of Nematology* 16, 240–245.

Rickard, D.A. and Barker, K.R. (1982) Nematode assays and advisory services. In: Riggs, R.D. (ed.) *Nematology in the Southern Region of the United States. Southern Cooperative Series Bulletin* 276, pp. 8–20.

Riggs, R.D. and Hamblen, M.L. (1962) Soybean cyst-nematode host studies in the family *Leguminosae. Report Series, Arkansas Agricultural Experiment Station,* No. 118, 18 pp.

Riggs, R.D. and Robbins, R.T. (1982) The variable soybean cyst nematodes. *Arkansas Farm Research* 31, 9.

Riggs, R.D. and Schmitt, D.P. (1988) Complete characterization of the race scheme for *Heterodera glycines. Journal of Nematology* 20, 392–395.

Riggs, R.D., Hamblen, M.L. and Rakes,L. (1981) Infra-species variation in reactions to host in *Heterodera glycines* populations. *Journal of Nematology* 13, 171.

Rinaudo, G., Alazard, D. and Mondiongui, A. (1988) Stem-nodulating legumes as green manure for rice in West Africa. In: *Green Manure in Rice Farming.* Manila, Philippines, pp. 97–109.

Ritchie, D.F. (1989) Improved peach tree longevity with use of fenamiphos in peach tree short-life locations. *Plant Disease* 73, 160–163.

Ritchie, D.F. and Bennett, M.H. (1986) Post-plant use of nematicides for ring nematode control and reduced tree death from peach tree short life complex, 1981–1985. *Fungicide and Nematicide Tests. American Phytopathological Society* 41, 76–77.

Rivoal, R., Person, F., Caubel, G. and Scotto la Massese, C. (1978) Méthodes d'évaluation de la résistance des céréales au développement des nématodes: *Ditylenchus dipsaci, Heterodera avenae* et *Pratylenchus* spp. *Annales de l'Amelioration des Plantes* 28, 371–394.

Rizaeva, S.M. (1983) [Occurrence of root-knot nematodes in new greenhouse nurseries.] *Uzbeksii Biologicheskii Zhurnal* No. 6, 64.

Robalino, G., Román, J. and Cordero, M. (1983) Efecto del nematicida-insecticida oxamil applicado al suelo y a las axilas de las hojas del bananero. *Nematropica* 13, 135–143.

Robbins, R.T. and Barker, K.R. (1973) Comparisons of host range and reproduction among populations of *Belonolaimus longicaudatus* from North Carolina and Georgia. *Plant Disease Reporter* 57, 750–754.

Roberts, P.A. (1987) The influence of planting date of carrot on *Meloidogyne incognita* reproduction and injury to roots. *Nematologica* 33, 335–342.

Roberts, P.A. and Greathead, A.S. (1986) Control of *Ditylenchus dipsaci* in infected garlic seed cloves by non fumigant nematicides. *Journal of Nematology* 18, 66–73.

Roberts, P.A. and May, D. (1986) *Meloidogyne incognita* resistance characteristics in tomato genotypes developed for processing. *Journal of Nematology* 18, 353–359.

Roberts, P.A. and Thomason, J. (1986) Variability in reproduction of isolates of *Meloidogyne incognita* and *M. javanica* on resistant tomato genotypes. *Plant Disease* 70, 547–551.

Roberts, P.A., Thomason, I.J. and McKinney, H.E. (1981) Influence of non hosts, crucifers and fungal parasites on field populations of *Heterodera schachtii*. *Journal of Nematology* 13, 164–171.

Roberts, P.A., Magyarosy, A.C., Matthews, W.C. and May, D.M. (1988) Effects of metam-sodium applied by drip irrigation on root-knot nematodes, *Pythium ultimum* and *Fusarium* sp. in soil and on carrot and tomato roots. *Plant Disease* 72, 213–217

Rodriguez Fuentes, M.E. and Caraballo Llosas, N. (1981) Influencia del regimen de riegos sobre los niveles poblacionales de *Rotylenchulus reniformis* Linford y Oliveira, 1940, en el boniato (*Ipomoea batatas* (L.) Lam.). *Centro Agricola* 8, 35–42.

Rodriguez-Kábana, R. and Ivey, H. (1986) Crop rotation systems for the management of *Meloidogyne arenaria* in peanut. *Nematropica* 16, 53–63.

Rodriguez-Kábana, R. and King, P.S. (1985) Evaluation of selected nematicides for control of *Meloidogyne arenaria* in peanut: a multi-year study. *Nematropica* 15, 155–164.

Rodriguez-Kábana, R., Jordan, J.W. and Hollis, J.P. (1965) Nematodes: biological control in rice fields: role of hydrogen sulfide. *Science* 148, 524–526.

Rodriguez-Kábana, R., Truelove, B. and King, P.S. (1976) A seed treatment method for control of plant parasitic nematodes. *Proceedings of the American Phytopathological Society* 3, 297.

Rodriguez-Kábana, R., Hoveland, C.S. and Haaland, R.L. (1977) Evaluation of a seed-treatment method with acetone for delivering systemic nematicides with wheat and rye. *Journal of Nematology* 9, 323–326.

Rodriguez-Kábana, R. King, P.S., Penick, H.W. and Ivey, H. (1979a) Planting time and post-emergence use of ethylenedibromide-chloropicrin mixtures for control of root-knot nematode on Florunner peanuts. *Highlights of Agricultural Research* 26, 19–20.

Rodriguez-Kábana, R., Penick, H.W. and King, P.S. (1979b) Control of nematodes on soybeans with planting-time applications of ethylene dibromide. *Nematropica* 9, 107.

Rodriguez-Kábana, R., Penick, H.W., King, P.S., Gray, F.A., Carden, E.L., McDaniel, N.R., Selman, F.B. and Ivey, H.W. (1979c) Control of nematodes on soybeans with planting-time applications of ethylene dibromide. *Nematropica* 9, 61–66.

Rodriguez-Kábana, R., Penick, H.W., King, P.S., Gray, F.A., Carden, E.L., McDaniel, N.R. and Selman, F.B. (1979d) Efficacy of planting-time applications of 1,3-dichloropropene for control of nematodes in soybean. *Nematropica* 9, 66–71.

Rodriguez-Kábana, R., King, P.S. and Pope, M.H. (1981a) Combinations of anhydrous ammonia and ethylene dibromide for control of nematodes parasitic on soybeans. *Nematropica* 11, 27–41.

Rodriguez-Kábana, R., King, P.S. and Pope, M.H. (1981b) Comparison of in-furrow applications and banded treatments for control of *Meloidogyne arenaria* in peanuts and soybeans. *Nematropica* 11, 53–67.

Rodriguez-Kábana, R., Weaver, C.F. and King, P.S. (1985) Combinations of 1,3-D and aldicarb for management of *M. arenaria* in peanuts. *Nematropica* 15, 93–106.

Rodriguez-Kábana, R., King, P.S., Robertson, D.G. and Weaver, C.F. (1988a) Potential of crops uncommon to Alabama for management of root-knot and soybean cyst nematodes. *Annals of Applied Nematology* 2, 116–120.

Rodriguez-Kábana, R., King, P.S., Robertson, D.G., Weaver, C.F. and Carden, E.L. (1988b) New crops with potential for management of soybean nematodes. *Nematropica* 18, 45–52.

Rodriguez-Kábana, R., Weaver, D.B., Robertson, D.G., Young, R.W. and Carden, E.L. (1990a) Sorghum in rotation with soybean for the management of cyst- and root-knot nematodes. *Nematropica* 20, 111–119.

Rodriguez-Kábana, R., Weaver, D.B., Robertson, D.G., Young, R.W. and Carden, E.L. (1990b) Rotations of soybean with two tropical legumes for the management of nematode problems. *Nematropica* 20, 101–110.

Rodriguez-Kábana, R., Robertson, D.G., King, P.S. and Wells, L. (1991a) American jointvetch and partridge pea for the management of *Meloidogyne arenaria* in peanut. *Nematropica* 21, 97–103.

Rodriguez-Kábana, R., Robertson, D.G., Wells, L. and King, P.S. (1991b) Soybean–peanut rotations for the management of *Meloidogyne arenaria* and *Sclerotium rolfsii*. *Nematropica* 21, 147–154.

Rodriguez-Kábana, R., Kokallis-Burelle, N., Robertson, D.G., King, P.S. and Wells, L.W. (1994) Rotations with coastal bermuda grass, cotton and bahia grass for management of *Meloidogyne arenaria* and Southern blight in peanut. *Journal of Nematology* 26 (4 suppl.), 665–668.

Roessner, J. (1987) Pilze als antagonisten von *Globodera rostochiensis*. *Nematropica* 33, 106–118.

Roffredi, D.M. (1775) Mémoire sur l'origine des petits vers ou Anguillules du blé rachitique. *Obs. Mem Phys. Hist. Nat.*, 5, 1–19.

Roffredi, D.M. (1776) Mémoire pour servir de supplément et d'éclarissement aux deux mémoires sur Anguillules du blé avorté et de la colle de farine. *Obs. Mem. Phys. Hist. Nat.* 7, 369–385.

Roget, D.K. (1988) Review of tillage and cereal root disease research in South Australia. *Plant Protection Quarterly* 3, 8–9.

Rohrbach, K.G. and Apt, W.J. (1986) Nematode and disease problems of pineapple. *Plant Disease* 70, 81–87.

Román, J., Rivas, X., Rodriguez, J. and Oramas, D. (1976) Chemical control of nematodes in plantains (*Musa acuminata* × *M. balbisiana*, AAB). *Journal of Agriculture of the University of Puerto Rico* 60, 36–44.

Román, J., Oramas, D., Green, J. and Torres, A. (1983) Control of nematodes and black weevils in plantains. *Journal of Agriculture of the University of Puerto Rico* 67, 270–277.

Romaşcu, E. (1977) Combaterea nematodului *Ditylenchus dipsaci* Kühn din culturile de usturoi printratamente chimice aplicate la sol (II). *Analele Institutului de Cercetări pentru Protecţia Plantelor* 13, 137–162.

Romaşcu, E., Lemeni, V., Popescu, I., Vasile, I. and Ciobami, E. (1977) Rezultate experimentale privind combaterea nematodului *Ditylenchus dipsaci* Kühn din culturile de usturoi I. Tratamente aplicate la materialul vegetal pentru saminta. *Analele Institutului de Cercetări pentru Proteçtia Plantelor* 12, 399–406.

Romero, M.D., Valdeslivas, A., Lacasta, C. and Duce, A. (1991) Evolucion de las polaciones di *Heterodera avenae* (Nematoda) y su efecto en el disarrolo y rendimiento de trigo en rotacion y monocultiva. *Suelo y Planta* 1, 323–334.

Routaray, B.N. and Das, S.N. (1988) Population management of root-knot nematode *Meloidogyne incognita* on papaya. *International Nematology Network Newsletter* 5, 25–27.

Routaray, B.N., Sahoo, H. and Das, S.N. (1987) Control of rice root nematode with carbofuran. *International Rice Research Newsletter* 12, 39–40.

Rovira, A.D., Brisbane, P.G., Simon, A., Whitehead, D.G. and Correll, R.L. (1981) Influence of cereal cyst nematode (*Heterodera avenae*) on wheat yields in South Australia. *Australian Journal of Experimental Agriculture and Animal Husbandry* 21, 516–523.

Roy, A.K. (1979) Effectiveness of rotation with non-host or fallow in reducing infestation of *Meloidogyne graminicola*. *Indian Journal of Nematology* 8, 156–158.

Roy, A.K. (1987) Rice diseases in Assam. *Indian Farming* 37, 25–27.

Roy, N.K., Prasad, D. and Taneja (1991) Nematotoxicity of 2-chloroethyl O-aryl O-methyl phosphonates. *Indian Journal of Nematology* 21, 85–90.

Rüdel, M. (1978) Bodenentsenchung zür Nematodenbekämpfung in Weinbau – Wirksamkeit und Rentabilität. *Weinberg und Keller* 25, 503–516.

Ruehle, J.L. (1972) Response of sand pine to parasitism by lance nematode. *Plant Disease Reporter* 56, 691–692.

Russell, C.C. and Perry, V.G. (1966) Parasitic habit of *Trichodorus christiei* on wheat. *Phytopathology* 56, 357–358.

Sagitov, A.O. (1971) [Use of nematicides in the control of the northern root-knot nematode in southern Kazakhstan.] *Vestnik Sel'skokozyaĭstvennoĭ Nauki*, Alma-Ata, No. 11, 44–46.

Sagitov, A.O. and Tulengutova, K.N. (1985) [Dependance between the population density of *Heterodera schachtii* and the yield of sugar beet.] *Doklady Vsesoyuznoi Akademii Sel'skok'hozyaĭstvennykh Nauk im v. I. Lenina* No 3, 27–28.

Sahu, S.C. and Chawla, M.L. (1986) A new virulent strain of rice root-knot nematode from Agartala, India. *International Nematology Network Newsletter* 11, 40.

Saigusa, T. and Aihara, T. (1977) [Ecology and control methods of root-lesion nematodes., *Pratylenchus* spp. in lily fields.] *Japanese Journal of Nematology* 7, 58–65.

Saigusa, T. and Matsumoto, Y. (1961) [On the root-knot nematode control by hot water treatment and its application to gladiolus corms.] *Research Bulletin of the Plant Protection Service, Japan,* No.1 30–35.

Saikia, D.K. and Phukan, P.N. (1986) Effect of crop rotation on the development of root-knot nematode, *Meloidogyne incognita* on jute crop. *Indian Journal of Nematology* 16, 110–111.

Saka, V.W. (1982) International *Meloidogyne* Project Report in Malawi. In: *Proceedings of the 3rd Research Planning Conference on Root-Knot Nematodes,* Meloidogyne *spp. Regions IV and V, 16–20 November, Ibadan, Nigeria,* pp. 31–36.

Sakhuja, P.K. and Sethi, C.L. (1986) Effect of some systemic nematicides on groundnut and multiplication of *Meloidogyne javanica*. *Indian Journal of Nematology* 16, 23–26.

Salas, J.A., Oyuela, R. and Stover, R.H. (1976) Effect of fallow on the burrowing nematode (*Radopholus similis*) of bananas. *Plant Disease Reporter* 60, 863–866.

Salawu, E.O. (1988) Effects of cassava root peeling and locust bean husk amendments in controlling *Meloidogyne incognita* on sugar cane (*Saccharum officinarum* L.). *Journal of Plant Protection in the Tropics* 5, 11–15.

Sampedro, J., Paneque, M. and Santos, Y.E. (1987) Obtencion de semillos de Platano libres de nematodos parasitos. *Ciencia y Técnica en la Agricultura, Protección de Plantas* 10, 105–113.

Sano, Z.I., Minagawa, N. and Ohshima, Y. (1980) [Comparison of control patterns and effects between DD and Telone II for *Meloidogyne incognita* and *Pratylenchus coffeae*.] *Proceedings of the Association for Plant Protection of Kyushu* 26, 156–158.

Santo, G.S. and Bolander, W.J. (1979) Control of *Meloidogyne hapla* in roots of red currants with foliar applications of oxamyl and phenamiphos. *Plant Disease Reporter* 63, 639–641.

Santo, G.S. and Qualls, M. (1984) Control of *Meloidogyne* spp. on Russet Burbank potato by applying metham sodium through centre pivot irrigation systems. *Journal of Nematology* 16, 159–161.

Santo, G.S. and Wilson, J.H. (1990) Evaluation of ethoprop and cadusafos for control of *Meloidogyne chitwoodi* on potato. *Nematropica* 20, 137–142.

Santo, G.S., Ponti, R.P. and Wilson, J.H. (1983) Control of *Meloidogyne chitwoodi* on potato with D-D, 1983. *Fungicide and Nematicide Tests. American Phytopathological Society* 40, 106–107.

Santo, G.S., Ponti, R.P. and Wilson, J.H. (1985a) Evaluation of deep placement of soil fumigants for control of *Meloidogyne chitwoodi* on potato, 1983. *Fungicide and Nematicide Tests. American Phytopathological Society* 40, 108.

Santo, G.S., Ponti, R.P. and Wilson, J.H. (1985b) Control of root-knot nematode on potato with Mocap and Dyfonate, 1983. *Fungicide and Nematicide Tests. American Phytopathological Society* 40, 105–106.

Santo, G,S., Ponti, R.P. and Wilson, J.H. (1985c) Control of *Meloiodogyne chitwoodi* on potato with soil fumigation alone and in combination with non fumigants, 1983. *Fungicide and Nematicide Tests. American Phytopathological Society* 40, 107.

Santo, G.S., Ponti, R.P. and Wilson, J.H. (1985d) Control of *Meloidogyne chitwoodi* on potato with Mocap and Temik, 1982. *Fungicide and Nematicide Tests. American Phytopathological Society* 40, 104.

Sarah, J.L. (1980) Utilisation de nématicides endothérapiques dans la lutte contre *Pratylenchus brachyurus* (Godfrey) (Nematoda, Pratylenchidae) en culture d'ananas. I Activité préventive et curative sur les infestations racinaires par application foliaire. *Fruits* 35, 745–757.

Sarah, J.L. (1983) Utilisation de nématicides endothérapiques dans la lutte contre *Pratylenchus brachyurus* (Godfrey) (Nematoda, Pratylenchidae) en culture d'ananas. IV Effets secondaires d'applications foliaires sur le développement des plants après le traitement d'induction florale et sur la maturation des fruits. *Fruits* 38, 523–540.

Sarah, J.L., Lassoudière, A. and Guérout, R. (1983) La jachère nue et l'immersion du sol: deux méthodes intéressantes de lutte integrée contre *Radopholus similis* (Cobb) dans les bananeraies des sols tourbeaux de Côte d'Ivoire. *Fruits* 38, 35–42.

Sarah, J.L., Kehe, M., Beugnon, M. and Martin, P. (1988) Expérimentation aves l'aldicarbe pour lutter contre *Radopholus similis* (Cobb) (Nematoda, Pratylenchidae) et *Cosmopolites sordidus* (German), (Coleoptera, Circulionidae) en bananeraies. *Fruits* 43, 475–484.

Sarakoski, M.L. (1977) Cleaning nematode-infested potatoes with sodium hypochlorite, NaClO. *Annales Agriculture Fenniae* 16, 137–140.

Sasser, J.N., Kirkpatrick, T.L. and Dybas, R.A. (1982) Efficacy of avermectins for root-knot control in tobacco. *Plant Disease* 66, 691–693.

Savary, A. (1952) Les nématodes nuisibles aux cultures de céréales. *Revue Romande d'Agriculture, de Viticulture et d'Arboriculture* 8, 35.

Saxena, M.C., Greco, N. and Di Vito, M. (1992) Control of *Heterodera ciceri* by crop rotation. *Nematologica Mediterranea* 20, 75–78.

Sayre, R.M.& Mountain, W.B. (1962) The bulb and stem nematode (*Ditylenchus dipsaci*) in southwestern Ontario. *Phytopathology* 52, 510–516.

Sayre, R.M., Patrick, Z.A. and Thorpe, H. (1965) Identification of a selective nematicidal component in extracts of plant residues decomposing in soil. *Nematologica* 11, 263–268.

Schacht, H. (1859a) Ueber einige Feinde der Rübenfelder. *Zeitschrift Ver. Rübenzucker-Industrie Zollver* 9, 175–179.

Schacht, H. (1859b) Ueber einige Feinde und Krankheiten der Zuckerrube. *Zeitschrift Ver. Rübenzucker-Industrie Zollver* 9, 239–250.

Schenk, P.K. (1972) Soil-borne pathogens in ornamental bulbs as related to quarantine problems. *Bulletin OEPP/EPPO* No. 7, 69–76.

Schiffers, B.C., Fraselle, J, Hubrecht, F. and Jaumin, L. (1984) Efficacité contre *Ditylenchus dipsaci* (Kühn) Fil. par nématicides incorporés dans l'enrobage de graines de féverole

(*Vicia faba* L.). *Mededeelingen van de Faculteit Landbouwwetenschappen, Rijksuniversiteit, Gent* 49, 635–641.

Schipke, L.G. and Ramsey, M.D. (1994) Control of banana burrowing nematode (*Radopholus similis*) by fenamiphos applied through micro-irrigation in North Queensland. *Australian Journal of Experimental Agriculture* 34, 109–114.

Schlang, J. (1985) Buchweizen, *Fagopyrum esculentum* Moench (Polygonaceae), eine feindpflanze gegen *Heterodera schachtii*. *Mitteilungen aus der Biologischen Bundesanstalt für Land- und Forstwirtschaft Berlin-Dahlem* No. 226, 104–114.

Schlang, J. (1991) Organische Pillierungszusätze zur Schadensabwehr von *Heterodera schachtii*. *Congrès d'Hiver – Institut International de Recherches Betteravières* No. 54, 197–203.

Schmitt, D.P. (1989) Effect of soil pH on nematicide efficacy on soyabean. *Journal of Nematology* 21(4 supplement), 615–618.

Schmitt, D.P. (1991) Management of *Heterodera glycines* by cropping and cultural practices. *Journal of Nematology* 23, 348–352.

Schmitt, D.P. and Imbriani, J.L. (1987) Management of *Hoplolaimus columbus* on cotton and soybean with tolerant soybean and nematicides. *Annals of Applied Nematology* (supplement to *Journal of Nematology*) 19, 59–63.

Schmitt, D.P., Corbin, F.T. and Nelson, L.A. (1983) Population dynamics of *Heterodera glycines* and soybean response in soils treated with selccted nematicides and herbicides. *Journal of Nematology* 15, 432–437.

Schmitt, D.P., Ferris, H. and Barker, K.R. (1987) Response of soybean to *Heterodera glycines* races 1 and 2 in different soil types. *Journal of Nematology* 19, 240–250.

Schneider, R.C., Zhang, J., Anders, M.M., Bartholomew, D.P. and Caswell-Chen, E.P. (1992) Nematicide efficacy, root growth and fruit yield in drip-irrigated pineapple parasitized by *Rotylenchulus reniformis*. *Journal of Nematology* 24, 540–547.

Schneider, R.C., Green, R.E. and Apt, W.J. (1993) Management of drip-applied nematicides in pineapple. *Acta Horticulturae* No. 334, 351–360.

Scholte, K. (1989) The effect of netted scab (*Streptomyces* spp.) and *Verticillium dahliae* on growth and yield of potato. *Potato Research* 32, 65–73.

Schuiling, M. and van Dinther, J.B.M. (1981) 'Red ring disease' in the Paricatuba oil palm estate, Para, Brazil. A case study. *Zeitschrift für Angewandte Entomologie*, 91, 154–169.

Schultz, O.E. and Cetas, R.C. (1977) Evaluation of granular nematicides for control of 'early maturity wilt' of potatoes in New York State. *Proceedings of the 9th British Insecticide and Fungicide Conference 21–24 Nov. 1977, Brighton* 2, pp. 491–498.

Scognamiglio, A. (1954) Esperimenti di lotta contro l'*Heterodera marioni* (Cornu) Goodey. *Annali della Facolta di Agraria di Portici della Universita di Napoli* 20, 140.

Scotto la Massese, C. (1965) Essais de destruction de *Tylenchulus semipenetrans* Cobb 1913 sur les scions [portes-greffes] de citrus issus de pépinières. *Comptes Rendus 1 ères Journées de Phytiâtrie et Phytopharmacie Circum-méditerranéennes*, Marseilles, 48–51.

Seinhorst, J.W. (1956) Population studies on stem eelworm (*Ditylenchus dipsaci*). *Nematologica* 1, 159–164.

Seinhorst, J.W. (1957) Phytonematology in Western Europe. *Technical Committee, Southern Regional Nematology Project (S-19)*, Alabama, USA, 39 pp.

Seinhorst, J.W. (1965) The relation between nematode density and damage to plants. *Nematologica* 11, 137–154.

Seinhorst, J.W. and Sauer, M.R. (1956) Eelworm attacks on vines in the Murray Valley irrigation area. *Journal of the Australian Institute for Agricultural Science* 22, 296–299.

Semikolenova, N.I. (1987) [Granular nematicides in the control of *Meloidogyne* on cotton.] In: *Paraziticheskie Nematody Rastenii Tadzhikistana*. Dushanbe, USSR, pp. 85–88.

Sethi, C.L. and Srivastava, A.N. (1986) Management of *Heterodera zeae* on maize through pesticidal seed dressing. *Indian Journal of Nematology* 16, 82–86.

Shagalina, L.M. (1979) [Use of lucerne in crop rotation of cotton on fields infected with *Meloidogyne acrita*.] *Izvestiya Akademii nauk Turkmenskoi SSR, Biologicheskie Nauki* No. 2, 91–92.

Shah, H.M., Patel, D.J. and Jaisani, B.G. (1985) Reactions of two Japanese [tobacco] varieties 'Okinawa-6' and 'Okinawa-12' to *Meloidogyne javanica* under Gujarat conditions. *Indian Journal of Nematology* 15, 108.

Sharma, A. and Trivedi, P.C. (1989) Influence of inoculum levels of fungus (*Paecilomyces lilacinus* (Thom.) Samson) on the biocontrol of root-knot nematode, *Meloidogyne incognita* (Chitwood). *International Nematology Network Newsletter* 6, 27–29.

Sharma, H.K. and Gill, J.S. (1993) Prevalence of races in root-knot nematode, *Meloidogyne incognita* in India. *Indian Journal of Nematology* 22, 43–49.

Sharma, N.K. and Sethi, C.L. (1975) Effects of initial inoculum levels of *Meloidogyne incognita* and *Heterodera cajani* on cowpea and on their population development. *Indian Journal of Nematology* 6, 117–123.

Sharma, R.K. and Tiagi, B. (1984) Studies on the effect of some amino acids and amino acid antimetabolite foliar sprays on the root-knot nematode infested pea plants. *Indian Journal of Nematology* 14, 61–62.

Sharma, S.B. and Swarup, G. (1984) *Cyst Forming Nematodes of India.* Cosmo Publications, New Delhi, 149 pp.

Sharma, S.B., Nene, Y.L., Reddy, M.V.I. and McDonald, D. (1993a) Effect of *Heterodera cajani* on biomass and grain yield of pigeon pea on vertisol in pot and field experiments. *Plant Pathology* 42, 163–167.

Sharma, S.B., Remanandan, P. and Jain, K.C. (1993b) Resistance to cyst nematode (*Heterodera cajani*) in pigeon pea cultivars and in wild relatives of *Cajanus*. *Annals of Applied Biology* 123, 75–81.

Sharma, S.B., Rupela, O.P. and Reddy, S.V. (1993c) Association of *Rotylenchulus reniformis* wih plant growth heterogeneity of pigeon pea. *Nematologica Mediterranea* 21, 129–132.

Sharma, S.K., Inderjit Singh and Sakhuja, P.K. (1980) Influence of different cropping sequences on the population of root-knot nematode, *Meloidogyne incognita* and the performance of the subsequent mung crop. *Indian Journal of Nematology* 10, 53–58.

Sharpe, R.R., Reilly, C.C., Nyczepir, A.P. and Okie, W.R. (1989) Establishment of peach in a replant site as affected by soil fumigation, rootstock and pruning date. *Plant Disease* 73, 412–415.

Shepherd, A.M. (1963) The emergence of larvae of *Heterodera goettingiana* Liebs. *in vitro* and a comparison between field populations of *H. goettingiana* and *H. rostochiensis* Woll. *Nematologica* 9, 143–151.

Shepherd, C.J. (1952) Tainting of tobacco by a dichloropropene–dichloropropane soil fumigant. *Nature,* 1073–1074.

Shepherd, J. and Way, J. (1986) Tobacco Research Board – Zimbabwe. *Nematological Society of Southern Africa Newsletter* No. 20, 9–10.

Shepherd, J.A. (1986) Effect of aldicarb on yield of flue-cured tobacco and root-knot nematode control in Zimbabwe. *Zimbabwe Journal of Agricultural Research* 24, 57–63.

Shepherd, R.L. (1988) Effect of root-knot nematodes alone and in combination with fungi on cotton seedling disease. In: *Proceedings of the Beltwide Cotton Production Research Conferences, USA.* Memphis, USA, p. 14.

Sher, S.A. (1959) A disease of carnations caused by the nematode *Criconemoides xenoplax*. *Phytopathology* 49, 761–763.

Sher, S.A. (1970) Revision of the genus *Nacobbus* Thorne and Allen, 1944 (Nematoda: Tylenchoidea). *Journal of Nematology* 2, 228–235.

Sher, S.A. and Munnecke, D.E. (1958) Nematodes and bacteria on rose. *California Agriculture* 12, 9–10.

Sher, S.A., Foot, F.J. and Boswell, S.B. (1959) A root-lesion disease of avocados. *Plant Disease Reporter* 43, 797–800.

Shimizu, K., Aiba, S. and Mitsui, Y. (1989) Nematodes. B. Soybean cyst nematode of soybean. *Extension Bulletin – ASPAC, Food and Fertilizer Technology* No. 302, 7–9.

Siddiqui, M.A. and Alam, M.M. (1987) Efficacy of seed dressing with extracts of neem and Persian lilac against *Meloidogyne incognita* and *Rotylenchulus reniformis. Nematologia Mediterranea* 15, 399–403.

Siddiqui, M.A. and Alam, M.M. (1988) Effect of seed dressing with plant latex on *Tylenchorhynchus brassicae* and plant growth of cabbage and cauliflower. *Pakistan Journal of Nematology* 6, 65–71.

Siddiqui, M.A. and Alam, M.M. (1989a) Control of stunt nematode by bare-root dip in leaf extracts of margosa and Persian lilac. *Pakistan Journal of Nematology* 7, 33–38.

Siddiqui, M.A. and Alam, M.M. (1989b) Effect of root exudates of neem and Persian lilac on plant parasitic nematodes. *Anzeiger für Schädlingskune, Pflanzenschutz, Umweltschutz* 62, 33–35.

Siddiqui, M.A. and Alam, M.M. (1989c) Possible utilization of a noxious weed in nematode control. *Biological Wastes* 28, 181–188.

Siddiqui, M.A. and Alam, M.M. (1990a) Further studies on the use of water hyacinth in nematode control. *Biological Wastes* 33, 71–75.

Siddiqui, M.A. and Alam, M.M. (1990b) Sawdusts as soil amendments for control of nematodes infesting some vegetables. *Biological Wastes* 33, 123–129.

Siddiqui, M.A. and Alam, M.M. (1990c) Control of root-knot, reniform and stunt nematodes by nimbin seed dressing. *Nematologia Mediterranea* 18, 19–22.

Siddiqui, Z.A. and Husain, S.I. (1990) Herbal control of root-knot and root diseases of chickpea. I. Effect of plant extracts. *New Agriculturist* 1, 1–6.

Siddiqui, Z.A. and Khan, A.M. (1975) Chemical control of the stunt nematode in the nursery beds of cabbage and cauliflower. *Indian Phytopathology* 27, 367–370.

Siddiqui, Z.A. and Mahmood, I. (1993a) The effect of ascorbic acid on *Rotylenchulus reniformis* and the rate of transpiration of tomato. *Nematologia Mediterranea* 21, 103–105.

Siddiqui, Z.A. and Mahmood, I. (1993b) Occurrence of races of *Heterodera cajani* in Uttar Pradesh, India. *Nematologia Mediterranea* 21, 185–186.

Siddiqui, Z.A., Khan, A.M. and Khan, M.W. (1979) Control of *Tylenchorhynchus brassicae* by the application of oil-cakes. *Indian Journal of Nematology* 6, 145–149.

Sikdar, A.K., Govindaiah, Madhava Rao, Y.R., Jolly, M.S. and Giridhar, K. (1986) Control of root-knot nematode disease of mulberry by Temik 10G and oil-cakes. *Indian Journal of Sericulture* 25, 33–35.

Silva, H.P., Monteiro, A.R. and Ferraz, L.C.C.B. (1987a) Tratamento hidrotémico de mudas de citros para a erradicação de *Tylenchulus semipenetrans. Nematologia Brasileira* 11, 143–152.

Silva, H.P., Monteiro, A.R. and Ferraz, L.C.C.B. (1987b) Controle químico de *Tylenchulus semipenetrans* em mudas de citros. *Nematologia Brasileira* 11, 176–183.

Singh, D.B., Reddy, P.P. and Syamasundar, J. (1984) Histological, histopathological and histochemical investigation on root-knot nematode resistant and susceptible lines of cowpea. *Nematologia Mediterranea* 12, 213–219.

Singh, I. and Prasad, S.K. (1973) Effect of some pesticides and growth regulators on *Anguina tritici* infesting wheat. *Indian Journal of Entomology* 34, 77–79.

Singh, I., Dalal, M.R. and Bhatti, D.S. (1989) Control of *Hirschmanniella oryzae* nematodes in rice. *International Rice Research Newsletter* 14, 34.

Singh, M. and Sharma, S.B. (1994) Temperature effects on development and reproduction of *Heterodera cajani* on pigeon pea. *Journal of Nematology* 26, 241–248.

Singh, R.S. and Sitaramaiah, K. (1966) Incidence of root-knot of okra and tomatoes in oil-cake amended soil. *Plant Disease Reporter* 50, 668–672.

Singh, R.S. and Sitaramaiah, K. (1967) Effect of decomposing green leaves, sawdust and urea on the incidence of root-knot of okra and tomato. *Indian Journal of Phytopathology* 20, 349–355.

Singh, R.S. and Sitaramaiah, K. (1971a) Control of root-knot through organic and inorganic amendments of soil – effect of oil cakes and sawdust. *Indian Journal of Mycology and Plant Pathology* 1, 20–29.

Singh, R.S. and Sitaramaiah, K. (1971b) Control of root-knot through organic and inorganic amendments of soil: effect of sawdust and inorganic nitrogen. *Indian Journal of Nematology* 1, 80–84.

Singh, R.S. and Sitaramaiah, K. (1973) Control of plant parasitic nematodes with organic amendments of soil. Final Technical Report. Effect of organic amendments, green manuring and inorganic fertilizers on root-knot of vegetable crops. *Research Bulletin, Experimental Station and College of Agriculture, G.B. Pant University of Agriculture and Technology, Pantnagar,* No. 6, 289 pp.

Singh, S.P. and Maheshwari, D.K. (1989) Effect of GA_3 on the phytotoxicity of aldicarb and carbofuran on seedling growth in *Capsicum frutescens* var. California Wonder and rate of root-knot infestation. *Journal of Phytopathology* 127, 158–168.

Singh, S.P. and Singh, V. (1988) Evaluation of oilcakes and nematicides for the control of *Meloidogyne incognita* infecting eggplants. *Indian Journal of Nematology* 18, 366–368.

Singh, S.P., Veena Pant, Khan, A.M. and Saxena, S.K. (1986) Changes in the phenolic contents, related rhizosphere mycoflora and nematode population in tomato inoculated with *Meloidogyne incognita*, as a result of soil amendment with organic matter. *Indian Journal of Nematology* 15, 197–201.

Singh, V.K., Rai, P.K. and Singh, K.P. (1993) Effect of simultaneous and sequential inoculatuion of *Heterodera cajani* and *Fusarium udum* in pigeon pea. *Current Nematology* 4, 129–133.

Sipes, B.S., Schmitt, D.P. and Oda, C.H. (1994) Comparison of single and double-chisel injection methods for the control of *Rotylenchulus reniformis* in pineapple. *Journal of Nematology* 25 (4 Suppl.), 773–777.

Sitaramaiah, K. and Pathak, K.N. (1981) Effect of growth regulators, phenolics and an aromatic acid on root-knot severity (*Meloidogyne incognita* and *M. javanica*) on tomato. *Zeitschrift für Pflanzenkrankheiten und Pflanzenschutz* 88, 651–654.

Sitaramaiah, K. and Singh, R.S. (1978) Effect of organic amendment on phenolic content of soil and plant and response of *Meloidogyne javanica* and its host to related compounds. *Plant and Soil* 50, 671–679.

Siti, E., Cohn, E., Katan, J. and Mordechai, M. (1982) Control of *Ditylenchus dipsaci* in garlic by bulb and soil treatments. *Phytoparasitica* 10, 93–100.

Sivakumar, C.V. (1987) Disinfestation of white tip nematode in rice seeds. *Indian Journal of Nematology* 17, 148–149.

Sivakumar, C.V., Meerzainudeen, M., Rajogopalan, P. and Kuppuswamy, S. (1973) Evaluation of certain systemic nematicides for the control of the root-knot nematode, *Meloidogyne incognita* and the reniform nematode, *Rotylenchulus reniformis* on okra. *Madras Agricultural Journal* 60, 530–533.

Sivakumar, C.V., Rajagopalan, P. and Meerzainudeen, M. (1974) Control of tomato root-knot by spot application of granular systemic nematicides. *South Indian Horticulture* 22, 5–7.

Sivakumar, C.V., Lakshmanan, P. and Balasubramanian, M. (1979) Interaction of a non-host plant (*Capsicum annuum* L.) and certain systemic nematicides in population decline of *Rotylenchulus reniformis*. *Revue de Nématologie* 2, 165–168.

Sivapalan, P. (1969a) Further developments in the control of the root-lesion nematode (*Pratylenchus loosi*) in tea nurseries. *Tea Quarterly* 40, 111–114.

Sivapalan, P. (1969b) Evaluation of pre-planting nematicidal treatments in young tea plantings. *Tea Quarterly* 40, 115–118.

Sivapalan, P. and Gnanapragasam, N.C. (1990) Nematode parasites of coffee, cocoa and tea. In: Luc, M., Sikora, R.A. and Bridge, J. (eds) *Plant Parasitic Nematodes in Subtropical and Tropical Agriculture.* CAB International, Wallingford, UK, pp. 404–430.

Sivapalan, P., Gnanapragasam, N.C. and Jebamalai, A.S. (1980a) Treatment of nursery soils with dazomet 98% granular ('Basamid') as an alternative method of control of the root-lesion nematode, *Pratylenchus loosi*, of tea. *Tea Quarterly* 49, 134–136.

Sivapalan, P., Gnanapragasam, N.C. and Jebamalai, A.S. (1980b) Chemical control of the root-lesion nematode, *Pratylenchus loosi* in new tea plantings. *Tea Quarterly* 49, 30–36.

Slack, D.A. and Hamblen, M.L. (1961) The effect of various factors on larval emergence from cysts of *Heterodera glycines*. *Phytopathology* 51, 350.

Slack, D.A., Riggs, R.D. and Hamblen, M.L. (1981) Nematode control in soybeans: rotation and population dynamics of soybean cyst and other nematodes. *Arkansas Experiment Station, Report Series* No. 263, 36 pp.

Smart, G.C. Jr. (1971) Population development of nematodes on bermuda grass treated annually with nematicides. *Nematropica* 1, 7–8.

Smart, G.C. Jr. and Perry, V.G. (1971) Effectiveness of fall applications of nematicides on bermuda grass. *Nematropica* 1, 15.

Smedley, E.M. (1936) The action of certain halogen compounds on the potato root eelworm, *Heterodera schachtii*. *Journal of Helminthology* 14, 11.

Smedley, E.M. (1939) Experiments on the use of isothiocyanates in the control of the potato strain of *Heterodera schachtii* (Schmidt). *Journal of Helminthology* 17, 31.

Smit, J.J. (1967) In: Coursey, D.G. (ed.) *Yams*. Longmans, London, pp. 115–120.

Smith, A.L. (1948) Control of cotton wilt and nematodes with a soil fumigant. *Phytopathology* 38, 943.

Smith. J. and Whitehead, A.G. (1975) The effect of dazomet on potato cyst-nematode (*H. rostochiensis* Woll.) and on King Edward potatoes grown in severely infested peaty loam. *Experimental Husbandry* 29, 89–96.

Smolik, J.D. (1978) Influence of previous insecticidal use on ability of carbofuran to control nematode populations in corn and effect on corn yield. *Plant Disease Reporter* 65, 95–99.

Song, X.S. (1991) [The study on control of root-knot nematode on peanut by using chloropicrin.] *Peanut Science and Technology* No.1, 19–21.

Soomro, M.H. (1994) Some observations on the survival and viability of *Meloidogyne graminicola* in the absence of any host. *Pakistan Journal of Nematology* 12, 137–140.

Southey, J.F. (1957) Observations on races of *Ditylenchus dipsaci* infesting bulbs. *Journal of Helminthology* 31, 39–46.

Southey, J.F. (1965) The incidence and location of stem eelworms on onion seed. *Plant Pathology* 14, 1–6.

Spaull, A.M. and Tones, S.J. (1986a) Benefits of applying nematicide for control of potato cyst-nematodes during stone separation. *Aspects of Applied Biology* No. 13, 349–353.

Spaull, A.M. and Tones, S.J. (1986b) Control of potato cyst-nematodes (*Globodera* spp.) by aldicarb incorporated during stone-windrowing or rotavation. *Proceedings of the 1986 British Crop Protection Conference, Pests and Diseases, 17–20 November, Brighton, UK*, 3, pp. 975–979.

Spears, J.F.,Mai, W.F. and Betz, D.O. (1956) A case history of the persistence of *H. rostochiensis* Wollenweber in a treated field. *Plant Disease Reporter* 40, 632.

Speich, J. (1982) Isazophos, A nematicide for the control of *Radopholus similis* in bananas. *Mededelingen van de Faculteit Landbouwwettenschappen Rijksuniversiteit Gent* 47, 715–719.

Spiegel-Roy, R., Vardi, A., Elhanati, A., Solel, Z. and Bar-Joseph, M. (1988) Rootstock selection from a Poorman orange × *Poncirus trifoliata* cross. *Proceedings of the International Society of Citriculture. VI International Citrus Congress, Middle-East.* Balaban Publishers,Tel Aviv, Israel, pp. 195–200.

Sprong, M.C. (1989) Inwerken van granulaten. *Landbouwmechanisatie* 40, 12–13.

Srivastava, A.N. (1980) Studies on the maize cyst nematode, *Heterodera zeae.* PhD Thesis, IARI, New Delhi.

Srivastava, A.N. and Sethi, C.L. (1984) Relationships of initial populations of *Heterodera zeae* with plant growth of maize and nematode reproduction. *Indian Journal of Nematology* 14, 110–114.

Srivastava, A.N. and Swarup, G. (1975) Preliminary studies on some graminaceous plants for their susceptibility to the maize cyst nematode *Heterodera zeae* Koshy *et al.*, 1970. *Indian Journal of Nematology* 5, 257–259.

Staniland, L.N. and Stone, L.E.W. (1953) Chlorphenol and related compounds as nematicides. *Journal of Helminthology* 27, 41–74.

Stanton, J.M. and O'Donnell , W.E. (1994) Hatching, mobility and infectivity of root-knot nematode (*Meloidogyne javanica*) following exposure to sodium hypochlorite. *Australian Journal of Experimental Agriculture* 34, 105–108.

Stapleton, J.J., De Vay, J.E. and Lear, B. (1991) Soil sterilization and field effects of ammonia-based fertilizers and soil solarization on pathogen survival and crop growth. *FAO Plant Production and Protection* Paper No. 109, 331–342.

Stark, F.L. Jr. (1948) Investigation of chloropicrin as a soil fumigant. *Memoirs, Cornell University Agricultural Experiment Station* No. 278, 61 pp.

Steele, A.E. (1964) Influence of prolonged association of sugar beet nematode and tomato on intensity of parasitism. *Journal of the American Society of Sugar Beet Technologists* 13, 170–176.

Steele, A.E. and Price, C. (1965) The effects of five legume crops on soil populations of the sugar beet nematode (*Heterodera schachtii* Schmidt). *Journal of the American Society of Sugar Beet Technologists* 13, 314–317.

Steiner, G. and LeHew, R.R. (1933) *Hoplolaimus bradys* n.sp. (Tylenchidae, Nematoda), the cause of a disease of yam (*Dioscorea* sp.). *Zoologischer Anzieger* 101, 260–264.

Stelter, H. (1970) Eine populationsdynamische Studie über Typ A des Kartoffelnematoden (*Heterodera rostochiensis* Wollenweber). *Nachrichtenblatt für den Deutschen Pflanzenschutzdienstes* Berlin, 24, 27–32.

Stelter, H. and Semmeler, W. (1967) Die Bekämpfung des Kartoffelnematoden in geschlossen Sanierungsgebieten. *Nachrichtenblatt für den Deutschen Pflanzenschutzdienstes*, 21, 234–237.

Stender, C., Glazer, I. and Orion, D. (1986) Effect of hydroxyurea on the ultrastructure of giant cells in galls induced by *Meloidogyne javanica. Journal of Nematology* 18, 37–43.

Stephan, Z.A. and Trudgill, D.L. (1983) Effect of time of application on the action of foliar sprays of oxamyl on *Meloidogyne hapla* in tomato. *Journal of Nematology* 15, 96–101.

Stettmeier, W. (1957) Bekämpfung des Wurzelgallenälchens im Gemüsebau mettels Natrium-methyldithiocarbamat. *Pflanzenschutz* 9, 183.

Steudel, W. (1972) Eight years' experience with monoculture of sugar beet in soil infested with *Heterodera schachtii. International Nematology Symposium (11th.) European Society of Nematologists, 3–8 September, 1972, Reading, UK*, pp. 70–72.

Steudel, W. (1985) Ergebnisse zur abundanzdynamik des rübennematoden (*Heterodera schachtii* Schmidt) und zum ertrag von zuckerrüben in einer zwanzigjähren monokultur auf dem institutsversuchsfeld in Elsdorf/Rhld. *Mitteilungen aus der Biologischen Bundesanstalt für Land-.und Forstwirtschaft. Berlin- Dahlem* No. 226, 115–128.

Steudel, W. and Rumpenhorst, H.J. (1978) Untersuchungen zur Populationsdynamik des Haferzystenälchens (*Heterodera avenae* Woll.) und zum Ertrag von anfälligam und resistentem Hafer in einiger Hafer-Dauerkultur. *Zeitschrift für Acker- und Pflanzenbau* 146, 90–108.

Steudel, W. and Thielemann, R. (1970) Weiter Untersuchungen zur Frage der Empfindlichkeit von Zuckerrüben gegen den Rübennematoden (*Heterodera schachtii* Schmidt). *Zucker* 23, 106–109.

Steudel, W., Schlang, J. and Müller, J. (1985) Untersuchungen zum Einfluss einiger zwischenfructe auf die abundanzdynamik des rübennematoden (*Heterodera schachtii* Schmidt) in verschiedenen bodentiefen. *Mitteilungen aus der Biologischen Bundesanstalt für Land- und Forstwirtschaft. Berlin-Dahlem* No. 226, 129–140.

Stewart, R.N. and Schindler, A.F. (1956) The effect of some ectoparasitic and endoparasitic nematodes on the expression of bacterial wilt in carnations. *Phytopathology* 46, 219–222.

Stirling, G.R. (1978) The role of *Dactylella oviparasitica* and other antagonists of nematodes in the biological control of root-knot nematodes (*Meloidogyne* spp.) on peach (*Prunus persica*). *Dissertation Abstracts International* 39B, 2045.

Stirling, G.R. (1981) Effect of temperature on infection of *Meloidogyne javanica* by *Bacillus penetrans*. *Nematologica* 27, 458–462.

Stirling, G.R. (1989) Organic amendments for control of root-knot nematode (*Meloidogyne incognita*) on ginger. *Australasian Plant Pathology* 18, 39–44.

Stirling, G.R. and Kerry, B.R. (1983) Antagonists of the cereal cyst-nematode *Heterodera avenae* Woll. in Australian soils. *Australian Journal of Experimental Agriculture and Animal Husbandry* 23, 318–324.

Stirling, G.R. and Wachtel, M.F. (1980) Mass production of *Bacillus penetrans* for the biological control of root-knot nematodes. *Nematologica* 26, 308–312.

Stirling, G.R. and Wicks, T.J. (1975) *Heterodera cruciferae* on cabbages in South Australia and its chemical control. *Plant Disease Reporter* 59, 43–45.

Stone, L.E.W. (1957) Observations on the control of potato-root eelworm (*Heterodera rostochiensis* Wollenw.) under glass by D-D and solubilized *p-m*-cresol. *Annals of Applied Biology* 45, 256–260.

Stone, L.E.W. (1961) Solubilized nematicides and trickle irrigation of tomatoes. *Journal of Horticultural Science* 36, 274–275.

Storey, G. (1984) The effect of oxamyl and thc growing of susceptible and resistant potato cultivars on the population dynamics of *Globodera rostochiensis* throughout the soil profile. *Annals of Applied Biology* 104, 131–141.

Stradioto, M.F., Babosa Ferraz, L.C.C. and Pitelli, R.A. (1983) Dinâmica populacional de *Pratylenchus brachyurus* em cultura de milho (*Zea mays* L.) infestada por plantas daninhas. In: *Trabalhos apresentados à VII Reunião Brasileira de Nematologia, Brasilia, DF, 21–25 de Fevereiro de 1983, Publicação* No. 7, pp. 99–115.

Streu, H.T., Jenkins, W.R. and Hutchinson, M.T. (1961) Nematodes associated with carnations *Dianthus caryophyllus* L., with special reference to the parasitism and biology of *Criconemoides curvatum* Raski. *New Jersey Agricultural Experiment Station Rutgers – The State University New Brunswick Bulletin* 800, 32 pp.

Sturhan, D. (1964a) Kreuzungsversuche mit biologischen Rassen des Stengelälchens (*Ditylenchus dipsaci*). *Nematologica* 10, 328–334.

Sturhan, D. (1964b) Zur Frage des Bekämpfung von *Longidorus maximus* (Bütschli).

Mitteilungen aus der Biologischen Bundesanstalt für Land- und Forstwirtschaft Berlin-Dahlem No. 111, 106–112.

Sturhan, D. (1965) Vergleichende Wirtspflanzen-untersuchungen und Stengelälchens (*Ditylenchus dipsaci*). *Nematologica* 10, 328–334.

Sturhan, D. (1969) Das Rassenproblem bei *Ditylenchus dipsaci*. *Mitteilungen aus der Biologischen Bundesanstalt für Land- und Forstwirtschaft Berlin-Dahlem* No. 136, 87–98.

Sturhan, D. (1971) Biological races. In: Zuckerman, B.M., Mai, W.F. and Rohde, R.A. (eds) *Plant Parasitic Nematodes*. Academic Press, New York, pp. 51–71.

Sturhan, D. (1985) Untersuchungen über verbreitung und wirte des nematodenparasiten *Bacillus penetrans*. *Mitteilungen aus der Biologischen Bundesanstalt für Land- und Forstwirtschaft Berlin-Dahlem* No. 226, 75–93.

Suatmadji, R.W. (1982) Control of root-knot nematodes, *Meloidogyne javanica*, in rooted stocks of grapevine, *Vitis vinifera*, by immersion in nematicides solutions at different temperatures and in hot water. *Nematologia Mediterranea* 10, 119–125.

Subramanian, P. and Velayutham, B. (1983) Control of rice root nematode. *International Rice Research Newsletter* 8, 15–16.

Subramaniyan, S. and Vadivelu, S. (1990) Effects of *Crotalaria spectabilis* extracts on *Meloidogyne incognita*. *International Nematology Network Newsletter* 7, 8–9.

Sud, U.C., Varaprasad, K.J., Seshadri, A.R. and Kher, K.K. (1984) Relationship between initial densities of *Rotylenchulus reniformis* and damage to cotton with fit to Seinhorst curves. *Indian Journal of Nematology* 14, 148–151.

Suett, D.L. and Jukes, A.A. (1988) Accelerated degradation of aldicarb and its oxidation products in previously treated soils. *Crop Protection* 7, 147–152.

Suga, T., Ohta, S., Munesada, K., Ide, N., Kurokawa, M., Shimzu, M. and Ohta, E. (1993) Endogenous pine wood nematicidal substances in pines, *Pinus massoniana, P. strobus,* and *P. palustris*. *Phytochemistry* 33, 1395–1401.

Suit, R.F. and Du Charme, E.P. (1953) The burrowing nematode and other parasitic nematodes in relation to spreading decline of citrus. *Plant Disease Reporter* 37, 379–383.

Suit, R.F. and Du Charme, E.P. (1957) Spreading decline of *Citrus*. *State Plant Board of Florida*, Vol.II, Bulletin 11, 24 pp.

Suit, R.F. and Feldman, A.W. (1965) Barriers for spreading decline control. *Proceedings of the Florida State Horticultural Society* 77, 52–56.

Suit, R.F. and Ford, H.W. (1950) Present status of spreading decline. *Proceedings of the Florida State Horticultural Society* 63, 36–42.

Suit, R.F., Du Charme, E.P. and Brooks, T.L. (1955) Effectiveness of the pull-and-treat method for controlling the burrowing nematode on citrus. *Proceedings of the Florida State Horticultural Society* 68, 36–38.

Sumner, D.R. (1967) Nematodes in bluegrass. *Plant Disease Reporter* 51, 457–460.

Sundaram, R. and Velayutham, B. (1988) Relative efficacy of some insecticides and neem cake in the control of *Rotylenchulus reniformis* and *Helicotylenchus dihystera* affecting garden bean. *Indian Journal of Nematology* 18, 329–331.

Sundaram, S. (1972) Tomatoes in India: need to control root-knot nematode. In: *Shell in Horticulture*, August 1972, Agricultural University, Coimbatore, India.

Sundararaj, P. and Mehta, U.K. (1990) Host status of some economic crops to *Pratylenchus zeae* and their influence on subsequent sugarcane crops. *Indian Journal of Nematology* 20, 165–169.

Sundararaju, P. and Koshy, P.K. (1982) Response of *Areca* collections against *Radopholus similis*. *Indian Journal of Nematology* 12, 404–405.

Sundararaju, P. and Koshy, P.K. (1986) Effect of different nematicides and neem oil cake in the control of *Radopholus similis* in yellow leaf disease affected arecanut palms. *Indian Journal of Nematology* 16, 44–47.

Swain, B.N. and Prasad, J.S. (1988) Influence of silica content in the roots of rice varieties on the resistance to root-knot nematode. *Indian Journal of Nematology* 18, 360–361.

Swanson, T.A. (1985) Root-knot nematode and *Fusarium* wilt diseases of cowpea and soybean. *Dissertation Abstracts International B. (Sciences and Engineering)* 45, 3389.

Swanson, T.A. and Van Gundy, S.D. (1984) Variability in reproduction of four races of *Meloidogyne incognita* on two cultivars of soybean. *Journal of Nematology* 16, 368–371.

Sykes, G.B. and Winfield, A.L. (1967) Studies on brassica cyst nematode *Heterodera cruciferae*. *Nematologica* 12, 530–538.

Szczygiel, A. (1967) [Preliminary estimation of the harmfulness of nematodes of the genus *Aphelenchoides* to strawberries in south Poland.] *Prace Instytutu Sadownictwa Skierniewicach* 11, 211–224.

Szczygiel, A. (1969a) Ocena skuteczności odkazánia sadzonek truskawek ciepla woda w zwalczaniu nicieni z rodzaju *Aphelenchoides*. *Prace Instytutu Sadownictwa Skierniewicach* 13, 133–145.

Szczygiel, A. (1969b) Ocena skuteczności niektórych środków chemicznych w zwalczaniu nicieni z rodzgju *Aphelenchoides* na truskawkach. *Prace Instytutu Sadownictwa Skierniewicach* 12, 309–321.

Szczygiel, A. (1969c) Dalsze badonia nad chemicznyni zwalcczaniem nicieni z rodzaju *Aphelenchoides* w sadzonkach truskawek. *Prace Sadownictwa Skierniewicach* 13, 147–155.

Szczygiel, A. (1970) [Trials on the control of *Aphelenchoides* spp. on strawberries with some granular organophosphates and Temik.] *Prace Instytutu Sadownictwa Skierniewicach* 14, 231–240.

Szczygiel A. (1980a) Zwalczanie *Aphelenchoides* spp. na metecznikach truskawek srodkami systemicznymi. Cz. I. Moczenie róslin przed sadzeniem. *Zeszyty Problemowe Postępów Nauk Rolniczych* No. 232, 69–75.

Szczygiel, A. (1980b) Zwalczanie *Aphelenchoides* spp. na metecznikach truskawek srodkami systemicznymi. Cz. II. Stosowanie środków na wiosne w okresie wefgetncji. *Zeszyty Problemowe Postępów Nauk Rolniczych* No. 232, 77–91.

Szczygiel, A. (1980c) Zwalczanie *Aphelenchoides* spp. na metecznikach truskawek srodkami systemicznymi. Cz. III. Stosowanie środków późna jesienia i wczesna wiosna. *Zeszyty Problemowe Postępów Nauk Rolniczych* No. 232, 93–100.

Szczygiel, A. (1981) Zwalczanie *Aphelenchoides* spp. na metecznikach truskawek srodkami sysemicznymi. Cz. IV. Stosowanie środków granulowanych pod korzenie rośiin podczas sadzenia. *Zeszyty Problemowe Postępów Nauk Rolniczych* No. 249, 101–107.

Szczygiel, A. and Rebandel, Z. (1982) Zwalczanie posozytniczych nicieni korzeniowych i werticyliozy truskawki za pomoca preparatow Di-Trapex i Benlate. *Roczniki Nauk Rolnczych, E (Ochrona Roślin)* 12, 163–174.

Szczygiel, A. and Rebandel, Z. (1990) Wplyw korzeniaka szkodliwego (*Pratylenchus penetrans*) na wzrost i plonowanie wiśni. *Prace z Zakresu Nauk Rolniczych* 65, 199–209.

Szczygiel, A. and Rebandel, Z. (1991) Skukeczność chemięznego odkażenia w ograniczaniu ujemnych skutkow zmęczenia gleby przu replantacji truskawki. *Roczniki Akademii Rolniczej w Poznaniu, Ogrodnictwo No. 19*, 149–159.

Tacconi, R. (1976) Prove di lotta contro il *Ditylenchus dipsaci* su fragola. *Informatore Fitopatologico* 26, 29–36.

Tacconi, R. (1979) Prove di lotta contro i nematodi fogliari della fragola (*Aphelenchoides fragariae* e *Ditylenchus dipsaci*). In: *Atti dell'Incentro Frutticolo S.O.I. (Societa Orticola Italiana) sulla Coltura della fragola svoltosi a Ferrara il 14 Novembre 1979, Italy*, 253–263.

Tacconi, R. (1980) Prove di lotta contro i nematodi epigei della fragola (*Aphelenchoides fra-*

gariae e *Ditylenchus dipsaci*). *Informatore Fitopatologico* 30, 3–8.

Tacconi, R. and Regazzi, D. (1990) Costi di coltivazione di colture intercalari resistenti a *H. schachtii*. *Informatore Fitopatologico* 40, 47–48.

Tacconi, R. and Santi, R. (1991) Effeto di avvicendamenti colturali su *Heterodera schachtii* (II contributo). *Informatore Fitopatologico* 41, 57–59.

Tacconi, R., and Venturi, G. (1991) Moyens agronomiques de lutte contre le nématode *Heterodera schachtii* en Italie. *Congrès d'Hiver. Institut International de Recherches Betteravières* No. 54, 221–243.

Tacconi, R., Giordani, G. and Toderi, G. (1988) Effeto di concimazione e di avvicendamenti su una populazione di *Pratylenchus thornei*. *Informatore Fitopatologico* 38, 47–50.

Tacconi, R., Biancardi, E. and Olimpieri, R. (1990) Effeto di avvicendamente colturali e di colture intercalari su *Heterodera schachtii* (II contributo). *Informatore Fitopatologico* 40, 47–51.

Tacconi, R., Mambelli, S. and Venturi, G. (1991) Effets de plantes-pièges sur *Heterodera schachti*. *Congrès d'Hiver. Institut International de Recherches Betteravières* No. 54, 263–275

Taconis, P.J. and Kuiper, K. (1964) Overdracht van het *Nicotiana* virus 5 door aaltjes van het geslacht *Trichodorus* in zaailingen van 5 gewassen. *Verslagen en Mededelingen van de Plantenziektenkundige Dienst te Wageningen* No. 141, 177–178.

Takaji, K. (1969) Nematodes confront tea plantations in Japan. *Japanese Agricultural Research Quarterly* 4, 27–29.

Takase, I., Oyama,H. and Fujishita, A. (1982) [Chemical control of pine wilting disease caused by *Bursaphelenchus xylophilus* and absorption and distribution of pesticides in the pine trees.] *Journal of Pesticide Science* 7, 473–480.

Tanaka, I. (1959) [Rice seed treatment test with rhodanate-acetic esters for the control of white-tip nematodes.] *Kyushu Agricultural Research* 21, 152–153.

Tarjan, A.C. (1961) Longevity of *Radopholus similis* in host-free soil. *Nematologica* 6, 170–175.

Tarjan, A.C. (1972) Response of *Pratylenchus*-infected citrus seedlings to stem applications of a paste of ethyl 4-(methylthio)-*m*-tolyl-isopropylphosphoramidate. *Plant Disease Reporter* 56, 626–627.

Tarjan, A.C. (1976) Applications of systemic nematicides to trunks of trees. *Journal of Nematology* 8, 303.

Tarjan, A.C. and O'Bannon, J.H. (1977) Nonpesticidal approaches to nematode control. *Proceedings of the International Society for Citriculture* 3, 848–853.

Tarté, R., Pinochet, J., Gabrielli, C. and Ventura, O. (1981) Differences in population increase, host preferences and frequency of morphological variants among isolates of the banana race of *Radopholus similis*. *Nematropica* 11, 43–52.

Taylor, A.L. and McBeth, C.W. (1941) Spot treatments with chlorpicrin and ethylene dichloride for control of root knot. *Proceedings of the Helminthological Society of Washington* 8, 53.

Taylor, C.E. and Murant, A.F. (1968) Chemical control of raspberry ringspot and tomato black ring viruses in strawberry. *Plant Pathology* 17, 171–178.

Templeton, G.E., Johnston, T.H. and Daniel, J.T. (1971) Benomyl controls rice white tip disease. *Phytopathology* 61, 1522–1523.

Teran, O.G., Perez, M. and Perez, J.A. (1986) Lucha química contra *Meloidogyne incognita* en semilleros de tabacco var. C-30. *Ciencia y Técnica en la Agricultura, Protección de Plantas* 9, 93–102.

Ternisien, E. and Melin, P. (1989) Étude des rotations culturale en bananeraie. I. Bilan des cultures de rotation. *Fruits* 44, 373–383.

Thakar, N.A. and Patel, C.C. (1985) Control of root-knot nematode, *Meloidogyne javanica*, on tomato by chemical bare-root dips. *Journal of Nematology* 15, 110–111.

Thakar, N.A. and Yadav, B.S. (1985a) Comparative pathogenicity of the reniform nematode *Rotylenchulus reniformis* on a susceptible and a resistant variety of pigeon pea. *Indian Journal of Nematology* 15, 167–169.

Thakar, N.A. and Yadav, B.S. (1985b) Phenolics in pigeonpea roots confer resistance to reniform nematode. *International Pigeonpea Newsletter* No. 4, 47–48.

Thakar, N.A. and Yadav, B.S. (1986) Role of total phenolics in pigeonpea resistance to reniform nematodes. *Indian Journal of Nematology* 16, 261–263.

Thames, W.H. and Heald, C.M. (1974) Chemical and cultural control of *Rotylenchulus reniformis* on cotton. *Plant Disease Reporter* 58, 337–341.

Thielemann, R. and Steudel, W. (1973) Neunjährige Erfahrungen mit Monokultur von Zuckerrüben auf mit *Heterodera schachtii* (Schmidt) verseuchtem Boden. *Nachrichtenblatt des Deutschen Pflanzenschutzdienstes* 25, 145–149.

Thomas, E. (1994) Veranderungen im Rassenspektrum von Kartoffelnematoden *Globodera rostochiensis* und *G. pallida* am Niederrhein und Uberlegungen zur resistenzzehtung. *Gesunde Pflanzen* 46, 270–274.

Thomas, E.E. (1923) The citrus nematode, *Tylenchulus semi-penetrans*. *California Agricultural Experiment Station Technical Paper* No. 2, 1–19.

Thomason, I.J. (1959) Influence of soil texture on development of the stubby-root nematode. *Phytopathology* 49, 552.

Thomason, I.J. and Fife, D. (1962) The effect of temperature on development and survival of *Heterodera schachtii* Schm. *Nematologica* 7, 139–145.

Thompson, A.K., Been, B.O. and Perkins, C. (1973) Nematodes in stored yams. *Experimental Agriculture* 9, 281–286.

Thompson, J.A., Miller, R.W., Conrad, N. and Carpenter, A.S. (1984) Evaluation of post-plant nematicides for control of the dagger nematode on apples, 1983. *Fungicide and Nematicide Tests. American Phytopathological Society* 39, 87.

Thorne, G. (1938) Alfalfa resistant to the stem nematode. *Proceedings of the American Society of Sugar Beet Technologists, 1938*, 71–72.

Thorne, G. (1939) Some factors governing the success of chemical treatment of soil for nematode control. *Proceedings of the Helminthological Society of Washington* 6, 60.

Thorne, G. and Giddings, L.A. (1922) The sugar beet nematode in the western states. *US Department of Agriculture. Farmers' Bulletin* No. 1248.

Thorne, G. and Scheiber, E. (1962) American dagger nematode (*Xiphinema americanum*) on coffee in Guatemala, with suggestions for nematode control in nurseries. *Plant Disease Reporter* 46, 857.

Thorne, G. and Schuster, M.L. (1956) *Nacobbus batatiformis* n.sp. (Nematoda: Tylenchidae) producing galls on the roots of sugar beets and other plants. *Proceedings of the Helminthological Society of Washington* 23, 128–134.

Tikhonova, L.V. (1966) [*Aphelenchoides besseyi* Christie, 1942 (Nematoda: Aphelenchoididae) of rice and its control.] *Zoologicheski Zhurnal* 45, 1759–1766.

Tikhornik, I.S. (1966) [Control measures against the stem nematode of onion.] In: [Protection of vegetable crops from pests, diseases and weeds.] Moscow: *Izdatelstvo 'Kolos'*, 65–70.

Timmer, L.W. and French, J.V. (1979) Control of *Tylenchulus semipenetrans* on citrus with aldicarb, oxamyl and DBCP. *Journal of Nematology* 1, 387–394.

Timofeev, N.N. and Ponin, I.Y. (1973) [Decrease of populations of the potato cyst nematode by applying small doses of picrolonic acid and growing forage grasses.] In: *Material Vsesoyuznogo Simpoziuma po Bor'be s Kartofel'noi Nematodoi, 3–5 Ijulya, 1973, Tartu.* Institut Zoologii i Botaniki Akademii Nauk Estonskoi SSR, p. 63.

Timofeev, N.N. and Ponin, I.Y. (1975) [The effect of various orders of anti-nematode rotation of crops with resistant potato varieties on the numbers of *Heterodera rostochiensis*.] *Byulleten 'Vsesoyuznogo Instituta Gel'mintologii im. K.I. Skryabina* No.15, 98–101.

Tiyagi, S.A., Ahmad, A. and Alam, M.M. (1990) Control of root-knot, reniform and stunt nematodes by root dip in leaf extracts of lemon grass. *International Pest Control* 32, 70–71.

Todd, E.H. and Atkins, J.G. (1959) White tip disease of rice. II. Seed treatment studies. *Phytopathology* 49, 184–188.

Todd, T.C. (1994) Soybean planting date and maturity effects on *Heterodera glycines* and *Macrophomina phaseolina* in southeastern Kansas. *Journal of Nematology* 25 (4 Suppl.), 731–737.

Toida, Y., Keereewan, S. and Puttirut, N. (1991) Assessment and prevention of mung bean damage caused by the root-knot nematode, *Meloidogyne javanica* in Thailand. *Japanese Journal of Nematology* 21, 6–10.

Tomminen, J. and Nuorteva, I.M. (1992) Pinewood nematode, *Bursaphelenchus xylophilus* in commercial sawn wood and its control by kiln-heating. *Scandinavian Journal of Forest Research* 7, 113–120.

Tomsett, A.A. (1975) The effect of pre-warming, pre-soaking and hot-water treatment on the growth of narcissus and control of stem eelworm (*Ditylenchus dipsaci* (Kühn) Filipjev). *Acta Horticulturae* No. 47, 165–170.

Toro, M.S. del and Mavrich, E. (1977) Control del nematode *Ditylenchus dipsaci* (Kühn) Filipjev en semilla de ajo con termoterapia. *Revista de la Facultad de Ciencias Agrarias, Universidad Nacional de Cuyo, Argentina* 21, 75–95.

Toro, M.S. del, Garay, J.A. and Colombino, M.A.A. (1988) Experiencias para el control de *Ditylenchus dipsaci* (Kühn) Filipjev, en ajo cv. Colorado, en San Luis, Argentina. *Revista de la Facultad de Ciencias Agrarias, Universidad Nacional de Cuyo, Argentina* 24, 35–47.

Townshend, J.L. (1963) The pathogenicity of *Pratylenchus penetrans* to strawberry. *Canadian Journal of Plant Science* 43, 75–78.

Townshend, J.L. (1989) Efficacy of oxamyl coated on alfalfa seed with a polymer sticker in *Pratylenchus* and *Meloidogyne* infested soils. *Journal of Nematology* 21, 242–246.

Townshend, J.L. (1990) Growth of carrot and tomato from oxamyl-coated seed and control of *M. hapla*. *Journal of Nematology* 22, 170–175.

Townshend, J.L. and Potter, J.W. (1979) Inhibitory infection of alfalfa seedlings by *Pratylenchus penetrans* by treating seed with oxamyl. *Canadian Journal of Plant Science* 59, 519–520.

Towson, A.J. and Lear, B. (1982) Control of nematodes in rose plants by hot-water treatment preceded by heat hardening. *Nematropica* 28, 339–353.

Triantaphyllou, A.C. (1975) Genetic structure of races of *Heterodera glycines* and inheritance of ability to reproduce on resistant soybeans. *Journal of Nematology* 7, 356.

Trudgill, D.L. (1979) Control of *Pratylenchus penetrans* and other replanting disorders in raspberries with dazomet. *Proceedings of the 1979 British Crop Protection Conference – Pests and Diseases – (10th British Insecticide and Fungicide Conference) 19–22 November, Brighton, UK* 1, pp. 211–214.

Trudgill, D.L. (1983) Raspberry re-planting disorders and soil sterilization. *Bulletin of the Scottish Crop Research Institute* No. 2, 21–30.

Trudgill, D.L. (1986) Effects of soil treatment for the control of *Pratylenchus penetrans* (Nematoda) on the growth and yield of raspberry (*Rubus idaeus*) in eastern Scotland. *Crop Research* 26, 89–109.

Trudgill, D.L. (1987) Effects of rates of a nematicide and of fertilizer on the growth and yield of cultivars of potato which differ in their tolerance of damage by potato cyst nematodes (*Globodera rostochiensis* and *G. pallida*). *Plant and Soil* 104, 235–243.

Trudgill, D.L. and Alphey, T.J.W. (1976) Chemical control of the virus-vector nematode *Longidorus elongatus* and of *Pratylenchus crenatus* in raspberry plantations. *Plant Pathology* 25, 15–20.

Trudgill, D.L. and Brown, D.J.F. (1978) *Pratylenchus penetrans*: a potential pest of raspberries in Scotland. *Plant Pathology* 27, 101.

Trudgill, D.L. and Cotes, L.M. (1983) Differences in the tolerance of potato cultivars to potato cyst nematodes (*Globodera rostochiensis* and *G. pallida*) in field trials with and without nematicides. *Annals of Applied Biology* 102, 373–384.

Trudgill, D.L., Mackintosh, G.M., Osborne, P. and Stewart, R.M. (1978) Control of potato cyst nematode (*Globodera rostochiensis*) by nematicides and a resistant potato cultivar at four sites in Scotland. *Annals of Applied Biology* 88, 393–399.

Trudgill, D.L., Holliday, J.M., Mathias, P.L., French, N., Mackintosh, G.M. and Tones, S.J. (1983) Effect of the nematicide oxamyl on the multiplication of *Globodera rostochiensis* and *G. pallida* on the haulm growth and yield of six potato cultivars with different levels of resistance and tolerance. *Annals of Applied Biology* 103, 477–484.

Truelove, B., Rodriguez-Kabana, R. and King, P.S. (1977) Seed treatment as a means of preventing nematode damage to crop plants. *Journal of Nematology* 9, 326–330.

Tülkikkala, K. (1981) Kemisk bekämpning av poatiscyst i Finland. In: *Nordisk växtskyddskonferens 1981, Skokloster, 17–19 mars. Del II. Zoologisk och nematologisk sektion, Uppsala, Sweden; Sveriges Landbruksuniversitet*, pp. 81–87.

Tülkikkala, K. (1991) Effect of crop rotation on *Globodera rostochiensis* and on potato yield. *Bulletin OEPP/EPPO* 21, 41–47.

Turner, G.O. (1963) Nellite nematocide used in irrigation waters increases cotton yields. *Biokemia* 1, 14–17.

Turner, S.J. (1989) New sources of resistance to potato cyst-nematodes in the Commonwealth Potato Collection. *Euphytica* 42, 145–153.

Turner, S.J. (1990) The identification and fitness of virulent potato cyst-nematode populations (*Globodera pallida*) selected on resistant *Solanum vernei* hybrids for up to eleven generations. *Annals of Applied Biology* 117, 385–397.

Uebayashi, Y., Amano, T. and Nakanishi, I. (1971) [On the chemical control of the rice white tip nematode, *Aphelenchoides besseyi* Christie, 1942.] *Bulletin of the Aichi-Ken Agricultural Experiment Station* No. 25, 50–70.

Upadhyay, K.D. and Dwivedi, B.K. (1987) Analysis of crop losses in pea and gram due to *Meloidogyne incognita. International Nematology Network Newzletter* 4, 6–7.

Vallotton, R. (1983) Le nématode à kyste *Heterodera carotae*, un dangereux ravageur de la carotte. *Terre Valaisanne* 32, 3.

Van Gundy, S.D. and Garabedian, S. (1982) Alternatives to DBCP for citrus nematode control. In: *Proceedings of the International Society of Citriculture, 9–12 November, 1981, Tokyo, Japan*, pp. 387–390.

Van Gundy, S.D. and Kirkpatrick, J.D. (1964) Nature of resistance in certain citrus rootstocks to citrus nematode. *Phytopathology* 54, 419–427.

Van Gundy, S.D., Perez, B.J.G., Stolzy, L.H. and Thomason. I.J. (1974) A pest management approach to the control of *Pratylenchus thornei* on wheat in Mexico. *Journal of Nematology* 6, 107–116.

Van Hoof, H.A. and Seinhorst, J.W. (1962) *Rhadinaphelenchus cocophilus* associated with little leaf of coconut and oil palm. *Netherlands Journal of Plant Pathology* 68, 251–256.

Varo Alcala, J., Tobar Jiménez, A. and Muñoz Medina, J.M. (1970) Lesiones causadas y reaciones provocadas por algumos nematodos en las raices de ciertas plantas. *Revista Ibérica de Parasitologia* 30, 547–566.

Vats, R. and Nandal, S.N. (1993a) Effect of different concentrations of leaf extracts of neem and *Eucalyptus* used as bare-root-dip treatment of tomato seedlings against *Meloidogyne javanica. Current Nematology* 4, 15–18.

Vats, R. and Nandal, S.N. (1993b) Preliminary studies on efficiency of aqueous plant parts extracts as bare-root-dip treatment of tomato seedlings against *Meloidogyne javanica. Current Nematology* 4, 181–185.

Vecht, J. van der (1950) Op planten parasiterende aaltjes (Tylenchidae). *Plagen van de Cultuurgewassen in Indonesie*, 16–42.

Vecht, J. van der and Bergmann, B.H.H. (1952) Studies on the nematode *Radopholus oryzae* (Van Breda de Haan) Thorne and its influence on the growth of the rice plant. *Pemberitaan Balai Besar Penjelidikan Pertanian Bogor, Indonesia*, No. 131, 82 pp.

Veech, J.A. (1977) A possible new approach to the chemical control of nematodes. *Journal of Nematology* 9, 184–185.

Veech, J.A. (1979) Terpenoid aldehydes in the resistance of cotton to the root-knot nematode. In: Brown, J.M. (ed.) *Proceedings of the Beltwide Cotton Production Research Conferences, 7–11 January, 1979, Phoenix, Arizona. National Cotton Council USA*, p. 16.

Vega, E. (1973) Tratamiento contra *Meloidogyne* (Nematoda: Heteroderidae) de barbados de vid. *Idia* No. 303, 15–20.

Veld, E.J.M.H. in 'T., Egberink, M.B., Pol, P.A. van de and S'Jacob, J.J. (1988) Fenamiphos as a dip treatment for rose rootstocks to control *Pratylenchus vulnus. Mededelingen van de Faculteit Landbouwettenschappen, Rijksuniversiteit, Gent* 53, 893–901.

Venkata Rao, C., Mani, A. and Rao, P.K. (1987) Use of herbicides as bare-root dip treatments for the control of the root-knot nematode, *Meloidogyne incognita*, on brinjal (*Solanum melongena* L.). *Tropical Pest Management* 33, 137–138.

Venkitesan, T.S. (1976) Studies on the burrowing nematode, *Radopholus similis* (Cobb, 1893) Thorne, 1949 on pepper (*Piper nigrum* L.) and its role in slow wilt disease. PhD Thesis, Department of Plant Pathology, University of Agricultural Sciences, Bangalore, India, 122 pp.

Venter, C., Waele, D. de and Meyer, A.J. (1992) Minimizing damage by *Ditylenchus destructor* to peanut seed with early harvest. *Journal of Nematology* 24, 528–532.

Vicente, N.E. and Acosta, N. (1992) Biological and chemical control of nematodes in *Capsicum annuum* L. *Journal of Agriculture of the University of Puerto Rico* 76, 171–176.

Vicente, N.E., Sanchez, L.A. and Acosta, N. (1991) Effect of granular nematicides and the fungus *Paecilomyces lilacinus* in nematode control in watermelon. *Journal of Agriculture of the University of Puerto Rico* 75, 307–309.

Viglierchio, D.R. (1971) Race genesis in *Ditylenchus dipsaci. Nematologica* 17, 386–392.

Viglierchio, D.R. (1987) Biological criteria in management considerations for *Meloidogyne chitwoodi. Nematologia Mediterranea* 15, 65–70.

Viglierchio, D.R., Maggenti, A.R., Schmitt, R.V. and Paxman, G.A. (1977) Nematicidal injection: targeted control of plant-parasitic nematodes of trees and vines. *Journal of Nematology* 9, 307–311.

Vik, F. (1968) Výsledky pokusů s ošetřenim semen cibule proti hádátku zhoubnému (*Ditylenchus dipsaci* (Kühn)). *Ochrana Rośtlin* 4, 221–230.

Vilardebó, A. and Guérout, R. (1976) A review of experiments on nematode control ethoprop (prophos), phenamiphos and carbofuran in French-speaking West Africa. *Nematropica* 6, 51–53.

Vilardebó, A., Guérout, R., Pinon, A. and Melin, P. (1972) La lutte contre les nématodes du bananier. Synthèse des études récentes avec les nématicides Némacur et Mocap. *Fruits* 27, 777–787.

Vilardebó, A., Boisseau, M., Lassoudière, A., Melin, P. and Ternisien, E. (1988) Expérimentation avec l'aldicarbe pour lutter contre *Radopholus similis* Cobb (Nematoda, Pratylenchidae) et *Cosmopolites sordidus* (Germar) (Coleoptera,

Curculionidae) en bananeraies. Première partie. Expérimentation réalisée en Martinique et au Cameroon. *Fruits* 43, 417–431.

Villanueva, L.M. and Davide, R.G. (1984) Evaluation of several isolates of soil fungi for biological control of root-knot nematodes. *Philippine Agriculturist* 67, 361–371.

Virville, J.D. and Person-Dedryver, F. (1989) Growth and respiratory activity of roots of various Triticinae tolerant or resistant to *Heterodera avenae* Woll., with or without infection by the nematode. *Revue de Nématologie* 12, 379–386.

Vladimirova, L.A. (1975) [Strawberry stem nematode in South-east Kazakhstan.] In: *Tematicheskii Sbornik Nauchnykh Rabot Instituta Plodovdstva i Vinogradarstva i Opytnykh Stantsii, 3*. Alma-Ata, Kazakh SSR, Kainar pp. 222–227.

Vorona, V.F. (1984) [Heterophos in the control of the potato stem nematode.] *Byulleten 'Vsesoyuznogo Instituta Gel'mintologii im K.I. Skryabina* No. 39, 56.

Vovlas, N. and Inserra, R.N. (1980) Consideerazioni sull'effeto dei trattamenti con nematocidi fumiganti e sistemici nella lotta contro *Ditylenchus dipsaci* su fragola nell' Italia meridionale. *Difesa della Pianti* 3, 163–170.

Vovlas, N. and Lamberti, F. (1976) Studies on the systemic action of some chemicals in the control of root-knot nematodes. *Nematologia Mediterranea* 4, 111–113.

Vovlas, N., Inserra, R.N. and Lamberti, F. (1978) *Ditylenchus dipsaci* su fragola nell Italia meridionale e relativi metodi di lotta. *Revista della Ortoflorofrutticoltura Italiana* 62, 253–268.

Vrain, T.C. (1987) Effect of *Ditylenchus dipsaci* and *Pratylenchus penetrans* on *Verticillium* wilt of alfalfa. *Journal of Nematology* 19, 379–383.

Vreeburg, P.J.M. (1981) De warmwaterbehandeling bij de teelt van rotstuin-narcissen. *Bloembollencultuur* 92, 177.

Walia, R.K. and Bajaj, H.K. (1986) Existence of host races of pigeon-pea cyst nematode, *Heterodera cajani* Koshy. *Nematologica* 32, 117–119.

Walia, R.K. and Bajaj, H.K. (1988) Further studies on the existence of races in pigeon pea cyst nematode, *Heterodera cajani*. *Indian Journal of Nematology* 18, 269–272.

Walia, R.K. and Seshadri, A.R. (1985) Chemical control of *Pratylenchus thornei* on chickpea through seed treatment. *International Chickpea Newsletter* No. 13, 32–34.

Walker, J.T., Specht, C.H. and Bekker, J.F. (1966) Nematocidal activity to *Pratylenchus penetrans* by culture fluids from actinomycetes and bacteria. *Canadian Journal of Microbiology* 12, 347–351.

Wallace, H.R. (1962) Observations on the behaviour of *Ditylenchus dipsaci* in soil. *Nematologica* 7, 91–101.

Wallace, H.R. (1971) The influence of the density of nematode populations on plants. *Nematologica* 17, 154–166.

Walter, J.M. and Kelsheimer, E.G. (1949) In-the-row application of soil fumigants. *Market Growers Journal* 78, 12 (5) 12, 33, 35–37; (6) 5, 37–39.

Walters, S.A. and Barker, K.R. (1994) Effect of *Paecilomyces lilacinus* in suppressing *Rotylenchulus reniformis* on tomato. *Journal of Nematology* 26 (4 suppl.), 600–605.

Wani, A.H., Ansari, A.P. and Alam, M.M. (1994) Nematicidal potential of latex extracts of *Calotropis procera* and *Euphorbia caducifolia* on root-knot nematodes infecting okra. *Annals of Plant Protection Sciences* 2, 72–74.

Warren, L.E. (1963) Control of sugar beet nematode with 1,3-dichloropropene in irrigation water. *Journal of the American Society of Sugar Beet Technologists* 12, 348–358.

Weaver, D.J. and Wehunt, E.J. (1976) Control of peach tree short life with hydrated lime and soil fumigation. *Proceedings of the American Phytopathological Society* 3, 270.

Weaver, D.B., Rodriguez-Kábana, R. and Carden, E.L. (1987) Soybean response to ethylene dibromide in a soil infested with *Meloidogyne arenaria* and *H. glycines*. *Annals of Applied Nematology* 1, 94–96.

Weaver, D.B., Rodriguez-Kábana, R. and Carden, E.L. (1988a) Multiple species nematode resistance in soybean: effect of genotype and fumigation on yield and nematode numbers. *Crop Science* 28, 293–298.

Weaver, D.B., Rodriguez-Kábana, R., Robertson, D.G., Akridge, R.L. and Carden, E.L. (1988b) Effect of crop rotation on soybean in a field infested with *Meloidogyne arenaria* and *Heterodera glycines*. *Annals of Applied Nematology* 2, 106–109.

Webley, D. P. and Jones, F.G.W. (1981) Observations on *Globodera pallida* and *G. rostochiensis* on early potatoes. *Plant Pathology* 30, 217–224.

Webster, J.W. (1964) The significance of biological races of *Ditylenchus dipsaci* and their hybrids. *Annals of Applied Biology* 59, 77–83.

Wehunt, E.J. (1984) Nematode parasites of peach and other tree crops. In: Nickle, W.R. (ed.) *Plant and Insect Nematodes.* Marcel Dekker Inc., New York, 435–455.

Wehunt, E.J. and Weaver, D.J. (1982) Effect of planting site preparation, hydrated lime and DBCP (1,2-dibromo-3-chloropropane) on populations of *Macroposthonia xenoplax* and peach tree short life in Georgia. *Journal of Nematology* 14, 567–571.

Wehunt, E.J., Horton, B.D. and Prince, V.E. (1980) Effects of nematicides, lime and herbicides on peach tree short life in Georgia. *Journal of Nematology* 12, 183–189.

Weingartner, D.P. and Shumaker, J.R. (1981) Comparisons of nematode control and potato yields in large plots treated with different combinations during 1977–1981. *Proceedings of the Florida State Horticultural Society* 94, 141–144.

Weingartner, D.P., Shumaker, J.R., Smart, G.C. Jr. and Dickson, D.W. (1976) A new nematode control program for potatoes grown in northeast Florida. *Proceedings of the Florida State Horticultural Society* 88, 175–182.

Weingartner, D.P., Shumaker, J.R., Dickson, D.W. and Littrell, R.C. (1978) Nematode control on Irish potatoes in northeast Florida using a soil fumigant and a non volatile nematicide both alone and in combination. *Journal of Nematology* 10, 301.

Westerdahl, B.B., Giraud, D., Radewald, J.D., Anderson, C.A. and Darso, J. (1994) Management of *Pratylenchus penetrans* on oriental lilies with drip and foliar applied nematicides. *Journal of Nematology* 25, (4 supp), 758–767.

Whitehead, A.G. (1965) Taxonomy, distribution and host–parasite relationships of the genus *Meloidogyne* Goeldi (Nematodea: Heteroderidae) PhD Thesis, University of London, 319 pp.

Whitehead, A.G. (1977a) Vertical distribution of potato, beet and pea cyst nematodes in some heavily infested soils. *Plant Pathology* 26, 85–90.

Whitehead, A.G. (1977b) Control of potato cyst-nematode, *Globodera rostochiensis* Ro1, by picrolonic acid and potato trap crops. *Annals of Applied Biology* 87, 225–227.

Whitehead, A.G. (1983) Loss of yield and quality in potatoes due to potato cyst-nematodes. *Proceedings of the 10th International Congress of Plant Protection, 20–25 November 1983*, Brighton, UK 1, p. 113.

Whitehead, A.G. (1984) Interaction of three lucerne cultivars and eleven English isolates of stem nematode (*Ditylenchus dipsaci*), 'lucerne race'. *Plant Pathology* 33, 33–37.

Whitehead, A.G. (1988a) Principles of granular nematicide placement for temperate field crops. In: Martin, T. (ed.) *Application to Seeds and Soil. British Crop Protection Council Monograph* No. 39 309–317.

Whitehead, A.G. (1988b) A vertical band–Dutch harrow technique for incorporating granular nematicides in soil to control potato cyst-nematodes, *Globodera rostochiensis* and *G. pallida*. *Annals of Applied Biology* 112, 613–616.

Whitehead, A.G. (1991a) The resistance of six potato cultivars to English populations of potato cyst-nematodes, *Globodera pallida* and *G. rostochiensis*. *Annals of Applied Biology* 118, 357–369.

Whitehead, A.G. (1991b) Selection for virulence in the potato cyst-nematode, *Globodera pallida*. *Annals of Applied Biology* 118, 395–402.

Whitehead, A.G. (1992a) Sources of resistance to stem nematode, *Ditylenchus dipsaci* (Kühn) Filipjev, in species of *Medicago* and *Trifolium*. *Annals of Applied Biology* 120, 73–81.

Whitehead, A.G. (1992b) Emergence of juvenile potato cyst-nematodes, *Globodera rostochiensis* and *G. pallida* and the control of *G. pallida*. *Annals of Applied Biology* 120, 471–486.

Whitehead, A.G. (1993) Integrated control of *Globodera pallida*. *Project Report. Potato Marketing Board, January 1993*, Oxford, UK, 28 pp.

Whitehead, A.G. (1994) Effect of cultivar, carbofuran and row spacing on the incidence of stem and leaf nematodes (*Ditylenchus dipsaci* and *Aphelenchoides ritzema-bosi*) in lucerne (*Medicago sativa*) and on lucerne yield. *Journal of Agricultural Science* 122, 225–230.

Whitehead, A.G. (1995) Decline of potato cyst-nematodes, *Globodera rostochiensis* and *G. pallida*, in spring barley microplots. *Plant Pathology* 44, 191–195.

Whitehead, A.G. and Bridge, J. (1985) Control of plant nematodes. In: Haskell, P.T. (ed.) *Pesticide Application: Principles and Practice*. Clarendon Press, Oxford, pp. 398–426.

Whitehead, A.G. and Hooper, D.J. (1970) Needle nematodes (*Longidorus* spp.) and stubby-root nematodes (*Trichodorus* spp.) harmful to sugar beet and other field crops in England. *Annals of Applied Biology* 65, 339–350.

Whitehead, A.G. and Nichols, A.J.F. (1992a) Control of potato golden cyst-nematode, *Globodera rostochiensis*, by nematicides applied once or twice in two rotations. *Journal of Agricultural Science* 119, 191–196.

Whitehead, A.G. and Nichols, A.J.F. (1992b) The effects of deep cultivation and oxamyl on control of potato cyst-nematode, *Globodera rostochiensis*. *Annals of Applied Biology* 120, 65–72.

Whitehead, A.G. and Tite, D.J. (1969) *Report of the Rothamsted Experimental Station for 1968, part I*, 157–158.

Whitehead, A.G. and Tite, D.J. (1972) Control of stem nematode (*Ditylenchus dipsaci*) attacking onions by aldicarb and dazomet in sandy loam. *Plant Pathology* 21, 89–92.

Whitehead, A.G. and Tite, D.J. (1987) Chemical control of stem nematode, *Ditylenchus dipsaci*, in field beans (*Vicia faba*). *Annals of Applied Biology* 110, 341–349.

Whitehead, A.G. and Tite, D.J. (1988) Effects of seed row and foliar applied pesticides on the yield of lucerne in clay soil, in the presence or absence of lucerne stem nematode, *Ditylenchus dipsaci*. *Journal of Agricultural Science* 110, 279–283.

Whitehead, A.G. and Westerdijk, C.E. (1987) Observations on the distribution of potato cyst-nematodes, *Globodera rostochiensis* and *G. pallida* on two fenland potato farms. *Annals of Applied Biology* 110, 571–578.

Whitehead, A.G., Fraser, J.E. and Greet, D.N. (1970a) The effect of D-D, chloropicrin and previous crops on numbers of migratory root-parasitic nematodes and on the growth of sugar beet and barley. *Annals of Applied Biology* 65, 351–359.

Whitehead, A.G., Tite, D.J. and Fraser, J.E. (1970b) The effect of small doses of nematicides on migratory root-parasitic nematodes and on the growth of sugar beet and barley in sandy soils. *Annals of Applied Biology* 65, 361–375.

Whitehead, A.G., Dunning, R.A. and Cooke, D.A. (1971) Docking disorder and root ectoparasitic nematodes of sugar beet. *Report of the Rothamsted Experimental Station for 1970, part II*, 219–236.

Whitehead, A.G., Fraser, J.E. and Storey, G. (1972a) Chemical control of potato cyst-nematode in sandy clay soil. *Annals of Applied Biology* 72, 81–88.

Whitehead, A.G., Tite, D.J., Fraser, J.E. and French, E.M. (1972b) Control of potato cyst-nematode, *Heterodera rostochiensis*, in peaty loam by D-D, aldicarb and a resistant variety of potato. *Annals of Applied Biology* 72, 307–312.

Whitehead, A.G., Tite, D.J. and Fraser, J.E. (1973a) Control of potato cyst-nematode, *Heterodera rostochiensis*, in sandy loam, by Du Pont 1410 (*S*-methyl 1-dimethylcarbamoyl) -*N* [(methylcarbamoyl) oxy] (oxythioformimidate) applied to the soil at planting time. *Annals of Applied Biology* 73, 325–328.

Whitehead, A.G., Tite, D.J., Fraser, J.E. and French, E.M. (1973b) Effects of D-D, Telone or dazomet applied to potato ridges in spring on potato cyst-nematode, *Heterodera rostochiensis*, in sandy loam and silt loam soils. *Annals of Applied Biology* 73, 105–111.

Whitehead, A.G., Tite, D.J., Fraser, J.E. and French, E.M. (1973c) Control of potato cyst-nematode, *Heterodera rostochiensis*, in sandy, peaty and silt loam soils by dazomet and Telone applied in different ways. *Annals of Applied Biology* 75, 257–265.

Whitehead, A.G., Tite, D.J., Fraser, J.E. and French, E.M. (1973d) Treating potato ridges in spring with aldicarb, D-D or dazomet to control potato cyst-nematode, *Heterodera rostochiensis*, in sandy clay and peat loam soils. *Annals of Applied Biology* 73, 203–210.

Whitehead, A.G., Tite, D.J., Fraser, J.E. and French, E.M. (1973e) Control of potato cyst-nematode, *Heterodera rostochiensis*, in silt and peat loams by ten pesticides applied to the soil at planting time. *Annals of Applied Biology* 73, 197–201.

Whitehead, A.G., Tite, D.J., Fraser, J.E. and French, E.M. (1973f) Control of potato cyst-nematode, *Heterodera rostochiensis*, in three soils by small amounts of aldicarb, Du Pont 1410 or Nemacur applied to the soil at planting time. *Annals of Applied Biology* 74, 113–118.

Whitehead, A.G., Tite, D.J., Fraser, J.E. and French, E.M. (1974) Control of pea cyst-nematode *Heterodera goettingiana*, by small amounts of aldicarb, Du Pont 1410, Ciba-Geigy 10576 or Dowco 275 applied to the soil at planting time. *Annals of Applied Biology* 78, 331–335.

Whitehead, A.G., Fraser, J.E., French, E.M. and Wright, S.M. (1975a) Chemical control of potato cyst-nematode, *Heterodera pallida*, on tomatoes grown under glass. *Annals of Applied Biology* 80, 75–84.

Whitehead, A.G., Tite, D.J., Fraser, J.E., French, E.M. and Smith, J. (1975b) Incorporating granular nematicides in soil to control potato cyst-nematode, *Heterodera rostochiensis*. *Annals of Applied Biology* 80, 85–92.

Whitehead, A.G., Bromilow, R.H., Tite, D.J., Finch, P.H., Fraser, J.E. and French, E.M. (1979a) Incorporation of granular nematicides in soil to control pea cyst-nematode, *Heterodera goettingiana*. *Annals of Applied Biology* 92, 81–91.

Whitehead, A.G., Fraser, J.E. and French, E.M. (1979b) Control of potato cyst-nematode, *Globodera pallida*, on tomatoes grown under glass, by applying steam or chemical nematicides to the soil. *Annals of Applied Biology* 92, 275–278.

Whitehead, A.G., Tite, D.J., Finch, P.H., Fraser, J.E. and French, E.M. (1979c) Chemical control of beet cyst-nematode, *Heterodera schachtii*, in some peaty loam soils. *Annals of Applied Biology* 92, 73–79.

Whitehead, A.G., Tite, D.J., Fraser, J.E. and French, E.M. (1979d) Control of stem nematode, *Ditylenchus dipsaci*, in onions, by granular nematicides applied to the soil. *Annals of Applied Biology* 93, 213–220.

Whitehead, A.G., Tite, D.J., Fraser, J.E. and French, E.M. (1980a) Control of potato cyst-nematode, *Globodera rostochiensis*, in a three course rotation. *Journal of Agricultural Science* 95, 293–304.

Whitehead, A.G., Tite, D.J., Fraser, J.E., French, E.M. and Short, L. (1980b) Effects of aldicarb and oxamyl in peaty loam soil on potato cyst-nematode, *Globodera rostochiensis*, and on resistant and susceptible potatoes. *Journal of Agricultural Science* 95, 213–217.

Whitehead, A.G., Tite, D.J. and Bromilow, R.H. (1981) Techniques for distributing non-fumigant nematicides in soil to control potato cyst-nematodes, *Globodera rostochiensis* and *G. pallida*. *Annals of Applied Biology* 97, 311–321.

Whitehead, A.G., Tite, D.J. and Fraser, J.E. (1983) Control of stem nematode, *Ditylenchus dipsaci* ('oat race') by aldicarb and resistant crop plants. *Annals of Applied Biology* 103, 291–299.

Whitehead, A.G., Tite, D.J., Fraser, J.E. and Nichols, A.J.F. (1984) Differential control of potato cyst-nematodes, *Globodera rostochiensis* and *G. pallida* by oxamyl and the yields of resistant and susceptible potatoes in treated and untreated soils. *Annals of Applied Biology* 105, 231–244.

Whitehead, A.G., Bromilow, R.H., Fraser, J.E. and Nichols, A.J.F. (1985) Control of potato cyst-nematode, *Globodera rostochiensis*, and root-knot nematode, *Meloidogyne incognita*, by organophosphorus, carbamate, benzimidazole and other compounds. *Annals of Applied Biology* 106, 489–498.

Whitehead, A.G., Fraser, J.E. and Nichols,A.J.F. (1987a) Variation in the development of stem nematodes, *Ditylenchus dipsaci*, in susceptible and resistant crop plants. *Annals of Applied Biology* 111, 373–383.

Whitehead, A.G., Tite, D.J., Fraser, J.E. and Nichols, A.J.F. (1987b) Effects of potato genotype, oxamyl and the numbers of potato cyst-nematodes, *Globodera rostochiensis* or *G. pallida* on tuber yields and nematode increase. *Annals of Applied Biology* 111, 161–172.

Whitehead, A.G., Tite, D.J., Fraser, J.E. and Nichols, A.J.F. (1987c) Control of beet cyst-nematode , *Heterodera schachtii*, by aldicarb applied to the soil by a vertical band-reciprocating harrow technique or by rotary cultivation. *Annals of Applied Biology* 110, 127–133.

Whitehead, A.G., Nichols, A.J.F. and Senior, J.C. (1991a) Control of potato pale cyst-nematode, *Globodera pallida*, with a granular nematicide and partially resistant potatoes. *Annals of Applied Biology* 118, 623–636.

Whitehead, A.G., Webb, R.M. and Beane, J. (1991b) Effect of rotation length and oxamyl on potato yield and the potato cyst-nematode, *Globodera rostochiensis*, in a sandy loam soil. *Annals of Applied Biology* 118, 371–380.

Whitehead, A.G., Nichols, A.J.F. and Senior, J.C. (1994) The control of potato pale cyst-nematode (*Globodera pallida*) by chemical and cultural methods in different soils. *Journal of Agricultural Science* 123, 207–218.

Wijnands, F.G., Kroonen-Backbier, B.M.A., Hofmeester, Y., Leeuwen-Haagsma, W.K.van, Boerma, J. and Dongen, G.J.M. van (1992) Themadag: bedrijfssystem en voor een akkbouw met toekomst. *Themaboekje – Proefstation en Informatie – en Kenniscentrum voor de Akkerbouw en de Groenteteelt in de Vollegrond* No. 14, 57–59, 111–118.

Willers, P. (1991) Control of root-knot nematode in ginger rhizome by dip treatment in fenamiphos (1991) *Inligtingsbulletin Navorsingsinstituut vir Sitrus en Subtropiese Vrugte* No. 230, 14–17.

Willers, P. and Meyer, A.J. (1979) Chemical dip treatment of strawberry propagation material for the control of *Meloidogyne hapla* and *Pratylenchus penetrans*. *Nematological Society of South Africa Newsletter* No. 11, 11–12.

Williams, C., Gilman, D.F., Jones, J.E. and Birchfield, W. (1983) Cotton-soybean rotation for reniform nematode control. *Louisiana Agricultural Experiment Station Bulletin* No. 741, 1–22.

Williams, T.D. and Beane, J. (1980) The effects of nematode resistant and susceptible spring oat cultivars and aldicarb on the cereal cyst nematode, *Heterodera avenae*, and yields in contrasting soil types. *Annals of Applied Biology* 95, 115–124.

Wilski, A. (1972) [The control of potato rot eelworm, *Ditylenchus destructor* Thorne, in potato tubers by dipping in Nemafos.] *Biuletyn Instytutu Ochrony Roślin* 4, 191–197.

Winchester, J.A. (1964) A root dip treatment for control of root-knot nematodes on celery transplants. *Plant Disease Reporter* 48, 782–783.

Winchester, J.A. (1968) Chemical control of nematodes in sugar cane. *Biokemia* No. 16, 15–17.

Winchester, J.A. and Burt, E.O. (1964) The effect and control of sting nematodes on Ormond Bermuda grass. *Plant Disease Reporter* 48, 625–628.

Windham, G.L. and Barker, K.R. (1986) Relative virulence of *Meloidogyne incognita* host races on soybeans. *Journal of Nematology* 18, 327–331.

Windrich, W.A. (1970) Control of *Ditylenchus dipsaci* (Kühn) in tulip bulbs by immersion in thionazin. *Horticultural Research* 10, 83–87.

Windrich, W.A. (1973) Bestrijding van stengelaaltjes bij narcisseteelt. *Bloembollencultuur* 84, 5.

Windrich, W.A. (1985) Control of stem nematode, *Ditylenchus dipsaci*, in tulip bulbs with aldicarb and oxamyl. *Crop Protection* 4, 458–463.

Winfield, A.L. (1965a) Potato root eelworm in Holland, Lincolnshire. *NAAS Quarterly Review* No. 67, 110–117.

Winfield, A.L. (1965b) Microplot experiments with pea cyst nematode. *Plant Pathology* 14, 158.

Winfield, A.L. (1970) Experiments on controlling stem eelworm (*Ditylenchus dipsaci* (Kühn) Filipjev) in narcissi and tulips. *Zeszty Problemowe Postępów Nauk Rolniczych* No. 92, 439–445.

Winfield, A.L. (1971) Control of bulb scale mite and stem nematode of narcissus, and reclaiming forced bulbs. *Plant Pathology* 20, 10–13.

Winfield, A.L. (1972) Observations on the control of the stem nematode *Ditylenchus dipsaci* of tulips with hot-water or thionazin treatment. *Journal of Horticultural Science* 47, 357–364.

Winfield, A.L., Murdoch, G. and John, M.E. (1971) Chemical control of stem nematode (*Ditylenchus dipsaci*) of bulb onions. *Proceedings of the 6th British Insecticides and Fungicides Conference* 1, 141–149.

Winslow, R.D. (1954) Provisional lists of host plants of some root eelworms (*Heterodera* spp.). *Annals of Applied Biology* 41, 591–605.

Winslow, R.D. (1955) The hatching responses of some root eelworms of the genus *Heterodera*. *Annals of Applied Biology* 43, 19–36.

Wood, F.H. and Foot, M.A. (1975) Treatment of potato tubers to destroy cysts of potato cyst nematode: a note. *New Zealand Journal of Experimental Agriculture* 3, 349–350.

Wood, F.H. and Foot, M.A. (1977) Decontamination of potato tubers grown in soil infested with potato cyst nematodes. *New Zealand Journal of Experimental Agriculture* 5, 315–319.

Wright, A.J., Nichols, D.B.R., Badcock, J.A. and French, N. (1986) Economics of applying nematicides to small populations of potato cyst nematodes *Globodera pallida* and *G. rostochiensis*. *Aspects of Applied Biology* No. 13, 355–362.

Wright, D.J. and Rowland,A.J. (1982) Susceptibility of different developmental stages of the root-knot nematode, *Meloidogyne incognita*, to the nematicide oxamyl. *Annals of Applied Biology* 100, 521–525.

Wright, D.J. and Womack, N. (1981) Inhibition of development of *Meloidogyne incognita* by root and foliar application of oxamyl. *Annals of Applied Biology* 97, 297–302.

Wright, D.J., Blyth, A.R.K. and Pearson, P.E.(1980) Behaviour of the systemic nematicide oxamyl in plants in relation to control of invasion and development of *Meloidogyne incognita*. *Annals of Applied Biology* 96, 323–334.

Wright, D.J., Birtle, A.J. and Roberts, I.T.J. (1984) Triphasic locomotor response of a plant-parasitic nematode to avermectin inhibition of the GABA antagonists bicuculline and picrotoxin. *Parasitology* 88, 375–382.

Wyss, U. (1971) Saugverhalten und Pathogenität von *Trichodorus* spp. in steriler Agarkultur. *Nematologica* 17, 501–507.

Xue, Z., Wu, J. and Chen, M. (1992) [Studies on habits of root knot nematodes of kenaf and their control.] *Acta Phytophylactica Sinica* 19, 117–121.

Yakimenko, G.F. and Efremenko, V.P. (1973) [Treatment of onion with nematicides before planting.] *Byulletin' Vsesoyuznogo Instituta Gel'mintologii im K.I Skryabina* No. 11, 98–100.

Yamashita, T.T. and Viglierchio, D.R. (1987) Field resistance to non fumigant nematicides in *Xiphinema index* and *Meloidogyne incognita*. *Revue de Nématologie* 10, 327–332.

Yassin, A,M. (1969) Glasshouse and laboratory studies on the biology of the needle nematode, *Longidorus elongatus*. *Nematologica* 15, 169–178.

Yassin, A.M. (1974) Root knot nematodes in the Sudan and their chemical control. *Nematologia Mediterranea* 2, 103–112.

Yoshida, K., Yoshimura, D. and Yokoyama, S. (1967) [Control of white tip disease of rice by granular pesticide application to seedbeds.] *Proceedings of the Association for Plant Protection of Kyushu* 13, 2–5.

Young, L.D. (1982) Reproduction of differentially selected soybean cyst nematode populations on soybeans. *Crop Science* 22, 385–388.

Young, L.D. (1984a) Effects of continuous culture of resistant soybean cultivars on soybean cyst nematode reproduction. *Plant Disease* 68, 237–239.

Young, L.D. (1984b) Changes in the reproduction of *Heterodera glycines* on different lines of *Glycine max*. *Journal of Nematology* 16, 304–309.

Young, L.D. (1994) Changes in reproduction of a *Heterodera glycines* race 5 isolate cultured on 'Cornell' and 'Bedford' soybeans. *Journal of Nematology* 26 (4 Suppl.), 653–655.

Youngson, C.R. and Goring, C.A.I. (1963) Nellite nematocide for root-knot nematode control. Laboratory and greenhouse studies. *Biokemia* 1, 2–8.

Yousif, G.M. and Badra, T. (1979) Effects of selected volatile, organophosphate and carbamate nematicides on hatching, infectivity and development of *Meloidogyne javanica*. *Nematropica* 9, 181–185.

Yuksel, H.S. (1960) Observations on the life cycle of *Ditylenchus dipsaci* on onion seedlings. *Nematologica* 5, 289–296.

Zacheo, G. and Bleve-Zacheo, T. (1984) Influence of different temperatures on resistance in tomato plants to *Meloidogyne incognita*. *Nematologia Mediterranea* 12, 221–230.

Zacheo, G. and Bleve-Zacheo, T. (1988) Involvement of superoxide dismutase and superoxide radicals in the susceptibility and resistance of tomato plants to *Meloidogyne incognita* attack.*Physiological and Molecular Plant Pathology* 32, 313–322.

Zacheo, G., Pricolo, G. and Bleve-Zacheo, T. (1988) Effect of temperature on resistance and biochemical changes in tomato inoculated with *Meloidogyne incognita*. *Nematologia Mediterranea* 16, 107–112.

Zaki, F.A. and Bhatti, D.S. (1986) Control of pigeon-pea cyst nematode, *Heterodera cajani* Koshy, 1967, by chemical seed treatment. *Indian Journal of Nematology* 16, 106–108.

Zawiślak, K., Sowa,S. and Gronowicz, Z. (1981) Wplyw matwikoodpornych odmian ziemniaka na zanikani w glebie populacji matwika ziemniaczanego (*Globodera rostochiensis* Woll.) i wydajnosc bulw. *Zeszyty Naukowe Akademii Rolniczo-Technicznij w Obzlijnie Rolnictwo* No. 31, 49–58.

Zehr, E.I. and Golden, J.K. (1986) Strip and broadcast treatments of dichloropropene compared for controlling *Criconemella xenoplax* and short life in a peach orchard. *Plant Disease* 70, 1064–1066.

Zehr, E.I., Miller, R.W. and Smith, F.H. (1976) Soil fumigation and peach rootstocks for protection against peach tree short-life. *Phytopathology* 66, 689–694.

Zehr, E.I., Lewis, S.A. and Gambrell, C.E. Jr. (1982) Effectiveness of certain nematicides for control of *Macroposthonia xenoplax* and short life of peach trees. *Plant Disease* 66, 225–228.

Zem, A.C., Rodrigues, J.A.S., Alves, E.J. and Lordello, L.G.E. (1981) Efeitos de nematicidas nas populações de nematoídes e produçao de bananeira 'Nanição'. In: *Trabalhos Apresentados à V Reunião Brasileira de Nematologia, 9–13 Fevereiro de 1981, Londrina, PR, Brasil. Publicação No.* 5 pp. 57–70.

Zhu, K.G., Yao, S.Y. and Zhang, J.Y. (1994) [Study on control of *Bursaphelenchus xylophilus* on pine by Nemacur.] *Journal of Zhejiang Forestry Science and Technology* 14, 16–19.

Zinovieva, S.V. and Chalova, L.I. (1987) Phytoalexins of potato and their role in the resistance to stem nematodes. *Helminthologia* 24, 303–309.

Zirakparvar, M.E. and Norton, D.C. (1981) Population characteristics of *Heterodera glycines* in Iowa. *Plant Disease* 65, 807–809.

Zou, J.J., Zhou,, C.A., Peng, J.C. and Yau, M.S. (1992) [A demonstration on the control effect of Rugby 10G against citrus root nematode.] *Plant Protection* 18, 42.

Zuckerman, B.M., Dicklow, M.B. and Acosta, N. (1993) A strain of *Bacillus thuringiensis* for the control of plant-parasitic nematodes. *Biocontrol Science and Technology* 3, 41–46.

Zuckerman, B.M., Matheny, M. and Acosta, N. (1994) Control of plant-parasitic nematodes by a nematicidal strain of *Aspergillus niger*. *Journal of Chemical Ecology* 20, 33–43.

Appendix

Crop plants recently reported resistant to *Meloidogyne* spp.

Crop plant	Cultivar
Aeschynomene americana	(a)
Amaranthus	(i) AC 13567, AC 13563 (Morgon), SI no.25, SI no.28, SI no.30, Bhubaneswar local, Kantei Khada, IHR-17-25, IHR-17-77, IHR-17-18, IHR-17-9, IHR-17-19, IHR-17-91.
Aubergine	(i) Gulla (1,2), Black Beauty, Vijaya, Banares Giant. (j) cv. 625.
Barley	(a) Bonus, CC89, CC163, Giza 121, Sahrawy. (i(1,3)) Bonus, CC163.
Bermuda grass (*Cynodon dactylon*)	(a,h,i,j) Coastal Coastcross 1, Tifton 44. (a,h,j) Tifton 78. (a (2), i (1,2)) Alice, Lancaster, Tifton 78.
Cabbage	(i) Leo # 80, K-K.
Capsicum annuum	(a, i, j) Naharia, Black India. (a,j) PM 217. (i(1,2,3,4)) Carolina Cayenne, Jwala. (i(1,3)) Pusa Jwala. (i(1)) KD-10.8, KD-30.8. (i) Malagachi, yellow 12–3-3, Jalandhari, Laichi-2, White Khandari, K-2, Pant Cl. (j) California Wonder
C. frutescens	(i(1) EL 21900.0, EL 38780.0, Dantsiga 0.0, PL 2289.0, PL 384750.0. (i) Webb.
Cassava	(i,j) Mandiocal, Retroz 1, Retroz 14, Para, Preta Legitima, Bahia Preta, Rosa Branca. (i) 73/138 (EBO 21 b/w), 75/668, 77/59 (Asiaka) 77/165, NR 7721, TMS 3001, U/41044.
Cotton (Sea Island) (*Gossypium barbadense*)	(a(1), i (2,3,4), j) Giza 68, Giza 70, Giza 75. (a, i(2), j) Giza 67, Bahlim 110. (a(1), i(2)) Giza 69.
Cotton (Upland) (*G. hirsutum*)	(a, i, ia, j) Deltapine 16. (a(1), i(2,3,4), j) Acala 4–42, Acala 67A. (a, i(2), j) Deltapine 16. (i(4)) Chuan 7273, Liaoyang 4, Jihe 328. (i) Georgia King, LA 887, M 315, Acala 1517–88, Stoneville LA 887, M-315/240, Aub 82 RNR, Aub 244 RNR, Aub 299 RNR, Aub 634, N 6072. (ia) 320-F, L55. (i – *Fusarium* complex) GaCot 79, Tamcot SP 37H Tamcot CAMD-E. (spp. – *Fusarium* complex) Auburn BR 2.

Crop plant	Cultivar
Colocasia esculenta (taro)	(i) Sree Rashmi, C 9.
Cowpea	(a,h,i,j) Amelia, Otilia, Pitiuba, Querenta – dias 1. (a, i(1,3,4)) cv. 64298. (a(1), i(3)) Colossus-80, California No. 5 Blackeye, Collossus, Mississippi Silver, Mississippi Purple, Hercules, Magnolia Blackeye, Worthmore (a,i) Erectset. (i,j) IC 19739, IC 20452, Iron, Colossus. (i(1,3)) California Blackeye 46. (a(1)) Pinkeye Purple Hull. (i(3) Azmerli. (i) Mississippi Silver, Barsati Mutant Iron, GC × Pusa Falguni 9, Co 3, C 152, P 426 Iron, Texas Pinkeye Purple Hull. (i(3)) Zipper Cream.
Crotalaria juncea	(i) K-12 Yellow.
C. spectabilis	(a,i,j)
C. anagyroides	(i)
C. paulina	(i)
Cauliflower	(i, j) Dania. (i) Flora Blanca, Early Patna, Late Patna. (j) Superial Maghi, 74–6C.
Chick pea	(i) K 1160, RSG 130, RSG 143, GCVT-Bold-K999, GIET PDG 85–18, GIET – GNG 317, BG 405.
Chloris gayana (Rhodes grass)	(j) Katambora.
Citrullus lanatus (watermelon)	(a(2), i(2), j) Mexico 74, Crimson Sweet, 11253, Gangpi 100, Zhong-Yu 1
Coffee (*Coffea arabica*)	(e) Rume Sudan, hybrid Sarchimor (i, j) Arussi S 11, Agaro Ginona, Kaffa CB-II, Hanar R2-CB-II, Geisha B, Pacas/MT, BK 13.
(*C. canephora*)	(e, i) Robusta.

In laboratory tests arabica coffee was immune to *M. hapla*, *M. incognita*, *M. javanica* and *M. thamesi* and robusta (*C. canephora*) coffee was immune to *M. hapla* and resistant to the other species.

Crop plant	Cultivar
(*C. congensis*)	(i(3)) Banguelan col. 5C104TP.
Grass *Panicum maximum*	(a,h,i,j) Umtali. (h,i) Natsukaze. (j) Guine, Petri.
Eragrostis curvula	(j) Ermelo.
Brachiaria brizantha	(j)
B. decumbens	(j)
Digitaria decumbens	(j) Pangola.

Crop plant	Cultivar
Groundnut	(a, i, ia, j) CES 101. (i(1,2,3)) IPRAL-67. (i(1)) Giza 4. (a(2)) Florunner (a) Mobwer Runner, R-33–1, C-41 (NRCG-31), NCAC-2196 (NRSG-1010), Local 256, Japtin-220–15, J 11. (h) Chatumbana.
Indigofera hirsuta (hairy indigo)	(a, i(1,3), j) FL 24, FL 101. (a,i,j) Flamingo (PI 538238).
Lucerne	(a,i,j) Florida 77. (h,i) Florida 77, Nevada Synthetic XX, Nevada Synthetic YY. (a) Hasawi NE-NAF-4, Hasawi SA 19, (h) UC Cibola, Nevada Synthetic XX (i) Cuf 101, Moapa 69, Cimarra. (j) Hasawi-Amer.
Maize	(a(1,2)) Northrup King 508, Pioneer, Brand 3147. (a) inbred Northrup King 508, Mp 313. (j) $IACH_s$ 1227, Ag-301, Ag-401, XL 678, IAC 100B, Contimax 322, Perola Piracicaba, $IACH_s$ 7777, IAC Maya Latente, N3, inbreds Mp 703, NC 246, Mp 307, F6, Mp704, T220, IAC Ip 365–4-1.
Oat	(a(1,2), i(3)) Coker 716. (a,i,j) FL 501, FL 502. (j) UFRG S1, UFRG S2, UFRG S3, UPF3.
Okra	(i) KS-381, KS-114, KSL-380.
Onion	(i(1)) Early Narrow Neck.
Papaya	(i) Pusa 22–3, Co-1, Ceylon C.S.C. Dwarf-1, Hybrid Peredeneya, Singapore Improved Sunrise.
Peach	(i,j) Nemaguard. (i) Shalil, Kanto-S, Stark Early Flame, China Flat, Japanese Double Bid, Compact Redhaven, Elberta, selections of open pollinated Okinawa and Okinawa × Swaton, *Prunus persica* var. *densa* Hybrid 420/1 (ex *P. davidiana* × *P. persica*). (j) Robbin, Flordasel, Loring, Red French.
Pigeon pea	(a,i,ia,j) Native. (i,j) Norman, ICP 11295, Pusa-33, Pusa-78, Pusa-85. (i) Pusa Ageti, P 3789. (j) PDA-1, LRG-30, UPAS 120, Pant A3, BDN 2.
Potato	(j) Ajax, Cardinal, Draga, Giant, Serana-28.

Crop plant	Cultivar
Rice	(g) MW 10, IR 36, Udaya, Daya, Palgar-1, Ilkorasali-2, Jahinga, Bonnet 73, 1014 mutant, CR_s-CR 1009, 1018, 292–5258, 280–5387, 146–7001, 146–4005, 401–7, 407–19, 407–6-2, 407–6-1, 407–4, 260–312, 254–43–1-6, CR 190–103, RP-2542–1639–50, RP-2633–59–9-7, RP-2362–110–54–27, RP-2151–192–2-5, RP-2151–178–1-2, WGL-47978, WGL-47998, OR-621–20–1, TM-10222, MTU-7029(B). (a(1), i(2,3), j) Philippine 24, IR 28, IR 459. (i(3)) A 95, Giza 171, Giza 172, Japonica 47. (i) Faro 27, Thabaung Mee, Gauk, IR 20. (j) Peta.
Sorghum (grain)	(a(2), i(3), j) Funk G-499 GBR, Funk- G 611, Funk G 522A Funk G-522DR, Coker 7723, Coker 7675, Coker 7623, Pioneer B 815, Pioneer 8222, Pioneer 8272. (a(1,2), i(3)) Cherokee, Northrup King 2260, Pioneer 8333.
Sorghum bicolor	(i)
Soybean	Maturity group when indicated, given in parentheses. (a) Sao Luix, FT-1, Industrial UFV-3, IAC-4, Sulina, Govan, Wright. (a, i, j) Perrin (VIII), Colquitt (VII), Tropical, Bragg, IAS-1, Santa Rosa, IAC-8. (a,i) Braxton, Cobb, Hampton. (i, j) BR 6, JC 5086, JC 5087, Doles. (h) Jangyeoskong. (i(4)) Hutton. (i) Choska (VI), Maxey (VIII), Rhodes, A 6785 (VI), A 6381 (VI), Sharkey (VI), PK-946, PK-948, JS-340, JS-2, KB-32, PBN-110, Forrest, Agrow 3307, FFR 398, Pioneer 9442, Amredo, Leflore, Centennial, Ring Around (RA) 680, RA 604, RA 702, Coker 485, Coker 596, Coker 627, Coker 368, Bedford, Jeff, Gordon, Foster, Hutton, Kirby, Lee 74, Eundaedu, Baegcheon, Dongbugtae, Danyeobkong, DS 74–20–1, DS 74–20–2, Bienville Coker's 4505, Hale-7, Hardee, Hill, IAS-2, IAS-3, IAS-4, IAS-5, Jackson, Louisiana, Parana, Mineira, Semmes, Doko. (iw) Braxton, Centennial, LA78–17717, LA 78–17908, LA 75–1794, LA 75–1799. (j) FT-Cometa, Decada, IAC-2, Tiaraju, Bossoer, FT-1, Coker 136, BR 6, Nova Bragg.
Strawberry	(h) Glima, Senga Sengana, Gorella, Cambridge Favourite, Madame Mantot, Surprise des Halles, Early Dawn. (i)

Crop plant	Cultivar
Sugarcane	(j) SP 72–4790, SP 70–1143, CB 40–13, CB 46–47, Co 740.
Sweet potato	(i) W 86, L 4–89, BP 4, Coastal Red, Norin 40, Norin 41, Topaz, Redglow, Sree Vardhini, Sree Nandini, H 268, Lanshu 88, Xushu 18, Shenglibaiho, Lanruishao. (i(3)) Sumor.
Tobacco	(i, j) McNair 135 (PCT 4-A). (i,ia) NC 95, Bel 4–30 (i(1,3)) NCD 60. (i(4)) Coker 209. (i) Speight G-28, CV 227, CV 347, Golden Gem, Coker 258, CV 116, CV 563, Mammoth Gold, Coker 86, Coker 51, Mammoth del Crest, Coker 254, Speight G-1469, Coker 2584, Coker 247, Speight G-33, Speight G-41, NC 2512, Virginia 770, Osibenbtchi.
Tomato	(a, i, ia, j) Hawaii 5229, Nematex, Small Fry, Cold Set. (a, h, i) KB 174, Nakhodka, Velikan, Bug^{66} r, Standard 69. (a, i, j) Nemadoro. (h, i, j) Piernita, Motabo Massol, Anahu, Rofoto, Monita, Gibrid 1 of Iordanova, Nemabrid FN. (i (1,2,3,4), j) Pusa-120, Calmart VFN, Panjabb NR-7, EC 173898, EC 173897 (Cal-Mart), EC 173896 (Kewalo), CLN 363 BC, VFN-Bush, VFN-8. (a,i) Namuna, Dar Surkhana, Surkhana 142, IPA-1, IPA-2, VFN-8. (i,j) Punjab NR-7, Nealain, Ultra Boy, T 1, T 5, Kristi, T 7, Kolhi, Monte Carlo, Breach, SL-12, 622 VFN, GC 204, Bonus VFN, Noria TMV$_2$ N, Terrific VFN, Vine Ripe, Atkinson, Small Boy, Nemared, Patriot, Rossol, Hawaii, NT-3, NT-8. (ia, j) Nematex, SL-120. (h) Nemared 341128. (i(1,2,3)) IPA-1. (i(1,2,4)) Pelican. (i) Hisar Lalit, Kewala, Matobo, Marsol, Uco 2 INTA, Uco 3 INTA, Uco 4 INTA, NT-12, SI-120, GS 27, Karnataka hybrid, VFN 77–177–1, 72 N58, Royal Chico, IPA-3, Roma VFN, VFN 77–92–2, Slumac, NTDR-1, Estafette, Ganita, INRA 518, Nemato, Diego, Mondial, Ife 1, Piersol, Pernita, Nemacross, Katala, G-370, H 7741, Multiset, Tomosa. (ia) Rossol, Piersol VFN. (j) Rubezh, Syuzhet, Carmelo, Barelli, Ace, Luxor, Energia VFN, Novia, F-187, F 193 D6, F 38 E2, F 455 D1, Ronita, Healani, NTDR-1, Roma, Better Bag, Ace Hy, Tobol 748, Piersol VFN, Nemasol.

Crop plant	Cultivar
Wheat	(a,i(1,3), j) Stork, Saka 8, Saka 61, Saka 69, Saka 80. (a, i(1,3)) Giza 155, Giza 157. (i(1)) Giza 160. (i) Kalyansona, Arthur 71, Coker 747, Coker 6815, Delta Queen, Double Crop, Florida 301, McNair 1813, McNair 1003, Southern Belle.

Letters in parentheses refer to species of *Meloidogyne* to which resistant: a = *arenaria*, c = *chitwoodi*, e = *exigua*, g = *graminicola*, h = *hapla*, i = *incognita*, ia = *incognita* (*acrita*), iw = *incognita* (*wartellei*) and j = *javanica* ; nos. in parentheses refer to race number, when known.

Root-stocks recently reported resistant to *Meloidogyne* spp.

Scion	Stock	
Aubergine	*Solanum torvum*	
Tomato	KNVF rootstock	(i)
	Bruinsma, Nematex	(i)
Grapevine	Kober 5BB (Berlandieri × Riparia)	(a(2), h, i(2))
	Fercal	(a,h)
	Dog Ridge, Salt Creek	(a,i)
	Harmony 504, Teleki 8B, 1103 Pauben, 1447 Pauben, 101–14, 3306, 1613 C.	(a)
	KM-151.	(h)
	Ramsey, Richter P5, Freedom, Harmony, 101–14 Mgt, Anka Kauchan, Black Sultan, Luglienga, Kandhari, Black Champa, cv. Criollo Negra, 1613 Couderc, 1616 Couderc.	(i)
	Masangwi.	(j)
Peach	Nemaguard, Okinawa	(i(1,2,3), j)
	Flordaguard	(i(1,3))
	Chico II, *Prunus davidiana*	(i(3))
	cvs Damasco 1869, GF 43, S Giuliano GF 655/2, Lovell, Halford, R 32–10, R 32–16, R 33–1, R 33–3, Rancho Resistant, Cadaman Avimag, S 37, S 60.	(i)
	Hansen 2168, Hansen 536, Nemared	(j)
	Ferciana Ishtara	(*M.* spp.)
Plum	Mariana 2624, Pixy, San Julian 655–2	(i)
	Ferciana Ishtara.	
Olive	*Prunus cerasifolia* × *P. spinosa*, *Vitis vinifera* × *V. rotundifolia*, *Juglans hindsii* × *J. nigra*.	(*M.* spp.)

Letters in parentheses refer to species of *Meloidogyne* to which resistant: a = *arenaria*, h = *hapla*, i = *incognita* and j = *javanica*; nos. in parentheses refer to race number, when known.

In general, the resistance of a rootstock is not affected by the susceptibility of the scion grafted onto it. For example, the resistance to *M. incognita* in *Solanum toxicarium*, *S. torvum* and *S. sisimbriifolium* was not affected when the susceptible tomato 'Kyouryokubeijo' was grafted onto them (Matsuzoe *et al.*, 1992).

It must be borne in mind that this appendix is a summary of what has been reported in recent literature. There is no guarantee that a resistant cultivar or rootstock resists all populations of the species or race indicated. Indeed, resistance can vary greatly between different populations of a race or species, so local assessment of the usefulness of a particular cultivar or rootstock is essential. Nevertheless this summary shows what may be useful nematode resistance. Cultivars with only moderate resistance have been omitted and no distinction is made between resistant and 'immune' plants.

General Index

Crop Plant Index